Deserts provide a harsh and inhospitable environment for plants and animals, and the ecosystem is correspondingly fragile and prone to disruption by a variety of external factors. The Mojave Desert is a winter-rainfall desert, experiencing drought in the summer months and occasional rain during the cooler winter months. For many years it has attracted the attention of ecologists and conservation biologists concerned to maintain the unique status of this region.

This book provides a broad overview of plant and animal ecology in the Mojave Desert, presented with a focus on data from Rock Valley, Nevada. The data from many research projects is organised into a synthesis describing community structure and dynamics in desert ecosystems.

ECOLOGICAL COMMUNITIES AND PROCESSES IN A MOJAVE DESERT ECOSYSTEM

Rock Valley, Nevada Test Site

ECOLOGICAL COMMUNITIES AND PROCESSES IN A MOJAVE DESERT ECOSYSTEM: ROCK VALLEY, NEVADA

PHILIP W. RUNDEL

&

ARTHUR C. GIBSON

Department of Biology and Laboratory of Structural Biology and Molecular Medicine
University of California at Los Angeles
Los Angeles, California

CAMBRIDGE
UNIVERSITY PRESS

CAMBRIDGE UNIVERSITY PRESS
Cambridge, New York, Melbourne, Madrid, Cape Town, Singapore, São Paulo

Cambridge University Press
The Edinburgh Building, Cambridge CB2 2RU, UK

Published in the United States of America by Cambridge University Press, New York

www.cambridge.org
Information on this title: www.cambridge.org/9780521465410

© Cambridge University Press 1996

This publication is in copyright. Subject to statutory exception
and to the provisions of relevant collective licensing agreements,
no reproduction of any part may take place without
the written permission of Cambridge University Press.

First published 1996
This digitally printed first paperback version 2005

A catalogue record for this publication is available from the British Library

Library of Congress Cataloguing in Publication data
Rundel, Philip W. (Philip Wilson)
Ecological communities and processes in a Mojave Desert ecosystem:
Rock Valley, Nevada / Philip W. Rundel and Arthur C. Gibson.
p.   cm.
Includes bibliographical references (p.   ) and index.
ISBN 0–521–46541–9 (hardback)
1. Desert ecology – Nevada – Rock Valley.   2. Adaptation (Biology) –
Nevada – Rock Valley.   3. Man – Influence on nature – Nevada – Rock
Valley.   I. Gibson, Arthur C.   II. Title.
QH105.N3R85   1996
574.5′2652′09793 – dc20   95–15896 CIP

ISBN-13 978-0-521-46541-0 hardback
ISBN-10 0-521-46541-9 hardback

ISBN-13 978-0-521-02141-8 paperback
ISBN-10 0-521-02141-3 paperback

Contents

Preface   *page* ix

Historical prologue on Rock Valley studies   xi

1 Introduction to the Mojave Desert   1
2 Physical geography of Rock Valley   21
3 Adaptations of Mojave Desert plants   55
4 Desert perennials of southern Nevada   84
5 Mojave Desert annuals   113
6 Adaptations of Mojave Desert animals   130
7 Mammals   155
8 Reptiles   174
9 Birds   197
10 Arthropods   214
11 Soil organisms and seed reserves   256
12 Nitrogen cycling   274
13 Human impacts on Mojave Desert ecosystems   290

References   318
Species index   350
Main index   357

Preface

The Mojave Desert is an expansive region of arid western North America, primarily in eastern California and southern Nevada. This is a winter-rainfall desert, which, like the Negev in Israel, experiences drought conditions during the very hot summer months and occasional rain during the cooler months in late fall and winter. Researchers, especially those from academic institutions with close access to the Mojave Desert, have utilized key research sites to investigate how animals and plants cope with the rigors of desert life. Of the several Mojave Desert research sites, none has been as productive as a creosote bush desert scrub located in Rock Valley on the Nevada Test Site (NTS). The Nevada Test Site, operated by the United States Department of Energy (DOE), occupies 350 000 hectares of arid and semiarid terrains in central-southern Nevada. The location of this large reservation is particularly interesting because NTS straddles the geographic boundary between the Great Basin, which is classified as a cold desert, and the Mojave Desert.

As discussed briefly in our Historical Prologue, the history of ecological research at NTS dates back to 1951, when UCLA researchers were first permitted access, especially to its disturbed lands. Under research contracts with the former Atomic Energy Commission (AEC), NTS eventually was designated as a base for investigating patterns and processes of desert ecosystems, with special emphasis on the communities of the Mojave Desert, found in the southern portions of the reservation.

Research facilities at Rock Valley have existed since the early 1960s, when the AEC agreed to utilize this relatively pristine basin for long-term studies on the effects of low-level gamma radiation exposure to desert populations of plants and animals. Having that history of intensive ecological study predisposed Rock Valley to be selected in 1970 as the Mojave Desert validation site for ecosystem studies under the Desert Biome program of the International Biological Program (IBP). Other desert IBP sites were Curlew Valley, Utah (Great Basin Desert), Silverbell, Arizona (Sonoran Desert), and Jornada Range, New Mexico (Chihuahuan Desert). With major funding by the National Science Foundation (NSF) to supplement the ongoing program of research sponsored by the AEC, six years of intensive ecological studies (1971–6) were carried out in Rock Valley. Ecologists from UCLA and several other institutions for the IBP-Desert Biome project collected voluminous biological data and important baseline measurements of environmental conditions and ecosystem processes.

Ecological research continued at NTS and in Rock Valley. The AEC, later the Energy Research and Development Administration (ERDA) and now DOE, sponsored basic ecological research there through contracts with the Laboratory of Biomedical and Environmental Sciences (LBES) at UCLA. Hence, after four decades of ecological research, Rock Valley clearly is the most intensively studied and best understood area of Mojave Desert ecosystem. Indeed, Rock Valley may arguably be considered one of the two or three best studied warm desert sites in the world. Both the breadth of research data available and the existence of relatively long-term records give this site critical significance for understanding the structure and function of desert ecosystems.

Despite the importance of the Rock Valley site and the large body of published literature on this and adjacent desert sites, there remain from the IBP studies useful blocks of data that have never been properly published outside the 'gray literature' research reports, which were prepared during the IBP-Desert Biome project. These reports are not widely available and contain occasional calculation errors, which resulted because those reports were prepared with too much haste.

The main objective in writing this book was to summarize the existing state of knowledge of the Mojave Desert ecosystem of Rock Valley. However, to understand that ecosystem the reader needs to appreciate the biological properties of the common residents in Rock Valley; hence explanations of their physiological, phenological, and behavioral adaptations have been included. No effort was made to cite every article or note published on desert organisms from the Mojave Desert communities, although we have cited a large corpus of literature. Unfortunately, we did not have the prerequisite data to publish here a comprehensive, comparative treatment on community ecology and population biology of the Mojave Desert with either the winter-rainfall deserts from different continents or the other desert regions of North America. Clearly the synthesis, which was begun with the IBP projects in North America, is still a goal for the future.

We are greatly indebted to Frederick B. Turner, who in the mid-1980s began to summarize the IBP findings in preparation for producing this book. Turner coordinated many research projects in Rock Valley since 1962, including the IBP studies, and consequently helped us to focus our attention on the most interesting observations. Many of the tables and much of the text concerning Rock Valley drew heavily on his analysis, and he assisted us in developing the historical prologue for the book.

Helen McCammon and the staff at the Ecological Research Division of the Office of Energy Research at DOE deserve heartfelt thanks for their support for ecological programs and the preparation of this volume. The basic research efforts summarized here were generously supported over the past three decades by DOE, NSF, the Environmental Protection Agency, and other private and public granting agencies. We also thank our UCLA colleagues at LBES and Barry Prigge in the Herbarium for assistance in researching and editing parts of the manuscript, and we are especially grateful to Philip A. Medica, Richard A. Hunter, and Paul Greger, research biologists at NTS, for enthusiastic support of the undertaking and for providing on-site expertise in the editing of many chapters.

Philip W. Rundel
Arthur C. Gibson

Historical prologue on Rock Valley studies

The first detonation of a nuclear fission device occurred at the Trinity site near Alamogordo, New Mexico, on July 16, 1945. Three weeks later a second and third device were detonated over Hiroshima and Nagasaki, Japan, bringing a quick end to World War II. These early devices were effective, but they were crude and unwieldy; therefore, further refinements in design and methods of construction and delivery, along with field testing, were necessary to ensure performance and reliability. Thus, the United States military needed suitable sites for continued nuclear testing.

Nuclear detonations in the atmosphere produced massive quantities of radioactive contaminants, which would ultimately reach the ground, especially not far downwind of the explosion. To minimize radiation exposure to populated areas from fallout, the first two series of postwar nuclear tests were conducted at remote Bikini Atoll (1946) and Eniwetok Atoll (1948) in the Pacific (Hines 1962). Conducting nuclear tests at the Pacific Proving Ground was financially constrained while the U.S. was engaged in the Korean War. Moreover, there were new political pressures in the balance of power following the detonation of a nuclear device by the Soviet Union on August 23, 1949, which resulted in the establishment in December, 1950, of a continental testing facility, the Nevada Test Site (NTS), on a large parcel of the Las Vegas–Tonopah Bombing and Gunnery Range. The Pacific Proving Ground was still used for testing large-yield nuclear devices during 1951, 1952, 1954, 1956, and 1958 while nuclear tests of smaller yield were conducted aboveground at NTS,

initially at Frenchman Flat, from January, 1951 until 1958.

The period of continental, atmospheric nuclear testing at NTS (1951–8) coincided with two unique and sharply contrasting biological programs. The Department of Defense established a military encampment at Mercury, Nevada, and systematically exposed soldiers to atomic blasts and post-explosion atmosphere, largely for the purpose of testing the psychological effects of these weapons. Also during the 1950s there was growing alarm about the deleterious effects of radiation, and thus the U.S. Atomic Energy Commission (AEC) funded investigations of the incorporation of fallout radionuclides into tissues of living organisms and potentially deleterious effects of these emitters.

The AEC was aware that aboveground testing in Nevada posed potentially grave risks. Above obvious public health issues loomed almost imponderable questions associated with effects of radioactive fallout on natural ecosystems. Fallout particles condense from superheated gases and contain complex mixtures of radionuclides with varying half-lives, each with its own physicochemical properties and biological availability. Some radionuclides pose little threat unless incorporated into biological tissues; others emit high-energy gamma radiations and, in sufficient amounts, produce acutely lethal exposure rates.

The Environmental Radiation Division of UCLA's Laboratory of Nuclear Medicine and Radiation Biology was responsible for mapping the geographical distribution, radiochemical composition, and intensity of tropospheric fallout resulting from aboveground testing in Nevada. The first test series so monitored was Operation Jangle in November, 1951 (Larson, Olafson, Mork, & Howton 1952). Similar efforts were continued during the 1950s, which concluded with the Plumbbob Series of 1957.

Tracking and measuring fallout from an aboveground test of large weapons was not easy. The approximate fallout pattern could be predicted from the yield and meteorological conditions that were expected at the time of detonation. Tests were postponed when wind directions were not realized, but even minor changes in wind speed and direction affected the pathways of fallout. At varying distances from ground zero fallout was collected at many stations as far east as western Utah. In addition to fallout particles, samples usually included soils, vegetation, jackrabbits, and rodents, and these were returned to Los Angeles for radiometric analysis, which was in itself a major undertaking.

The studies of the 1950s at NTS were principally an attempt to meet a social responsibility and not a series of scientific experiments done in the hypothetico-deductive mode. Results should be realistically viewed as an attempt to provide descriptive analyses of a staggeringly complex set of physicochemical and meteorological interacting processes. There was no real opportunity to assess biological effects because measurements were made in areas where exposure doses to organisms were considerably lower than acutely lethal levels. The ultimate justification for the enormous effort of environmental radiation research came many years later, when, in a new sociopolitical context, suits were brought against the U.S. government on behalf of residents of southwestern Utah and northern Arizona who had allegedly died or become ill as a result of exposure to fallout. It was then that the data described above played, and are still playing, a part in the evaluation of these claims.

Partly as the result of political concern, but more as the reaction to findings on the deleterious biological effects of radioactive fallout and pollution of the world's atmosphere, the Soviet Union and United States, along with six other nations, in May, 1958, agreed to cease aboveground testing of nuclear weapons. That first moratorium on testing took effect on October 31, 1958, although the last Soviet test was conducted on November 3. However, in 1961, during the high tensions of the Cold War the Soviet Union resumed aboveground testing on July 29 with the detonation of a 50- to 60-megaton device. At NTS the United States resumed testing in 1962 with the Sedan event, but shifted to underground testing, which very greatly reduced the release of radioactive materials to the atmosphere. Then after much diplomacy and posturing, the United States, Soviet Union, and United Kingdom signed the Limited Ban Treaty of 1963, which effectively banned atmospheric testing of nuclear weapons in the atmosphere, outer space, and underwater for these nations. China was not bound to that treaty and conducted its first atmospheric test of an atomic bomb in 1964 and a hydrogen bomb in 1967.

Aboveground nuclear tests at NTS were routinely detonated on the barren, closed basins of Yucca Flat and Frenchman Flat. Much of the early biological research was therefore concentrated around those sites and in Mercury Valley, location of the military community. However, the amount of general information on the ecosystems there was very limited, and thus the AEC began to sponsor research at NTS to investigate broader ecological aspects of the affected communities. From 1957 to 1963 UCLA researchers at NTS were joined by investigators from New Mexico Highlands University, who studied the effects of physical disturbance and residual radiation from nuclear bomb detonation on plants, algae, and fungi. Scientists from Brigham Young University were contracted (1959–1966) to collect data, especially in Yucca Flat and Frenchman Flat, on species diversity, geographical distribution, population structure, and seasonal phenology of animals. The publications and reports to the AEC that resulted from those and the earlier studies (1951–1966) have been referenced elsewhere (Schultz 1966; Wallace & Romney 1972; O'Farrell & Emery 1976).

The multiple effects of aboveground detonations on desert scrub communities included not only nuclear radiation, but also thermal radiation and fire, ground shock, air blast, and dust, and it was impossible to separate the effects of any single factor and certainly not any particular form of radiation. At NTS the research effort of the 1950s largely described the geographical extent and intensity of radioactive fallout in portions of the Great Basin and amounts of various radionuclides in soils and organisms of NTS desert ecosystems. These measurements were essentially descriptive and did not provide any understanding how the eco-system functioned as a unit. The fallout-contaminated areas provided opportunity to investigate both immediate and delayed radiation effects. However, the radiation levels beyond a blast-disturbed zone decreased so rapidly that it was difficult to establish necessary ecological baseline information on biotic components and to develop methods for documenting long-term radiation effects.

The nuclear moratorium gave researchers the opportunity to plan new types of studies for federally funded programs. Research had to depart from the experiments that used either acute, one-time or fractionated doses to test survivorship of individuals; instead studies were needed that made a distinction between immediate lethality versus long-term local extinction owing to radiation-induced sterility. For example, animals might survive the immediate effects but, because of dispersion of radioactivity in fallout, organisms would be exposed to long-lived emitters, for example, ^{90}Sr and ^{137}Cs, and the accumulated doses to reproductive organs might be sterilizing. Little research had used sublethally irradiated animal populations to study changes in plant or animal populations under natural conditions. Such experiments were needed to diagnose the impacts of continuous, low-dose radiation.

An interesting experiment was begun in 1961 on Long Island, New York at the Brookhaven National Laboratory, where researchers continuously exposed an oak–pine forest and an oldfield system to gamma irradiation (Woodwell 1962, 1963, 1965, 1967; Woodwell & Sparrow 1962; Sparrow & Woodwell 1963; Woodwell & Oosting 1965; Turner 1975). Each site had a centrally

located tower with an unshielded gamma emitter, i.e., one emitting only electromagnetic radiation. This was a successful design for analyzing effects on perennial plants, which showed dramatic close-in effects of continuous irradiation and very little effect beyond 100 m from the source. Arthropods showed some direct lethality from the radiation and indirect effects via changes in the vegetation (Brower 1964). However, the experimental design did not expose animals to a uniform dose, so that interpreting results on mobile animal populations was not possible.

In fall, 1959, Norman R. French at UCLA conceived a similar experiment for research on animal populations at NTS (French 1964). Until then research had been concentrated around the detonation sites, but those valleys were unreasonable candidates for studies to obtain the needed baseline ecological data because they contained residual effects of disturbance and contamination. French appraised the NTS resources to find a suitable, undisturbed desert scrub community for long-term ecological studies. That site was to be Rock Valley.

Rock Valley (frontispiece), which covers about 130 km^2, is located adjacent to the southern boundary of NTS and about 20 km west of Mercury. Based on a proposal from French, the AEC agreed to set aside this unused valley for long-term ecological studies, especially as an example of lower bajada desert scrub (*Larrea tridentata–Ambrosia dumosa* community) for the northernmost Mojave Desert. This site had easy access to a paved highway and had received very little radioactive contamination from the aboveground nuclear tests. In March, 1960, French initiated preliminary studies on a high-density population of heteromyid rodents in Rock Valley at a site having an elevation of 1000 m and a gentle slope of 3 to 4 percent. He established a grid of 400 live-traps covering a 9-ha circular plot to safely capture and release rodents at monthly intervals, and from this he began accumulating data to evaluate the changing age structure of the population.

French proposed to establish four permanent, 9-ha (20 acre) circular plots, each with a diameter of 165 m, and to enclose three of these (A–C) by erecting a strong fence that would prevent the immigration and emigration of rodents during experiments. The novelty of French's proposal was that fenced plot B would be used as a site for a controlled experiment to expose the plants and animals of the community to chronic, continuous but *relatively uniform*, low-level gamma (ionizing) irradiation, to be compared with organisms in unfenced plot D, which experienced only natural (background) irradiation. Radiation was to be generated from a 33 600 curie ^{137}Cs source mounted on a 15-m tower that would be erected in the center of the circular plot, and this radiation source could be lowered and then stored in a lead cask once per month (a five-day period) to permit field work. An advantage of using ^{137}Cs was that there was no residual contamination, which meant that safe levels of radiation could be achieved instantaneously whenever personnel had to enter the enclosure and study area. A ^{137}Cs source of that size was used to avoid any heating problem. The fence would prevent escape of nonflying irradiated animals, so that investigators could follow the effects of radiation that may require a long time to become expressed as some life-history parameter.

It was critical for the design of the Rock Valley experiment to have fairly uniform dosage of radiation throughout the enclosure, so that investigators could sample a relatively homogeneous treated population. Elsewhere, for example, at the Brookhaven studies, fixed radiation sources had not controlled for uniform dosage of the entire animal population. To achieve that goal, the cesium source at Rock Valley was subtended by a differentially thickened lead shield disc to reduce exposure directly beneath the source, so that similar dosages would be received from the center to the fence. Measurements during the experiment showed that the ground at the center of the plot received 750 mR hr^{-1} while the periphery received 75 mR hr^{-1}, only a tenfold difference, whereas at Brookhaven the unshielded source produced more than a 250-fold difference, and differential shielding by the tree canopies reduced transmission of gamma radiation from the tower and further complicated analyses.

French tested four fence designs and selected one that prevented rodents from passing through, over, and under it (Figs. P.1–P.2). Fences were installed on the circular, 9-ha plots in summer, 1962 (plot A) and fall, 1963 (plots B and C). Each was constructed of hardware (wire) cloth (3 × 3 mesh) having openings that were approximately 6 × 6 mm in size. Wire cloth (1.2 m wide) was supported by steel fence posts installed at 3-m intervals. The lower 0.3 m of the wire cloth was buried below ground, and the soil was firmly packed against the cloth. A continuous piece of sheet-metal flashing, in the form of a roof having overhangs on both sides, was fastened to the top of the wire cloth. The flashing was riveted, and segments were welded at junctions.

In preparation for the broad, long-term ecological and radiation studies, the AEC–UCLA program brought to Rock Valley additional personnel to study the terrestrial vertebrates and identify shrub patterns on the plots. The research staff was in place by 1962, when initial sampling gave strong indication that this would be a very good location

Figure P.1 Nature of fenced enclosure for circular 9-ha research area in Rock Valley.

Figure P.2 Closer view of enclosure fence, showing hardware cloth buried in ground and having a galvanized metal cover that prevents escape of most terrestrial organisms.

for the proposed population studies. Thus, the biotic surveys were completed prior to January, 1964, when the tower was erected on plot B.

Special microthermoluminescent dosimeters had to be implanted on all tagged, censused rodents to know precisely what exposure of gamma radiation each of these mobile individuals would receive. For Rock Valley the original dosimeters were sealed glass capsules 0.8×6.0 mm with 0.6 mg of lithium fluoride. Lithium fluoride had an advantage over earlier designs because its response to gamma radiation was constant over a wide range of exposures and temperatures. These dosimeters showed that active animals received 2–10 R d^{-1} from the cesium source, and cumulative exposure over 1636 days (five years) for stationary

objects (plants) was 29 kR at 20 m from the tower and 3 kR at 170 m. Through 1976 research with ionizing radiation continued with plants and lizards on plot B, and companion studies were established on the other plots in Rock Valley, especially within plots A and C, where rodent trapping occurred regularly to monitor population parameters and studies on reptiles were aggressively pursued by herpetologists.

Before radioecology studies were begun in the early 1960s at Brookhaven and Rock Valley, researchers had determined that mammals were more radiosensitive to acute doses than other groups of animals (Odum 1959). Hence, those who awaited the results at Rock Valley expected that *Perognathus formosus* would show the greatest effects of chronic exposure to low-dose irradiation. For pocket mice, French and coworkers (French, Maza, Hill, Aschwanden, & Kaaz 1974) found instead very slight differences in survivorship curves between irradiated versus control plots, no difference in lethality, and no perceived sterility (Chapter 13). Quite unexpected was the discovery in spring, 1969, by Joseph R. Lannom, jr. that female lizards of *Crotaphytus wislizenii* lacked reproductive color changes and, subsequently, that they were permanently sterile (Turner, Licht, Thrasher, Medica, & Lannom 1973). In fact, four species of lizards showed marked female sterility following two years of continuous gamma irradiation under natural conditions (Chapter 13). Thus, contrary to conventional wisdom when the experiment began, lizards and not mammals showed serious long-term effects of radiation, probably because they have poorer molecular repair capacity than mammals.

Whatever the reader's feelings are about purposely exposing living organisms to unnatural levels of radiation, the experiment was the best of its kind at that time to obtain real values on potential lethal or damaging levels of radiation to organisms under otherwise natural conditions, and hence a way to predict the potential dangers to them. That issue was one of the formidable questions when the era of nuclear-testing began. But in addition to the applied data for radiation biology, the study established a standard of excellence for monitoring age–class mortality rates and life expectancy for mobile animals (pocket mice and lizards) living in essentially natural settings, and from this would come important observations on the nature of change in populations of desert animals.

By the time that the gamma radiation experiments were completed on rodents in spring, 1969, the worldwide issue of aboveground testing had also been resolved by treaties and resolutions. In fact, by 1970 public interest had shifted from concerns about nuclear attack to that of extremely low-level exposure from nuclear power plants and alternative forms of energy production.

During the 1960s a U.S. national committee, operating under the International Union of Biological Sciences, was set up to organize an American contribution to the International Biological Program (IBP). One component of this program was to be a detailed study of desert ecosystems. The specific object of this Desert Biome project was to improve understanding of the dynamics of arid ecosystems to a point where ecologists and environmental scientists would have a firm basis of theory and data on which to advise land managers on these natural resources (Norton 1975). A major premise in selecting one or more representative sites was that biological relationships elucidated from studies of undisturbed systems would allow projections of the impact from disturbance at other sites.

Rock Valley had a sustained biological and climatological record since 1962 and for this reason was chosen to be a validation site for the Desert Biome program of the International Biological Program (IBP), representing the Mojave Desert (Chapter 2). The validation site was the area where researchers would make sequential measurements of state variables, for example, densities, standing crop, biomass, etc., over several years, and use these values to 'validate' models to be tested for desert systems from process studies.

The IBP validation site covered 0.46 km^2 and was located directly north of plots A and C. It was a five-sided, roughly trapezoidal parcel that had six somewhat recognizable vegetational groupings (El-Ghonemy, Wallace, Romney, & Valentine 1980). Within that area investigators did mostly nondestructive types of sampling in an elaborate set of plots, and additional destructive sampling was done just off the validation site in comparable vegetation. One could argue – and some did – that Rock Valley was not the most typical form of *Larrea–Ambrosia* desert scrub for the Mojave Desert as-a-whole, but nowhere could the organizers find an equivalent and well-protected parcel with so much long-term ecological data already accumulated. Consequently, from 1971 to 1976 Rock Valley was used by dozens of researchers, who expanded studies on rodents, lizards, and shrubs and included especially massive harvesting efforts to quantify productivity and biomasses of shrubs, winter annuals, cryptogams, and arthropods and to lesser degrees other represented groups of organisms.

The IBP Project Coordinator for the Mojave

Desert validation site was Frederick B. Turner from UCLA. He had served since 1962 as a principal investigator on studies of reptiles in Rock Valley, and steered the ecological programs at Rock Valley after French left the project, upon completing analysis of irradiated rodents.

Whereas the IBP program in general was perhaps only marginally successful in attaining some of its objectives, in particular those related to the modeling efforts (National Academy of Sciences 1977), overall the effort convincingly demonstrated the effectiveness of multidisciplinary approaches to research on complex problems. At the time the U.S. IBP program was the largest research effort ever launched in biology, involving 1800 American scientists and more than 57 million dollars in federal grants.

Since 1976 ecological studies at Rock Valley, particularly within plots A and C, have been con-ducted primarily by resident ecologists at NTS. In 1976 there was a shift in emphasis for ecological research to evaluate forms of energy other than radioactivity. Since 1976, research at Rock Valley has been maintained, but at a relatively low level; primary emphasis has been on biology of desert tortoise (*Gopherus agassizii*), now listed as a threatened species, and desert annuals. The most recent studies have been especially valuable in recording the degree of change in this desert ecosystem since the IBP data were collected.

Clearly, Rock Valley is one of the best-studied desert sites in the world, and the breadth and depth of past studies, described in this volume, has had a significant impact on desert research throughout the world. Much remains to be done, of course, but the long-term record of ecological communities and processes at Rock Valley represents an invaluable and unique resource.

1 Introduction to the Mojave Desert

Biological research in deserts has long stimulated the keen interest of natural scientists. In part this is because we are fascinated by any living organism, including our own type, that is capable of surviving the harsh physical conditions of deserts, and in part it is because knowledge gained from desert biology aids our understanding of the physiology and ecology of nondesert species and of how their systems might respond to stressful environmental conditions like those imposed by deserts.

Fortuitously, around the world there are many different types of desert environments, ranging from very hot to cold, extremely arid to semiarid, winter to summer precipitation, and having clear skies to having cloud or fog cover. Within a desert, soils can vary from sands or gravels to clay. Each desert environment offers researchers a different set of environmental parameters under which plant and animal responses can be studied and analyzed for comparisons with research results at other localities. So it is that desert research at the Nevada Test Site (NTS) has been a key for understanding the environmental biology of the Mojave Desert.

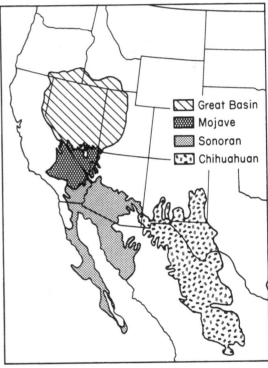

Figure 1.1 Deserts of North America.

THE MOJAVE DESERT REGION

The Mojave Desert is situated between the Great Basin Desert to the north and the Sonoran Desert to the south (Fig. 1.1). It is located in eastern California on the inland side of the coastal Transverse Ranges and the Sierra Nevada between 34°N and 37°N latitude and in the southern one-sixth of Nevada, along the Virgin River drainage near St. George in southwestern Utah, and near Kingman

(Mohave County) in the northwestern corner of Arizona (Fig. 1.2).

Of the four North American desert regions, the Mojave Desert is the smallest and in many ways a transitional zone between its larger desert neighbors (Bender 1982). Total area within the Mojave Desert Region (MDR) is approximately 124 000 km², this including some nondesert, higher elevation zones of isolated mountain

[1]

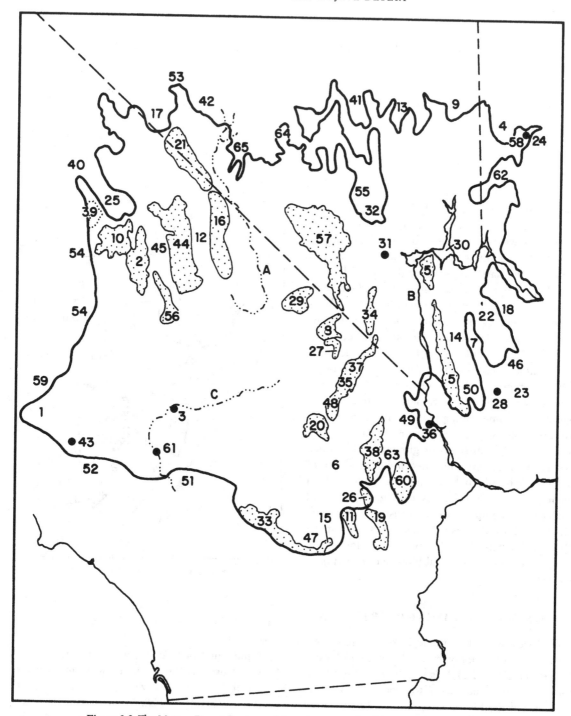

Figure 1.2 The Mojave Desert Region and its major geographical locations and important cities.

ranges. The Mojave Desert therefore constitutes about one-tenth of the total desert area in the United States and Mexico and in size is roughly equivalent to the state of New York. Overall it is a much lower desert than the Great Basin Desert; the elevations of Mojave Desert valley basins are usually above 600 m and seldom exceeding 900 m, as compared with the mean valley elevation of 1200 m for the Great Basin. The Colorado subdivision of the Sonoran Desert also has many low basins, frequently below 400 m and as low as −67 m at the Salton Sea in southern California, hence its recognition as the 'low desert'. Unlike the Sonoran Desert, where most drainage is ulti-mately into the Gulf of California, most drainage of the Mojave Desert and southern Great Basin is internal within the basin and range topography. Exceptions to this are the Muddy and Virgin rivers and several other smaller drainages (for example, Sacramento Wash and Detrital Wash), which drain portions of the eastern Mojave into the Colorado River.

BOUNDARIES AND TOPOGRAPHIC RELIEF

Landforms of the Mojave Desert are characteristic basin and range structures, formed by isolated

Key to locations

Location			
1	Antelope Valley	34	McCullough Mountains
2	Argus Range	35	Mid-Hills
3	Barstow, California	36	Needles, California
4	Beaver Dam Mountains	37	New York Mountains
5	Black Mountains	38	Old Woman Mountains
6	Cadiz Valley	39	Owens Dry Lake
7	Cerbat Mountains	40	Owens Valley
8	Clark Mountain Range	41	Pahranagat Range
9	Clover Mountains	42	Pahute Mesa
10	Coso Range	43	Palmdale, California
11	Coxcomb Mountains	44	Panamint Range
12	Death Valley	45	Panamint Valley
13	Delamar Mountains	46	Peacock Mountain
14	Detrital Valley	47	Pinto Basin
15	Eagle Mountain	48	Providence Mountains
16	Funeral Mountains	49	Sacramento Mountains
17	Gold Mountain	50	Sacramento Valley
18	Grand Wash Cliffs	51	San Bernardino Mountains
19	Granite Mountains, Riverside County	52	San Gabriel Mountains
20	Granite Mountains, San Bernardino County	53	Shoshone Mountain
		54	Sierra Nevada
21	Grapevine Mountains	55	Sheep Range
22	Hualapai Valley	56	Slate Range
23	Hualapai Mountain	57	Spring Mountains
24	Hurricane Cliffs	58	St. George, Utah
25	Inyo Mountains	59	Tehachapi Mountains
26	Iron Mountain	60	Turtle Mountains
27	Ivanpah Mountains	61	Victorville, California
28	Kingman, Arizona	62	Virgin Mountains
29	Kingston Range	63	Ward Valley
30	Lake Mead	64	Yucca Flat
31	Las Vegas, Nevada	65	Yucca Mountain
32	Las Vegas Range	A	Amargosa River
33	Little San Bernardino Mountains	B	Colorado River
		C	Mojave River

mountain ranges oriented mostly along a north–south axis (Fig. 1.2). Most of the ranges are fairly narrow, and in the MDR the highest point is Charleston Peak (3652 m), located west of Las Vegas, Nevada, in the Spring Mountains. Extensive alluvial fans, termed bajadas (Fig. 1.3), form around the perimeter of these mountains and extend outward to low-elevation intervening basins, which commonly contain dry lake beds termed playas (Fig. 1.4). Playas receive water runoff and fine sediments from the surrounding mountains and desert and may contain standing water during the rainy season, but playas are also sites where large volumes of water evaporate, leaving behind high concentrations of salts in playa soils. Death Valley, having the lowest point (−86 m) in the United States and several playas, lies between the Amargosa Range and the Panamint Mountains in eastern California (Hunt 1975). Other well-known

playas of the MDR are China Lake and Rogers Dry Lake, the United States space shuttle landing site on Edwards Air Force Base near Lancaster, California.

Boundaries

The MDR can be delimited by a set of encircling mountain ranges. The western margin of the southern Mojave Desert is bounded by the southern Sierra Nevada and the Tehachapi Mountains in Kern County, California. The western portion of the southern boundary is clearly defined by the Transverse Ranges with the Mojave Desert, skirting the northern slopes of Liebre Mountain, Sierra Pelona, the San Gabriel Mountains, and the San Bernardino Mountains. At Morongo Valley, between the San Bernardino and Little San Bernardino Mountains, the Mojave Desert

Figure 1.3 A typical bajada with Mojave Desert south of Barstow, California, supporting desert scrub of creosote bush (*Larrea tridentata*) and white bursage (*Ambrosia dumosa*).

Figure 1.4 An unusual sight when Soda Lake near Baker, California, normally a large dry playa, is filled with rainwater during an exceptionally wet rainy season.

extends southward, thus including most of the Little San Bernardino Mountains and all of the Eagle Mountains above an elevation of about 600 m.

The eastern portion of the southern boundary, between the Mojave and Sonoran deserts, is ill-defined and admittedly arbitrary. Unlike the steeper slopes in the western portion of the Mojave Desert, where species characteristic either of the Mojave Desert or the neighboring region appear or disappear within a relatively short distance, the long, gentle slopes of the valleys in the southeastern portion result in the gradual loss of Sonoran Desert species over many kilometers. There is no obvious sudden appearance of a diagnostic Mojave Desert species, such as Joshua tree (*Yucca brevifolia*), hence the difficulty of demarcating the southern boundary between the two deserts. From Eagle Mountain the southern boundary extends northeast to about the southern tip of Nevada. As drawn it crosses the eastern end of Pinto Basin and skirts the northern end of Coxcomb Mountains of eastern Joshua Tree National Monument, passing between the Iron and Granite mountains, north along Ward Valley, excluding

Dandby Lake and elevations below about 300 m, then south and around the east slope of the Turtle Mountains to about 600 m elevation. From the Turtle Mountains the boundary between the Mojave and Sonoran deserts continues north to Nevada following the 600 m contour on the eastern slopes of the Turtle, Stepladder, Sacramento, and Dead mountains or, wherever elevations are less than 600 m, along the crest of these mountains.

In Arizona the southern boundary also is currently ill-defined and arbitrary. A narrow finger extends southeastward from Nevada; this sector includes the Black Mountains above the 600 m contour and Sacramento and Detrital valleys between about 600 and 1350 m. Another finger extends southeastward into the Hualapai Valley, where there is a long, gradual transition to the Sonoran Desert. Many characteristic elements of both deserts are found at the same localities. Thus in the vicinity of Wickiup, Arizona, there are Joshua trees of MDR and Sonoran elements such as saguaro (*Carnegiea gigantea*), and at Congress Junction there is an isolated patch of Mojave Desert vegetation.

The eastern boundary of the Mojave Desert is well defined because of the abrupt elevational changes that occur along the western slopes of the Grand Wash Cliffs and Virgin Mountains of Arizona and the Hurricane Cliffs of Arizona and Utah. The only area that is perhaps poorly defined is within the Grand Canyon, where Mojave Desert vegetation extends for about 150 km along the lower slopes of the canyon.

The northern boundary across Nevada shows a transition into the Great Basin Desert. Where there are mountains, the delimitation of the Mojave Desert is easy, being defined by the abrupt changes in vegetation, but in the broad north-south valleys the transition is more gradual and sometimes unclear. In these valleys the creosote bush (*Larrea tridentata*), which is the primary index species of the three warm deserts of North America, drops out many kilometers south of the northern limit of Joshua tree, and here blackbrush (*Coleogyne ramosissima*) and sagebrush (*Artemisia tridentata*) appear.

Proceeding from the northeastern corner of the Mojave Desert at the Hurricane Cliffs near La Verkin, Utah, the northern boundary follows the Virgin River drainage, including the lower slopes of Beaver Dam Mountains, along the southern base of the Clover and Delamar mountains and the South Pahroc, Hiko, Pahranagat, and Desert ranges. The Mojave Desert extends up the valleys between these ranges, thus extending beyond Delamar Lake in Delamar Valley, to Alamo in the Pahranagat Valley, and to the 1300 m contour in Tikaboo Valley. From Tikaboo Valley the boundary passes south of the Groom Range, down Emigrant Valley to Nye Canyon, along the base of the Halfpint Range, across the northern end of Frenchman Flat at NTS, and then along the 1300 m contour on the southern base of the Shoshone, Yucca, and Timber mountains, Pahute Mesa, and Stonewall Mountain, crossing near Sarcobatus Flat, along the south face of Gold Mountain, across the north end of Death Valley, over the lower slopes of the Last Chance Range, and the north end of Eureka Valley to the Inyo Mountains in California. Other familiar ranges near Death Valley are the Grapevine and Funeral mountains of the Amargosa Range and the Panamint Mountains.

From the Inyo Mountains the boundary follows the east slope of the range down Eureka Valley, around the higher slopes of the Saline Range and back to the Inyo Mountains, continuing south along Saline Valley, and then including the higher slopes of the Nelson Range, Hunter Mountain, and the Argus and Coso ranges. From the Coso Range the Mojave Desert forms a narrow finger into southern Owens Valley, where it extends up the eastern side to the valley, barely beyond Owens Lake. The boundary contacts the Sierra Nevada again just south of Owens Lake.

Topographic relief

Topographic relief is relatively small in the southwestern Mojave, with elevations commonly ranging from 625–1300 m. Extensive urbanization has taken place here in recent years as the influence of the Los Angeles suburban area and farming enterprises have extended into the Antelope Valley around Palmdale, Lancaster, Pearblossom, and Littlerock, Rose Valley, and Fremont Valley.

The central and eastern regions of California include more than 20 mountain ranges having different areal extent and elevation. From northwest to southeast these include the Kingston Range (643 km², 2232 m maximum elevation), Mesquite Mountains (272 km², 1573 m), Clark Mountain Range (403 km², 2416 m), Ivanpah Range (incl. Mescal Range, 316 km², 1840 m), New York Mountains (556 km², 2296 m), Mid Hills (359 km², 1954 m), Providence Mountains (349 km², 2148 m), and Granite Mountains (one of three so named in the Mojave Desert; 251 km², 2069 m). Near the southern limit of the Mojave Desert and to the north of Joshua Tree National Monument and the Granite Mountains are the extensive Kelso Dunes at a basin elevation of 950 m and covering 142 km² (Thorne, Prigge, & Henrickson 1981). This strip of desert and mountain ranges in the Mojave Desert of California constitutes its biologically most diverse region.

In Nevada the Spring Mountains and Sheep Range are two of the major mountain systems within the MDR. These Nevada ranges have limestone parent materials, which through erosion have been sculptured into a number of spectacular caverns. Similar limestone and dolomitic rocks characterize most ranges within the MDR.

Lakes and rivers

During the pluvial periods of the Pleistocene, an extensive drainage system of rivers and lakes existed within the Mojave Desert (Fig. 1.5), as well as in much of the Great Basin. Both Searles Lake and Panamint Lake, which are now playas, were very large at one time, having areas of 1000 and 700 km² and maximum depths of 195 and 275 m, respectively. The Owens River drainage in the Owens Valley now terminates at Owens Lake to the south, but in pluvial periods it formed drain-

age connections with Death Valley through a chain of lakes to the south and east. Lake Manly in Death Valley was fed both from the Owens River, via Searles Lake and Panamint Lake, as well as the interconnected drainages of the Amargosa and Mojave rivers (Hubbs & Miller 1948; Hunt 1975). Tecopa Lake and Pahrump Lake were the largest Pleistocene lakes of the Amargosa River system.

Now the entire MDR has little riverine activity. The Amargosa River system flows southward for 160 km from above Beatty in southern Nye County, Nevada before turning north again into Death Valley. In years of heavy winter runoff, surface water may still flow as far as the playa basin of Lake Manly, but there have been no discharges in historical times into Bad Water, which marks the northern limit of the Pleistocene drainage system. The Mojave River system begins on the northern slopes of the San Bernardino Mountains and winds northward. Once beyond its mountain tributaries and in the desert environment, flow in the Mojave River is intermittent, but some water is able to flow eastward for 190 km into the Mojave Desert before all flow disappears. This is a classic example of an allogenic drainage, which originates outside the desert area and disappears within it (Cooke & Warren 1973; Neal 1975; Duffy & Al-Hassan 1988). Three large but shallow lakes, Mojave, Little Mojave, and Manix lakes, formed in this drainage during the Pleistocene (Hubbs & Miller 1948), but today the playa Soda Lake is the principal salt-encrusted basin (Fig. 1.4).

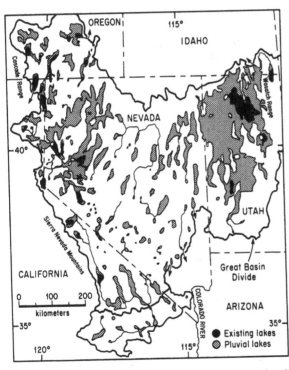

Figure 1.5 Pluvial lakes and drainages during the height of the Pleistocene in the Mojave Desert Region and Great Basin. The southern area separated by a dark line has external drainage. The present-day occurrence of lakes is shown as dark area. From Dobrowolski, Caldwell, and Richards (1990).

CLIMATE GRADIENTS AND VARIABILITY

Precipitation

High mountains to the west and south of the Mojave Desert provide a rain shadow that effectively intercepts moist air masses from the Pacific Ocean. The stable subtropical high pressure over the northeastern Pacific Ocean, the same system that provides the mediterranean-type climate for coastal California, is the source of the predominant westerly and southwesterly winds that affect the MDR. Airmass subsidence from high pressure systems brings a predominance of clear and dry conditions. High-pressure systems formed intermittently during the fall and winter months over the Intermontane Region result in easterly winds that warm adiabatically as they descend from the higher elevations in the Intermontane Region to further deflect approaching storms. These winds are known as Santa Ana winds when they reach southern California.

As in southern California, winter and spring precipitation results when the northeastern Pacific High is displaced to the south, and low pressure troughs form over the western United States. Under these conditions, large cyclonic storms from the Gulf of Alaska may bring in moist and unstable air masses. In the latitudes of central and southern California most of the water content of the air is precipitated against the western slopes of the Sierra Nevada or Transverse Ranges. As these air masses then cross over the Mojave Desert, they are greatly reduced in moisture, especially as compressional heating of air descending into the desert basins further reduces relative humidity and increases stability. Occasionally, however, a storm or rapid succession of storms sweep into the MDR and bring light to moderate autumnal and winter precipitation for one to several days. Gusty winds and cool temperatures usually accompany such storms, and in the coolest sites precipitation may occur as snow.

In the Mojave Desert infrequent events of summer precipitation arise from totally different weather systems. These events occur when the subtropical high is weakened, resulting in the monsoonal influx of moist air from either the Gulf of Mexico or the Gulf of California, bringing air flow from the south and east into the eastern MDR. Thermal heating of these humid air masses produces strong convectional storms, which can be enhanced by convergence or topography. Such convectional storms can bring brief but intense precipitation bouts to local areas. Rarely tropical hurricanes, called chubascos, form off the western coast of Mexico, move northward, and penetrate into the southern MDR, bringing widespread torrential rains in August or September, and these may be the principal rains in the driest regions of

the Colorado subdivision of the Sonoran Desert in Imperial County, California, and adjacent Mexico. Storm fronts that originate over the Gulf of Mexico bring summer rains to the Chihuahuan Desert, mainland Mexico, and eastern Sonoran Desert starting in mid-summer, and occasionally they carry precipitation to westernmost Arizona and California when the subtropical high is weakened.

The above description of seasonal precipitation defines the rainfall pattern of the three warm desert regions found within the United States (Fig. 1.6; see also MacMahon & Wagner 1985). Far to the east, the Chihuahuan Desert Region receives most precipitation from the Gulf of Mexico in late summer months, whereas in California the Mojave Desert receives primarily winter rainfall. Between these is the Sonoran Desert, whose eastern portion receives the majority of the precipitation from the summer storms that have already passed across the Chihuahuan Desert, and the western portion receives some of the leftover precipitation from the winter storms from the Pacific Ocean that dampened the Mojave Desert. In a central zone around Tucson, Arizona, both summer and winter rainfalls are significant, and thus a biseasonal rainfall regime is present. Late summer chubascos from the Gulf of California also reach as far east as Tucson and westward along the international borderland to the southernmost Mojave Desert.

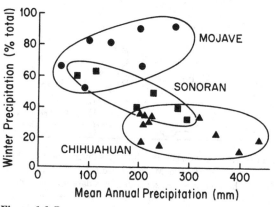

Figure 1.6 Precipitation pattern of desert regions in western North America. After MacMahon and Wagner (1985).

The Great Basin Desert, which is a cool desert, often receives measureable precipitation in each season (Turner 1982a), some winter precipitation from the fronts that produce the Mojave Desert pattern, some from the summer rains from the Gulf of California, and some at other times from the storm fronts that provide precipitation to the Intermontane Region.

Visualizing the major sources of rainfall helps to explain the patterns of precipitation within the MDR. The western and central portion, including Palmdale, Lancaster, and Bishop, have mean annual precipitation of about 135 mm, with only about 5 mm on average of this total falling in summer because chubascos and westerly storms from Arizona are very uncommon. Summer precipitation hardly ever reaches the Owens Valley. Trona and Inyokern have only 70–75 mm yr^{-1} precipitation (Turner 1982b), with the same low frequency of summer rain. In the southeastern portion, adjacent to the Coloradan Sonoran Desert, summer precipitation increases in frequency and amount due mostly to chubascos, thus producing an irregular biseasonal precipitation regime. Twenty-nine Palms at the southern margin of the Mojave Desert has virtually equal long-term mean winter and summer precipitation (85 mm total), consisting of fairly predictable winter rains and snows, but intense summer thunderstorms in some years. Along the eastern Mojave Desert from Needles, California northward through Kingman, Arizona, and Searchlight, Boulder City, and Las Vegas, Nevada, about one-third or more of their long-term mean precipitation comes from the east in summer after mid-July. The northern Mojave Desert, including NTS, commonly receives two to four times more winter than summer precipitation, a pattern of regular winter rains and highly irregular occurrence of summer storms. Habitats of Great Basin Desert in NTS and northward tend to receive fairly even precipitation in all four seasons of the year.

Total mean precipitation levels in the Mojave Desert show a strong correlation with elevation. This general pattern can be seen graphically in data from 63 weather stations in southern Nevada (Fig. 1.7). The relationship appears to be curvi-

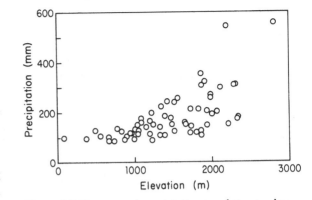

Figure 1.7 Mean annual precipitation in relation to elevation for 63 weather stations throughout the Mojave Desert Region.

linear, with stations below 1000 m having a similar precipitation level of 90–120 mm yr^{-1}. Above 1000 m elevation there is a general increase in precipitation, although local rain-shadow effects cause considerable scatter in those data.

Long-term climatic records from three stations in southern Nevada emphasize the high variability from year to year. Mina (mean 109 mm yr^{-1}), Beatty (159 mm yr^{-1}), and the Las Vegas Airport (101 mm yr^{-1}) all have 60-year records that range from virtually zero precipitation in some years to more than double long-term mean values (Fig. 1.8). Maximum annual precipitation at any Mojave lowland locality rarely exceeds 250 mm yr^{-1}.

Temperature

Of the four major desert regions in North America, the Sonoran Desert is the most subtropical because even the most northern sites rarely receive snow or subfreezing winter temperatures. The Chihuahuan Desert has a much higher base elevation and experiences many subzero winter temperatures. To the west the Mojave Desert is typically warm, but may experience cool Arctic air masses during the winter rainy season and hence receives some snow. To the north the Great Basin experiences cold winter temperatures and regularly receives snow covering. As before, areas of the MDR adjacent to another desert are also transitional in these climatic parameters.

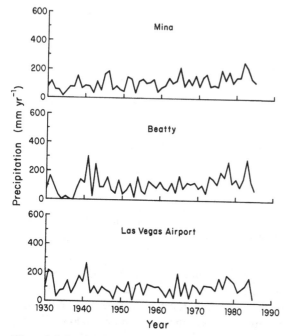

Figure 1.8 Long-term precipitation records at Mina, Beatty, and Las Vegas, Nevada, located in northern Mojave Desert Region.

Both seasonal and diurnal extremes of temperature are large in the Mojave Desert. Beatty, a typical but high Mojave Desert site at 1085 m elevation, has mean summer high (daytime) temperatures reaching 37°C in July, but mean nighttime temperatures drop sharply to 16°C, due to cloudless night skies. In December and January mean low temperatures at Beatty drop to −2 to −3°C at nighttime and rise to means of 13°C during the day. Each year Beatty has an average of 88 days with temperatures below 0°C. In the southern MDR Lancaster and Palmdale have mean daily maximum temperatures in July of 35–36°C and mean daily minimum temperatures in January of 0 to −1°C. Similarly dramatic daily and seasonal temperature differences are recorded at sites in the 'lower' Colorado Desert of southern California, but here freezing is rare and summer temperatures are greater; they have mean daily minimum temperatures in January that are several degrees above zero and mean daily maximum tem-

peratures in July that are 5–6°C higher (Turner 1982b).

Greater temperature extremes are observed elsewhere within the Mojave Desert. In Death Valley mean July maximum temperatures reach a remarkable 46°C. Sarcobatus, Nevada, just north of Beatty, experiences 144 subzero nights per year, whereas Las Vegas averages only 46 such nights (Houghton, Sakamoto, & Gifford 1975).

As shown for Beatty (Fig. 1.9), diurnal temperature changes of 20–25°C are common in MDR during summer months, but smaller ones occur during the winter. Cold air drainage is an important component of large diurnal temperature oscillations in the basin-and-range portions of this desert; dense, cold air masses flow downslope at night, bringing temperature inversions into the basins and warmer air displaced upward on the slopes, thus affecting local distributions of the species in the biota.

BIOTA

The biota of the lowland Mojave Desert is one that is neither clearly defined on the basis of endemic species nor distinctive in having a high diversity of perennial plants or resident animals. Such characteristic plant and animal species of the southeastern MDR are those also present in the western Sonoran Desert, which shares a similar

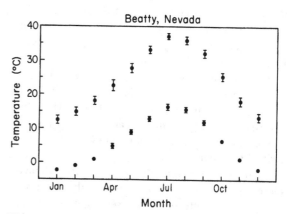

Figure 1.9 Mean monthly high and low temperatures at Beatty, Nevada.

but somewhat warmer climatic pattern and common physiographic features. Likewise, the northernmost sites in Nevada possess typical perennial plant species from the cold-adapted Great Basin, and transitional biotic communities are located at NTS between the two desert provinces (MacMahon 1979).

Vascular plants

Endemic plants mostly occur along the California–Nevada border of MDR, especially in the mountain ranges of Inyo County, California (Raven & Axelrod 1978). In particular, the Inyo region has at least 200 species that do not occur elsewhere in California, and many plants of Inyo County do not occur nearby at NTS. The flora of the Mojave Desert is fairly rich in winter annuals, having over 230 species, of which at least 80 are endemic. None of the 17 species of summer annuals is endemic to the MDR (Johnson 1976).

The upper elevational, western, and northern boundaries of Mojave Desert communities are indicated by the disappearance of creosote bush. In the Mojave Desert the Joshua tree is the most distinctive species, and in most localities this species also disappears near the upper elevational limits of the desert, where fairly cold winter temperatures occur and pinyon pines (especially *Pinus monophylla*) and junipers (*Juniperus* spp.) replace Joshua tree as the dominant of the community. Actually creosote bush is usually absent about 150 m below the pinyon–juniper woodland. Along the southern and southeastern boundary of MDR, ironwood (*Olneya tesota*), a distinctive leguminous tree of desert washes in the Sonoran Desert, and the green-stemmed perennial chuparosa (*Justicia californica*), both abruptly disappear, apparently because they cannot tolerate subzero night temperatures. Similarly, foothill palo verde (*Cercidium floridum*) also terminates its very broad range just outside the Mojave Desert. The eastern boundary of MDR is vaguely defined, and in Arizona Joshua tree, a Mojave element, occurs with some of the westernmost specimens of the giant saguaro (*Carnegiea gigantea*). In the northern portion of the Mojave Desert, blackbrush forms a transitional boundary between stands of sagebrush, which typify Great Basin desert scrub, and Mojave creosote bush scrub.

In the isolated mountains of the MDR topographical features are more diverse than in typical desert lowland habitats and precipitation is much greater, so that the biota is decidedly richer, and species composition is very different. Animals that require daily water and dense cover find refuge in the mountains from the more open and drier lowlands. Mountain habitats also harbor calciphiles, plants that tend to be restricted to high limestone soils.

Floristic treatments

Perhaps because the MDR is poorly delimited as a biotic province, there have been no definitive regional treatments of the species. Lacking a floristic survey of the MDR, one can none the less account for most species by combining several published floras, beginning with the broad regional treatments for California (Munz & Keck 1959; Munz 1974; Hickman 1993), floras that cover nearly two-thirds of the province, the flora of southern Nye County, Nevada, including NTS, by Beatley (1976a), and floras of Arizona (Kearney & Peebles 1960; MacDougall 1973) and Utah (Welsh, Atwood, Goodrich, & Higgins 1987; Albee, Shultz, & Goodrich 1988). The flora of the northern MDR of California was surveyed by DeDecker (1984), and the vascular plants of the Charleston Mountains, Nevada, by Clokey (1951). Additional California regional floras were completed for the eastern Mojave Desert mountain ranges (Thorne et al. 1981) and Antelope Valley and Indian Wells in Kern County (Twisselman 1967). In all the MDR flora probably contains about 1900 native and naturalized species, including the nondesert, montane taxa.

Floristic composition and endemics

Rowlands, Johnson, Ritter, and Endo (1982) provided a numerical summary of vascular plant families in the MDR of California, including 1836 taxa. Of these there were 1680 native and 156 intro-

Table 1.1. *The twenty most diverse families of vascular plants of the Mojave Desert in California (after Rowlands, Johnson, Ritter, and Endo, 1982). These families constitute 74% of the flora.*

Family	Number of species	
	Native	Introduced
Apiaceae	30	2
Asteraceae	270	17
Boraginaceae	63	0
Brassicaceae	61	17
Cactaceae	23	0
Chenopodiaceae	37	11
Cyperaceae	34	2
Euphorbiaceae	36	2
Fabaceae	124	11
Hydrophyllaceae	55	0
Lamiaceae	29	1
Loasaceae	32	0
Nyctaginaceae	24	0
Onagraceae	37	0
Poaceae	139	47
Polemoniaceae	67	0
Polygonaceae	81	3
Rosaceae	29	4
Scrophulariaceae	80	1
Solanaceae	23	4

duced species belonging to 575 genera in 109 families. Of the native species, 28 (1.7%) were ferns or fern allies, 18 (1.1%) were gymnosperms, 252 (15.0%) were monocotyledons, and 1382 (82.3%) were dicotyledons. Species of Asteraceae were the most numerous, with 287 species (Table 1.1), followed in diversity by Poaceae (186 spp.), Fabaceae (135 spp.), Polygonaceae (84 spp.), and Scrophulariaceae (81 spp.). The 10 most diverse plant families together comprised 977 native and 107 introduced species (Table 1.1), thus 58% of the native flora and 59% of the total vascular plant flora. Introduced species have become widely naturalized, even very undesirable weeds, and five families included two-thirds of the total: Poaceae (47 spp.), Asteraceae (17 spp.), Brassicaceae (17 spp.), Fabaceae (11 spp.), and Chenopodiaceae (11 spp.).

Raven and Axelrod (1978) identified seven endemic Mojave plant genera, all to the Inyo Region: *Dedeckera* and *Gilmania* (Polygonaceae), *Swallenia* (Poaceae), *Oxystylis* (Capparaceae), *Hecastocleis* (Asteraceae), *Scopulophila* (Caryophyllaceae), and *Arctomecon* (Papaveraceae). Endemic species here occur in montane habitats, for example, wedged in limestone crevices and outcrops, and lowland habitats, such as alkali basins.

Vertebrates

The vertebrate fauna has not been summarized for the MDR, but checklists of birds and terrestrial animals have been published for the principal federal reserves, NTS (Hayward, Killpack, & Richards 1963; Jorgensen & Hayward 1963, 1965; Tanner & Jorgensen 1963; Allred, Beck, & Jorgensen 1963; Tanner 1969; O'Farrell & Emery 1976), Joshua Tree National Monument (Miller & Stebbins 1964; Rowlands et al. 1982), and Death Valley National Monument (Rowlands et al. 1982). Brown (1982) published lists of the common animals in each of the biotic communities of the southwest, and the data showed that the Mojave Desert vertebrate fauna is largely that of the western Sonoran Desert. The most unexpected vertebrates are the remarkable desert pupfishes (*Cyprinodon*), which inhabit underground pools, such as Devils Hole, Nevada, and are thermophilic and eurythermic (Brown 1971).

As with plants, the terrestrial vertebrate fauna of MDR has low species diversity and very low endemism, and the species are those shared either with the Sonoran Desert to the east or the Great Basin to the north. One Mojave Desert endemic is the Mojave fringe-toed lizard (*Uma scoparia*), which inhabits sand dunes. Some of the vertebrates, particularly rodents and reptiles, have distinctive Mojave Desert geographic subspecies or races, such as the Mojave shovel-nosed snake (*Chionactis occipitalis occipitalis*), Mojave sidewinder (*Crotalus cerastes cerastes*), the Panamint rattlesnake (*C. mitchelli stephensi*), and the desert night snake (*Hypsiglena torquata deserticola*), but the majority of the Mojave forms are generally not recognized as distinct. No species of resident birds is endemic to MDR.

VEGETATION PATTERNS

Vegetation types of lowland desert habitats are strongly related to patterns of geomorphic landforms. Distinctive plant communities and their faunas are characteristic for bajada slopes, playa basins, lowland arroyos, natural springs, and mid-elevational mountain slopes. In addition, environmental factors that affect temperature (elevation and slope exposure) and water availability (elevation, slope exposure, and seepage zones) as well as soil properties (soil texture, substrate parent material, and salinity) are important for determining presence or absence of species in each type of Mojave Desert community.

Many attempts have been made to devise a workable classification of plant communities for the Mojave Desert, and there is no consensus. Table 1.2 shows a range of opinions from several widely used treatments of California vegetation. Munz (1974), Brown, Lowe, and Pase (1979), Vasek and Barbour (1977), Kuchler (1977), and Turner (1982b) recognized few, broadly defined plant communities or series having many variants. Other workers have recognized the distinctiveness of numerous variants and have emphasized more narrowly defined plant associations (Rowlands et al. 1982; Thorne 1982). Beatley (1974a, 1976a) focused instead on plant associations of central-southern Nevada and NTS. These physiographic habitat types include: bajadas, arroyos, playa basins, natural springs, and mid-elevational mountain slopes.

Bajadas

Creosote bush scrub (*Larrea tridentata* dominant; Fig. 1.3) is not only the characteristic and widespread vegetation community on bajadas of the Mojave Desert, covering nearly two-thirds of the total area, but also the dominant vegetation of the Sonoran Desert to the east (Mabry, Hunziker, & DiFeo 1977). *Larrea* may occur in relatively pure stands or with a variety of codominant shrubs, in the Mojave Desert the most common being white bursage (*Ambrosia dumosa*), a dwarf shrub. Highest measured density of *Larrea* has been 1790

shrubs ha^{-1}, but densities usually range from 200 to 700 shrubs ha^{-1} (Vasek & Barbour 1977). The densities of shrubs and the relative importance of associated species are controlled by climate and soil physical properties (Beatley 1974a, 1975, 1976a) as well as topographic diversity and relative community age (Johnson, Vasek, & Yonkers 1975; Vasek, Johnson, & Brum 1975; Vasek, Johnson, & Eslinger 1975; Vasek, & Barbour 1977).

Creosote bush scrub occurs from below sea level in Death Valley to above 1500 m on south-facing exposures in the northern Mojave Desert, but typical creosote bush scrub in this desert occurs below 1200 m in elevation (Beatley 1976a, 1979; Turner 1982b). In lowland desert habitats *L. tridentata* typically does not occur on saline soils, but instead grows best on adjacent well-drained sandy soils. Lunt, Letey, and Clark (1973) have noted that fine-textured basin soils may not provide adequate oxygen for the roots.

Mean annual precipitation in bajada communities of the Mojave Desert range from 40 to 218 mm yr^{-1} (Vasek & Barbour 1977), and of central-southern Nevada from about 120–160 mm yr^{-1} (Beatley 1974a). However, the variance of these mean levels is very high. At 24 bajada sites on NTS, from 1962 to 1972 annual levels of precipitation ranged from 49 to 324 mm (Beatley 1976a).

There are several different forms of creosote bush scrub within the Mojave Desert (Beatley 1979; Turner 1982b). On bajadas throughout the region the typical codominants are creosote bush and white bursage. Common woody associates include desert thorn (*Lycium andersonii*) and bladder sage (*Salazaria mexicana*; often in washes); at NTS and near other Great Basin habitats hopsage (*Grayia spinosa*) and shadscale (*Atriplex confertifolia*) are very typical associates (Beatley 1976a). Another common species is cheesebush (*Hymenoclea salsola*), which often is very abundant along roadsides and washes. In northern MDR of Nevada, where soils are deep, loose sands, lacking a surface pavement, more common shrubs are *Psorothamnus arborescens*, *Krameria erecta* (*Pima rhatany*), *Ephedra nevadensis*, *Ceratoides lanata* (winter fat), and, in some areas,

Table 1.2. *Comparison of plant community treatments for the Mojave Desert by Beatley (1976), Brown et al. (1979), Kuchler (1977), Rowlands et al. (1982), Thorne (1982), Turner (1982b), and Vasek & Barbour (1977). Abbreviations: assoc = association, f = forest, GB = Great Basin, Moj = Mojave, sc = scrub, wd = woodland.*

Vasek & Barbour 1977	Turner 1982b	Beatley 1976	Brown et al 1979	Thorne 1982	Rowlands et al 1982
Creosote bush sc	Creosotebush series	Mojave Desert	Warm temperate desertlands	Desert sc	Desert sc complex
	Larrea-Ambrosia assoc	Bajada assocs	Moj desert sc	Creosote bush sc	GB subcomplex
Saltbush sc	*Larrea-Atriplex confertifolia* assoc	*Larrea-Ambrosia*	Creosotebush series	Desert psammophytic sc	Sagebrush sc
Shadscale sc	*Larrea-Lycium-Grayia*	*Larrea-Atriplex*	*Larrea*	Stem-succulent sc	Blackbrush sc
Blackbush sc	Saltbush series	*Larrea-Lycium-Grayia*	*Larrea-Ambrosia*	Desert wash sc	Hopsage sc
	Xerophytic phase	Arroyos	*Larrea-Yucca*	Desert rupicolous sc	Moj-Colorado Desert sc subcomplex
Joshua tree wd	Halophytic phase	Mountains	Blackbrush series	Desert calcicolous sc	Creosote bush sc
	Shadscale series	*Atriplex*	*Coleogyne*	Mixed desert sc	Cheesebush sc
	Blackbush series	Springs & Seepage	*Coleogyne-Yucca*	Low-desert wds	Succulent sc
	Joshua tree series	*Ash-Screwbean-Baccharis*	Joshua tree series	Desert microphyll wd	Saline-alkali sc subcomplex
		Atriplex	*Yucca brevifolia-Acamptopappus-Larrea*	Desert oasis wd	Allscale-alkali sc
		Atriplex-Haplopappus	*Yucca brevifolia-Coleogyne*	Desert riparian wd	Desert-holly sc
			Yucca brevifolia-Larrea	Alkaline sc & meadow	Shadscale sc
		Mesquite	Bladder-sage series	Saltbush sc	Moj saltbush-allscale sc

Transition desert	Mesquite series	Shadscale sc	Desert microphyll wd complex
Lower bajada assocs	Catclaw series	Allscale sc	Mesquite microphyll wd subcomplex
Grayia-Lycium	Saltbush series	Fourwing saltbush sc	Desert-semidesert sc-steppe complex
Lycium pallidum-Grayia	*Suaeda torreyana*	Desert-holly sc	Desert-semidesert sc-steppe subcomplex
Lycium shockleyi-Atriplex	*Atriplex*	Gypsicolous sc	Indian rice grass sc-steppe
Upper bajada assocs	Cold temperate desertlands	Alkali sink sc	Desert needle grass sc-steppe
Larrea-Grayia-Lycium	GB scrublands	Alkaline meadow & aquatic	Big galleta sc-steppe
Coleogyne	Blackbush series	Transition communities	Galleta-blue grama sc-steppe
GB Desert	Sagebrush series	Joshua tree wd	Desert alkali grassland subcomplex
Basin assocs	*Artemisia tridentata*	Blackbush sc	Saltgrass meadow
Atriplex confertiflora	*Artemisia-mixed sc-grass*	Sagebush sc	Desert saxicole subscrub complex
Atriplex-Ceratoides	*Artemisia nova*	GB sagebrush sc	Calciphyte saxicole subsc
Atriplex-Kochia	Shadscale series	Pygmy sagebrush sc	
Atriplex-Sarcobatus	*Atriplex confertifolia*	Montane f & wd	
	A. confertifolia mixed sc	Low montane	

(Table cont. overleaf)

Table 1.2. – (cont.)

Vasek & Barbour 1977	Turner 1982b	Beatley 1976	Brown et al 1979	Thorne 1982	Rowlands et al 1982
		Bajada assocs	Forest & wd formation	Pinyon-juniper wd	Noncalciphyte saxicole subsc
		Atriplex canescens	GB conifer wd	Yellow pine f	Desert psammophytic complex
		Artemisia tridentata	Pinyon-juniper series	Upper montane	Xeric-conifer wd/f complex
		Mountain assocs	Pinus monophylla	White fir-pinyon wd	Xeric conifer wd subcomplex
		Artemisia nova	Pinus monophylla-Juniperus californica	Limber pine f	Utah juniper-one leaf pinyon wd
		Artemisia-Pinyon-Juniper	Pinus monophylla-Juniperus osteosperma Juniperus osteosperma Juniperus californica	Bristlecone pine f	California juniper-one leaf pinyon wd
		White fir Artemisia-Cercocarpus	P. monophylla-Juniperus californica-chaparral	Freshwater meadow, marsh & aquatic	Desert montane f subcomplex
			Boreal forest & wd	Streamside marsh	White fir f enclave
			Bristlecone pine-limber pine series	Lake, pond & quiet stream aquatic	
			Strand formation	Reservoir & tank subaquatic	Subalpine f
			Mojavean inland strand		Streamside & oasis wd complex
			Mojavean inland submergents		Streamside wd subcomplex
					Cottonwood-willow-mesquite bottomland
					Cottonwood-willow streamside wd
					Desert oasis wd subcomplex
					Washington fan palm oasis

Menodora spinescens and *Acamptopappus shockleyi*. Perennial grasses, such as *Oryzopsis hymenoides* (also *Achnatherum hymenoides*, Indian rice grass), are also characteristic at the highest elevations.

On bajada slopes above 1100 m, where soils have an imperfectly developed surface pavement, creosote bush occurs with *Lycium andersonii* and *Grayia spinosa*; these soils generally lack a caliche hardpan and have a sandy-loam texture with rock fragments scattered throughout the profile (Beatley 1976a). Other species present are the associates of the lower bajada community, listed earlier, plus *Haplopappus cooperi*. In the broad transition in southern Nevada from Mojave Desert to Great Basin Desert, creosote bush diminishes in abundance and then drops out of the community, so that from 1300 to 1600 m closed drainage basins typically have *Grayia* and *L. andersonii* as codominants (Beatley 1976a). At these elevations, characteristic Great Basin shrubs, such as *Ceratoides lanata*, *Tetradymia axillaris* (horsebrush), *Atriplex canescens* (fourwing saltbush), and *Artemisia spinescens*, frequently occur.

A third bajada association is the creosote bush/white bursage series with the addition of shadscale. Beatley (1976a) has observed that this type dominates on lower elevation bajadas with calcareous soils derived from limestones or dolomites. Surface pavements are generally well developed in these soils, and the general matrix is either finely textured sand or silt loam. Gravels or rock fragments are usually present in the root zone, and the depth of the roots is limited by a relatively shallow caliche zone. In the limestone zones where this plant formation occurs, potential associates include *Psorothamnus arborescens*, *Krameria erecta*, *Ephedra torreyana* or *E. funerea*, *Lycium shockleyi* or *L. pallidum*, and *Yucca schidigera*. Among the cacti, *Opuntia ramosissima* is restricted to this association and *O. basilaris* is frequently present. Overall, this subtype of creosote bush scrub has the highest species diversity of perennial plants.

On gently sloping bajadas above 800 m elevation, where winter temperatures may be relatively cold and soils are loose, one often finds stands of Joshua tree (Fig. 1.10). *Yucca brevifolia* appears at lower elevations in creosote bush scrub, but changes into a recognizable community, Joshua tree woodland (Thorne 1976; Vasek & Barbour 1977) at higher elevations. Although this tree is very conspicuous, actually codominant shrubs have greater cover. In the southern Mojave Desert common perennials are *Salazaria mexicana*, the bunch grass *Hilaria rigida*, *Haplopappus cooperi*, *Hymenoclea salsola*, *Ephedra nevadensis*, *Coleogyne ramosissima*, the buckwheat *Eriogonum fasciculatum*, and a variety of others, including succulent chollas (*Opuntia*) and *Y. baccata*. Where creosote bush drops out and junipers (*Juniperus*) and pinyon pines (*Pinus*) appear is the transition from desert vegetation to semiarid, nondesert pinyon–juniper woodland. In Nevada codominants of Joshua tree woodland frequently are *E. nevadensis*, *S. mexicana*, *Hilaria rigida*, *Y. baccata*, *O. ramosissima*, *C. ramosissima*, *Eriogonum fasciculatum*, and *Menodora spinescens* (Vasek & Barbour 1977). Distributions of *Y. brevifolia* and *S. mexicana* are nearly restricted to the Mojave Desert.

As the northern Mojave Desert merges with the southwestern portions of the Great Basin Desert, other desert transitions are present (Beatley 1975, 1976a). The most common of these is an association strongly dominated by blackbrush. This is a transition zone between creosote bush scrub and the open drainage basins at middle elevations dominated by Great Basin sagebrush scrub (*Artemisia tridentata*). *Coleogyne* grows as low shrubs in nearly pure stands, with up to 50% ground cover. Beatley (1974a, 1975) interpreted the entrance and eventual dominance of *Coleogyne* over *Larrea* as a function of increased winter precipitation rather than responses to low temperature extremes. *Coleogyne* coexists with creosote bush on upper bajadas having an annual mean precipitation of about 160 and becomes dominant at 225-240 mm yr^{-1}. Several other desert transitions in southern Nevada, including one with the Nevada endemic *Lycium shockleyi* and no *Larrea*, appear to be correlated with low nocturnal temperatures.

Figure 1.10 Joshua tree woodland (*Yucca brevifolia*) north west of Kelso, California, in the central Mojave Desert.

Playa basins

Shadscale scrub, with *Atriplex confertifolia* as dominant, is a widespread plant community in the deserts of western North America. In the southern Mojave Desert this vegetation type occurs on steep slopes with heavy, rocky soils, but in northern portions and the Great Basin it often occurs on alkaline, pluvial lake basins (Young, Evans, & Major 1977). In general authors agree with Billings (1949) that shadscale is a distinct zone between Mojave creosote bush scrub and Great Basin sagebrush scrub. The shadscale community, which typically has *Atriplex confertifolia* and *Artemisia spinescens*, usually occurs where rainfall is low and soluble salts are relatively high. Sometimes *Atriplex canescens*, *Ceratoides lanata*, and *Yucca brevifolia* also occur in this community (Vasek & Barbour 1977; Rowlands et al. 1982).

Saltbush scrub (Vasek & Barbour 1977), also called alkali sink, is a series of common plant associations within the deserts of North America. In the Mojave Desert at least two phases have been distinguished, a xerophytic formation on dry soil and having plants with limited salt tolerance, and a halophytic formation growing in fine-textured soils with available ground water and salt concentrations up to 6% (Vasek & Barbour 1977; Turner 1982b; Rowlands et al. 1982). The xerophytic phase, also called desert holly scrub, is codominated by *Atriplex polycarpa*, *A. confertifolia*, and *A. hymenelytra*, for which the community is named. The halophytic phase, also called iodinebush-alkali scrub or allscale-alkali scrub (Rowlands et al. 1982), has succulent species of Chenopodiaceae, especially iodinebush (pickleweed, *Allenrolfea occidentalis*), *Nitrophila occidentalis*, and species of *Salicornia*, *Suaeda*, and sometimes *Sarcobatus*. Gradients between the two phases can be seen around playas and dry lakes. Examples of such communities can be found in Death Valley (Hunt 1966), the Amargosa Valley, and the Ash Meadows area along the California–Nevada border (Beatley 1976a). *Distichlis spicata*, a dioecious, halophytic perennial grass, may form open meadows on seasonally saturated saline soils. Along lower elevation playas or in sand dune areas, where ground water is available at depth, the phreatophytic mesquite (*Prosopis glandulosa* var. *torreyana*) may form dense stands. This species can utilize both fresh and moderately saline ground water.

Arroyos

Arroyo or desert wash communities draw their floras from elements of surrounding bajada and mountain slope habitats, with distinctive assemblages in washes of limestone and volcanic parent materials. Frequently wash dominants are phreatophytic species, which have roots anchored deep below the surface in or near the water table; consequently, these are the localities where the largest trees and shrubs, other than the succulent Joshua tree, occur in the Mojave Desert. The desert riparian woodland of the Colorado River and its tributaries is atypical of the surrounding desert and harbors numerous introduced species, including such undesirables as saltcedar (*Tamarix*). On floodplains and bajadas of the Mojave Desert arroyo vegetation contains fewer trees than in the adjacent ones of the Sonoran Desert, but such trees as desert-willow (*Chilopsis linearis*), smoke tree (*Psorothamnus spinosus*), and mesquites (*Prosopis* spp.) may occur with arrowweed (*Pluchea sericea*), other native shrubs, and naturalized saltcedar.

Natural springs

Natural springs are rare in the lowlands of the Mojave Desert, but two worthy exceptions are Ash Meadows and Oasis Valley on the California–Nevada border. Ash Meadows was formed by several dozen springs, drawn from deep underground sources in limestone parent materials. Oasis Valley is also spring-fed from the Amargosa River. Soils at both localities are extremely fine textured and are high in salt concentration (Beatley 1976a).

There are a number of narrowly restricted or endemic plants in Ash Meadows as well as groves of velvet ash (*Fraxinus velutina* var. *coriacea*), screwbean (*Prosopis pubescens*), cottonwood (*Populus fremontii*), and typical willow-leaved native shrubs of *Baccharis*, *Pluchea*, and, of course, willow itself (*Salix*).

In a few localities near the southern border of the Mojave Desert occur outposts of California fan palm (*Washingtonia filifera*), a community called desert oasis woodland (Thorne 1976), because these arborescent palms only occur where there is a natural spring. However, this community is in reality one of the warmer Sonoran Desert, because the fan palm is sensitive to freezing temperatures (Brown, Carmony, Lowe, & Turner 1976).

Mid-elevational mountain slopes

Within the MDR many of the ranges have limestone or dolomite parent material with shallow, rocky soils. Especially interesting are the woody perennials, some of which are endemic to those mountain ranges and others have outposts, long disjunctions from ranges on the California coast or southeastward in the Chihuahuan Desert. There also are widely scattered gypsum outcrops, on which one observes a suite of gypsophilous endemic species (Meyer 1986).

Beatley (1976a) described a lack of characteristic plant associations in the southern Nevada mountains and noted instead relatively individualistic distributions of many woody and herbaceous species. Most woody shrubs of the bajada associations extend their ranges to some of the lower mountain rocky slopes.

SUMMARY

The Mojave Desert is a relatively small desert that in many ways is transitional between the Colorado subdivision of the Sonoran Desert in southern California and western Arizona and the Great Basin Desert of central Nevada and Utah. Most of the Mojave Desert occurs between 600 and 900 m elevation as a basin-and-range topography consisting of isolated mountain ranges with extensive bajadas, broad valleys, and barren playas. The southern boundary is conveniently defined by the elevation above where freezing is a limiting factor for typical trees of the Sonoran Desert, and the highest and northern localities of the Mojave Desert are delimited where creosote bush (*Larrea tridentata*), a primary index of Mojave desert scrub, is replaced by upland species, typically at an elevation of 1300 m. Many of the isolated ranges within the Mojave Desert Region have nondesert, montane vegetation. There is little riverine activity today and no large body of permanent water within the region.

The Mojave Desert has a winter rainfall pattern, receiving storms that generate over the Pacific Ocean and sweep across California between November and March. Occasionally a storm system may also move across the desert bringing summer and fall precipitation, principally from the east or south, hence the eastern portion of the Mojave Desert tends to have a greater percentage of summer rainfall. Rainfall from year to year is highly variable, but often ranges from 85 to 160 mm, depending on location, and locally rarely exceeds 250 mm yr^{-1}. Subzero minimum winter temperatures are fairly common for lowland desert scrub communities.

Neither the flora nor the fauna of the Mojave Desert is sharply delimited or distinctive, because species commonly have distributions ranging into either of the neighboring deserts. For plants the Inyo Region has the highest amount of endemism, and there are seven genera of angiosperms that are endemic to the Mojave Desert Region. For animals endemism is uncommon, in large part because their preferred habitats are widely distributed in the Mojave and Sonoran deserts, in particular creosote bush communities on bajadas and communities of *Atriplex* around playas. In the Mojave Desert arroyos typically have only shrub associations, although native and introduced trees grow where ground water is present.

2 Physical geography of Rock Valley

The Nevada Test Site, located between 36° 35′ and 37° 15′N latitude and 115° 55′and 116° 35′W longitude, occupies 350 000 hectares of arid and semiarid terrain in southern Nye County, Nevada (Fig. 2.1). Geological structure of this region is diverse and extremely complex, consisting of mountain ranges separated by broad, linear valleys and, in the northern sector, a large volcanic plateau. Rock Valley (frontispiece), which covers about 130 km², is located adjacent to the southern boundary of NTS and about 20 km west of Mercury, Nevada. This unused valley was set aside in 1960 for long-term ecological studies, especially as an example of lower bajada desert scrub (*Larrea tridentata–Ambrosia dumosa* community) for the northernmost Mojave Desert. The site chosen (Fig. 2.2) had an average elevation of 1040 m and a gentle slope of three to four percent.

GEOLOGICAL HISTORY AND TOPOGRAPHY

From Precambrian times through most of the Paleozoic, the Mojave Desert Region and adjacent Great Basin were under shallow water, a continental shelf environment within which tremendous sedimentary processes were taking place, resulting in thick deposits of silicates, limestones, and dolomites. Approximately one-third of the parent material of exposed upland areas at NTS possesses mixtures of Paleozoic rocks, with an aggregate thickness up to 12 000 m (Johnson & Hibbard 1957), and nearly 80% of these deposits are limestones and dolomites ranging from Lower Cambrian to Upper Permian in age (Kistler 1968). During the Mesozoic Era, intense crustal deformation, folding, thrust faulting, and normal faulting shaped these strata; rocks were intruded and compressed on an east-southeast, west-northwest axis (Johnson & Hibbard 1957; Ekren 1968; Blankennagel & Weir 1973). Further deformation occurred during the Cenozoic in response to extensive tectonic activity; the Cenozoic rocks are mostly volcanic extrusives and basin-filling sediments. The present-day topography of NTS includes alternating ranges and sediment-filled valleys on the east and normally faulted volcanic

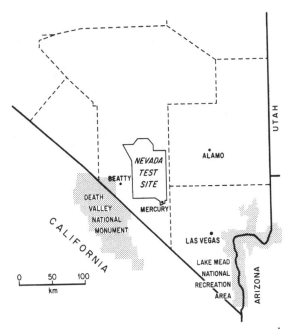

Figure 2.1 Location of the Nevada Test Site in central-southern Nevada.

[21]

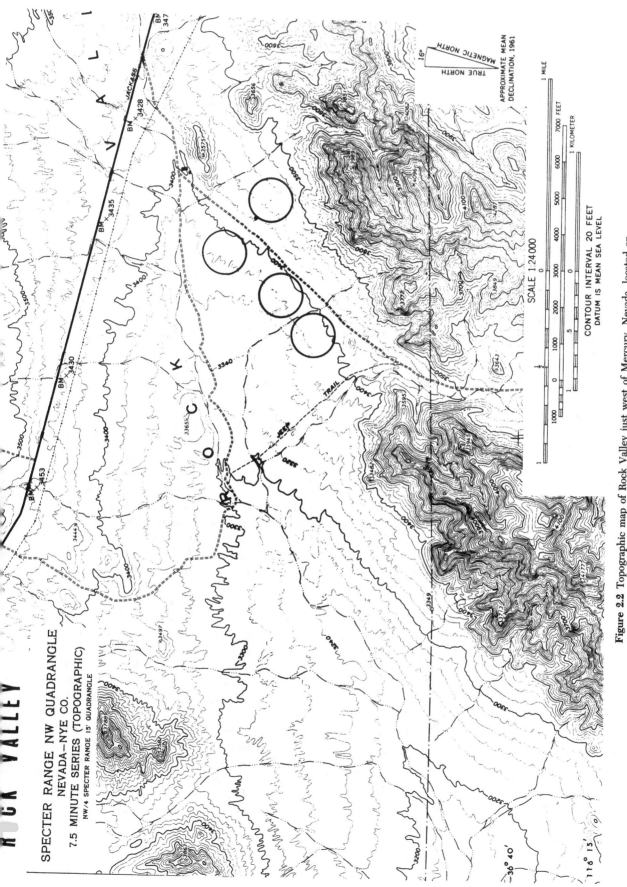

Figure 2.2 Topographic map of Rock Valley just west of Mercury, Nevada, located on the Nevada Test Site. Base map from U.S. Geological Survey.

plateaus to the west (Ekren 1968; Zoback & Zoback 1980; Sinnock 1982).

Over much of the NTS region the Lower Paleozoic sedimentary parent materials have been uplifted and folded several times across a north–south axis. Structural deformation began no earlier than Upper Carboniferous or Lower Permian and continued into the Miocene (Johnson & Hibbard 1957). Much of the uplifting was related to plate

tectonics along this, the former western coastline of North America (Brown & Gibson 1983).

Today the topography of NTS is dominated by three large valleys, Yucca Flat, Jackass Flats, and Frenchman Flat, bordered by mountains and ridges of moderately steep relief (Fig. 2.3). Jackass Flats is an open basin with an outlet to the southwest, the Forty-mile Canyon drainage, ultimately flowing into Death Valley as the Amargosa Valley

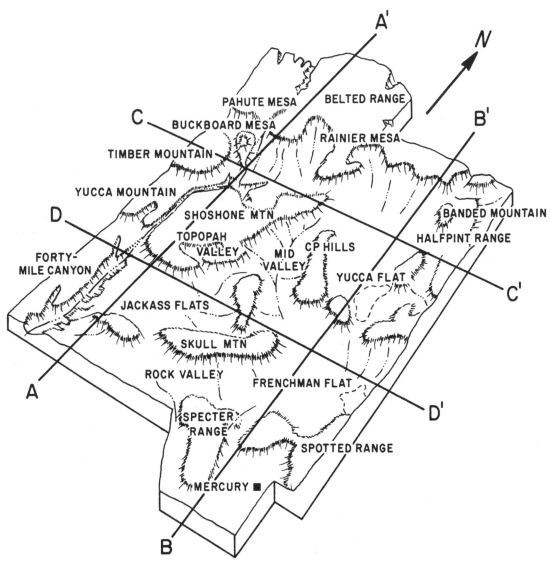

Figure 2.3 Geomorphic landforms at the Nevada Test Site. Geologic sections along four transect lines are shown in Figure 2.4. Adapted from O'Farrell and Emery (1976).

drainage. This outlet drops in elevation to 819 m at the margin of NTS and is the lowest elevation present on the preserve (O'Farrell & Emery 1976). The other two major basins, Yucca Flat and Frenchmen Flat, are closed basins, both lacking an outlet for surface runoff and having no gravitational outflow of air at night. These two valleys also each contain a large playa.

Four smaller basins, Mercury Valley, Rock Valley, Mid-Valley, and Topopah Valley, are also present at NTS, and all have open drainage to the southwest or south.

Two massive mountains, Rainier Mesa and Pahute Mesa, along with the smaller Buckboard Mesa dominate the northern and northwestern portions of NTS. Rainier Mesa reaches to 2341 m, the highest point at NTS. The western margin of the preserve is bordered by an old volcanic caldera, Timber Mountain, and the smaller Yucca Mountain. Relatively small mountain ranges and hills border the major basins. Jackass Flats is bordered to the west by Yucca Mountain, to the north by Shoshone Mountain, and to the south by Skull Mountain. Yucca Flat is circumscribed by Rainier Mesa and the Eleana Range to the north and west, the Halfpint Range to the east, and the CP Hills to the south. South of the CP Hills lies Frenchman Flat, which has the Spotted Range along its southern edge. Rock Valley has Skull Mountain to the north and the Specter Range to the south.

The Yucca Flat formation is a Cambrian unit, which outcrops widely along the northeastern margin of the valley, where it forms many steep hills and ridges. Banded Mountain takes its name from alternating beds of dark limestones and light-colored dolomites in this formation. Other large areas of mountains surrounding Yucca Flat and those around Frenchman Flat were formed by the Pogonin limestone of Ordovician age. The Ranger Mountains and Red Mountain south of Frenchman Flat are composed of Eureka quartzites, also of Ordovician age. Thrust faults were important in producing some of these mountains, particularly in the area around Yucca Flat, where evidences of four separate thrusting events can be seen, leaving older Paleozoic sediments positioned over younger Paleozoic sediments. These thrust faults

are thought to have occurred mostly in the Cretaceous (Johnson & Hibbard 1957).

In addition to these thrust faults, almost every part of NTS has been affected by normal faults (Fig. 2.4). Like those of the Great Basin, normal faults in the northern half of NTS have a southerly trend and have influenced the orientation of mountain ranges, but in the southern half faults trend gradually westward. South of NTS the direction of normal faults and mountain ranges resumes a southerly trend. Normal faulting began as early as the Miocene, following completion of major structural deformation in the region, and has continued until Recent time (Johnson & Hibbard 1957). The most prominent structural fault at NTS is the Yucca Fault, which extends more than 25 km from the northern end of Yucca Flat into Frenchman Flat to the south. Yucca Fault shows evidence of 300 m of movement along its length and in northern Yucca Flat forms a scarp over 25 m in height. Recent faulting across the playa basin in Frenchman Flat has produced linear belts of sagebrush (*Artemisia tridentata*) growth along the rift line, where fruits have collected and soil moisture conditions have been favorable for germination and seedling establishment.

Alluvial and colluvial deposits of Quaternary age cover the major valleys at NTS and 30% of the total area. The majority of this detritus has been transported from bedrock areas by intermittent streams, which flow out of the surrounding mountains when torrential rainstorms occur. In the valleys this material has formed deposits 250–400 m in depth, and in parts of Yucca Flat and Frenchman Flat the deposits exceed 1000 m (Cornwall 1972). Slope gradients and amounts of precipitation have determined the size of particles and quantity of sediments that have been moved. Steep alluvial slopes along the margins of the valleys may have elevational gradients of 50–100 m km^{-1}, whereas at the valley centers the gradients are only 2–3 m km^{-1}. Removal of particles from the mountains and their deposition on the bajadas below is a continual and continuing process. Detrital material that fills the valley basins generally consists of interfingering lenses of unconsolidated alluvium.

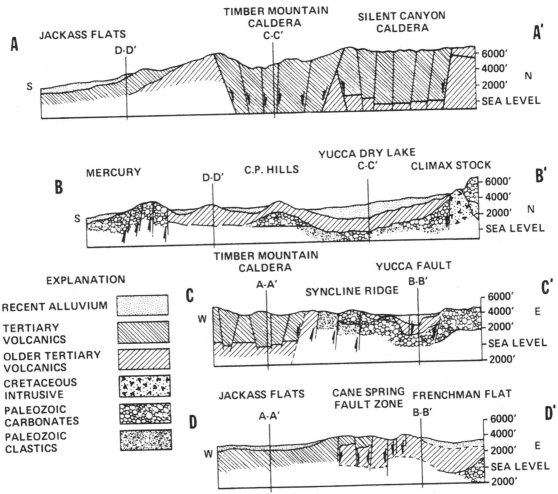

Figure 2.4 Geological structures for cross sections through the Nevada Test Site, as shown in Figure 2.3. From O'Farrell and Emery (1976).

Tertiary volcanic rocks, most notably ash-flow tuffs, form parent materials on 70% of the outcrops at NTS (Fig. 2.4). The oldest volcanics are the Monotony Valley tuffs, dated about 26.5 MY bp, which form the most widespread ash flows in the United States (Table 2.1; Ekren 1968). Five volcanic centers formed in and around NTS, beginning with the Belted Range Center about 14 MY bp. The caldera forming the Belted Range tuff is now buried beneath Pahute Mesa in the northwest sector of NTS. The Paintbrush tuff

extended widely 12 MY bp from a caldera north of the current town of Beatty (Fig. 2.5), but two million years later this was itself covered by ash flow from the Timber Mountain caldera to the east. About 7 MY bp the Thirsty Canyon tuff was extruded from the Black Mountain caldera, and this material covered portions of the older Belted Range and Timber Mountain tuffs.

Rock Valley consists of deep Quaternary alluvium and colluvium overlying Tertiary volcanics. Small local areas of eolian sands occur along slopes

Table 2.1. *Principal volcanic formations of the Nevada Test Site and vicinity (data from Ekren 1968; Nobel 1968; Kistler 1968).*

K/Ar Date, MY bp	Unit	Lateral extent, km	Volume, km³
–	basalt lavas	–	–
7	Thirsty Canyon tuff	80	50
11	Timber Mountain tuff	100	500
12	Paintbrush tuff	70	120
14	Belted Range tuff	60	70–100
14	Kane Wash tuff	60	200
17	Fraction tuff	120+	>500
	dactite and latite lavas	–	–
21	White Blatch Spring tuff	120+	>500
26	Monotony Valley tuff	120+	1000+

of the Specter Range to the south, which itself consists mostly of Paleozoic carbonates. Also in the southern sector are Little Skull Mountain and Skull Mountain, which were formed from Tertiary basalts and tuffs.

HYDROLOGY

Hydrologic flow systems at NTS are comparable to those found elsewhere in the Mojave Desert and Great Basin (Rush 1971; Young 1972; Blankennagel & Wier 1973). Commonly precipitation over mountain ranges exceeds that over intermontane basins. Runoff channels and arroyos transport surface water flow laterally, but infiltration also takes place. During heavy rainfall, streams and arroyos can carry large amounts of water, and these flows may form temporary shallow lakes in the playa basins, lasting as long as several months. Permanent surface water is limited to a few, relatively small springs in the upland areas. All desert surface channels and arroyos are ephemeral and flow only after rains. Permanent or seasonally intermittent flows are important resources for wildlife and vegetation. On NTS the only permanent water are the springs, for example, Cane Spring, Tippipah

Springs, Topopah Springs, and White Rock Springs.

Surface flow systems are strongly influenced by geologic substrates. Clastic and crystalline rocks act as barriers to water movement and thus frequently form the boundaries to flow systems. Where present this rock type can result in a high-perched water table and small, geographically controlled groundwater basins. Only the deepest montane channels that are underlain by such rocks maintain permanent streams and small springs. Springs and seepage zones occur in some of the deep channels, where they represent discharge from groundwater pools. Carbonate rocks, derived from the massive limestone and dolomite deposits, are commonly fractured and cavernous and therefore highly permeable to water flow. Mountains with these substrates appear much drier, with relatively less surface runoff and few small springs, and groundwater instead appears in relatively large springs at the foot of the mountains or in nearby valleys. The broad and flat water tables produced in carbonate terrains may extend for long distances with little control by surface topography. Interbasin flows of groundwater resulting from such conditions are common for eastern and southern Nevada, including NTS.

Patterns of hydrological flows on NTS and adjacent areas have been described in considerable detail (Winogard 1962; Winogard & Thordarson 1968, 1975; Maxey 1968; Young 1972). Paleozoic limestones and dolomites, which underlie NTS, allow broad hydrologic basins to form at depths of 90–150 m beneath the valley floors of the southern portions of the property. Along this surface water flows from north to south along two major groundwater systems (Fig. 2.6). The Pahute Mesa system is fed from input to fractured volcanic rocks of Pahute Mesa and Timber Mountain in northwestern NTS, and this water flows southward into the Amargosa Desert area. The Ash Meadows groundwater system moves beneath the eastern two-thirds of NTS, collecting moisture of Rainier Mesa and other ranges to the north, as well as seepage from Yucca Flat and Frenchman Flat, before discharging into Ash Meadows and the southeastern Amargosa Desert.

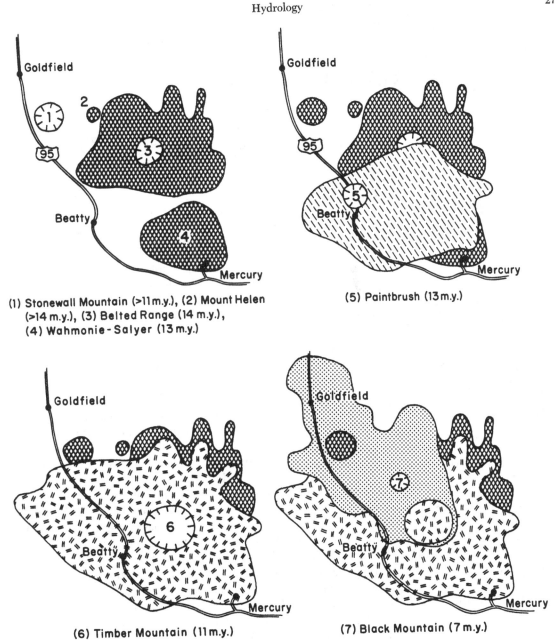

(1) Stonewall Mountain (>11 m.y.), (2) Mount Helen (>14 m.y.), (3) Belted Range (14 m.y.), (4) Wahmonie-Salyer (13 m.y.)

(5) Paintbrush (13 m.y.)

(6) Timber Mountain (11 m.y.)

(7) Black Mountain (7 m.y.)

Figure 2.5 Geologic development of seven volcanic centers and five ash-flow tuff sheets in the vicinity of the Nevada Test Site. Adapted from Ekren (1968).

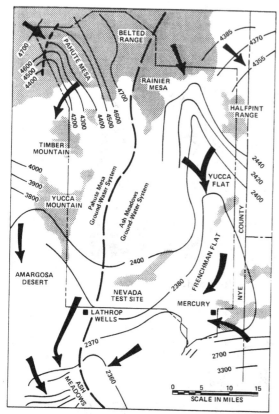

Figure 2.6 Elevational level and direction of flow for groundwater pools at the Nevada Test Site. The stippled areas are upland mountain regions. From O'Farrell and Emery (1976).

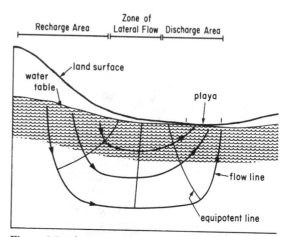

Figure 2.7 Schematic representation of groundwater flow along line of equipotential flow from mountain areas to valley bottoms. After Maxey (1968).

Groundwater 'mounds' are formed beneath mountain ranges, as the porous soils in the mountains are recharged through downward movements of water. At low montane elevations sufficient groundwater may not collect to allow lateral flow belowground or surface discharge of groundwater. When groundwater recharge is sufficiently high, lateral subsurface flow occurs at right angles to equipotential lines (Fig. 2.7), and such flows may cross large areas, where surface runoff and precipitation are too low to recharge soil moisture. Surface discharge of groundwater occurs where aquifers are formed along fault lines or other favorable flow paths.

SOILS

Soil types

There has been no official soil classification work done at NTS, although several series names were suggested from limited surveys conducted by the Environmental Protection Agency in specific localities (Leavitt 1970; Leavitt & Mason 1971). Considerable field and laboratory characterizations of soil profiles were also done by other workers in the northern Mojave Desert areas on NTS (Romney, Hale, Wallace, Lunt, Childress, Kaaz, Alexander, Kinnear, & Ackerman 1973; Romney, Wallace, Kaaz, & Hale 1980; Case 1984). As expected, chemical and physical characteristics of local soils vary according to the source of the parent materials, and these in turn are derived from the mountain masses; hence, the detritus and alluvium consist of limestone, dolomites, quartzite conglomerate, shale, tuff, and rhyolite (Johnson & Hibbard 1957). On the bajadas the basin floors consist of transported alluvial materials up to 1000 m in thickness (Cornwall 1972).

Soils of the lower basins at NTS belong to Thermic families, whereas those at higher elevations fit into the Mesic family classification. Extensive areas have soils that would be classified as Entisols, a soil order having only an ochric epipedon and showing no pedogenic horizons. These soils are found on relatively recent alluvium in the flattest zones of the valleys. In contrast, on the gently to moderately sloping bajadas Aridisols are found; these have ochric epipedons and petrocalcic or calcic subsurface horizons. Argillic horizons are weakly developed on some of the bajadas but are commonly absent, and, because these horizons are not well represented at NTS, this indicates that the suborder classification for much of the Aridisols would be Orthids. On the steepest bajadas, where diagnostic subsurface horizons are lacking, Entisol is the dominant soil order found. At higher elevations certain soils may meet the requirements of the Alfisol order.

Residual soils that were formed from weathering products of underlying rock are nearly always shallow. They occur on the more gentle slopes of the mountains and foothills and on mesas and other areas of undulating terrain. On the steeper mountain slopes, where the processes of weathering and erosion are extensive, bare rock outcrops are commonly the prevailing substrate. The weakly developed niche soils, if present at all beneath, around, and below rock outcrops and 'flatrock' areas, are sites at which certain rare plant species have been found (Beatley 1976a).

Alluvial soils on the bajadas are formed on materials that were sorted along a particle size gradient at the time of deposition. Only a crude sorting of rock debris has occurred on the margins of the valleys. Boulders, cobbles, and gravels characterize parent materials of the upper bajadas and become less frequent in a gradient downslope, whereas smaller particles are transported along with excess salts to the playas of the valleys, where there are fine-textured sediments composed mostly of clay and silt particles. Where the finer materials have been removed from the surface by wind and water, larger fragments comprise a pavement, which is highly resistant to further erosion.

The playa basins in Yucca Flat and Frenchman Flat are geologically distinct from the other valley sediments. These playas are covered with relatively impermeable, fine silts and clays, which are well sorted and uniform both laterally and vertically. The thickness of the playa sediments are thought to be variable, but probably up to 50 m (Farnham, in Johnson & Hibbard 1957).

Carbonates

On the bajadas below the limestone mountain ranges can be found soils with high carbonate concentrations. These calcareous soils are readily permeable to water and typically brought to field capacity to depths of one meter with recharge from annual precipitation. In fact, the subsurface of such desert soils rarely becomes air dry. Soil profiles that are relatively low in carbonate usually have developed on alluvial sediments of mixed volcanic and limestone origin. These soils are located on the northern bajadas of Frenchman Flat and in Jackass Flats, western Yucca Flat, and Rock Valley.

Pedogenic calcretes (syn. caliche, kankar, and kundar), are widely distributed in arid region soils (Reeves 1976; Goudie 1985; Phillips, Melmes, & Foster 1987) and at NTS are almost always coincident with calcareous soils on gently sloping to moderately sloping bajadas near limestone formations. Calcrete is an indurated horizon of calcium carbonate and silica, a restrictive carbonate–silicate hardpan enveloping other soil materials that were originally present in the profile. At NTS the upper surfaces of calcrete horizons are found at depths of 0.3–0.6 m and may be 0.5–1.0 m in thickness. However, depth of hardpan development generally is associated with the depth to which water-carried salts have consistently penetrated year after year. In some stream-dissected terrains several outcroppings of calcrete are found.

Calcretes at NTS frequently overlie a discrete textural boundary in the soil profile of coarse sand. This suggests that formation began just above the coarse sand layer. Hypothetically, downward movement of water would be halted at the coarse sand layer until the water potential rose to the point that would allow entry into the pores of the sand, according to the formula:

$$\psi = \frac{2\gamma}{r}$$

where ψ is the water potential, γ is the surface tension value for water, and r is the radius of the soil pores. As water is withdrawn from the system, either by plant transpiration or other water depletion processes, precipitation of $CaCO_3$ would begin and then be repeated in other wetting cycles.

'Desert pavement' (Figs. 2.8–2.9) is a characteristic desert landscape feature (Dorn & Oberlander 1981; Watson 1989) and often occurs at locations on NTS where the terrain is relatively flat. Desert pavement has a surface composed entirely of varnished gravel- or stone-sized materials (Fig. 2.9), and pavement areas typically are devoid of plant growth, except at 'clump' locations. Gravel is mostly 20–40 mm in depth, and frequently the gravel surface layer is underlain by soil material that is highly vesicular and would at the outset appear to be permeable to water. However, Musick (1975) found that desert pavement soils in Yuma, Arizona have unusual characteristics. These soils tend to be sodic, i.e., they have an exchangeable sodium percentage greater than 15%, or saline-sodic, and water infiltration is extremely slow because deflocculation of the soil colloids occurs either because of the high levels of exchangeable sodium or the introduction to the system of low-salt rainwater. In some desert pavement soils the vesicular surface layer becomes jellylike when wetted and can actually flow down a gentle slope while the lower soil remains dry. Consequently, only a fraction of the precipitation may in some circumstances infiltrate desert pavement and become available for plant use. Low water availability, due to poor infiltration of rain, probably is the major factor limiting plant success, given that halophytes do not colonize saline-sodic desert pavement. In addition, high salinity and sodium levels probably inhibit seedling establishment. Musick speculated that desert pavement is maintained because after a rain shower evaporation from the surface would draw salts back to the surface soil, which underlies the 'pavement'.

At numerous locations at NTS, surface rocks that are half buried have a white undercoating (Figs. 2.10–2.11), consisting of a mixture of calcium carbonate and silica. The coating, which may be as much as ten millimeters in thickness, probably develops in the following manner. During unusual weather conditions upon cooling at night the temperature on the underside of a rock decreases to below the dew point. Water in the vapor phase is transferred to, and condenses on, that underside, and liquid flow from the rock moistens the adjacent soil. Subsequently, when the rock warms, water is evaporated, thus causing liquid phase flow to reverse, now flowing toward the rock and carrying with it minerals, which are precipitated on the underside of the rock.

Soil structure at NTS

At NTS most of the 79 soil profiles that were analyzed by UCLA investigators occurred on alluvial deposits and contained unconsolidated parent materials of both sedimentary and volcanic origins (Romney et al. 1973; Table 2.2). These soils displayed typical characteristics of nonsaline desert soils: coarse texture, low organic matter content, low carbon/nitrogen ratios, and calcic or calcrete horizons. These soils, which showed little, if any, evidence of leached upper horizons or profile development, are young, and much of the site-to-site diversity reflects their mixed alluvial origins. Where B horizons were found, the sites were fairly level.

Most soils on NTS are readily permeable to water. Soils with relatively high proportions of silts and clays along with a well-developed desert pavement retain moisture significantly longer than do sandy soils without a pavement. Runoff is highly variable, depending upon the nature of the soil surface, slope gradient, and rainfall intensity. Wherever desert pavement is well developed and rainfall is relatively high, most runoff is lost as sheetwash, producing rapid accumulations of water in the drainage courses. Fortunately at NTS, most rainfall is not of high intensity, except when associated with summer thunderstorms.

Particle size distribution and soil cations

As previously indicated, most soil locations that were studied at NTS had a low clay content (Table

Figures 2.8–2.11 Rock features on desert pavement. **Figure 2.8**. (Top left) General aspect of desert pavement in Rock Valley. **Figure 2.9**. (Top right) Varnish on the upper surface on a desert pavement rock. **Figure 2.10**. (Bottom left) Unvarnished rock of desert pavement in Rock Valley, which when turned over (**Figure 2.11**, bottom right) had a white calcareous undercoating.

Table 2.2. *Properties of bare-ground, nonsaline desert soils on bajadas in the southern Nevada Test Site (data from Romney et al. 1973). Profile study sites had slopes of 1–2% and mixed volcanic and limestone parent materials: Jackass Flats (elev. 1097 m), Mercury Valley (915 m), and Rock Valley (1030 m).*

Horizons	Jackass Flats (Site 64)			Mercury Valley (Site 10)				Rock Valley (Site 50)		
	A12	C1	C2	A12	C1	C2	C3	A2	B	C1
Profile properties										
Depth, cm	0–11	11–48	48–100	0–15	15–33	33–44	44–	0–12	12–21	21–28
Particle size, %										
sand	88.0	85.4	86.9	93.0	96.7	97.2	98.0	86.4	77.5	89.0
silt	26.4	10.2	5.8	4.2	2.4	2.0	1.4	92	10.8	6.8
clay	4.6	4.4	7.3	2.8	0.9	0.8	0.6	4.4	11.7	4.2
pH (paste)	8.3	8.4	8.3	8.4	8.4	8.6	8.6	8.4	8.0	8.4
Organic-matter, %	0.18	0.20	0.04	0.15	0.18	0.12	0.10	0.20	0.17	0.23
Lime, %	1.2	2.3	0.6	13.0	15.0	15.0	19.0	3.4	3.6	3.4
Exchangeable ions, meq/										
Na^+	0.48	0.48	4.70	0.52	0.38	0.52	0.61	0.50	0.41	0.80
K^+	1.68	1.86	0.85	0.40	0.45	0.27	0.28	0.60	0.31	0.28
Ca^{++}	6.60	2.64	3.96	3.96	3.86	2.64	3.96	5.28	3.96	3.90
Mg^{++}	1.50	2.76	1.44	1.44	2.10	2.82	0.09	2.88	0.09	1.50
Cl^-	0.20	0.00	0.30	0.73	0.49	0.61	0.33	0.00	1.60	0.86
NO_3	0.00	0.00	0.00	10.0	0.00	0.00	0.00	0.00	0.00	0.00
$SO_4^=$	0.10	0.13	0.10	0.50	0.50	0.03	0.60	0.13	0.03	0.05
Micronutrients, ppm										
Iron	0.3	0.3	0.6	0.1	0.1	0.1	0.1	0.3	0.5	0.5
Zinc	0.47	0.78	0.99	0.40	0.22	0.10	0.50	0.52	0.92	0.40
Copper	0.15	0.20	0.30	0.10	0.10	0.05	0.10	0.20	0.30	0.15
Manganese	1.55	1.25	1.15	0.75	0.55	0.60	0.45	0.95	0.65	0.85
Boron	1.80	0.50	4.40	1.04	0.60	0.92	0.80	0.96	0.10	0.60
Phosphorus (NaHCO₃ extract) ppm	1.40	1.40	0.00	0.40	0.24	0.20	0.20	0.36	0.36	0.36

2.2). Higher clay concentrations occur in closed drainage basins, such as in the playa of Frenchman Flat. Because of the presence of much volcanic glass, ash, and pumice at NTS (Ekren 1968), it is reasonable to assume that these materials contribute to the presence of noncrystaline clays, such as allophane and imogolite. Because of the low degree of weathering, smectites, micas, and probably kaolinitic clays are to be expected. The low clay content generally found in the various soil horizons is one of the striking characteristics of these desert soils. Stratifications of higher clay concentration were found occasionally in some horizons of profiles developed on alluvial sediments, but only a few sites, specifically those located in the closed-drainage basin of Frenchman Flat, had high concentrations of clay throughout the soil profile.

Loess, consisting of silt-sized material, blankets a major portion of the study areas. Most loess probably originated from the most recent volcanic ash falls, which have been further eroded, mixed, and transported by wind (Pye 1982).

Cation exchange capacities vary considerably between sites and reflect clay contents of the soils. In Mercury Valley, Rock Valley, and Yucca

Table 2.3. *Profile characteristics of soil developed on limestone parent materials at Rock Valley Site 54 beneath a shrub and in bare ground between shrubs (data from Romney et al. 1973).*

Profile properties	Shrub horizons			Bare horizons		
	A1	C1	C2	A2	C1	C2
Depth, cm	0–6	6–24	24–70	0.8	8–16	16–63
Color, wet	10YR4/3		10YR5/4		10YR5/4	
Coarse sand, %	20.2	12.2	37.4	18.2	14.8	31.0
Fine sand, %	67.3	73.8	49.4	31.5	67.7	59.1
Silt, %	6.6	8.8	7.7	12.9	11.0	5.6
Clay	5.9	5.2	5.3	7.4	6.5	4.3
pH, paste	7.9	8.3	8.2	8.0	8.3	8.3
pH, saturated extract	8.6	8.9	8.7	8.8	8.7	8.8
Moisture retention, %						
Saturation	42.9	37.0	20.0	22.9	28.1	37.5
1/3 bar	13.5	17.1	23.0	15.7	16.6	17.1
1 bar	10.2	10.1	15.8	10.8	11.0	11.4
15 bar	9.8	6.5	8.8	7.5	7.4	6.4
Carbonate, %	10.4	15.3	10.2	12.2	19.4	28.0
Saturated extract soluble cations and anions						
Sodium, m.e. l^{-1}	1.75	0.28	0.61	0.45	0.37	0.40
Potassium, m.e. l^{-1}	7.75	5.58	1.55	0.54	0.53	0.67
Calcium, m.e. l^{-1}	33.00	8.28	2.64	3.28	2.64	2.64
Magnesium, m.e. l^{-1}	14.25	7.88	4.56	4.17	2.84	2.76
Chloride, m.e. l^{-1}	1.60	0.05	0.05	0.02	0.02	0.02
Sulfate, m.e. l^-	0.63	0.10	0.10	0.10	0.08	0.08
Boron, ppm	2.10	2.50	1.50	0.50	0.80	1.00
Organic carbon, %	2.42	0.52	0.27	0.20	0.17	0.20
Organic nitrogen, %	0.23	0.05	0.04	0.03	0.02	0.02
Phosphorus (NaHCO$_3$ extract), ppm	6.60	2.31	0.05	0.05	0.05	0.05
Cation exchange capacity, m.e. 100 g^{-1}	15.0	12.8	10.3	12.2	12.0	10.0

Flat soils are low in exchangeable sodium but high in calcium and magnesium, whereas soils from Jackass Flats are low in sodium, calcium, and magnesium and high in potassium. Highest cation exchange capacity and exchangeable potassium are found in soils from Frenchman Flat; here potassium concentrations sometimes exceed the exchangeable calcium and magnesium content.

The pH values of soil profiles at NTS generally ranged from 8 to 9, but occasionaly pH values from 7.5 to 8.0 occurred in A horizons beneath shrub canopies (Table 2.3). Conductivity of the saturation extracts was higher in A and B horizons beneath shrub clumps than in exposed areas.

Soil salinity

Soils are classified as nonsaline-nonsodic whenever electrical conductivity is less than 0.4 mmhos mm^{-1} and exchangeable sodium percentage is less than 15. Nonsaline soils are typical of profiles within root zones of perennial desert scrub vegetation. Saline conditions have been measured in deeper horizons in Rock Valley and Jackass Flats, but saline soils occurred mostly in and around the playas. The highest levels of soluble salts were found in deeper soil horizons on Frenchman Flat. Soluble sulfates were relatively low in all profiles. Levels of micronutrients, such as manganese, iron, copper, and zinc, were low

Table 2.4. *Variations in saline properties from soil profiles beneath shrubs at Site 38 near the French-man Flat drainage playa (data from Romney et al. 1973).*

Profile properties	Horizons					
	A1	A2	C1	C2	C3	C4
Depth, cm	0–12	12–23	23–66	66–75	75–90	90–94
pH, saturated extract	8.9	8.7	8.1	7.8	7.5	8.0
Anions, m.e. l^{-1}						
Cl^-	5.90	2.40	13.40	32.30	11.10	9.30
NO_3^-	0	0	75.00	387.50	156.20	150.06
$SO_4^=$	0.15	0.15	1.81	1.16	1.26	1.71
Exchange sodium, %	3.6	5.7	11.4	14.7	12.9	17.7
Phosphorus (NaHCO₃ extract), ppm	1.08	1.16	0.24	0.20	0.24	0.40
Cation exchange capacity, m.e. l^{-1}	14.4	13.3	11.9	13.8	9.4	8.1

and variable between sites; however, shrubs showed no obvious signs of micronutrient deficiencies, and therefore micronutrient cycling probably was taking place. Soluble boron contents varied considerably between soils from various sites, and the highest levels were measured in Frenchman Flat. Boron, nitrates, and chlorides tended to accumulate more in profiles beneath shrub clumps than in those from bare surfaces.

Another example of variations in saline properties of soil is given in Table 2.4 for one soil profile in the Frenchman Flat playa, illustrating the degree to which soluble salts tend to accumulate at greater depths in the closed drainage basins. Of interest in this particular location are the higher concentrations of nitrate in deeper parts of the profile. The low level of sulfates was due to precipitation by calcium.

The playas are areas where silt and clay accumulate, as well as salts, and the patterns of accumulation are interesting. For example, at Frenchman Flat silt is higher than clay at all depths that were measured. This supports the conclusion that weathering processes are not far advanced for the region, otherwise clay levels in soils would not be so low. Also notable are the relatively high levels of exchangeable potassium both in the playa and bajada soils. Lack of plant growth on playas is certainly due to high salt levels at the soil surface as well as extremely harsh abiotic features of those habitats.

Effects of vegetation on soil formation

There is clear evidence that soil-forming processes are accelerated beneath desert shrubs, demonstrating an important role for perennials in desert ecosystems. Soils that are not directly associated with vegetation can be described as parent material that is weakly modified. Profile development is minimal; horizons are indistinct; organic matter accumulation is negligible; eluviation is slight; color usually resembles that of parent materials; and structure is weakly developed. In contrast, soil profile development beneath long-established perennial grasses and shrubs is often well defined: horizons are often recognizable, accumulation of organic matter is relatively high, color differentiations occur, and structural development can be seen (Romney et al. 1973).

One of the most consistent features of NTS soils is that the highest concentrations of organic carbon, organic nitrogen, and available phosphorus occur in the upper horizons beneath shrubs (Table 2.3; Romney, Wallace, Kaaz, & Hale 1980). Shrubs intercept wind-borne materials, which are added to coarser alluvial material at the plant bases. Salts also accumulate there, probably from recycling through leaf fall and litter decomposition as well as trapping of air-borne materials. Beneath shrubs there tends to be some decomposition of subsurface hardpans and a better developed A horizon. Carbon/nitrogen ratios of

Figures 2.12–2.13 Coppice mounds in Mojave desert scrub near Barstow, California. **Figure 2.12**. (Top) Fairly low, broad mound beneath the canopy of creosote bush (*Larrea tridentata*); mounds of this type tend to have relatively high soil nitrogen and other nutrients and therefore support abundant winter annuals. **Figure 2.13**. (Bottom) A prominent mound formed at the base of an old white bursage (*Ambrosia dumosa*).

organic matter are usually between 12–15. Below the A horizon the carbon/nitrogen ratios are around 10. Nitrates may be high. The A horizon is the only one in which organic carbon content may exceed 1%. As stated earlier, beneath shrubs soil pH also tends to be lower and conductivity of saturation extracts is higher.

A soil-forming feature that impacts upon the desert ecosystem is the accumulation of wind-blown materials at the bases of shrubs and grass tussocks, as well as other discontinuities on the land surface (Figs. 2.12–2.13). This accumulation of suspended soil particles and litter may build beneath each shrub into a 'coppice mound'. Resuspended soil particles may be trapped in the canopy of vegetation and subsequently deposited on the ground by washing action of rainfall or by association with leaf fall. Once again, the importance of mounds – or islands of nutrient accumulation – is that higher biological activity occurs in the mounds because soluble plant nutrients are concentrated there (Table 2.5). Species with clonal patterns of reproduction, such as the vegetatively spreading individuals of creosote bush (*Larrea tridentata*), may have very old mounds, and studies in the Mojave Desert have shown that clones may be as old as 10 000 years (Sternberg 1976; Vasek 1980). Coppice mounds are favored sites for shaded entrances of animal burrows.

An interesting spin-off from studies in radioactive fallout-contaminated areas at NTS has been the opportunity to use radionuclide 'tags' to follow their involvement in this wind-blown resuspension and redeposition process, and in the subsequent formation and growth of mounds beneath shrub clumps (Romney et al. 1973; Essington & Gilbert 1977; Gilbert & Essington 1977; Tamura 1975). Romney et al. (1973) showed that mounds contained higher levels of the long-lived ^{137}Cs isotope than did adjacent soil in fallout areas downwind from earlier aboveground nuclear tests. Contaminants that accumulated in these mounds were and continue to be incorporated in the nutrient cycle of desert vegetation and small animals, and they probably will become more biologically active over long periods of time.

The actual genesis and dynamics of mound for-

Table 2.5. *Range in ratios of selected soil properties measured in mound soil beneath shrubs and in adjacent bare-ground soil from 79 profile study sites (data from Romney et al. 1973).*

Profile properties	Range of shrubs/ bare-ground ratios
Cation exchange capacity	0.9–1.6
Conductivity (E_{c25})	3.6–9.4
Organic carbon	1.9–8.1
Soluble nutrients	
Na^+	2.5–14
K^+	2.5–19
C^{++}	2.6–13
Mg^{++}	3.9–30
Cl^-	1.9–40
NO_3^-	34–730
SO_4^-	4.5–48
Phosphorus ($NaHCO_3$ extract)	2.0–23

mation is not well understood. Wallace and Romney (1972) attempted to age-date coppice mounds by analyzing numbers of mature shrubs, annual growth rings, and organic matter turnover. Estimated ages of mature, stable mounds ranged from 100 to 300 years, but many mounds were much younger and in various stages of development.

Soil profiles of Rock Valley

For long-term ecological studies a team from AEC–UCLA established four permanent, 9-ha (20 acre) circular plots (A–D; Fig. 2.14), each with a diameter of 165 m, and a strong fence was erected to enclose plots A–C, preventing the immigration and emigration of rodents during experiments. Sustained climatological records were begun in 1962. The IBP validation site, used 1971–6, covered 0.46 km^2 and was located directly north of plots A and C. It was a five-sided, roughly trapezoidal parcel (Fig. 2.15) that had six somewhat recognizable vegetational groupings (El-Ghonemy, Wallace, Romney, & Valentine 1980).

Soils on the Rock Valley validation site are derived from a heterogeneous, highly calcareous

Figure 2.14 Aerial view of circular experiment plots C (upper), B (left of the road), and A (beneath C), 9-ha parcels of Mojave desert scrub enclosed by a strong fence that prohibited rodents from leaving or entering the plots. Plot D was an unfenced parcel to the north (right) of irradiated plot B, across the road that traversed Rock Valley. From Turner (1975).

Table 2.6. *Soil profiles made in Rock Valley at Sites 59–62 (Romney et al. 1973), located at the four corners of the IBP validation site. Each profile was established beneath a creosote bush (Larrea tridentata). Soil consistency was friable, mostly nonsticky, and either loose or soft, and all layers were violently effervescent. For a chemical analysis of approximately the same nonsaline soil, see data for Site 50 in Rock Valley (Table 2.2).*

	Depth, cm	% sand	% clay	% lime	% organic carbon	pH (paste)	Conductivity (Ec 25), mmhos cm^{-1}
Site 59 (SE)							
Horizon A1	0–6	92.4	3.7	11.4	0.87	8.3	1.37
A2	6–12	93.1	2.7	5.5	0.49	8.7	0.61
C1	12–23	88.2	4.6	15.0	0.38	8.8	0.44
C2	23–34	88.8	4.1	16.0	0.32	8.8	0.42
C3ca	34–57	90.6	3.8	36.5	0.30	8.7	0.42
Site 60 (SW)							
Horizon A1	0–5	89.5	4.7	5.0	1.07	7.9	1.21
A2	5–18	75.6	7.9	13.5	0.24	8.6	0.45
A3	18–38	85.1	6.2	21.7	0.34	8.7	0.45
C1	38–63	87.1	6.3	19.8	0.27	8.6	0.62
Site 61 (NE)							
Horizon A1	0–9	91.8	4.6	3.1	1.95	8.1	2.27
A2	9–19	72.5	9.5	5.5	0.25	8.6	0.69
B	19–37	86.1	5.6	3.2	0.21	8.7	0.53
C1	34–47	87.9	3.8	3.6	1.18	8.7	0.53
Site 62 (NW)							
Horizon A1	0–9	91.5	3.8	4.0	1.18	8.3	1.93
A2	9–21	78.3	7.3	6.5	0.42	8.7	0.70
C1	21–32	83.7	6.1	6.4	0.36	8.7	0.49
C2	32–50	90.1	2.4	20.0	0.28	8.7	0.65

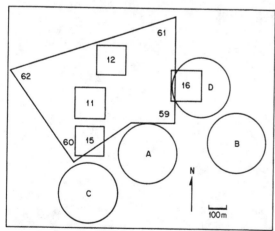

Figure 2.15 Research areas established in Rock Valley for long-term ecolgical studies. Circular plots A–D were fenced during the early 1960s for the first ecological studies, and the 5-sided IBP site was established in 1971. Squares numbered 11, 12, 15, and 16 were major sampling plots for plants and animals, and 59–62 were sites of soil analyses for the IBP.

alluvium that is composed primarily of Cambrian limestone with some tuff and basalt. A well-developed desert pavement covers the soils.

Romney et al. (1973) analyzed soil profiles from 13 sites within Rock Valley from elevations of 1015–1110 m, including all four corners of the IBP validation site (Sites 59–62) and on the periphery of the ecological study plots (A through D) that were used during the 1960s. Soil pits 0.74 m × 2.0 m in size were dug by hand in October, 1970, and the ones on the corners of the validation site were taken beneath creosote bush (*Larrea tridentata*). They were strikingly similar in horizon characteristics (Table 2.6). Soil textures were loamy sand in the uppermost profile and gravelly loamy sand below (20–45% gravel). Soil pH ranged from 7.9 to 8.3 in the A1 horizon to 8.8 in the C horizon, and when treated with acid all soils were violently effervescent because the parent material was limestone. Soil permeability was moderate to rapid. The surface was 30–70% gravel.

At eight sites within Rock Valley Romney et al. (1973) compared each soil profile beneath a shrub to that of an adjacent bare area. In all cases the A1 horizon was present beneath shrubs, but was absent in the corresponding bare area, and soils beneath shrubs, typically were deeper than in the bare areas, but differed little in their physical or chemical properties, being calcareous and sand to loamy sand with much gravel in the C horizon. Although one soil had a depth of 110 cm, other soils were significantly thinner and were underlain by cemented gravel, caliche, or other cemented pan variations.

Several special soil characteristics were measured at the validation site during June, 1972 by Porter and Mitchell from the University of Wisconsin (Table 2.7). These attributes have potential importance for developing transient heat models.

In October, 1972, Stolzy and Mehuys from the University of California at Riverside collected soil samples from the upper 0.50 m of a soil profile near the validation site. They determined that the >2 mm size fraction composed about 62% of the total weight of their samples (including stones). Bulk density (g cm^{-3}) of soil lacking stones was 1.50 (Mehuys 1973). Samples were also used to make laboratory measurements of hydraulic conductivity (Mehuys 1973; Mehuys, Stolzy, Letey, & Weeks 1973), taken 25 mm from the drying end of the soil column at varying levels of water potential (Table 2.8). These measurements were made using the >2 mm soil fraction and with unaltered samples (i.e., including stones and larger particles). Further information may be found in Mehuys et al. (1973) and Weeks and Richards (1967).

Soil water for plant growth

In desert scrub ecosystems at NTS annual precipitation normally recharges surface soil to a maximum depth of 1 meter, primarily as the result of gentle rainfall and snowfall from broad-area storms during winter and early spring. At other times of the year rainfall rarely dampens the soil. Light

Table 2.7. *Some physical attributes of Rock Valley soils (data from Warren Porter and John Mitchell taken in June, 1972).*

Soil attribute	Measurement
Solar absorptivity	0.79
Thermal conductivity	0.05 cal cm^{-1} min^{-1} °C^{-1}
Density-specific heat product	0.5 cal cm^3 °C^{-1}
Surface roughness	
0–60 cm	0.9 cm
61–200 cm	3.7 cm
Overall approximation	3.3 cm

Table 2.8. *Laboratory measurements of hydraulic conductivity of a Rock Valley soil (data from Mehuys et al. 1973).*

Water potential, −MPa	Hydraulic conductivity, mm s^{-1}	
	>2 mm fraction	with stones and larger particles
0.01	–	1.7×10^{-4}
0.02	1.7×10^{-6}	1.9×10^{-6}
0.03	4.2×10^{-7}	2.8×10^{-7}
0.10	2.2×10^{-8}	1.7×10^{-8}
0.30	2.2×10^{-9}	1.7×10^{-9}
1.0	2.8×10^{-10}	1.7×10^{-10}
3.0	1.9×10^{-11}	2.8×10^{-11}

rains evaporate rapidly from the surface soil, and there is high runoff during high intensity, short duration summer thunderstorms, resulting in sheet flow that moves rapidly into drainage channels. As a result, vegetation is dependent primarily upon the cool season annual recharge within the top meter of the soil.

Moisture retention on the bajadas ranges from moderate to low, but vegetative growth in the spring, coupled with drying environmental conditions, rapidly deplete soil water, and for 6–8 months, sometimes more, the soil water potential stays below −3.0 MPa. In fact, during summer months and drought periods soil water potentials frequently fall below −9.0 MPa, i.e., a dry soil.

At water balance study plots near Mercury and a low-level waste repository site in Frenchman Flat, during summer drought researchers observed an upward flow of water vapor from deeper underground storage areas into the shallow soil, where shrub roots occurred. This upward movement of water vapor into dry soil was measured by a combination of psychometric and resistance sensors as well as by tracking the movement of tritium through the soil atmosphere from sources buried deeper than 30 m. Desert shrubs have the capability of utilizing this moisture, resulting from underground vapor phase transport (Hunter 1991).

CLIMATE AND ROCK VALLEY MICROCLIMATE

Typical lowland Mojave Desert habitats in southern Nevada and adjacent Death Valley, California, are some of the most arid desert localities within North America. Numerous weather stations below 1000 m have mean annual precipitation less than 120 mm (Chapter 1), and these low values, when coupled with high spring and summer daytime temperatures, nearly cloudless skies, and extremely low water vapor concentration in the atmosphere and hence high evaporation, produce an extremely arid environment for perennial plants and summer-active animals. Rarely annual precipitation at a particular lowland site exceeds 250 mm, but opposing that may be several successive years with precipitation values significantly less than the mean. Under such parching conditions, desert soils may lack available moisture for months at a time, and plants and animals must be opportunistic in their use of water, when it is available.

Precipitation

As described in Chapter 1, the climate of southernmost Nevada and the southern Nevada Test Site is typical of the central Mojave Desert, whereas the climate northward is transitional to the Great Basin (Schaeffer 1968; Houghton 1969). Winter storms (November through April) may bring widespread precipitation, usually as rain, from the Pacific Basin. In summer few storms ever reach southern Nevada, but sometimes storms originating from the Gulf of Mexico and passing westward bring rain to NTS; these storms occur as isolated, local showers. Occasional tropical storms (chubascos) may also bring heavy but infrequent and scattered rains to southern Nevada during September and early October (Quiring 1968).

Long-term precipitation records for southern Nevada

French (1983) has summarized long-term precipitation records for many weather stations in southern Nevada. Because the majority of those stations were moved frequently during the recording period, some care must be taken in interpreting the long-term mean values at many of those sites, but stations at Boulder City, Ada, and Mina were permanent. Throughout NTS more than 30 precipitation stations were also maintained between 1962 and 1980, and these are considered to be very reliable data. From NTS data mean values ranged from 86 to 310 mm yr^{-1}, and there was a general linear correlation of precipitation increasing with elevation for stations located from 700 to 2300 m elevation (Fig. 2.16). Consequently, the low elevation Mojave Desert localities are drier than the higher transitional desert and Great Basin desert scrub communities.

On NTS long-term records of precipitation, intensity and duration have been analyzed in cen-

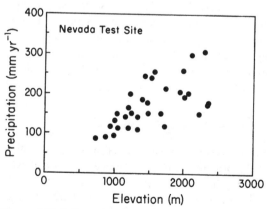

Figure 2.16 Relationship of mean annual precipitation to elevation at the Nevada Test Site. Data from French (1983).

Table 2.9. *Storm frequency, duration, and intensity for Well 5B, central Frenchman Flat, Nevada Test Site (adapted from French 1983). Data were taken from 15 years of records (1963–1979). Winter storms were those occurring from October through April, whereas summer storms were those from May through September.*

Storm duration, hr	Storm frequency, %	Precipitation frequency, %	Number of winter storms	Winter storm intensity, mm storm^{-1}	Number of summer storms	Summer storm intensity, mm storm^{-1}
0–1.9	61	34	183	1.8	122	2.8
2–3.9	19	19	63	3.8	31	4.8
4–9.9	14	28	58	7.9	11	8.9
9.9	6	19	29	11.9	2	20.3
Total storms			333		166	

tral Frenchman Flat at Well 5B (942 m elevation), a location comparable to Rock Valley. Based on 16 years of records, Well 5B had a mean annual precipitation of 117 mm, having an average of 72% of the precipitation falling in winter and spring (French 1983). Data for 499 storms during a 15-year period indicated that greater than 60% of the precipitation events lasted two hours or less, whereas only 6% lasted longer than ten hours (Table 2.9). However, short duration storms (less than two hours) accounted for only one-third of total precipitation, but long duration ones (greater than ten hours) comprised nearly 20% of total rainfall. Two-thirds of all storms occurred during winter, October through April. Although mean summer storm intensity was higher than that of winter storms in every category of storm duration, the high proportion of short duration summer storms strongly influenced the winter bias (72%) of total precipitation.

Rock Valley precipitation records

The Rock Valley precipitation station, at an elevation of 1040 m, over 16 years of records had a mean annual precipitation of 150 mm (French 1983), and during the 6 years of the IBP (1971–6) annual precipitation ranged from 62.5 mm in 1975 to 223 mm in 1976 (Fig. 2.17).

Monthly precipitation values, obtained during

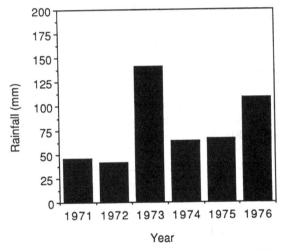

Figure 2.17 Annual precipitation at Rock Valley from 1971 to 1976 during the IBP study.

the IBP, are presented in Table 2.10 for comparisons with 14-year monthly means from 1963 to 1976. Standard deviations of monthly and annual precipitation values over the record years were both high and seasonally variable, although relative standard deviation, calculated as the monthly standard deviation divided by monthly mean precipitation, was relatively consistant at 1.0–2.0. Such variation is typical of many desert regions of the world (Solbrig 1979), and this promotes an

Table 2.10. *Monthly rainfall (mm) recorded at the IBP validation site as compared with the 14-year monthly means of rainfall obtained at Well 5B in Frenchman Flat (data from French 1983).*

Month	14-year	Standard	1971	1972	1973	1974	1975	1976
January	13.1	20.1	0	0	23.1	34.0	1.0	0
February	25.3	36.0	8.1	0	47.8	1.0	4.0	98.0
March	13.3	20.0	1.3	0	76.7	6.4	26.9	1.5
April	11.0	16.1	0	1.8	14.9	0	9.9	15.5
May	4.7	5.6	19.3	0	10.4	0.3	8.1	7.1
June	3.8	4.5	0	11.2	4.2	0	0	0
July	10.4	14.2	0.5	0	0	25.6	0	13.5
August	12.4	15.0	33.3	20.1	2.3	1.6	1.8	0
September	7.9	15.5	0	8.4	0	0	5.3	51.0
October	8.9	12.5	0	32.0	3.8	25.9	0.5	34.8
November	13.8	14.4	0.8	29.7	14.0	2.5	4.8	0.3
December	13.7	19.0	40.6	0	10.9	32.8	0	1.3
Total	138.1	162.5	103.9	103.2	208.1	130.1	62.5	223.0

unpredictability in availability of limited moisture resources. Under these circumstances the concept of mean precipitation values has little utility. The IBP years were somewhat unusual in having 2 very wet winters, in 1973 and 1976, out of 6, and hence a relatively high value of mean annual precipitation (Fig. 2.18).

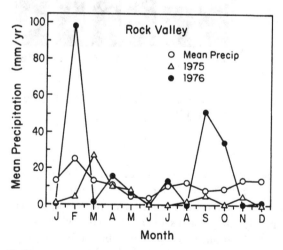

Figure 2.18 Monthly pattern of precipitation at Rock Valley compared with rainfall in 1975 and 1976.

Individual rainfall events in Rock Valley

Between 1974 and 1976 individual rainfall events were recorded in Rock Valley. Tables 2.11 to 2.13 give the measurements at Station 1 for 1974–6, respectively, on duration of events, amount of precipitation, and rainfall intensity. Rainfall events of 7 mm or more occurred 6 times (16%) in 1974, 3 times in 1975 (12%), and 7 times (15%) in 1976. A 7-mm rainfall is considered to be one that will effectively moisten a dry desert soil to a depth of 0.08–0.10 m (Nobel 1976a), where many roots of the plants occur. Actually only two events, one in July, 1974, and the other in May, 1975, probably reached that critical soil depth during the three summers. About 40% of the events yielded 1 mm of rainfall or less, which is considered a 'trace', and in 1975, a dry year (62.5 mm), the root zone of plants in dry soil probably received a good supply water about four times, in March, April, May, and September.

Mean rainfall intensity (Tables 2.11–2.13) from 1974 to 1976 was 1.2 mm hr^{-1}, and maximum intensity was 11.3 mm hr^{-1} in September, 1976. Rainfall intensities for six intervals were analyzed to determine whether that parameter varied sea-

Table 2.11. *Individual rainfall events and duration and intensity of each event in Rock Valley during 1974.*

Date of rainfall event		Duration, hr	Rainfall during event, mm	Rainfall intensity, mm hr^{-1}
January	1	2	0.76	0.38
	6	10	10.92	1.09
	7	22	14.99	0.69
	8	1	0.25	0.25
	16	1	1.02	1.02
	17	1	0.25	0.25
	20	2	0.51	0.25
	21	1	3.05	3.05
	21	1	2.29	2.29
February	19	1	0.66	0.66
	28	1	0.41	0.41
March	3	5	3.30	0.66
	8	3	0.76	0.25
	8	2	2.90	1.45
	8	1	0.25	0.25
May	19	1	0.25	0.25
July	22	2	3.05	1.52
	23	7	17.78	2.54
	30	3	4.06	1.35
August	3	1	1.27	1.27
October	2	1	2.54	2.54
	2	2	0.51	0.25
	3	2	7.35	3.67
	7	2	0.51	0.25
	8	1	0.25	0.25
	26	3	3.56	1.19
	27	1	0.25	0.25
	27	1	0.25	0.25
	28	1	0.76	0.76
	28	9	9.13	1.01
	31	2	1.02	0.51
November	2	1	0.27	0.27
	2	2	2.32	1.16
December	4	6	17.78	2.96
	4	2	3.30	1.65
	28	4	5.84	1.47
	29	4	5.59	1.40

sonally or between different years (Table 2.14). A 2×3 factorial analysis of variance showed a distinct effect owing to years ($F = 6.7$, $F_{.01} = 4.82$), but not to differences between seasons ($F = 2.3$, $F_{.05} = 3.94$). Hence, at least for the study period considered, winter rains were not significantly more intense than rains during the remainder of the year. On the other hand, rains between November, 1974, and October, 1975, a period of low rainfall (93 mm), were less intense than other rains, suggesting that intensity and amount of rainfall may be directly related.

Spatial variability of local rainfall was measured at Rock Valley in 1972 and 1973 by using four additional stations placed at the corners of the 0.46 km^2 validation site. These data (summarized

Table 2.12. *Individual rainfall events and duration and intensity of each event in Rock Valley during 1975.*

Date of rainfall event		Duration, hr	Rainfall during event, mm	Rainfall intensity, mm hr^{-1}
January	28	2	0.76	0.38
	29	1	0.51	0.51
February	2	1	0.25	0.25
	3	4	1.27	0.32
	14	6	2.29	0.38
	15	3	0.76	0.25
March	5	5	5.08	1.02
	5–6	14	9.65	0.69
	8	15	8.64	0.58
	11	1	0.51	0.51
	13–14	5	3.56	0.71
April	10	2	1.78	0.89
	11	4	2.54	0.64
	11	9	3.56	0.40
	11–12	7	3.05	0.44
May	20	16	8.64	0.54
August	18	1	1.30	1.30
	21	2	1.27	0.63
September	8	1	0.25	0.25
	9	3	4.83	1.61
	10	1	0.25	0.25
	10	1	0.76	0.76
October	7	1	0.25	0.25
	12	1	0.25	0.25
November	28	2	3.81	1.90
	29	2	1.02	0.51

monthly by Turner 1973; Turner & McBrayer 1974), showed only small amounts of such variation.

Temperature

Typical lowland sites in the Mojave Desert have high spring and summer daytime temperature (Table 2.17). At the extreme is Death Valley, which has a mean annual temperature of 24.7°C (MacMahon & Wagner 1985).

Rock Valley temperature records

During the IBP study at Rock Valley continuous air temperatures were recorded at 2.4 m above the ground surface at the same NOAA climatolo-gical station (National Oceanographic and Atmospheric Administration) used for precipitation records. During 1972 and 1973 air temperatures were also recorded at the 0.30 m height, both in the open and under a shrub canopy. Mean monthly air temperatures from 1971 to 1976 ranged from 5.5°C in January to 31.1°C in July (Table 2.15). Standard deviations of monthly mean temperature for this period were lowest in summer and fall months and highest in the spring, reflecting the unpredictability of temperatures as well as precipitation during the growing season.

Ranges of daily minimum and maximum air temperatures at Rock Valley are shown in Tables 2.16 and 2.17. Freezing temperatures were recorded during 7 months of 1971 and 5 or 6 months during the other IBP years. If data for

Table 2.13. *Individual rainfall events and duration and intensity of each event in Rock Valley during 1976.*

Date of rainfall event		Duration, hr	Rainfall during event, mm	Rainfall intensity, mm hr^{-1}
February	5	2	3.1	1.5
	6	11	33.5	3.0
	6	2	5.6	2.8
	7	3	5.6	1.9
	7	1	6.3	6.3
	8	1	1.3	1.3
	8	1	1.3	1.3
	8	5	12.7	2.5
	8	3	12.7	4.2
	9	3	5.6	1.9
	9	1	3.1	3.1
	9	1	3.6	3.6
	10	5	3.8	0.8
March	3	1	1.5	1.5
April	13	4	7.6	1.9
	13	3	4.8	1.6
	14	1	2.3	2.3
	15	2	0.8	0.4
May	6	2	3.8	1.9
	6	1	1.3	1.3
	6	1	0.8	0.8
	6	1	0.3	0.3
	7	1	0.2	0.2
	7	1	0.5	0.5
	7	1	0.2	0.2
July	16	1	0.3	0.3
	16	6	5.1	0.9
	16	2	1.8	0.9
	25	3	1.0	0.3
	26	1	0.5	0.5
	26	2	0.5	0.3
	27	3	0.5	0.2
	30	1	0.8	0.8
	31	2	2.5	1.2
September	5	1	1.8	1.8
	8	1	0.5	0.2
	10	2	3.1	1.5
	10	3	33.8	11.3
	10	1	1.0	1.0
	11	2	5.1	2.6
	25	1	1.0	1.0
	25	3	2.3	0.8
	30	1	0.8	0.8
October	1	10	23.9	2.4
	1	9	10.9	1.2
November	15	1	0.3	0.3
December	30	1	0.5	0.5

Table 2.14. *Number of rainfall events and mean rainfall intensity in Rock Valley between November, 1973, and October, 1976.*

Interval	Number of events	Mean rainfall intensity, mm hr^{-1}	Range of rainfall intensity, mm hr^{-1}
November, 1973–March, 1974	18	1.04	0.25–3.05
November, 1974–March, 1975	17	0.85	0.25–2.96
November, 1975–March, 1976	16	2.38	0.51–6.30
April–October, 1974	16	1.12	0.25–3.67
April–October, 1975	13	0.63	0.25–1.61
April–October, 1976	31	1.36	0.35–11.30

Table 2.15. *Mean monthly air temperatures (°C; at 2.4 m height) in Rock Valley between 1971 and 1976. No data were recorded for October and November, 1975.*

Month	1971	1972	1973	1974	1975	1976	Mean	Standard deviation
January	8.8	6.5	3.1	2.7	4.3	7.7	5.5	2.5
February	6.1	11.7	6.9	7.4	5.7	8.9	7.8	2.2
March	9.0	19.3	7.0	12.3	7.6	9.9	10.9	4.5
April	12.0	16.1	14.4	14.4	9.4	13.1	13.2	2.3
May	14.9	20.8	22.5	22.6	19.5	22.8	20.5	3.0
June	24.3	27.1	28.0	29.5	26.6	25.0	26.7	1.9
July	20.1	34.1	31.5	30.1	30.6	30.4	31.1	1.5
August	28.3	28.3	29.5	28.7	28.8	27.1	28.5	0.8
September	22.1	23.9	24.5	27.5	26.8	23.1	24.7	2.1
October	16.7	14.2	18.0	17.5	N/A	16.7	16.6	1.5
November	6.1	7.9	9.1	10.2	N/A	12.2	9.1	2.3
December	4.6	3.8	6.5	3.8	8.1	7.6	5.7	1.9

Table 2.16. *Range of daily minimum air temperatures (°C; at 2.4 m in height) in Rock Valley between 1971 and 1976. No data were recorded for October and November, 1975.*

Month	1971	1972	1973	1974	1975	1976
January	−3–9	−3–4	−8–7	−10–5	−13–7	−7–8
February	−8–5	−6–12	−1–7	−5–9	−8–8	−2–7
March	−10–12	4–14	−1–5	−4–12	−7–9	−1–15
April	−4–11	0–17	1–14	0–10	−6–10	−2–12
May	1–15	2–19	3–19	2–20	−1–19	7–20
June	1–20	13–22	8–27	12–23	7–24	8–27
July	14–27	17–32	12–28	10–27	15–28	14–29
August	16–26	15–26	11–29	16–24	16–32	12–26
September	3–23	8–21	9–19	16–26	14–26	12–23
October	−4–18	−4–18	2–16	1–19	N/A	3–16
November	−5–4	−2–10	−4–10	0–10	N/A	−4–13
December	−7–9	−9–4	−3–6	−9–6	−7–7	−3–6

Table 2.17. *Range of daily maximum air temperatures (°C; at 2.4 m in height) in Rock Valley between 1971 and 1976. No data were recorded for October and November, 1975.*

Month	1971	1972	1973	1974	1975	1976
January	1–22	8–18	1–19	2–16	3–20	4–24
February	4–22	7–27	7–17	8–20	2–22	7–23
March	−1–27	19–33	2–19	6–28	3–21	2–25
April	8–26	10–20	13–31	16–30	7–24	11–30
May	14–31	16–36	14–40	15–41	13–36	19–36
June	21–39	28–44	24–45	30–45	25–39	25–40
July	33–42	34–49	35–45	29–43	33–44	27–42
August	31–42	24–43	31–43	33–38	28–42	28–44
September	16–41	25–37	26–38	30–38	28–38	22–43
October	4–33	9–33	17–34	6–33	N/A	17–31
November	3–19	8–21	6–25	10–23	N/A	4–29
December	2–14	−3–19	7–19	0–16	5–23	12–19

October and November were available for 1975, there would likely have been 7 or 8 months with freezing temperatures. The lowest temperature reached was −13°C in January, 1975. Summer low temperatures in July and August ranged widely from 10 to 32°C, with variations up to 20°C in the nightly values for a single month. In most months ranges of daily low temperatures commonly were 10–15°C (Table 2.16).

Summer high temperatures commonly were 30–45°C, and the high daily temperature at Rock Valley was 49°C during July, 1972 (Table 2.17). High temperatures above 40°C were recorded in 2 to 4 months each year. Despite having low winter minimum temperatures in December and January, daily maximum temperatures of these months commonly reached 10–15°C. During the IBP study the lowest daily maximum reading was 3°C in December, 1972.

Overall seasonal patterns of temperature variation appear in the IBP data. Periods of relatively cold temperature occurred through most of 1971 and in the early part of 1975. Even early in 1972 temperatures were generally high in comparison with other years. Periods of unusual temperatures were often, but not invariably, associated with atypical patterns of precipitation. For example, the elevated temperatures in spring, 1972 were associated with a period of very low rainfall, whereas in early 1973 abnormal cold weather

coincided with heavy rains. Unexplained, however, was the unusually cold weather in 1971 that was unrelated to heavy rainfall.

To assess the impact of shrub canopy on air temperatures, measurements were taken every 4 hours for one entire month per season, this for 1972 and 1973 (Table 2.18). Under shrubs temperatures were statistically lower throughout the year at midday. During April and July at sunrise canopy temperatures were statistically higher than those in the open, and in general the modifying effects of shrub canopy on air temperatures were least pronounced during fall and winter. Variances of temperature measurements in the open were always greater than those taken beneath the shrub canopy.

Rock Valley relative humidity

Because Rock Valley is not located near agricultural enterprises or natural sources of surface water, it can be used as a fairly typical site for describing daily and seasonal profiles of relative humidity for the northern Mojave Desert.

Relative humidity was recorded in the southeastern corner of the Rock Valley validation site with Wescor hygrothermographs, which were positioned 0.15–0.25 m above ground surface. Readings were taken daily at 6-hr (1972–4) or 2-hr intervals (1975–6). Chart readings at less than 5%

Table 2.18. *Mean air temperatures (°C) in the shade and open at four times of day during four 30-day periods at Rock Valley during 1972 and 1973. Asterisks indicate significantly different pairs of means.*

Period	Year	In the open				In the shade			
		0600	1200	1800	2400	0600	1200	1800	2400
January 12–February 10	1972	3.5	15.3*	6.2	4.6	3.8	12.8	6.6	4.9
	1973	4.3	9.8*	8.3	8.1*	4.5	7.9*	4.5	7.7*
April 1–30	1972	10.0*	24.0*	17.6	9.9	11.9*	22.2*	17.1	10.4
	1973	10.0*	24.2*	18.8*	16.1*	11.6*	17.5*	15.7*	14.6*
July 1–30	1972	20.4*	38.3	37.1	25.1	23.3*	37.7	36.1	25.7
	1973	18.9*	42.5	35.5*	30.9*	22.1*	37.9*	32.2*	29.4*
October 1–30	1972	9.4	21.4*	15.5	10.0	9.7	19.6*	15.1	10.1
	1973	11.0	26.4*	22.1*	21.2	11.1	21.8*	20.7*	21.1

Table 2.19. *Highest and lowest relative humidities (percent) recorded at Station 1 in Rock Valley between 1972 and 1976.*

Month	1972		1973		1974		1975		1976	
	Low	High	Low	High	Low	High	Low	High	Low	High
January	10	86	30	89	20	81	8	86	5	70
February	5	87	30	92	8	77	11	78	5	91
March	11	69	28	97	6	76	6	88	5	88
April	5	82	21	93	5	49	5	84	5	95
May	5	69	7	93	5	48	5	78	5	95
June	5	83	5	77	5	39	5	59	5	68
July	5	43	5	41	5	52	5	62	5	90
August	5	77	5	77	5	80	5	84	5	75
September	13	99	5	53	5	57	5	90	5	99
October	15	92	5	69	5	85	5	82	5	99
November	22	92	13	83	14	76	5	89	9	95
December	22	89	5	81	9	80	6	80	5	85

were arbitrarily set at 5%, and at best readings were accurate to within ±5% but were prone to highest error at low humidities. For a month of n days, the mean relative humidity was computed as the average of $4n$ measurements (6-hr intervals) or $12n$ measurements (2-hr intervals). Maximum and minimum monthly humidities were simply the highest and lowest readings, respectively, during the month.

Over the five years of measurements mean monthly relative humidity ranged from 43% in January and February to 17% in June and July (Table 2.19). The highest monthly mean was 65%

in November, 1972, and the low was 13% in July, 1972 and June, 1974. Extremely low relative humidities of 5% or less occurred throughout the year, and mean monthly values were high during the during the rainy months, but were more highly correlated with mean monthly air temperature ($R = -0.71$) than with monthly rainfall ($R = 0.60$). Probably the relatively poor correlation between humidity and rainfall was because only monthly measures were used for analysis; correlations on a short-term basis would have been higher. When average monthly relative humidity values(H) for 1972–6 were examined in terms of monthly air temperatures (T) and monthly rainfall totals (R), the relationship was:

$$H\ (\%) = 0.32R - 0.92T + 44.0$$

This model had a multiple R of 0.82 ($R^2 = 0.67$).

Solar irradiance

The general pattern of insolation for southern Nevada may be inferred from measurements of percent of maximal (clear-day) sunshine realized in nearby Las Vegas (Table 2.20). Individual annual totals (1969–75) ranged from 81.5% in 1973 to 87.3% in 1975, resulting in an overall annual mean about 85%. Although year-to-year differences were slight, percent of maximal sunshine for the months of January through April ranged from as low as 74.4% in 1973 to 91.8% in 1972, reflecting the degree of cloud cover related to rainfall during those periods, wet and dry seasons, respectively.

Rock Valley irradiance

For Rock Valley, the theoretical maximum irradiance per year, assuming cloudless conditions, is about 1000 W m^{-2}. Theoretical maximum daily fluxes of incident solar energy range from about 42 W m^{-2} in June to about half that value during December and early January. Measurements of actual solar radiation at Rock Valley were made between October 11 and December 31, 1971, throughout 1972 (Fig. 2.19), and between January 5 and July 29, 1973. During 1972, the total incident solar energy was 73% of theoretical maximum. An analysis of the amounts of measured radiation relative to the maximum for the 53 time periods showed that the slope of the regression did not differ significantly from zero ($F = 0.80$, 51 d.f.). At Rock Valley for the first five months of years 1972 and 1973 measured incident radiation amounted to about 20% of theoretical irradiance under cloudless conditions.

Wind patterns at Rock Valley

As part of long-term weather monitoring at the NOAA site in Rock Valley, wind speeds and direc-

Table 2.20. *Typical radiative environment for 1969 to 1975 in Las Vegas, Nevada.*

Month	Average percent of possible sunshine	Range	Standard deviation
January	76.0	61–92	13.5
February	79.9	62–92	12.4
March	82.3	70–97	10.6
April	90.3	84–97	5.0
May	89.4	83–95	5.0
June	92.3	86–98	4.6
July	88.4	78–95	6.7
August	88.1	78–94	5.4
September	93.9	85–100	5.4
October	83.4	55–95	13.3
November	78.6	70–87	6.1
December	76.4	63–82	10.8

tions were continuously recorded during the six years of IBP research. Annual wind patterns were similar throughout that period, and they are probably applicable to many similar lowland sites within the Mojave Desert.

Monthly mean wind speeds (the means of daily means) ranged from 9 to 15 km hr^{-1}, with highest means in March or April and lowest speeds between July and October (Table 2.21). Average monthly minimum wind speeds (*ca.* 3 km hr^{-1})

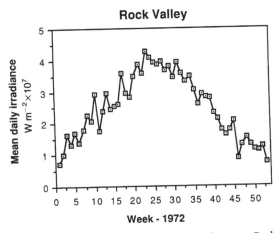

Figure 2.19 Seasonal changes in solar irradiance at Rock Valley from October, 1971 to July, 1973. Points are mean values for each week.

Table 2.21. *Mean monthly wind speeds (km hr⁻¹) in Rock Valley between 1971 and 1976. No measurements were made during October, 1975.*

Month	1971	1972	1973	1974	1975	1976	6-year mean	Standard deviation
January	12.2	12.5	13.1	11.4	11.8	12.5	12.3	0.60
February	10.6	11.3	12.7	12.7	11.5	10.7	11.6	0.93
March	12.2	11.6	11.4	11.7	12.8	13.1	12.1	0.69
April	12.6	14.0	13.7	13.5	13.1	11.1	13.0	1.05
May	13.1	11.9	11.4	11.9	12.9	11.2	12.1	0.78
June	11.4	11.7	12.2	11.5	12.6	12.1	11.9	0.46
July	10.8	12.6	10.5	11.9	11.6	10.8	11.4	0.81
August	10.8	10.7	11.0	10.6	12.4	11.4	11.2	0.67
September	13.4	11.9	11.8	10.8	12.1	9.2	11.5	1.41
October	14.8	9.4	12.2	11.1	N/A	10.7	11.6	2.03
November	11.5	11.0	11.5	11.4	12.2	11.1	11.5	0.42
December	12.7	14.4	11.8	10.8	12.7	12.7	12.5	1.19
Overall mean	12.2	11.9	11.9	11.6	12.2	11.4	11.9	

Table 2.22. *Mean annual wind speeds (km hr⁻¹) from eight directions in Rock Valley between 1971 and 1976.*

Direction	1971	1972	1973	1974	1975	1976	6-year mean	Standard deviation
N	10.9	11.4	9.5	9.3	10.1	9.9	10.2	0.8
NE	12.8	11.6	11.2	11.1	11.7	11.0	11.6	0.7
E	11.2	11.2	11.2	10.7	10.4	10.6	10.9	0.4
SE	7.0	7.0	8.3	7.8	7.7	7.4	7.5	0.5
S	13.5	14.1	14.6	12.7	12.1	10.4	12.9	1.5
SW	12.0	11.3	10.5	10.5	11.4	10.1	11.0	0.7
W	8.5	8.3	7.9	8.5	8.4	7.6	8.2	0.4
NW	10.1	11.1	11.3	12.7	11.0	11.2	11.2	0.8

almost always occurred during the summer, and the highest average monthly maximum speeds (23 km hr⁻¹) typically occurred between April and June. March through June also tended to have the greatest variations in daily means within a given month. For each year the highest hourly wind (37–42 km hr⁻¹) occurred in five different months.

Wind direction generally changed during each day, as it does in most typical desert valleys. Night winds most often blew from the east to the northeast, and day winds switched to come from the south to southwest. There were also strong seasonal trends, with dominant winds from the east to northeast during the winter and from the south

to southwest during warmer months. Wind blew either from the east-northeast or south-southwest 76–90% of the time, depending on the year and season. Between September and March winds predominantly were from the north and east (54–71%), and April to August winds were characteristically from the south-southwest.

Table 2.22 summarizes mean wind speeds at Rock Valley measured from eight compass directions. Dominant winds, discussed above, were generally stronger, whereas winds from the southeast and west were light.

Run of the wind for each month was computed by multiplying the number of hours the wind blew

Table 2.23. *Monthly run of the wind (km) in Rock Valley between 1971 and 1976. No data were available for December, 1972 and October, 1975.*

Month	1971	1972	1973	1974	1975	1976	6-year mean	Standard deviation
January	4958	7754	9810	8489	9080	9315	8234	1755
February	7110	4595	8552	8507	7706	5877	7058	1563
March	9057	6125	7347	8707	9643	9772	8442	1430
April	9056	8040	9828	9722	9569	7749	8994	897
May	9738	8882	8452	8877	9527	8332	8968	564
June	8190	8419	8756	8250	8981	6507	8184	876
July	8027	8707	7836	8206	8811	8050	8273	396
August	8058	8002	8135	7860	9212	8457	8287	495
September	9325	8578	8517	7781	5800	6648	7775	1324
October	6368	7002	9107	8269	N/A	7944	7738	1074
November	5261	7921	8293	8198	5248	7376	7050	1426
December	7819	N/A	8774	8031	9489	6421	7981	1221
Annual mean	7695	7639	8617	8408	8460	7704	8087	

from a given direction times the mean wind speed (for that direction and summing the resulting products; Table 2.23). Relatively little season variation was found in mean run, and annual run of wind varied less than 5% of the 6-year annual mean (8087 km). Most months with low wind runs occurred during the autumn-winter rainy season, but from 1971 to 1976 the variability of this measure was relatively high from September to March and low for the hot months.

The conclusions from these data are that Rock Valley has winds principally from the east to south quadrant (90–180°) but that they are generally not strong enough to have obvious mechanical drag effects on either plants or animals. Air movement can influence the microenvironments of plants and animals (Nobel 1981a), but because the diel and seasonal patterns of wind speed and direction were consistent, it is unlikely that wind has played a significant role in year-to-year variations in abundances of plant and animal species of Rock Valley.

Rock Valley is not subjected to major dust storms, which are characteristic of many low elevational desert scrub habitats (Goudie 1978, 1983).

SOIL MICROCLIMATE
Rock Valley studies

In Rock Valley gravimetric measurements of soil moisture were made throughout 1971 in the open at two stations, sampled at depths of 0.15 and 0.30 m. Soil psychrometers were installed to monitor soil water potential, beginning in October, 1971, with Wescor MJ-55 psychrometric voltmeters and switching in 1974 to Wescor Model HR33T (see Rundel & Jarrell 1989). The model MJ-55 psychrometer could not measure potentials less than about 5.0 MPa, and all lower readings were set at that value. The model HR33T was more sensitive and could register soil water potential to −7.5 MPa (1974–6). Psychrometric measurements between 1971 and 1973 were made at four stations, both between shrubs and in the open and each with sensors at 0.15 and 0.30 m. Measurements between 1974 and 1976 were made at only two stations, one in the open and the other beneath shrubs, at depths of 0.03, 0.15, 0.30, and 0.45 m.

Experiments conducted in 1971 with soils from two stations showed that the relationship of percent soil water content (W) and moisture potential (B, in -MPa) could be expressed as follows:

$$W = 0.944 - 4.23(\log B) + 0.26(\log B)^2$$

The saturation level of these soils was about 12% water by weight.

Moisture potentials of soils at Station 1 were measured for the six-year period, with 1971 values converted from gravimetric data. The profiles differ qualitatively because methods were not constant, for example, depths of 0.15 and 0.30 m in some instances, but 0.03 and 0.30 m in others. However, these differences are instructive in that they illustrate important distinctions in moisture dynamics at different soil levels. Measurements made at 0.45 m were not plotted because they did not differ in any significant way from those made at 0.30 m. Changes in measurements at 0.45 m were over less amplitude and somewhat retarded relative to those taken at 0.30 m.

Except in 1976, soils at 0.3 m, either in the open or beneath a shrub canopy, were at or near field capacity during December. This was also true of soils at 0.15 and 0.30 m, except in 1975, and soil field capacity continued in those years where winter storms periodically added soil moisture. During the extremely wet period in early 1973, soils were still at field capacity as late as March. This soil at depth of 0.03 m experienced the greatest variations in moisture content, often having rapid changes from extremely dry to saturated conditions following 'light rain', presumably several millimeters. After the cessation of winter rains, soils beneath shrubs typically dried more rapidly than those in the open, and a comparison of drying rates of 0.03-m soil in open and beneath shrub canopy were often dramatic. As expected, soil moisture potentials at 0.15 and 0.30 m changed much less rapidly than those of surface soils. Once soils beneath shrubs were dry, they often remained so at times when soil in the open showed measureable increases in moisture content. These differences apparently reflect the intense biological activity centered around the 'fertile islands' of vegetation (Wallace, Romney, & Hunter 1980a) but may, in part, be caused by less

rainfall beneath shrubs due to interception by the canopy.

Near-by studies

Annual pattern of soil CO_2 levels was monitored in two Mojave desert communities, lowland *Larrea-Ambrosia* desert scrub (elev. 840 m) and *Coleogyne* scrub (elev. 1400 m), northwest of Las Vegas, Nevada (Amundson, Chadwick, & Sowers 1989). In the upper 0.5 m of those soils, CO_2 concentrations were very low and changed little throughout the sampling period (mid-January to early October) and increased about a partial pressure of 1.0 KPa only in mid-April at the creosote bush site. These levels were significantly lower than upland sites, which had cooler and moister microclimates, and reflects how desert soils have low levels of organic matter and decomposition.

POTENTIAL EVAPOTRANSPIRATION IN ROCK VALLEY

Abiotic measurements made in Rock Valley during the IBP study did not include all variables necessary to estimate potential evapotranspiration (PET) using the Penman equation (Penman 1956). However, using mean monthly air temperature and relative humidity values (Olivier 1961) some crude estimates of PET can be made for the years 1972–6. Each pair of monthly mean values can be used to estimate wet bulb depression values (WBD) from a standard psychrometric table. Monthly WBD values can then be multiplied by values of W that are appropriate for 36.7°N latitude (Olivier 1961). Rock Valley values of W for January through December were 0.30, 0.40, 0.59, 0.73, 0.83, 0.84, 0.84, 0.70, 0.67, 0.48, 0.32, and 0.25, respectively. These products were an estimate of PET (mm d^{-1}) and when multiplied by the number of days in the month can yield estimates of monthly PET.

For the years 1972–6 estimated PET in Rock Valley ranged from 1977 to 2084 for calendar years and 1797–2255 for hydrologic years (Table 2.24). These estimates are about 11% higher than those

Table 2.24. *Estimated potential evapotranspiration (PET) in Rock Valley for 1972 to 1976. Rain-year is defined as September (y n − 1) through August (year n).*

Year	PET, mm Calendar-year	PET, mm Rain-year
1972	2016	–
1973	1977	1797
1974	2280	2255
1975	2094	2096
1976	2054	2190
Mean	2084	2085
Standard deviation	118	202

computed by Webb, Lauenroth, Szarek, and Kinerson (1983) for Rock Valley in 1972 (1789 mm) and 1973 (1765 mm). Of course, these values emphasize the very dry nature of this desert valley, which directly relates to the relatively sparse perennial plant life of the IBP validation site. For areas similar to Rock Valley, estimates developed by the Department of Water Resources, State of California (1975), places potential evapotranspiration at approximately 2 meters per year.

SUMMARY

The Nevada Test Site has had a complex geologic history of volcanism and sedimentation, also affected by regional and local faulting events, but the Mojave Desert portion of NTS is dominated by a series of valleys, only two of which (Yucca Flat and Frenchman Flat) have a central playa. Rock Valley lacks a playa and has open drainage to the south and southwest, and the ecological study site (elevation 1040 m) is located on a lower bajada with a very gentle slope covered with deep Quaternary alluvium and colluvium.

Permanent surface water is absent in lowland desert scrub habitats on NTS. Limestones and dolomites underlying the valleys allow hydrologic basins to form at depths of 90–150 m and groundwater to flow from north to south. Soils have formed, therefore, under very dry climatic conditions.

In Rock Valley soils are those typical of gentle slopes. These are shallow, calcareous alluvial soils with poor definition of horizons and having a fairly well developed desert pavement and calcrete at the base of the profile. These soils are loamy sand with pH ranging from 7.9 to 8.3 in A1 horizons to 8.8 in the C horizon. Like most nonsaline desert soils, such lower bajada soils have indistinct profiles and negligible accumulation of organic matter. Highest concentrations of organic carbon, nitrogen, and available phosphorus occur in the upper horizons beneath shrubs, which also tend to form coppice mounds by accumulating loosely consolidated debris around the plant base.

The northern Mojave Desert of southern Nevada has a winter-rainfall pattern, with a mean annual precipitation of about 120 mm, and the region receives approximately 70% of its precipitation from November through March. During the IBP studies Rock Valley had two unusually wet winters (1973 and 1976) and consequently a relatively high mean annual precipitation (150 mm) during the six years of research. Standard deviations of monthly and annual precipitation values are high and seasonally variable, and this produces an unpredictability in availability of limited soil moisture. In Rock Valley during three summers (1974–6) there were probably only two rain showers that moistened soil to the rooting depth of shrubs (0.08-0.10 m). During wet winter rainy seasons, soil remained saturated as late as March, but once soil dried in late winter or spring, it stayed dry, excepting after a rare and unpredictable heavy summer storm, until the next winter season of rains. Soil beneath shrubs often remained dry while that of open areas was wetted by light rain, probably because plant canopies intercepted the rain.

In Rock Valley summer high temperatures were mostly 30–45°C, often exceeding 40°C and with a maximum of 49°C, and winter lows were mostly 10–15°C, on some days below freezing and with a minimum of −13°C. Throughout the year relative humidity frequently was 5% or less, and very low monthly means of 13% occurred during the summer, whereas mean relative humidity was significantly higher (to 65%) during rainy winter months. The site tended to have the highest mean wind speeds in April and lowest ones during the hot, dry season, especially from July to October. For the IBP site, 76–90% of the time wind blew from the east-northeast (mostly September and March) or south-southwest (especially April through August); mean annual wind run was 8087 km and showed little yearly variance. The combination of dry air, high daytime summer temperatures, and mean monthly wind speeds of 9.2– 14.4 km hr^{-1}, create the characteristic high potential evapotranspiration of this desert region.

This climatic regime describes a desert in which the most benign conditions for biological activity can occur only in warm spring months. Summer becomes a time when perennial plants must deal with dry soils and desiccating ambient conditions, and animals must deal with little or no performed water and daytime conditions that are not only desiccating, but also so hot that they can be lethal.

3 Adaptations of Mojave Desert plants

Environmental stresses of low and unpredictable precipitation, low relative humidity with desiccating winds, and high summer temperatures characterize climates of deserts (Chapter 2) and, coupled with low nutrient availability, produce severe limitations on plant growth. Despite such stresses, desert scrub communities often contain surprisingly large amounts of plant biomass and may possess a remarkable diversity of perennial growth forms. Additionally, annuals and short-lived herbaceous perennials irregularly contribute a small but important biomass component, one that is crucial for maintaining faunal diversity of desert communities, and in favorable years net primary production by desert annuals may approach half that of perennials. Species that inhabit desert scrub of the Mojave Desert have adaptations permitting substantial carbon assimilation while coping with the shortage of available water and high, potentially lethal temperatures. Biologists have just begun to examine and advance our understanding of the physiological, morphological, and phenological adaptations that are used by plants of the Mojave Desert to grow successfully in their peculiar desert environment.

ADAPTIVE STRATEGIES

In ecology the assumption is generally made that survivorship and reproductive success of plants are highly correlated with traits that maximize net carbon gain. The manner by which net carbon gain is maximized depends on a complex interaction of factors determining the photosynthetic capacities of leaves and entire canopies and on how that

carbon and critical nutrients are allocated to tissues within the plant. These factors include biochemical characteristics of the photosynthetic pathway (C_3, C_4, and Crassulacean acid metabolism or CAM), physiological traits of whole-tissue response and source/sink relationships, morphological forms of photosynthetic tissues and organs, and phenological features of plant growth.

Conceptual models

One simple model relating net carbon gain to tolerance of environmental stress during the growing season is shown in Figure 3.1. In this model, modified from Ehleringer (1985), each curve rep-

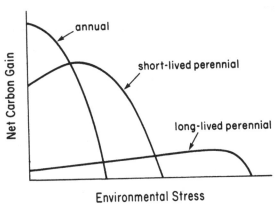

Figure 3.1 Adaptational model relating net carbon gain to plant growth form and environmental stress. Each curve represents a suite of physiological, morphological, and phenotypic responses of net carbon gain to increasing severity of environmental stress during the period of growth. Adapted from Ehleringer (1985).

resents a specific and different life history strategy, with high rates of net carbon gain occurring at the expense of lower tolerance to environmental stress. Long-lived perennial shrubs have relatively low net carbon gain per unit time, but are buffered against environmental stress. Annuals, at the other extreme, have high rates of net carbon gain, but restrict their active growth to relatively short periods, especially when soil moisture is highest and usually when environmental stress is lowest. Short-lived perennials that have drought deciduous leaves are intermediate in response.

A similar model has been used to suggest how morphological expressions of growth form may have selective advantage in desert environments by maximizing net carbon gain (Orians & Solbrig 1977; Solbrig & Orians 1977; Solbrig 1982). With desert plants, however, it is important to consider the possibility that selection for maximizing net carbon gain may be less important than adaptations to optimize efficiency of water use.

Levitt (1972, 1980) analyzed how plants deal with environmental stresses as belonging to two strategies, avoidance and tolerance. Avoidance is used by organisms to maintain the 'normal' physiological condition, whereas tolerance is used by organisms to withstand extremes in the internal environment. Each species possesses ways to avoid and to tolerate environmental stresses. Woody perennials of deserts physiologically may be avoiders of water stress by storing water (succulents) to nonavoiders, for example, those species that can tolerate considerable cellular dehydation. Desert annuals were classically treated as intolerant avoiders, but many old conclusions about these ephemerals, presumably avoiders or 'escapers' and 'evaders' of desert stresses, have been changed after observing that numerous species have highly specialized features for tolerating the hottest and driest summertime conditions (Björkman, Pearcy, Harrison, & Mooney 1972; Mooney, Ehleringer, & Berry 1974; Mooney, Troughton, & Berry 1974; Ehleringer, Mooney, & Berry 1977; Mulroy & Rundel 1977; Downton, Berry, & Seemann 1984; Seemann, Berry, & Downton 1984; Smith & Nobel 1986). Another type of conceptual model interprets

desert plant strategies in terms of photosynthetic adaptations to a seasonally fluctuating thermal environment (Mooney 1980; Fig. 3.2). Some species produce photosynthetic organs that are suitable for only a narrow temperature range, for example, either cool spring conditions or hot summer conditions; some plants acclimate, i.e., adjust, the thermal optimum of photosynthetic organs to more closely match the temperature regime; and some maintain a steady or homeostatic condition of a photosynthetic organ regardless of changes in the thermal regime.

General accounts of desert plant adaptations have emphasized how desert plants, especially their photosynthetic organs, are modified to tolerate the desiccating environment and thereby reduce transpiration and conserve water (Warming 1925; Shields 1950, 1951; Orshan 1954; Oppenheimer 1960; Fahn 1964; McCleary 1968; Evenari, Shanan, & Tadmor 1971). However, the key to interpreting desert plant adaptations in a physiological sense is to understand transpiration as only one of several important factors of leaf physiology, including uptake of carbon dioxide, control of leaf temperature, and capture of photosynthetically active radiation or PAR (Ehleringer 1985; Smith & Nobel 1986).

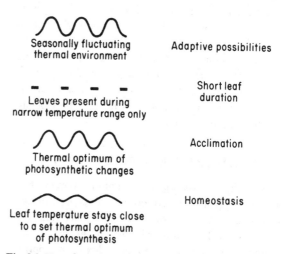

Fig. 3.2 Hypothetical models of the adaptation of photosynthetic activity to a seasonally fluctuating warm desert environment. Adapted from Mooney (1980).

Energy balance and leaf morphology

The most obvious structural feature of desert plants is that typical leaves are small and narrow, a condition termed microphylly. Many writers have treated this as an adaptation to reduce the transpiring surface area of plant canopies, hence a way to limit water loss. Gates (1968; Gates, Alderfer, & Taylor 1968; Gates & Papian 1971) provided a biophysical approach to model the energy balance of leaves under hot and dry desert conditions. This energy balance approach has proved very valuable in understanding evolutionary selection for leaf size and form.

In most plants, several physiological processes begin to break down when cell temperatures exceed 42–45°C, which are common midday summer temperatures in the desert (Björkman, Boynton, & Berry 1976; Björkman & Badger 1977; Schreiber & Armond 1977; Björkman, Badger, & Armond 1980; Kappen 1981). Broad leaves absorb much short wave and infra-red solar radiation and therefore can greatly exceed ambient temperatures, thereby attaining lethal temperatures, whereas narrow leaves, kept under identical conditions, remain close to ambient temperature and thereby may stay below lethal temperatures. This is clearly demonstrated in two computer simulations for cloudless, midday summer conditions of high solar irradiance (1000 W $^{-1}$ s^{-1}), ambient temperature of 40°C, and relative humidity of 50% (Fig. 3.3). In relatively still air (a wind speed of 0.1 m s^{-1}) very broad leaves (200 mm × 50 mm) with closed stomates (no transpiration) can experience internal temperatures exceeding 55°C, intermediate leaves (50 mm × 50 mm) have temperatures around 48°C, whereas narrow leaves (10 mm × 50 mm) with closed stomates have temperatures around 44°C (Fig. 3.3). Temperatures of broad leaves can be lowered substantially below lethal temperatures only by using high rates of transpiration, i.e., by evaporative cooling, but leaf temperatures of narrow leaves are close to ambient temperature without transpirational cooling. Under conditions with a slightly higher wind speed (1.0 m s^{-1}), nontranspiring, very broad leaves still exceed lethal leaf temperatures at 48°C

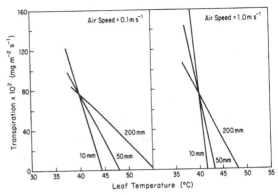

Figure 3.3 Computer simulations of leaf temperature in response to leaf size and wind speed at an ambient temperature of 40°C, irradiance of 1000 W m^{-2}, and relative humidity of 50%. Adapted from Gates (1971).

whereas the narrow ones are well below lethal and within one degree of ambient temperature (Fig. 3.3). Hence, the first adaptive advantage of a narrow desert leaf would be to survive, i.e., avoid, lethal high temperatures.

The second adaptive advantage of a narrow leaf in a hot environment is that having a cooler leaf (i.e., one close to ambient temperature) transpirational water loss is greatly decreased. The concentration of water vapor in air is very strongly dependent on air temperature, so that warm air holds much more water vapor than cool air. By decreasing internal leaf temperature by several degrees, water vapor concentration in intercellular air spaces also decreases substantially, thus producing a smaller gradient to the dry atmosphere relative to a leaf with a higher internal temperature.

Leaves of most desert plants, especially the leafy shrubs and summer-active annuals, are significantly less than 10 mm in width, and leaf temperatures are therefore very close to ambient. In addition, numerous species have dimorphic leaves, relatively broad ones, which develop in any season having abundant soil moisture, and very narrow ones, which are formed with the onset of drought. The common example of leaf dimorphism in a shrub from California deserts is *Encelia farinosa* (brittlebush).

Field studies have demonstrated that very high

rates of transpiration can cause broad leaves to have internal temperatures significantly below that of ambient temperature, due to the cooling effect of increased transpiration and in agreement with the predictions of the Gates computer simulations (inversion points in Fig. 3.3; Lange 1959; Smith 1978). Broad leaves are shed when water supply decreases so much that leaves cannot be maintained below lethal temperatures. Narrow leaves, which may avoid lethal temperatures, often abscise when water deficits become critical for maintaining leaf turgor (Szarek & Woodhouse 1977; Ehleringer 1982; Meinzer, Rundel, Sharifi, & Nilsen 1986).

Rare instances of Mojave Desert plants having very broad leaves, such as viney species of *Cucurbita*, have below ground storage with high capacitance to provide some water for high transpiration rates to lower leaf temperature by evaporative cooling and produce high root pressures to refill shoot vessels (Lange 1959; Rundel & Franklin 1991). Widest leaves of Mojavean species are those of an herbaceous perennial *Rumex hymenosepalus*, which typically are greater than 100 mm wide and are active in the spring as long as soil moisture is adequate. *Datura wrightii* and *Asclepias erosa*, herbaceous perennials with wide leaves, are spring and summer active, but are restricted to roadsides and sandy washes, where the plants appear to tap moist soil. Other wide leaves are several woody phreatophytes along arroyos and washes or at springs, where roots definitely use plentiful groundwater. On the southernmost edge of the Mojave Desert occurs the California fan palm (*Washingtonia filifera*), a plant with extremely large evergreen leaves that can only be cooled by having access to permanent groundwater. On the other hand, many species of phreatophytes have narrow leaves, for example, *Hymenoclea salsola* (cheesebush), which grows along washes and controls leaf temperature by utilizing differing transpiration rates (Ehleringer 1985; Comstock & Ehleringer 1988; Comstock, Cooper, & Ehleringer 1988).

Leaf orientation and reflectance

Leaf orientation and spectral characteristics have served as other parameters to consider for conceptual models of leaf energy balance and to relate leaf structure and physiological processes of photosynthesis and transpiration with biotic factors in desert environments (Ehleringer & Werk 1986). As shown in Fig. 3.4, leaf absorptance and leaf angle each act to control the amount of solar irradiance reaching the photosynthetic machinery of a leaf. Orientation of a desert leaf, including both its angle and azimuth, controls aspects of solar radiation uptake in three manners. These variables affect daily integrated radiation load, peak irradiance, and diurnal distribution of irradiance. The significance of such orientation can be seen in data for evergreen shrubs from the Mojave and Sonoran deserts (Ehleringer & Werk, 1986). Evergreen leaves of *Larrea tridentata* (creosote bush), *Simmondsia chinensis* (jojoba), and *Atriplex hymenelytra* (desert holly) all tend to be more steeply inclined than leaves of drought-deciduous shrubs growing with them. Steeper leaf angles, combined with nonrandom leaf azimuths, can reduce interception of solar radiation at midday during summer months, when problems of leaf energy balance are most severe (Neufeld, Meinzer, Wisdom, Sharifi, Rundel, Neufeld, Goldring, & Cunningham 1988).

Leaf orientation is not always fixed in desert

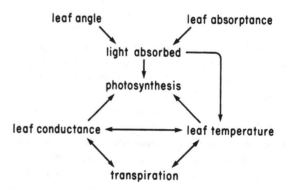

Figure 3.4 Leaf energy balance and photosynthesis in relation to leaf angle and leaf absorptance. Adapted from Ehleringer and Werk (1986).

species. Numerous desert annuals and a smaller number of perennials exhibit leaf solar tracking, in which leaf laminae remain perpendicular to the direct rays of the sun throughout the day (Mooney & Ehleringer 1978; Ehleringer & Forseth 1980; Ehleringer 1985). Leaves exhibiting this property may receive as much as 35% more direct solar radiation than a fixed leaf with a horizontal orientation, and from this there are major increases in net rates of production (Ehleringer & Forseth 1980). The frequency of leaf tracking in desert annuals is inversely related to length of the growing season, with 75% of summer annuals in the Sonoran Desert exhibiting some degree of solar tracking (Ehleringer & Forseth 1980). Solar tracking by leaves under conditions of drought stress may have a different nature, wherein leaves may avoid high leaf temperatures by minimizing exposure to direct sunlight with tracking to maintain laminae parallel with rays of the sun (Mooney & Ehleringer 1978; Forseth & Ehleringer 1980, 1982, 1983a,b).

Another version of orientation is the cladode, the flattened and often vertically oriented photosynthetic stem of certain cacti (Opuntia). Computer models and field studies have demonstrated that at each location cladodes on a plant tend to be oriented to maximize PAR interception integrated over the entire growing season (Nobel 1982a, b; Gibson & Nobel 1986). Likewise, for species with nonsucculent, cylindrical, photosynthetic stems, the arrangement of branches is designed to maximize PAR interception (Neufeld et al. 1988).

It has been known for many years that leaves of desert plants commonly have leaf reflectances higher than those of plants in less xeric environments (Billings & Morris 1951). Leaf absorptance to PAR typically is 80–85% in mesic communities (Ehleringer & Werk 1986). However, surveys of leaf absorptances in the Mojave and Sonoran deserts have found ranges of 60–85% in most growth forms, but as low as 29% in Encelia farinosa (Ehleringer 1981; Fig. 3.5). Decreases in leaf absorptance may result from a variety of morphological traits of the leaf surface, but especially thickness and narrow or dead, air-filled trichomes or, on succulents, epicuticular wax.

Many accounts of desert plants formerly concluded that trichomes are present as part of a leaf strategy to lower transpiration rates, presumably by influencing water vapor conductance across a thicker boundary layer. Whereas many desert leaves have dead trichomes, and in species with dimorphic leaves trichomes are more pronounced on pre-drought narrow leaves (Cunningham & Strain 1969), there is serious doubt now that typical trichomes on leaves produce any physiologically significant difference in boundary layer conductance. Instead, extensive research on species of Encelia, including selections with different trichome densities, has found that dead trichomes function primarily as structures that reduce leaf heat load by reflecting infra-red radiation (Ehleringer, Björkman, & Mooney 1976; Ehleringer & Björkman 1978a, b; Ehleringer & Mooney 1978; Ehleringer 1980, 1981, 1982, 1983a). The immediate benefit of lower leaf temperature is, as before, a lower transpiration rate because the water vapor gradient is decreased. On the other hand, trichomes also block incoming PAR and therefore have a negative effect, when present, of decreasing photosynthetic rate of the mesophyll. In the Mojave Desert common species with shielding trichomes include Ambrosia dumosa (white bursage), extremely common in most lowland desert scrub. More importantly, leaves of Larrea (Fig. 3.6), the ubiquitous dominant shrub, are coated with thick resin deposits (Fig. 3.7), which are secreted by the glandular trichomes, and this resin layer has a positive effect in maintaining a favorable vapor pressure deficit or VPD (Meinzer, Wisdom, Gonzalez-Coloma, Rundel, & Shultz 1990).

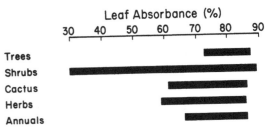

Figure 3.5 Leaf absorptance to solar radiation in the 400–700 nm waveband for dominant growth forms in the Mojave and Sonoran deserts. Adapted from Ehleringer (1981).

Figure 3.6 Creosote bush, *Larrea tridentata*.

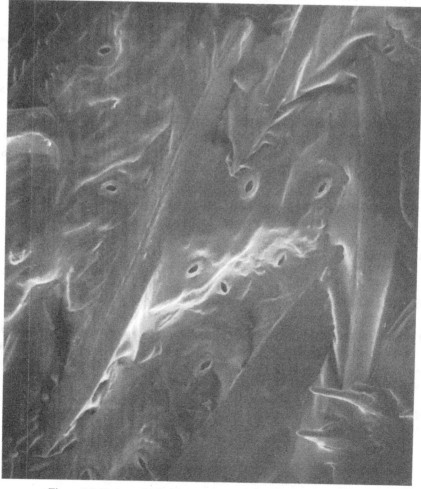

Figure 3.7 Resinous leaf surface of creosote bush, *Larrea tridentata*.

Physiological acclimation

In addition to structural adjustment of leaf size and protection by trichomes, species may also have adaptations to deal with changes of the thermal environment via a physiological strategy involving seasonal changes in the biochemical responses of enzyme systems to temperature (Mooney 1980). This ability of a plant to change temperature optimum for photosynthesis in an adaptive manner to seasonal thermal fluctuations is termed acclimation. For desert shrubs acclimation was first tested in four common North American species, *Larrea tridentata* and *Encelia farinosa* of desert scrub communities and *Chilopsis linearis* (desert-willow) and *Hymenoclea salsola* of desert washes (Strain & Chase 1966; Strain 1969). However, of those only plants of *L. tridentata* showed photosynthetic acclimation, an adjustment upward, when grown at high temperatures. *Encelia farinosa* and *H. salsola* exhibited only a slightly elevated temperature optima when grown at high temperature, so they utilized the homeostatic strategy. *Chilopsis linearis* had its optimal temperature for photosynthesis at low temperatures, thereby showing that this phreatophyte does not have an interesting photosynthetic adaptation to the desert environment.

In a study of *Larrea* from Mojave populations (Mooney, Björkman, & Collatz 1977), optimum temperature for CO_2 assimilation changed from about 22°C in March to 32°C in September (Fig. 3.8). Changes in stomatal conductance and dark respiration played no role in those acclimations to different temperatures; instead, the shifts in temperature optima for photosynthesis involved intrinsic properties of the photosynthetic apparatus. Researchers detected differences in fluorescence properties of chloroplasts and chlorophylls when acclimated and control plants of *Larrea* were compared (Armond, Schreiber, & Björkman 1977; Schreiber & Armond 1977). Seasonal acclimation of this general type has been reported in the C_4 halophyte *Atriplex lentiformis* (Pearcy 1977), the perennial C_4 bunchgrass *Hilaria rigida* (Nobel 1980), and a C_3 desert fern, *Notholaena parryi* (Nobel 1978).

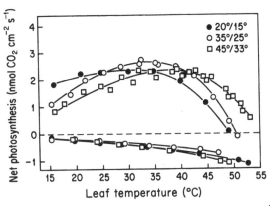

Figure 3.8 Seasonal acclimation in temperature response of photosynthesis for leaves of *Larrea tridentata*. Adapted from Mooney, Björkman, and Collatz (1978).

Leaf anatomy and photosynthetic pathways

Typical leaves of lowland desert perennials, in addition to being small and narrow, exhibit a syndrome of anatomical features not characteristic of plants in other biomes (Gibson 1996). Leaves may be soft or firm but have few or no cells with thick, lignified cell walls (excluding lignified trichomes). Some notable exceptions are deciduous leaflets of many legumes, which tend to have well-developed primary phloem fibers, and those species with evergreen leaves, especially *Mortonia utahensis* (sandpaper bush), and *Simmondsia chinensis* and less so *Larrea tridentata*. Collenchyma is mostly absent except in species with prominent veins, e.g., Lamiaceae. Cuticle coats the leaf to prevent water loss through the epidermal cell walls, but, contrary to popular accounts, this waxy layer is not noticeably thickened, certainly not like that for plants of chaparral habitats. In certain genera, for example, *Haplopappus*, *Chrysothamnus*, and *Larrea*, resins coat the outer epidermal cell walls. All leaves of warm desert plants are amphistomatic, meaning that stomata occur on both leaf surfaces (Mott, Gibson, & O'Leary 1982); moreover, stomata are small and present at fairly high densities, but typically they are not sunken. Mesophyll is composed either exclusively of cylindrical palisade parenchyma cells (Fig. 3.9) or of palisade layers on the adaxial (upper) and abaxial (lower) side and a poorly defined, thin central zone of

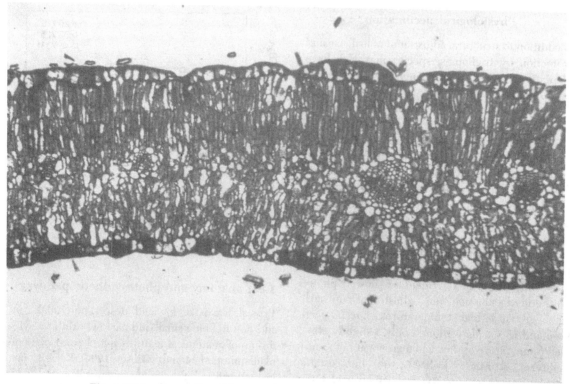

Figure 3.9 Leaf transection of *Larrea tridentata*, showing isolateral arrangement of meso-phyll palisade cells.

spongy mesophyll (isolateral leaves; Shields, 1951). Because the palisade parenchyma cells tend to be extremely narrow, there are many cells per surface area and the leaf is compact, which also means that desert leaves tend to have extremely high internal mesophyll cell surface area relative to v/A ratios and also many chloroplasts per unit of leaf tissue (Nobel 1976b; Smith & Nobel 1978; Longstreth, Hartsock, & Nobel 1980). At the same time, the palisade cells in a layer sparingly touch each other, so that intercellular air spaces, although smaller than in mesophytic leaves, are still well developed.

The structural design of the typical leaf can be interpreted primarily as a strategy to maximize photosynthetic rate at times when plants are still capable of opening their stomata (Gibson 1996). A high number of chloroplasts per leaf surface area enables the leaf, living in a very high-light environment, to capture much PAR, and high A^{mes}/A signifies that much internal surface area is available for diffusion of CO_2 into the cells (Smith & Nobel 1978; Longstreth et al. 1980; Nobel 1991a). At first glance the presence of many stomates on both sides of a leaf seems counterintuitive if saving water is the strategy. However, researchers have concluded that amphistomaty allows maximum CO_2 uptake and may permit significantly more CO_2 uptake than for leaves having the same number of stomates, but only on the abaxial side, because a second boundary layer conductance is added while stomatal conductance remains the same (Mott et al. 1982). Thus, the typical desert leaf shows adaptations for maximal use of PAR, high CO_2 through the stomates, and rapid diffusion of CO_2 in the liquid phase into the chlorophyll-rich cells. Desert leaves tend to have fairly high rates of transpiration when the stomates are open (Maximov 1929), hence the major adaptations for water conservation are a sensitive regulation of stomatal apparatus, an effective cuticle to prevent unwanted transpiration when stomates

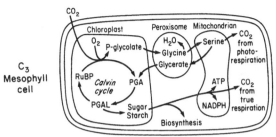

C₃ Mesophyll cell

Figure 3.10 Metabolic pathway of C₃, Calvin-Benson cycle photosynthesis.

are closed, and a lowering of leaf temperature, by reducing leaf size, to decrease the water vapor concentration gradient to the atmosphere (Schulze, Lange, Buschbom, Kappen, & Evenari 1972; Lange, Schulze, Kappen, Buschbom, & Evenari 1975; Schulze & Hall 1982).

The great majority of desert species have leaves with Calvin-Benson or C_3 photosynthesis (Fig. 3.10; for details see Salisbury & Ross 1991; Taiz & Zeiger 1991), in which the first product of CO_2 fixation is phosphoglycerate (PGA), a compound with three carbon atoms. Within the chloroplast, where PGA is produced via a reaction catalyzed by the enzyme Rubisco (ribulose-1,5-biphosphate carboxylase), the synthesis of sugar also occurs. Over 90% of the species in warm deserts have this type of metabolism, including perennial plants that occur in virtually every North American desert microclimate, from the driest creosote bush desert scrub to the phreatophytic habitats of washes. Leaves of herbaceous species, including short-lived annuals, have a similar overall design.

In sharp contrast are leaves of C_4 species, which include certain shrubs in the Chenopodiaceae and Amaranthaceae and herbs in several additional families (Mulroy & Rundel 1977). Leaves of C_4 species are amphistomatic, but the mesophyll is arranged around the veins in a rosette-like pattern around a distinctive ring of large bundle sheath parenchyma cells (Fig. 3.11). This mesophyll design is termed Kranz anatomy.

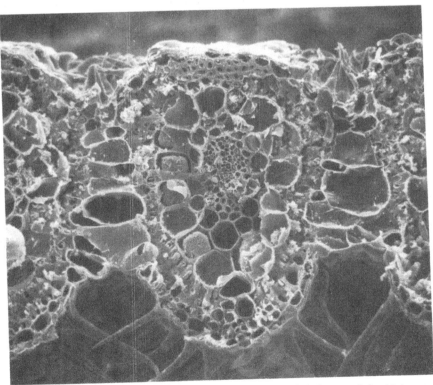

Figure 3.11 Kranz anatomy in *Hilaria rigida*, a common bunchgrass of the Mojave Desert. Large parenchyma cells form a prominent bundle sheath around each vascular bundle.

The C$_4$ biochemical pathway (Fig. 3.12a; for a recent treatment, see Salisbury & Ross 1991; Taiz & Zeiger 1991) is named from the first product of CO$_2$ fixation is oxaloacetate (OAA), a compound with four carbon atoms. In plants with C$_4$ metabolism, two separate carboxylation processes take place during daytime carbon assimilation. The first step utilizes phosphoenolpyruvate (PEP) carboxylase to fix carbon into OAA and immediately into the organic acids malate or aspartate, and this process occurs entirely within the mesophyll cells adjacent to the open stomates. Malate and aspartate, both C$_4$ organic acids, are then transported via plasmodesmata to the specialized bundle sheath cells. Within bundle sheath cells, which have relatively thick cell walls and very prominent, agranal chloroplasts, the organic acids

are decarboxylated to release CO$_2$, which is refixed into PGA using the C$_3$ type of carbon fixation and then into sugars, which are loaded directly into the phloem for prompt export from the bundle sheath. Hence there is spatial separation between the place where CO$_2$ is initiated fixed (general mesophyll) and where it is ultimately used to make sugar. Both C$_3$ and C$_4$ species have uptake of CO$_2$ exclusively during daytime.

Warm deserts typically have many species that are either stem or leaf succulents, although the transitional Mojave Desert, which has some freezing winter temperatures, has comparatively few. Many succulents have a biochemical pathway for photosynthesis called Crassulacean acid metabolism (CAM), which has nocturnal uptake of CO$_2$ (Fig. 3.12b; Kluge & Ting 1978; Osmond 1978; Ting & Gibbs 1982; Winter 1985; Ting 1985; Nobel 1985, 1988, 1991b; Gibson & Nobel 1986; von Willert, Eller, Werger, Brinckmann, & Ihlenfeldt 1992). The name was derived from Crassulaceae, a widespread family of succulents in which this system was discovered, but in the Mojave Desert species of Cactaceae are the most common examples. Anatomy of photosynthetic organs of CAM plants is entirely different from the other two biochemical pathways in that the chlorophyll-bearing tissue consists of very large mesophyll cells, rectangular to isodiametric, each having an extremely large central vacuole (Gibson 1982).

In CAM plants there is a temporal separation of CO$_2$ uptake, which occurs in the dark, and the manufacture of sugar, which occurs during the day, but both processes are present within the same cell. When stomates open at night, CO$_2$ diffuses into the cytosol of a cell and is combined with PEP into organic acids, primarily malate, which is loaded into large vacuoles. When sunlight appears, the stomates close and the chloroplasts intercept PAR for photosynthesis. Malate moves out of the vacuole and is decarboxylated, forming CO$_2$ and PEP, and the CO$_2$ is subsequently used via typical C$_3$ photosynthesis to make PGA enroute to sugars. In CAM plants there are major diurnal changes in pH as organic acids are made and stored in vacuoles in the dark and then broken

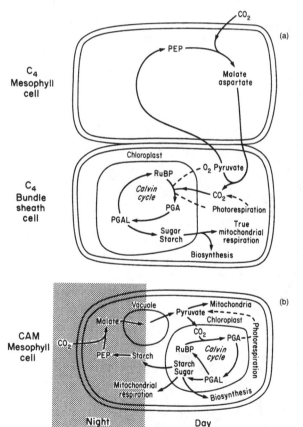

Figure 3.12 Metabolic pathways in C$_4$ photosynthesis and Crassulacean acid metabolism (CAM).

down the next day by decarboxylation to release the CO_2 for C_3-type photosynthesis.

Detailed explanations and comparisons of the three photosynthetic pathways have been published elsewhere (for example, Salisbury & Ross 1991; Taiz & Zeiger 1991), and these discuss the significance of C_4 photosynthesis and CAM as physiological strategies for desert life. Plants having C_4 metabolism typically possess adaptations for high photosynthetic capacity in environments with high temperatures and high light intensities, whereas CAM plants have adaptations for conserving and storing water and tolerating very high internal temperatures. Under natural conditions, CAM plants tend to have low photosynthetic capacity and greatest productivity during the cool months.

In a C_3 leaf a significant fraction of the CO_2 that is fixed during photosynthesis is lost at the same time by a process called photorespiration. Rate of photorespiration is especially high under conditions of high temperatures and high oxygen concentration. In a C_4 leaf the site where photorespiration occurs is within the bundle sheath, yet here concentration of CO_2 is exceedingly high, because this is where organic acids are decarboxylated, and concentration of O_2 is low, because oxygen is not being generated by the light reaction in these cells. Consequently, in a C_4 leaf photorespiration is virtually absent and physiologically insignificant because any CO_2 that is generated is used immediately to make sugar. All other things being equal, C_4 leaves can normally fix more CO_2 per unit time than C_3 leaves. This is not true at very low temperatures, where C_3 plants have higher net photosynthesis than C_4 plants, but at high temperatures C_4 species often show remarkably high levels of photosynthesis. Moreover, under optimal water potential and temperature conditions many C_4 desert herbs and even some annual species with C_3 metabolism are not PAR-saturated at full sunlight (2000 umol photons m^{-2} s^{-1}), whereas leaves of most desert plants are PAR-saturated typically at 25–60% full sunlight (Smith & Nobel 1986).

Many desert C_4 species are active during extreme midsummer environmental stress. For example, in *Tidestromia oblongifolia*

(Amaranthaceae), which is a common herb in Death Valley and other parts of the Mojave Desert, maximum photosynthesis takes place at 47°C under well-watered conditions, and solar irradiance remains limiting even at midday (Björkman, Mooney, & Ehleringer 1975). It has been suggested, therefore, that the principal adaptation of C_4 metabolism is to allow species to maximize photosynthetic rates during periods of the summer when ambient temperatures are supraoptimal for C_3 plants. This pattern of adaptation is consistent with the ecological relationships of many herbaceous C_4 taxa of subtropical grasses and desert summer annuals and ephemerals (Mulroy & Rundel 1977; Nobel 1980). However, species of desert shrubs, such as many species of *Atriplex*, also exhibit C_4 metabolism with an adaptive significance of this metabolic system different than for the herbaceous taxa. The maximal rates of photosynthetic assimilation in these shrubs is no higher than those of competing C_3 shrubs. Instead, they utilize a relatively high efficiency of water use, another property of C_4 plants (grams of CO_2 fixed in photosynthesis per kilogram water vapor lost by transpiration) by maintaining relatively low internal CO_2 concentrations, and thus remain physiologically active during summer drought.

Water-use efficiency is high in CAM succulents (Gibson & Nobel 1986; Nobel 1976a, 1988, 1991b). They have temporal separation of CO_2 uptake, which occurs at night, and use of CO_2 in photosynthesis, which occurs during the day, consequently transpiration occurs at night when the stomates are open and the air temperature is lower, not during the hot daytime. As discussed earlier, having lower leaf temperature decreases the water vapor concentration gradient between leaf tissues and the air, thus there are much lower rates of transpiration if stomates are open only at night. During the day water loss is essentially eliminated by closing the stomates.

CAM plants have greater energy requirements for fixation per molecule of CO_2 than either C_4 or C_3 metabolic systems (Taiz & Zeiger 1991; Nobel 1991b), and productivity tends to be relatively low for succulents growing under conditions of low soil water and partial shading of photosynthetic organs. Productivity is also highest when night temperatures are low, which promotes uptake of CO_2 and storage of malate in the vacuoles, thus providing large amounts of CO_2 via decarboxylation during the following light period (Gibson & Nobel 1986; Nobel 1988). When plants are exposed to high summer temperatures, photosynthetic rates are low and the plant becomes a heat sink and must tolerate internal temperatures above those that are lethal for nonsucculent species. The biochemical nature of high-temperature resistance in CAM tissues is not well understood.

PLANT WATER RELATIONS

Despite realizing the importance of water as a limiting resource for plant growth in desert ecosystems, scientists know surprisingly little about the relationship between plant productivity and water consumption. The importance of water as a spatially and temporally variable resource undoubtedly is a driving force via natural selection in producing much of the diversity of growth forms that occur in many desert regions. Because water is differentially available within soil profiles, given irregular inputs of precipitation and differential features of water runoff and soil infiltration, no single morphological or physiological strategy of water uptake is more successful than other strategies. Rooting architecture, leaf canopy area, stomatal response, and other physiological variables may all affect plant water relations and thus the degree to which desert plants either avoid or tolerate the extremes of summer drought.

Tradeoff between transpiration and photosynthesis

Transpiration is an unavoidable loss of water vapor when stomates open to allow uptake of CO_2 for photosynthesis (Fig. 3.13; Kramer 1969; Nobel 1991a). Like all gases, carbon dioxide and water vapor diffuse toward regions of lower concentration. Concentration of CO_2 is high in the atmosphere (currently about 360 ppm), whereas in typical leaf cells, where CO_2 is used to make sugars,

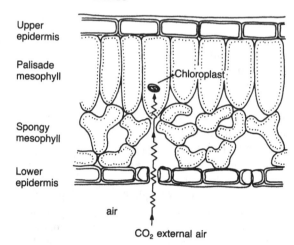

Figure 3.13 Schematic diagram of leaf gas exchange for CO_2 and water vapor. See text for discussion.

the concentration is much lower. Carbon dioxide diffuses into the leaf through stomates and to the photosynthetic cell walls, where it enters a liquid phase for diffusion into the cell protoplast and specifically chloroplasts. Conversely, water is very abundant within the cells, typically being greater than 90% of the volume; water evaporates from internal cell surfaces leading to an atmosphere within the intercellular air spaces that is essentially saturated with water vapor. Water vapor diffuses along a concentration gradient, or vapor pressure gradient, from saturated in the leaf intercellular air spaces through the stomates and unstirred boundary layer surrounding the leaf to the dry ambient air. Because both gases utilize the same exchange pathway through stomates, modifications in the pathway will proportionately affect both gases, hence rates of both photosynthesis and transpiration.

Many evolutionary models of optimal function in desert plants (for example, Johnson 1975; Orians & Solbrig 1977; Solbrig & Orians 1977), while recognizing the tradeoff between photosynthetic assimilation and transpirational water loss, have nevertheless focused on maximizing primary production rather than on the significance of water-use efficiency or WUE. Water-use efficiency is a ratio of dry matter production to water consumption, and this probably is more important in determining survival and reproductive success. Variations in WUE have been noted since the early part of the century (Briggs & Shantz 1914), and now ecologists are increasingly focusing on the significance of WUE in explaining adaptive strategies of desert plant water relations (Nobel & Smith 1986).

The significance of differing adaptive strategies in regulating plant water relations can be seen in midday water potentials of 11 species of desert perennials growing in midsummer at the same site in the Sonoran Desert of California (Fig. 3.14; Nilsen, Sharifi, & Rundel 1984). These species exhibit several mechanisms for both drought avoidance and tolerance. At one extreme were shallow-rooted desert evergreens, such as *Larrea tridentata* and *Simmondsia chinensis*, which at this site experienced afternoon leaf water potentials more negative than −6.0 MPa (more negative water potentials indicate greater amounts of environmental water stress). Drought-deciduous shrubs, such as *Encelia farinosa* and *Justicia* (*Beleperone*) *californica*, which also have shallow roots, were experiencing more positive leaf water potentials at −3.5 to −4.5 MPa, the same as measured for deep-rooted phreatophytes, such as *Prosopis glandulosa* and two other woody legumes,

which had access to ground water. For the woody phreatophytes, their large leaf areas produce sufficient transpirational fluxes to lower midday leaf water potentials beyond what might be expected. Relatively high afternoon leaf water potentials (relatively low stress) were also seen in the wash plants *Chilopsis linearis* (desert willow) and *Hyptis emoryi* (desert lavendar) that were completely leafless during the summer or those of *Psorothamnus* (*Dalea*) *spinosus* (smoke tree) that relied on photosynthetic stems. Then at the other extreme were herbaceous perennials that utilize large capacitance (storage) in underground roots or shoots to buffer aboveground tissues from large diurnal shifts in water stress while the plants still act as phreatophytes with roots drawing water from deeper soil layers. Such large-leaved perennials can maintain very high leaf water potentials, such as −1.5 MPa for *Cucurbita digitata*, a desert gourd, which experienced the same temperature regime.

Strategies for avoiding or tolerating low water potentials

Drought avoidance for desert plants often involves morphological and phenological strategies of changing the form and amount of leaf tissues during the summer months of maximum stress. This is most easily seen in drought-deciduous growth forms or in the concentrated growth activity under moderate spring conditions of desert ephemerals.

Physiological strategies to minimize drought stress are also commonly utilized by desert plants, in particular via stomatal control of water loss. Low stomatal conductances conserve water and thus lower plant water stress (Monson & Smith 1982; Nilsen et al. 1984). When ambient temperature is very high and air is very dry, VPG is very steep and transpirational loss can be exceedingly high if stomates remain wide open. Thus, many desert plants limit their transpiration rates during periods of summer water stress by having narrower (less open) apertures and limiting stomatal opening to the most favorable times of a day when leaf water potentials are the highest.

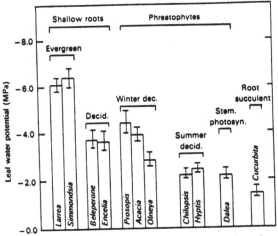

Figure 3.14 Comparative midday water potentials in late summer for 11 species of desert perennials living along Nude Wash in the Sonoran Desert of California. Adapted from Nilsen, Sharifi, and Rundel (1982).

Most desert leaves possess the ability for stomates to open even when plant water potential is less than -3.0 MPa, often -4.0 to -5.0 MPa, and in one case to -12.0 MPa (Smith & Nobel 1986). These values are significantly lower than those tolerated by mesophytes and signifies that typical desert leaves endure lower water content, experience stomatal opening under drought conditions, and can draw water from relatively dry soils as long as leaf water potential is lower than that of the soil.

Both short-term and long-term changes in plant water status are thought to be involved in the balance between transpiration rate and leaf water potential (Schulze & Hall 1982). When transpiration rates are low, such as during the night, leaf water potentials approach an equilibrium with soil water potential, and after sunrise, as leaf conductance and transpiration rates increase, leaf water potentials decrease following a curvilinear response (solid lines of model in Fig. 3.15). Under given conditions of soil moisture availability, a plant may operate along one of these lines over short-term changes in plant water status. As drought conditions set in, however, pre-dawn water potentials decrease as soil moisture becomes more limiting, and resistances to hydraulic flow within the plant may increase. These long-term changes in plant water status commonly act to

move the slope of the transpiration/water potential relationship along the dotted line in Figure 3.15. The dependence of plant water potential on seasonal and diurnal changes in transpiration rate have been shown for desert shrubs (Kappen, Oertli, Lange, Schulze, Evenari, & Buschbom 1975; Nilsen, Sharifi, Rundel, Jarrell, & Virginia 1983).

Drought tolerance as a physiological strategy for desert plants requires adaptations for maintaining positive turgor potentials and thus preventing stomatal closure and cessation of growth when soil moisture would normally become limiting. There are several interrelated physiological, biophysical, and morphological mechanisms that can act to maintain such turgor in shrubs that remain physiologically active during drought even though their shallow roots are in dry soil. One such mechanism is seasonal osmotic adjustment, whereby there is a decrease in tissue osmotic potentials resulting from a net increase in inorganic or organic solutes in the cell protoplast (Turner & Jones 1980). In this way, a favorable water potential gradient along the soil-root-shoot continuum can be maintained so that positive turgor conditions persist despite decreasing soil water potentials. Comparative patterns of seasonal osmotic adjustments have been described in several desert perennials (Bennert & Mooney 1979; Monson & Smith 1982; Nilsen, Meinzer, & Rundel 1989). An alternative to large seasonal changes in osmotically active solutes is by lowering the tissue osmotic potential through decreases in the cell fraction of osmotically active water (Calkin & Pearcy 1984). Changes in this cell fraction can be achieved by seasonal changes in the elasticity of cell walls. *Larrea tridentata* utilizes this mechanism to maintain relatively constant levels of turgor over a wide range of leaf water potentials (Meinzer, Rundel, Sharifi, & Nilsen 1986). The advantage of changes in cell wall elasticity over seasonal osmotic adjustment are not fully understood as drought-tolerating strategies, but perhaps active uptake of inorganic solutes or synthesis of organic solutes in cytoplasm may be more energetically expensive than sequestering the same amount of solutes in smaller cell volumes.

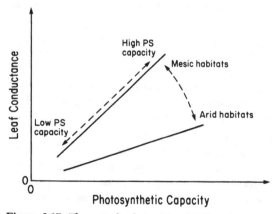

Figure 3.15. Theoretical relationship of stomatal conductance and assimilation rate in a mesic and an arid environment. Adapted from Schulze and Hall (1982).

Variability in water relations according to habitat

The water relations of an individual species varies not only temporally, but also spatially with the nature of the habitat in which a plant grows and the location of the roots. Wash habitats, for example, frequently store more soil moisture than adjacent bajadas or rocky slopes. This effect can be observed easily by noting that plants growing along washes are more robust and retain their leaves longer, sometimes even year round, when nonwash plants are drought deciduous. This effect can be quantified by measuring the variations in pre-dawn water potentials, as in *Haplopappus cooperi*, a low desert shrub, which was studied at Yucca Mountain, Nevada (Smith & Nowak 1990). In early August, under extreme conditions of summer drought stress, midday leaf water potentials reached -7.5 MPa on bench habitats, but in the adjacent wash the minimum values were only about -4.0 MPa.

The response of many desert shrubs to habitat gradients of increasing water stress, as from wash to transition habitats to dry bajadas and rocky slopes, is a change in WUE as a response to lowered leaf water potentials (Ehleringer & Cooper 1988). The highest mean WUE among desert shrubs, as measured using stable carbon isotope ratios (for a summary of the ecological significance of changes in carbon isotope ratios in plant leaves, see Rundel, Ehleringer, & Nagy 1988) generally occurs in long-lived species, whereas relatively short-lived, opportunistic species have lower WUE (Ehleringer & Cooper 1988). The slower growth rates, which one would expect to be correlated with lower WUE, may well be an important correlate of longevity.

Root systems

Prior to the pioneering early studies by Cannon (1911) at the Carnegie Laboratory in Tucson, Arizona, conventional wisdom suggested that the survival of desert plants was dependent on extensive root systems that could penetrate great depths in the soil. This form of rooting architecture was con-sidered to be essential for woody desert plants to take up moisture during long dry seasons (Schimper 1903). Cannon was able to show that root systems of desert shrubs were highly variable with regard to the extent of lateral roots and the presence or absence of a prominent taproot. Many schemes of classification for rooting architecture have been developed (see Cannon 1949; Zohary 1961), but none of these has been widely accepted.

It is perhaps best to utilize simple ecological approaches to root architecture and designate broad groups of growth forms. Woody perennials tend to have superficial laterals and moderately deep taproots or vertically descending laterals. Many common Mojave Desert shrubs, such as *Larrea tridentata* and species of *Lycium* and *Encelia*, have a well-developed system of lateral roots as well as at least one central root that penetrates lower soil horizons. However, lateral roots are largely absent in *Ephedra* and shrubs in wash habitats, where surface water drains rapidly from the coarse soils. Shallow-rooted perennials with little or no development of a taproot comprise another set, of which shrubby cacti and *Yucca* spp. are good Mojave Desert examples, but numerous low nonsucculent shrubs and subshrubs, such as *Krameria* (ratany), fit here also. Deeply rooted phreatophytes are species with primary development of taproots, which extend deeply to reach permanent or semi-permanent groundwater pools. *Prosopis glandulosa* and other woody legumes are phreatophytes of the Sonoran Desert, and *Prosopis* extends its range into the Mojave Desert. Exotic species of *Tamarix* have successfully invaded many phreatophytic locations. Whereas early ecological paradigms presented the expectation that root-to-shoot biomass ratios in warm desert ecosystems would be high, this is not the case. Field studies of shrubs in warm deserts from North America, the Middle East, and Australia have generally reported root/shoot ratios of 0.5–1.0 (Rundel & Nobel 1991). These values are comparable to those found for shrubs in temperate forest and chaparral habitats. Root/shoot ratios of desert annuals are much lower, commonly 0.1–0.3 (Bell, Hiatt, & Niles 1979; Forseth, Ehleringer, Werk, & Cook 1984).

Because water is a critical limiting factor for desert ecosystems, there is reason to expect that coexistence among species may be due in part to differences in water uptake strategies expressed as variations in species–species patterns of rooting architecture. Cody (1986a) determined that highest levels of community diversity of perennials in the Mojave Desert occur on coarse alluvial fans, where water infiltrates rapidly and is available in deep soil horizons. Thus, the degree of competition between species may be a function of the competition for water at specific levels within the soil profile. Species with similar rooting architectures would be expected to compete more than would species with differing patterns of rooting, and thus be more likely to occur together. Cody (1986a, b) interpreted spatial patterns of nonsucculent and succulent shrubs in the eastern Mojave Desert as examples of this type of competition.

As pointed out by Barbour (1973), evidence from spatial patterning that purports to demonstrate competition of desert shrubs in warm deserts is inconclusive, so that manipulation studies of plant spacing have been used to monitor the physiological effects of competitive interactions between individuals of the same species and of different species (Fonteyn & Mahall 1978, 1981; Phillips & MacMahon 1981; Robberecht, Mahall, & Nobel 1983; Ehleringer 1984). Some interactions have been shown to be significant. Undoubtedly we have much to learn about belowground mechanisms of competition between desert plants for water and nutrients. A different approach is needed to understand the mechanism of how seedlings for each species become established (Vasek 1979/80; Hunter 1989). Advances will come from improved knowledge of variations in rooting architecture, experimental manipulation studies, and theoretical models of competition for limited resources (Caldwell & Richards 1986; Tilman 1988).

PLANT GROWTH FORM AND PHENOLOGY IN THE MOJAVE DESERT

By understanding how plants respond in structural and physiological ways to the abiotic conditions of

desert environments, an investigator can then look at the different life history strategies of desert plants. In the deserts of the world there are numerous successful plant growth forms within the desert environment, and many of these occur in the Mojave Desert.

Evergreen shrubs

Larrea tridentata, (creosote bush; Fig. 3.6), is the dominant shrub over huge areas of the Chihuahuan, Sonoran, and Mojave deserts. It frequently occupies the driest and hottest sites and in the Mojave Desert occasionally may be found in pure stands on sandy or gravelly bajada slopes (Fig. 1.3). *Larrea* is adapted to grow in many types of desert soils, but usually is absent from saline soils (Benson & Darrow 1981).

Creosote bush is also remarkable because it is a true evergreen; individual leaves, which have two leaflets, commonly do not live for more than 8–14 months, but a canopy of photosynthetically active leaves is present throughout the year (Sharifi, Meinzer, Nilsen, Rundel, Virginia, Jarrell, Herman, & Clark 1988). Under conditions of extreme drought, leaf biomass may be greatly reduced, but the shrubs are never leafless and generally exhibit positive net photosynthesis throughout the year. It is expected that during the summer months herbivores would eat the evergreen leaves, which have a high protein content, but, in fact, concentrations of resins on the leaf surface and within the epidermis deter most insects and vertebrates from consuming the foliage. None the less, the few insects that eat leaves of *Larrea* tend to be monophagous, giving *Larrea* a specialized fauna (Schultz 1981). Jackrabbits feed on young shoots, but only after excising the leaves (Chew & Chew 1965).

Larrea is used as the model system for investigating the adaptive strategy of evergreen shrubs, but there are other species in the warm deserts of North America that have nonsucculent leaves during every month of the year (Table 3.1). *Mortonia scabrella* (sandpaper bush) is a codominant species in parts of the Chihuahuan Desert, and the related species *M. utahensis* occurs in limestone mountain ranges, especially of the MDR. *Sim-*

Table 3.1. *True evergreen shrubs occurring in the warm deserts of North America. Leaves are present on shoots year-round, although the canopy may have few or even no leaves after successive years of drought.*

Larrea tridentata, creosote bush.	Common throughout all warm deserts at all elevations.
Simmondsia chinensis, jojoba.	Sonoran Desert, upper bajadas in easter region, sometimes at low elevation sites in California and Baja California.
Mortonia, sandpaper bush.	A genus of evergreen shrubs restricted to limestone outcrops; *M. scabrella* occurs in widely scattered localities in the Chihuahuan Desert, and *M. greggii* and *M. palmeri* are more narrowly restricted in northern Mexico; *M. utahensis* primarily occurs in limestone mountains of the Mojave Desert and Utah, but occasionally occurs in desert habitats as low as 900 m.
Viscainoa geniculata.	A monotype endemic to the Viscaíno Region of the Sonoran Desert in Baja California.
Berberis, barberry.	Three species, *B. fremontii*, *B. haematocarpa*, and *B. trifoliolata*, occasionally found in upper desert scrub associations but more commonly occur in scrub and woodland communities at higher elevations.
Eriogonum fasciculatum ssp. *polifolium*.	A subspecies of the common California buckwheat of coastal sage scrub and chaparral habitats and one that occurs fairly frequently in Joshua tree woodland and adjacent desert scrub in southern California.
Coleogyne ramosissima, blackbrush.	A common shrub of upper bajadas in the desert transition vegetation of the Mojave Desert with more mesic habitats, occurring northward in the Groom Ranges, Nevada. This may not be a true evergreen, because its leaves in winter appear to be inactive, and leaves of the previous season abscise when a new flush of leaves is produced in the spring.
Atamisquia emarginata.	An occasional shrub in thorn scrub desert in the Sonoran Desert of Mexico, but also native to a disjunct desert region in Argentina.
Purshia glandulosa.	Shrub of chaparral, pinyon–juniper woodland, and sclerophyllous woodland, but occasionally occurring in desert scrub and Joshua tree woodland above 1200 m in elevation.
Quercus, scrub oak.	Shrubs of chaparral and sclerophyllous woodland associations, but *Q. dunnii* and *Q. turbinella* rarely appear in adjacent desert scrub communities.
Garrya flavescens var. *pallida*, silk-tassel.	A chaparral plant that may occur in the adjacent desert transition zone.

mondsia chinensis is a species of the Sonoran Desert, where it often occurs in the eastern region at relatively high elevation sites, but may also occur at low elevation sites in western localities with *Larrea*. *Atriplex hymenelytra* (desert holly), a shrub that occurs mostly on limestone soils and along washes, also typically has leaves year-round (Mooney, Ehleringer, & Björkman 1977). Other evergreen species are absent from typical desert scrub communities, but appear peripherally at higher elevations, where desert scrub adjoins other semiarid vegetation types, such as chaparral, coastal sage scrub, sclerophyllous woodland, pinyon–juniper woodland, and, in Mexico, tropical thorn scrub.

Comparatively few studies of long-term population dynamics have been carried out for warm desert shrubs, especially in the Mojave Desert, but it is well known that individuals of *Larrea* have the potential to attain great ages. Ramets (individual shoots) continue to be produced on the perimeter of the canopy while older shoots die near the center (Vasek & Barbour 1977). In the Sonoran Desert, using growth rings, old stems of *Larrea* have been dated at about 65 years (Chew & Chew 1965), and photographic records from Arizona have provided evidence that large individuals may be more than 100 years old (Shreve & Hinckley 1937). In seven populations of *Larrea* from the Mojave Desert with growth-ring studies of seedlings Vasek (1980) showed that plants having stem radii of 15–35 mm had mean ages of 19–60 years. Large clonal rings of creosote bush in the southern Mojave Desert appear to have been produced by a single genetic individual that grew at the peripheral and died within the center, this over very long periods of time. Growth analyses and radiocarbon dating suggest that certain large clones may be thousands of years old (Sternberg 1976; Vasek 1980).

Population studies of *Larrea tridentata* are interesting because investigators have noted that stands may have regular, clumped, or random distributions of individuals (Barbour 1969, 1973). Studies in California by Woodell, Mooney, and Hill (1969) found a significant linear correlation between mean annual rainfall and *Larrea* density in 12 populations of the Sonoran and Mojave deserts, but there was no such correlation for other perennials. Detailed studies by Beatley (1974a, 1975, 1976b) at NTS defined the effects of precipitation and temperature on the distribution of the evergreen *Larrea*.

Larrea tridentata, the representative long-lived desert evergreen shrub, has as predicted (see Fig. 3.1) low net carbon gain per unit time, but superb physiological adaptations to endure environmental stresses (Oechel, Strain, & Odening 1972; Odening, Strain, & Oechel 1974; Solbrig 1979). Field measurements of *Larrea* from numerous desert sites show that the evergreen leaves have positive carbon gain during every month of the year, and maximum photosynthesis at each temperature regime only varies up to 10% (Mooney, Björkman, & Collatz 1978). The highest observed rate of CO_2 uptake was a respectable 26 mol CO_2^{-2} s^{-1}, but this was lower than several other codominant leafy shrubs and the herbs occupying open areas (Smith & Nobel 1986). Creosote bush leaves are PAR-saturated at 50% full sunlight and rarely achieve that because the canopy is self-shading, even though the plants are V-shaped and somewhat open, and leaves are oriented more or less vertically rather than perpendicular to direct incoming solar radiation. PAR is blocked from mesophyll by scattered trichomes and a resin that coats the entire leaf and gives the plant a distinc-

tive aroma and a varnished appearance. Research has shown that presence of the resin, although it limits photosynthesis, drastically reduces transpiration (Meinzer et al. 1990). *Larrea* can maintain positive uptake of CO_2 when leaf water potential is −6 MPa, a point at which most shrubs are totally leafless or at zero turgor (Odening et al. 1974).

Atriplex hymenelytra (Fig. 3.16) is a very different type of evergreen shrub, one that occurs mostly in low elevation, dry alkali sink habitats, such as Death Valley, where it survives very high summer temperatures. Like *Larrea*, *A. hymenelytra* has a mean upper thermal stability of photosynthesis above 50°C (Downton, Berry, & Seemann 1984), but unlike creosote bush, which has poor stomatal control, desert holly is a C_4 species that can maintain tight stomatal control when soils become dry (Sanchez-Diaz & Mooney 1979). Its

Figure 3.16 Desert holly, *Atriplex hymenelytra*.

leaves are polymorphic and highly reflective due to the presence of vesicular trichomes that shield the mesophyll. This species maintains a fairly constant WUE throughout the year (Mooney, Ehleringer, & Björkman 1971; Pearcy, Björkman, Harrison, & Mooney 1974), and although rate of maximum photosynthetic assimilation is only moderate, no higher than those of sympatric C_3 shrubs, favorable WUE allows a higher level of season productivity.

Species of *Atriplex* are commonly associated with halophytic conditions. Plants growing under saline conditions must maintain a high concentration of osmotically active substances to allow a favorable water potential gradient to exist from soil to leaf tissues. Whereas the increase in osmotic potential of leaf tissues is primarily accomplished by accumulation and tolerance of inorganic salt ions, a similar result may be achieved by accumulation of large amounts of organic acids. For C_3 species, osmotic adjustment using organic acids is usually too expensive metabolically to be a viable strategy; however, in these C_4 species organic acid synthesis can be used for this purpose (Osmond, Winter, & Ziegler 1982). *Atriplex* excretes excess salts by vesicular trichomes (salt glands), but this requires considerable investment of energy.

Drought-deciduous shrubs

Many shrubs of the Mojave Desert are characterized by having a growth strategy with drought-deciduous leaves, nonsucculent stems, and fibrous, nonsucculent roots. Leaf production mostly occurs in early spring, from mid-February to April in the Mojave Deert (Ackerman, Romney, Wallace, & Kinnear 1980; Chapter 4), and a leaf canopy persists until drought conditions become extreme in summer and shoot water potential no longer is sufficient for daily stomatal opening. Leaves may be polymorphic, wherein the degree of xeromorphism corresponds to the condition of shoot water potentials at the growing tips when the leaf blades were formed, so that leaves formed under declining water potentials have stomatal response, surface area, and spectral characteristics

appropriate for increasing aridity (Ehleringer 1980; Mooney 1980).

These perennials range from woody shrubs with long-lived aboveground stems with well-developed canopies to low, suffrutescent plants (subshrubs) consisting of shoots that arise from below ground. In semi-woody, suffrutescent species, individual shoots may die back when drought is very severe or after the shoot has produced terminal reproductive structures, hence the name 'branch shedders' (Evenari et al. 1971).

Ambrosia dumosa is the most common drought-deciduous perennial within Mojave desert scrub and at NTS. Typically these low plants are leafless from mid- to late summer until new leaf production in the following spring, but in years when infrequent summer storms rewet the soil, either spring leaf flushes are maintained, if senescence has not yet occurred or new shoots are produced from axillary buds.

Leaves of *A. dumosa* have dense trichomes on both surfaces, giving this the common name white bursage. This species has very distinctive isolateral leaves; abaxial and adaxial palisade parenchyma are demarcated by a wide spongy mesophyll that has secondary cell walls, so that the two halves of the leaf appear to be physiologically separated (Fig. 3.17). An individual plant of white bursage has a relatively short life span; growth ring studies at NTS have suggested that individuals commonly attain ages of no more than 15–25 years (Wallace & Romney 1972). None the less, seedling establishment of *A. dumosa* is much more common than that of long-lived shrubs.

Larger woody shrubs with drought-deciduous phenology are exemplified by the genus *Lycium*. Lyciums are common in most nonsaline desert scrub communities of the Mojave Desert, and three species occur at NTS. In some localities *Lycium* is mostly restricted to washes (Wallace, Romney, & Hunter 1980), where robust specimens may attain a height of 2 m, but typically they are waist high. One group of species has narrowly spatulate, thin leaves, as represented by *L. pallidum* at NTS, and another by small, semi-succulent terete leaves, as represented by *L. andersonii*. Mesophyll of both leaf types is unusual because

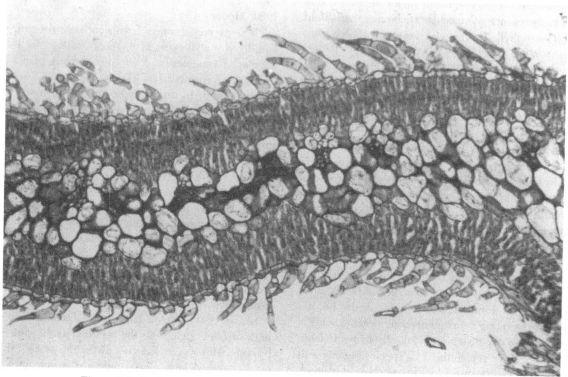

Figure 3.17 Leaf transections of white bursage, *Ambrosia dumosa*, having isolateral organization of mesophyll and abundant hairs covering both surfaces.

the palisade cells tend to be fairly large and the cells are therefore not compact.

Lycium also has a very common shoot design called long shoot-short shoot organization. A young shoot forms with significant internodes between subsequent leaves on a shoot (the long shoot), but after the leaves on the original long shoot abscise, leaving the stem bare, the new sets of leaves arise, for example, in the spring, from the axillary buds without formation of internodes, so that they appear to form in clusters (the short shoots). On desert plants with short shoots, leaf emergence is extremely rapid following spring rains (Humphrey 1975), and long shoots typically form if summer rains occur. Another property of long shoot–short shoot organization is that the plant does not use valuable materials to produce stems for new leaves, which has the added and perhaps greater benefit of creating space between

shoot branches so that PAR can penetrate an open canopy to improve photosynthetic rates. When species of *Lycium* produce long shoots, typically when spring growing conditions are very favorable (Wallace & Romney 1972), many of the axillary buds are activated to form branches, but as water stress develops the shoot tips die and harden as spines.

Seedlings of *Lycium* that germinate following fall or winter rains rarely survive drought conditions of summer; however, those that germinate following occasional summer rains have a longer period to become established and thereby higher rates of survival (Went 1955). This pattern of seedling success is true for many woody species.

Krameria erecta (Pima rhatany) is an ecologically important, intricately branched, low-growing shrub, typically 0.2–0.6 m in height, with narrow, drought-deciduous leaves. Observations at Rock

Valley showed that spring growth is initiated at least four weeks later than for other species, and leaves persist for up to two months longer during the summer drought (Ackerman et al. 1980). Species in the genus *Krameria* are considered to be root parasites, at least as young plants forming haustorial connections with the roots of neighboring plants to use their nutrient and water resources (Kuijt 1969). No studies have demonstrated this biotic association for species of the Mojave Desert, either for adult plants or for young plants as they become established.

Although *Ambrosia*, *Lycium*, and *Krameria* are widespread and locally abundant in the Mojave Desert, the ecophysiology of these species is virtually unknown, being limited to a few early studies (Bamberg, Kleinkopf, Wallace, & Vollmer 1975; Clark, Letey, Lunt, Wallace, Kleinkopf, & Romney 1980; Kleinkopf, Hartsock, Wallace, & Romney 1980), which reported relatively high maximum photosynthetic rates exceeding and noted sharp declines in CO_2 uptake with fairly small decreases in leaf water potential. Many Sonoran Desert shrubs with drought-deciduous leaves, such as *Encelia farinosa*, *Hyptis emoryi*, and *Ambrosia deltoidea* (triangle-leaved bursage) have quite high rates of leaf photosynthesis (Nobel 1976b; Szarek & Woodhouse 1976; Smith & Nobel 1977; Ehleringer & Björkman 1978b). Studies of three drought-deciduous species from Yucca Mountain, Nevada, occurring in desert transition communities like those in Rock Valley, illustrate how leaf area index declines as predawn water potentials decrease (Fig. 3.18; Smith & Nowak 1990).

Many low shrubs and subshrubs of the Mojave Desert utilize stem photosynthesis to allow low levels of carbon assimilation to take place in stem tissues, where water-use efficiency is relatively high, and thereby allow possible carbon gain to occur during severe drought at the end of the growing season. The contrasting photosynthetic behavior of leaves versus photosynthetic stems of one very common Mojave drought-deciduous species, *Hymenoclea salsola* (Fig. 3.19), has been described (Comstock & Ehleringer 1988). This species has relatively long but narrow, linear

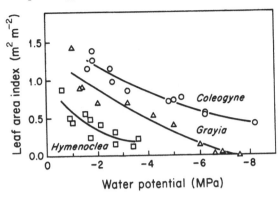

Figure 3.18 Relationship of leaf area index to predawn water potential in three species of drought-deciduous shrubs at Yucca Mountain, Nevada. From Smith and Nowak (1990).

(essentially cylindrical) leaves, which gradually abscise through the summer and sometimes into early autumn. These plants tend to grow along washes or in disturbed sites, where rain runoff accumulates. Comstock and Ehleringer reported that leaves have very high photosynthetic rates as compared with leaves of other C_3 shrubs and much higher under well-watered conditions than photosynthetic stems of *Hymenoclea*. However, young stems also have quite high photosynthetic rates relative to leaves of other species, and they operate with a favorable WUE. During periods of low water stress, leaves are more productive than stems, but under periods of water stress, for example, at shoot water potentials of -2.3 MPa, superiority of the leaf is lost. None the less, the photosynthetic strategy of this species combines the productivity of both organs in a very complex way for optimizing CO_2 uptake until the plant experiences very high water stress, when leaves are shed and the stems contribute little to productivity. A comparable usage pattern of leaf and stem photosynthesis is present in many suffrutescent desert perennials (Comstock et al. 1988).

In lowland desert scrub of the MDR the majority of C_3 shrub species become dormant during summer drought, for example, species of *Acamptopappus* and other diminutive shrubs of

Figure 3.19 Cheesebush, *Hymenoclea salsola*.

Asteraceae and *Psorothamnus arborescens* (Fabaceae).

Grayia spinosa (hopsage), is a species characteristic of the higher elevations of Mojave desert scrub and adjacent parts of the Great Basin as well. *Grayia* is generally among the earliest shrubs to initiate growth following winter rains, for example, beginning in February at Rock Valley, but it also becomes leafless and dormant in early summer and does not respond to summer rains with new vegetative growth, and only rarely responds in the fall (Ackerman et al. 1980). Summer pre-dawn water potential of *G. spinosa* has been measured below −8.0 MPa in the vicinity of NTS (Smith & Nowak 1990). Growth ring studies suggest that *G. spinosa* has a life-span at NTS of at least 20 years (Wallace & Romney 1972).

Little has been reported on the ecophysiology of *Coleogyne ramosissima* (blackbrush), an important semi-evergreen shrub of transition desert communities. Although Beatley (1976a) considered blackbrush communities as a characteristic unit of Mojave desert scrub, the communities show their best development in the Great Basin along the valleys of the Colorado River and lower Green River in southeastern Utah (Cronquist, Holmgren, Holmgren, & Reveal 1972).

The most comprehensive ecological discussion of *Coleogyne* was presented by Bowns and West (1976). *Coleogyne* has a relatively shallow and diffuse root system, and pre-dawn water potentials of plants in central-southern Nevada may reach to −7.5 MPa (Smith & Nowak 1990) even though it occurs at sites with higher rainfall than *Larrea tridentata* (Beatley 1976b, 1979). Leaf structure of *Coleogyne* is more typical of chaparral than that of desert shrubs (Fig. 3.20), having a totally different internal design than described earlier in this chapter. The ecological significance of that variation for a desert environment is not known. Because

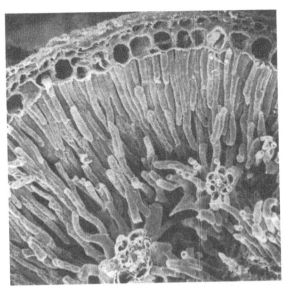

Figure 3.20 Leaf transection of blackbrush, *Coleogyne ramosissima*, which has anatomical features more typical of chaparral plants than desert shrubs.

Coleogyne tends to grow in very dense stands, it has been suggested that root splitting may result in clonal propagation (Manning & Groeneveld 1990).

Two other shrubs are important elements in the perennial vegetation of Rock Valley, the C_4 *Atriplex confertifolia* (shadscale) and C_3 *Ceratoides lanata* (winter fat). Both species are widespread throughout desert and semi-desert grasslands of western North America, but are more characteristic of the cooler Great Basin (Caldwell 1985). Whereas *C. lanata* is deciduous in summer, *A. confertifolia* continues to produce leaves into mid- and late summer and maintains a semi-evergreen growth habit. Caldwell (1985) suggested that the value of this evergreen-like strategy may be in having leaves available for photosynthesis during early spring, when moisture conditions are highly unpredictable.

Detailed ecological and physiological studies of shadscale and winter fat were conducted during the IBP program of the Great Basin and therefore results characterized the nature of adaptive strategies under cooler conditions than in the Mojave Desert. High photosynthetic capacities were not found; maximum rates of net CO_2 assimilation

were only 14 μmol m^{-2}s^{-1} for *A. confertifolia* and 19 μmol m^{-2}s^{-1} for *C. lanata*, rates that were similar to those of big sagebrush (*Artemisia tridentata*), a C_3 shrub and the dominant species of the Great Basin desert scrub (Depuit & Caldwell 1975; Caldwell, White, Moore, & Camp 1977). Such restricted photosynthetic capacities appear to be an evolutionary accommodation to the relatively cool temperatures that prevail during the spring growing season. The majority of the net annual carbon gain occurred in the spring and early summer. These C_4 species show a marked decrease in WUE during the dry summer months.

Aphyllous C_3 shrubs

A significant number of woody desert shrubs and trees are essentially leafless (aphyllous) throughout much or all of the year, and therefore rely heavily or exclusively on stem photosynthesis (Gibson 1983). The most thoroughly studied taxa of this type are Sonoran Desert species of *Cercidium* (palo verde) and *Psorothamnus spinosus* (smoke tree), leguminous trees (Adams, Strain, & Ting 1967; Adams & Strain 1968, 1969; Szarek & Woodhouse 1978; Nilsen, Meinzer, & Rundel 1989). However, these arborescent legumes are absent from the Mojave Desert. Several aphyllous shrubs are fairly common in Mojave desert scrub, most notably the genus *Ephedra*, with *E. nevadensis* (Fig. 3.21) and *E. funerea* at NTS and in Rock Valley. This genus of gymnosperms occurs in arid and semiarid environments worldwide, and its success in desert regions is likely attributable in part to the presence of vessels in its wood (typically not present in gymnosperms but present in dicotyledons) and its root system, which in some species exceeds 4 m in depth (Cannon 1911). Growth rates of *E. nevadensis*, as sampled in Rock Valley, vary with seasonal soil moisture conditions (Ackerman et al. 1980). On lower bajada slopes and plains, where much sand and loess are blown by the wind, this species commonly forms extensive clumps through rooting of underground shoots (Wallace & Romney 1972).

Other examples of aphyllous shrubs of the Mojave desert scrub are *Senna armata*, a low leg-

Figure 3.21 Mormon tea, *Ephedra nevadensis*.

uminous shrub most common in the southern MDR, *Thamnosma montana* (turpentine bush), a species that prefers upland slopes and washes, *Menodora spinescens*, a low, mat-forming shrub that is fairly common in the northern MDR, especially on and near NTS, and *Castela emoryi* (crucifixion thorn), a large, spinescent shrub of the western Mojave and Sonoran deserts. Few detailed ecophysiological studies have been done on any of these aphyllous species.

Stem and leaf succulents

Succulent desert species, typically utilizing Crassulacean acid metabolism, are widespread in arid and semiarid regions of the world. Succulent biomass is commonly greater in warm than in cold desert regions and in summer rainfall deserts over those with predominantly winter rainfall. Nevertheless, cacti, especially several species of chollas

(*Opuntia*), and *Yucca* are highly visible succulent components of some lowland desert scrub regions in the Mojave Desert. Platyopuntias (*Opuntia*), which have large, flattened stems called cladodes, are more common on the eastern and southern zones of MDR and Colorado subdivision of the Sonoran Desert, but *O. basilaris* (beavertail cactus; Fig. 3.22) is an exception that occurs throughout the Mojave Desert. A branched barrel cactus, *Echinocactus polycephalus*, and several low cactus growth forms are more common in rocky slopes than on the desert floor.

Yucca brevifolia (Joshua tree; Fig. 1.10), which is a C_3, not a CAM, species, is the characteristic tree at the outer and upper elevational limits of the Mojave Desert (Smith, Hartsock, & Nobel 1983), where it may achieve a height of 12 m. This species needs some cold temperatures during the winter in order to produce its magnificant inflorescences. *Yucca schidigera* is locally very common

Figure 3.22 Beavertail cactus, *Opuntia basilaris*.

on bajadas and has some CAM features (Nobel 1988). New growth of *Y. schidigera* occurs with the production of a set of 6 new leaves at a time, 2–4 sets per year. If such a phenological pattern is consistent, then the large clumps of this species in Mercury Valley have ages up to 200 years. Wallace and Romney (1972) found for *Y. brevifolia* that 2–8 sets of leaves were produced annually, depending on soil moisture conditions, so that large trees in Yucca Flat at NTS may be 200 years old. For Joshua tree, annual CO_2 uptake for its evergreen leaves has been estimated at 22 mol CO_2 m^{-2} yr^{-1}, which is a moderately high rate of photosynthesis, and the species is also noteworthy for displaying tight control of water loss during the dry season (Smith et al. 1983). Its leaves form rosettes with high leaf area indexes and fairly even distribution of PAR throughout the canopy, including both leaf surfaces. Net photosynthesis of *Y. brevifolia* occurs under a moderate day/night growth regime 31°C/17° and is significantly decreased at typical summer temperatures; none

the less, the temperature optimum for photosynthesis shows acclimation of approximately 9°C from the coldest to hottest growth regimes (Smith et al. 1983).

For cacti, the physiological and ecological aspects of CAM have been described earlier in this chapter and have been the subject of several recent reviews (Nobel 1985, 1988; Gibson & Nobel 1986; Smith & Nobel 1986). Desert succulents have an obvious ecological adaptation in their extensive water storage tissue, which provide a considerable volume of water behind every unit of transpiring surface area, but CAM itself provides considerable efficiency in conserving limited water resources. Nobel (1985) has calculated that under typical conditions of atmospheric water vapor content for a warm desert region, transpiration rates for a CAM plant at night (10°C) should be one-sixth that under day conditions (30°C).

Bunchgrasses

Perennial bunchgrasses are an intermediate growth form between woody desert shrubs and the ephemeral annuals and herbaceous perennials, which lack aboveground living tissues during extremes of environmental stresses. Bunchgrasses are widespread throughout western North America and sometimes are important members of desert scrub communities, particularly *Oryzopsis* (*Achnatherum*) *hymenoides* and *Hilaria rigida*. Studies of perennial bunchgrasses in the Great Basin have observed that rooting depths are generally less than 2 m, shallower than that of sympatric woody shrubs (Weaver 1919; Dobrowolski, Caldwell, & Richards 1990).

In deserts, species from different genera are often morphologically similar, having culms and leaf sheaths that have active photosynthetic surfaces during the late spring and summer, but they are physiologically diverse, some having C_3 (*Oryzopsis*) and others being C_4 photosynthesis (*Hilaria*). An ecologically widespread species on sandy plains, dunes, and bajadas of the Mojave and Sonoran deserts is *H. rigida*, a C_4 species. Its extremely high photosynthetic capacity (67 μmol CO_2 m^{-2} m^{-2} s^{-1}) and high WUE may be important factors in the successful establishment and growth of this species at many arid sites (Nobel 1980). This species has a relatively high temperature optimum for CO_2 uptake and undergoes a major physiological acclimation of temperature curve responses for photosynthesis changing from 29°C for a winter temperature regime to 43°C for a midsummer regime, thus enabling *Hilaria* to grow well in cool and hot months (Nobel 1980). Because shoots are competing with each other for water and nutrients, clump size and distance between intraspecific clumps strongly influence growth rates within this species (Nobel 1981b; Robberecht, Mahall, & Nobel 1983). Comparable data are needed on native perennial *Stipa* of the Mojave Desert.

Annuals

Two major groups of desert annuals can be recognized, winter annuals and summer annuals, separated by the phenology of their seasonal growth activity (Mulroy & Rundel 1977). Winter annuals are those species that germinate following fall or winter rains (typically October through mid-December) that measure 25 mm or more and complete the life cycle during the spring (Beatley 1974b). Most winter annuals remain in a vegetative rosette or tuft until stem elongation begins in March or early April, when air temperature increases (Beatley 1969b). In this way, vegetative rosettes take advantage of microenvironmental temperatures at the ground surface, which may be 10–15°C warmer than those at 10 cm above the surface (Mulroy & Rundel 1977). Leaves of winter annuals typically are narrow, amphistomatic, and isolateral, like those of the shrubs, but they appear to have lower values of A^{mes}/A. Growing seasons for development of winter annuals may be as long as five to eight months in a year with early germination and as short as six weeks in an unfavorable year. Considerable levels of developmental plasticity are present because the conditions of the growing season are unpredictable; species that in a favorable climate grow to a height of 50–100 cm or more may grow to only 5 cm in a year where drought follows the period of plant establishment. Winter annuals commonly produce a few flowers and fruits early in their period of vegetative development (Went & Westgaard 1949; Shreve 1951; Mulroy & Rundel 1977), probably a developmental strategy to allow minimal reproductive success so that some seeds of the species can be added to the soil seed reserve for the future.

Summer annuals germinate following a heavy summer rain and complete the life cycle in late summer or early fall (typically July or August). They may be as abundant as winter annuals where biseasonal rainfall patterns are present, such as in the eastern Sonoran Desert (Chapter 2). In upland desert areas of Arizona and on the plains of Sonora, Mexico, summer annuals may carpet large expanses and be very robust plants with fairly large leaves. West of the Colorado River, where

Table 3.2. *Summary of physiological characteristics of the common angiosperm life forms of the northern Mojave Desert. Adapted from Smith and Nobel (1986).*

Parameter	Evergreen shrubs	Drought-deciduous shrubs	Aphyllous C$_3$ shrubs	Stem and leaf succulents	Bunchgrasses	Winter annuals	Summer annuals
Maximum photosynthetic rate	low to moderate	moderate to high	low to moderate	low	moderate to very high	moderate to very high	high to moderate
Minimum seasonal water potential	low	moderate to low	moderate to low	high to moderate	moderate	high	high to moderate
Water-use efficiency in dry season	low to moderate	moderate	low	low to high	moderate	low	moderate
Temperature optimum	moderate to high	moderate	moderate	low	moderate	low to high	high to moderate
Seasonal temperature acclimation	generally high	low to moderate	low	high	moderate	low to high	low to high

summer thunderstorms are infrequent, abundance and diversity of summer annuals drops markedly, so that the ephemeral flora is largely that produced by winter rains. The Mojave Desert flora is notably poor in summer annuals (Mulroy & Rundel 1977).

Many summer annual species exhibit C$_4$ metabolism – as characterized by Kranz anatomy in the leaves – which is well suited to their phenological pattern of growth. High rates of net photosynthesis are present under the high temperature and irradiance conditions that occur with summer rains (Ehleringer 1983b, 1985). In this way the metabolic system of these annuals, along with the large surface area of individual leaves or leaflets, allows them to maximize their growth rates and complete the life cycle in a relatively short time (Mulroy & Rundel 1977). The basal rosette, typical of winter annuals, is generally absent in summer-active species, probably because temperatures at ground level are suboptimal or even lethal.

Whereas C$_3$ species are found in summer rainfall areas of the Sonoran Desert, particularly in moist and disturbed microhabitats, C$_4$ annuals are the dominant system. For the Mojave Desert, all of the small number of native summer annuals have C$_4$ metabolism (Fig. 3.12a). However, despite the metabolic advantages of C$_4$ metabolism for attaining high rates of CO$_2$ assimilation, for example, *Amaranthus palmeri* with 82 μmol CO$_2$ m^{-2} s^{-1} (Ehleringer 1983b), a number of C$_3$ winter annuals have maximum rates nearly as high as those of typical C$_4$ annuals (Mooney, Björkman, Ehleringer, & Berry 1976; Forseth & Ehleringer 1982; Ehleringer 1985). In comparison with summer annuals, winter annuals also tend to have much lower thermal optima for photosynthesis, lower, often very low, lower water stress tolerance, and lower WUE (Table 3.2; Smith & Nobel 1986).

Comparisons of photosynthetic characteristics

Table 3.2 summarizes photosynthetic characteristics of the various growth forms that are found in the Mojave Desert. Excluded was any review of spring-active herbaceous perennials, including mostly geophytes, on which essentially no ecophysiological has been done, although it is assumed that they are similar to winter annuals.

Photosynthetic capacity tends to be low in evergreens and succulents, which fix carbon in all months of the year, and highest in ephemerals, which have adaptations for taking advantage of short periods with high soil moisture content. Correlated with high photosynthetic capacity is the character to need high levels for light saturation. Drought-deciduous shrubs, as expected, are intermediate with moderate values when conditions are favorable, whereas perennial grasses are expected to have high photosynthetic capacity. For production ephemerals have the highest levels (on a leaf area basis) but are good for all categories except the slow-growing succulents of this desert. Water stress tolerance is highest for evergreen shrubs, intermediate for most other perennials, and generally low for ephemerals and succulents. On the other hand, those succulents with CAM and C_4 annuals and perennials have high values of WUE.

Although ecophysiological studies have been conducted on several characteristic species of the Mojave Desert, and those adaptive designs are now fairly well understood, the majority of species, including many widespread perennials, have never been investigated along these lines of research. Many of the generalizations discussed here may need to be refined or revised as new data are gathered from warm desert ecosystems.

SUMMARY

Typical leaves of warm desert plants are small and narrow, a design that enables leaf temperature to be near ambient temperature even when the stomates are closed, so that the leaf can avoid lethal temperatures during summer drought. If leaf temperature is lowered, transpirational water loss will also decrease because the water vapor concentration decreases in intercellular air spaces of the leaf; by having a lower concentration in the leaf there is a smaller diffusion gradient to the surrounding dry air, thus an important driving force of transpiration is reduced. Under dry-soil conditions, broad leaves overheat unless cooled via massive transpirational water loss, which typically

is not an environmental option without availability to abundant deep soil water. Some desert plants display leaf polymorphism, whereby broad leaves are formed when soil moisture is high and narrower leaves follow as that water is depleted.

Early analyses of desert plant adaptations emphasized that their principal adaptive strategy is to minimize water loss. Obvious examples are the water-storing succulents with Crassulacean acid metabolism (CAM), which open stomates at night and thereby experience substantially reduced transpiration than if stomates were open during the heat of the day, and these plants have very high water-use efficiencies. However, adopting an explanation that emphasizes water conservation is inconsistent with many properties of desert plants, and has caused most features of desert plants to be misinterpreted. The typical leaf must be viewed as one designed to maximize carbon gain whenever conditions are favorable. The great majority of nonsucculent desert species have C_3 photosynthesis. As one expects if net carbon uptake is the goal, their leaves are amphistomatic with high densities of stomates on each surface to maximize CO_2 uptake when stomates are open, have stomates that can open at low to extremely low leaf water potentials for prolonging the period for photosynthesis to occur, and have isolateral mesophyll with high internal cell surface area for rapid diffusion of CO_2 into palisade parenchyma cells. Such leaves also have a relatively thin cuticle, and stomates are not sunken. Most summer annuals and a few species of shrubs, most notably *Atriplex*, have C_4 photosynthesis with amphistomatic leaves but the characteristic Kranz anatomy.

Leaf orientation enables some species to reduce interception of infra-red radiation during midday of summer months, and leaf suntracking has been employed either to minimize high leaf temperatures or to maintain high solar irradiance for maximizing photosynthesis. Desert leaves may be covered with dense trichomes, which absorb and reflect infra-red radiation and thus reduce leaf temperature and, correspondingly, transpiration. Some plants of warm deserts exhibit physiological acclimation whereby temperature optima for photosynthesis change to more closely match the prevailing seasonal thermal regime.

Water is a spatially and temporally variable resource that has produced many growth forms in deserts. Shallow-rooted evergreen shrubs and some drought-deciduous shrubs tolerate leaf water potentials lower than -6.0 MPa. Woody phreatophytes, with deep roots anchored in ground water, and shrubs in sandy washes have relatively high transpirational fluxes at midday during drought. A few large-leaved perennial vines can maintain very high leaf water potentials at -1.5 MPa also by tapping water from deeper soil layers, but having a large storage root to buffer aboveground tissues from diurnal shifts in water stress. Ephemerals, which are rooted only in the upper soil horizons, are heavily dependent on having adequate soil moisture, but the C_4 summer annuals have rapid growth rates and good water-use efficiency. Other physiological adaptations for water management include reducing stomatal apertures during less favorable times and mechanisms to maintain positive turgor potentials in leaves under water stress that should cause stomates to remain closed. It is reasonable to expect that coexistence among species is due in part to the varied strategies of water uptake and management. Competition for water between codominants of desert scrub have been demonstrated.

In Mojave desert scrub, located in valleys and on lower bajadas, one observes a variety of different growth forms. *Larrea tridentata* is the obvious example of the widespread evergreen shrub that generally exhibits positive net photosynthesis throughout the year and tolerates extremely low water potentials. Other evergreens, such as *Simmondsia chinensis*, *Mortonia utahensis*, and *Atriplex hymenelytra*, a C_4 species, have more restricted distributions based on soil preferences. Evergreens tend to have reasonably good rates of CO_2 uptake, but lower than some codominant leafy shrubs.

Many shrub and subshrub species are drought deciduous, most notably *Ambrosia dumosa*, which is a low-growing and extremely abundant codominant with *L. tridentata* in the Mojave Desert. Species of *Lycium* are drought-deciduous shrubs with long shoot–short shoot organization. *Krameria erecta* and several species common along washes have photosynthetic stems and leaves, and these are somewhat transitional to those species that have ephemeral leaves and rely almost entirely on aphyllous, photosynthetic old stems for carbon uptake. Examples of widespread aphyllous species are such shrubs as *Ephedra* spp., *Senna armata*, *Thamnosma montana*, and *Menodora spinescens* and the arborescent legumes of *Cercidium* and *Psorothamnus*. Locally common in Mojave desert scrub may be succulents, especially cacti, CAM species, and *Yucca* spp., which include both CAM and C_3 species. *Yucca brevifolia* is a C_3 species with moderate rates of CO_2 uptake. It grows best under moderate day/night thermal regimes.

Among herbaceous species, bunchgrasses can take hold and be important components of desert scrub communities. Some species have C_3 and others C_4 photosynthesis, and of the C_4 species *Hilaria rigida* has an extremely high photosynthetic capacity and a high WUE, two factors that probably contribute to its success. The Mojave Desert also has a diverse winter and summer flora of annual species. Winter annuals, which mostly germinate from October through mid-December with late fall and early winter rains are adapted to grow best under cool temperature regimes, whereas the summer C_4 annuals have very high carbon assimilation rates at high temperature and full irradiance. Summer and winter annuals from the desert have among the highest rates of CO_2 assimilation, and the leaf anatomy shows typical anatomy to favor high uptake of CO_2.

4 Desert perennials of southern Nevada

The Nevada Test Site occupies the zone of transition between the Mojave Desert Region (MDR) flora to the south and the Great Basin Desert flora to the north. At NTS one can find associations of desert plants that are clearly assignable to each of the two desert floras, as well as localities with mixed assemblages, transitions where the warm desert species decrease in abundance and disappear, while cold desert species appear with increasing frequency. By sampling species abundance, productivity, and phenology from sites within this diversity of habitats, researchers at NTS began an important ecological program to obtain long-term records of how environmental factors influence plant growth in desert ecosystems.

DESERT VEGETATION AT THE NEVADA TEST SITE

Beatley (1974a, 1975, 1976a, c, 1979) clearly described the typical and transitional types of perennial vegetation in central-southern Nevada, with a primary focus on those at NTS. She not only produced a flora based on thousands of personal collections and herbarium materials, but also used a network of 68 permanent study sites, most located within the eight major drainage basins at NTS, to obtain long-term environmental data and measurements of plant community characteristics for descriptions and analyses of desert vegetation. Consequently, the vegetation on NTS and its borderlands have been carefully mapped (Fig. 4.1), and plant ecologists have a very valuable storehouse of observations for investigating growth and productivity of perennial plants as related to abiotic factors.

Mojave desert scrub on the Nevada Test Site

Typical stands of Mojave desert scrub prevail in the southern portion of Jackass Flats, most of Frenchman Flat, and most of Rock Valley and Mercury Valley at the Nevada Test Site. One dominant shrub in these communities is *Larrea tridentata* (creosote bush; Fig. 1.3, 3.6), which is the most widespread and ecologically important shrub of the warm desert regions of North America (Benson & Darrow 1981). The lowest sites (940–1100 m; Sites 1, 7, 8, 22, 28, 32 in Fig. 4.1) often have *Ambrosia dumosa* (white bursage; Fig. 4.2) as codominant. This vegetation occurs on sandy limestone or dolomitic soils (Chapter 2) that lack desert pavement. At 1000–1100 m elevation on NTS (Sites 3, 4, 6, 10, 15, 20, 23–25) abundance of *Ambrosia* sharply decreases while that of *Grayia spinosa* (hopsage), *Lycium andersonii* (desert thorn), and *L. pallidum* (box thorn) increases (Sites 5, 11, 12, 16, 17, 21, 27). Eventually *Larrea* also becomes a minor member of the vegetation and then finally is absent at elevations above 1325 m. This absence of *Larrea* typically marks the boundary of MDR vegetation.

On certain calcareous bajadas having fine sandy or silt loams and desert pavement (Site 29) one encounters *Larrea* as a codominant with *Atriplex confertifolia* (shadscale). *Larrea* occurs elsewhere northward in southern Nevada, but other characteristic species of the MDR, *A. dumosa*, *Krameria*

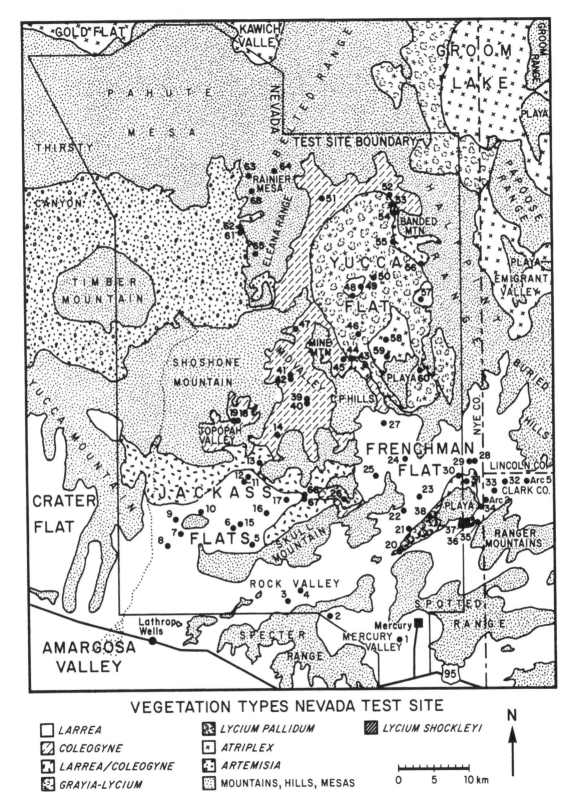

Figure 4.1 Vegetation type map of the Nevada Test Site and surrounding areas. Permanent vegetation plots established by Beatley are located by numbers. Redrawn from Beatley (1976a).

Figure 4.2–4.5 Figure 4.2 (Top left) White bursage, *Ambrosia dumosa*. Figure 4.3 (Top right) Desert thorn, *Lycium andersonii*. Figure 4.4. (Bottom left) Blackbrush, *Coleogyne ramosissima*. Figure 4.5 (Bottom right) Box thorn, *Lycium pallidum*.

erecta (Pima rhatany), and *Psorothamnus arborescens* (Mojave indigo bush), have their northern limits at NTS in Yucca Flat.

Mojave Desert arroyos on NTS typically lack trees, and are habitats for shrubs such as *Hymenoclea salsola* (cheesebush), *Salazaria mexicana* (bladder sage), *Chrysothamnus paniculatus*, *Encelia virginensis*, *Bebbia juncea* (sweet bush), *Senna armata* (Spiny senna), *Prunus fasciculata* (desert almond), and *Thamnosma montana* (turpentine bush). The only arborescent species that occurs within lowland desert scrub at NTS is *Yucca brevifolia* (Joshua tree; Fig. 1.10), and it is virtually endemic to the MDR (Chapter 1). *Lycium andersonii* (Fig. 4.3) is often associated with small washes (Wallace, Romney, & Hunter, 1980b).

Great Basin communities

The Great Basin Desert habitats are mostly found in the volcanic mountains and mesas of the northern NTS, typically above 1550 m elevation. Great Basin communities cover Pahute and Rainier mesas, the Belted Range, Timber Mountain, and the northern part of Forty-Mile Canyon, and northeast to north of the NTS boundary in Emigrant Valley, Groom Lake, and Gold Flat. One characteristic vegetation type is a semiarid scrub dominated by *Atriplex confertifolia*, sometimes in nearly pure stand, but often with *Ceratoides lanata* (winter fat; Site 58). However, *A. confertifolia* is also widespread in the warm deserts to the south and is especially common near the playas of closed drainage basins (Site 31), for example, with *Sarcobatus vermiculatus* (greasewood) or *Kochia americana* (green molly; Site 59).

Another Great Basin desert vegetation type, this on very sandy soil, is scrub composed of *A. canescens* (four-winged saltbush; Site 30), which commonly is the only species of woody perennials at a site. Beatley sampled a community of *A. canescens* at 1676 m elevation (Site 65). Fairly dense stands of *Artemisia tridentata* (big sagebrush; Site 62) or *A. nova* (black sagebrush; Site 61) occur at similar elevations, and in some of these stands sagebrush contributes 95% of total shrub cover (Beatley 1976a). *Artemisia tridentata* is a characteristic species throughout the Great Basin (West 1983), but *A. nova* is especially common on NTS in the Forty-Mile Canyon, often associated with *Chrysothamnus viscidiflorus* ssp. *puberulus* and several species of cacti. At sites (63, 64, 68) above 1800 m in elevation, sagebrush desert scrub is replaced by semiarid pinyon–juniper (*Pinus monophylla–Juniperus osteosperma*) woodland; these are habitats where annual precipitation exceeds 250 mm, most falling as winter snow, and mean minimum temperatures that range from −4.1 to −1.5°C (Beatley 1975). However, either *A. tridentata* or *A. nova* often dominates in mixed, ecotonal scrub–woodland associations.

In the southern half of NTS, where most of the ecological studies have been conducted, some Great Basin plant communities occur, particularly those of *Atriplex*. In Yucca Flat northward along a transect from the playa margin one can observe *A. confertifolia* at lowest elevation with *Kochia americana* (Site 59), then with *Ceratoides lanata* (Site 58), and higher yet with *Grayia spinosa* and *Lycium andersonii* (Site 50). In southern Yucca Flat the *Atriplex–Ceratoides* association occurs with *Grayia*, *Lycium*, and *Artemisia spinescens*.

Transition desert associations

Several transition desert associations occur between the vegetation zones of Mojave and Great Basin deserts. The most important of these transitions are dominated by species of *Lycium* (mostly 1250–1400 m elevation; Fig. 4.3) or *Coleogyne ramosissima* (blackbrush; Fig. 4.4) on slightly higher slopes (1250–1525 m). A transect through the vegetational zones of Jackass Flats can illustrate this pattern (Table 4.1). *Larrea-Ambrosia* Mojave desert scrub occurs below 1000 m and near the playa (elevation 930 m). On the gently rising slopes from 1000 to 1100 m, *Lycium andersonii*, and *Grayia spinosa* are increasingly common, and from 1100 to 1250 m, but rarely to 1350 m elevation, the communities are dominated by *Larrea*, *Lycium*, and *Grayia*. From 1250 to 1350 m elevation *Coleogyne* joins the association (Site 13). *Larrea* is largely absent from the association by 1350 m, and above 1425 m (Sites 14, 18) *Coleogyne* forms the overwhelming dominant of the scrub vegetation. Where *Coleogyne* appears in the vegetation, absolute shrub cover increases from lows of 9.3–23.1% with *Larrea* dominant to a high of 50% where *Coleogyne* occurs in essentially pure stand. The *Grayia–Lycium andersonii* association is very common elsewhere on lower bajadas, especially on the northern and western slopes of Yucca Flat (Sites 46, 50) extending southward to Mid Valley and Topopah Valley. In this community *Ephedra nevadensis* now appears to be increasing in importance (Hunter, pers. comm.). The *Coleogyne* associations are found primarily throughout Mid Valley, Topopah Valley (Site 18), northern and eastern Jackass Flats, and northern and western Yucca Flat (Site 51). On the northern

Table 4.1. *Vegetational zonation in Jackass Flats and nearby Mid Valley and Tonopah Valley, showing elevational range, mean height of vegetation (m), annual precipitation (mm) based on 10 years of data, and percent shrub cover. Data from Beatley (1979).*

Vegetation Association	Elevation, m	Mean height of vegetation, m	Annual precipitation, mm	Percent shrub cover
Bare playa	*ca.* 930			0.0
Larrea-Ambrosia	<1000	0.27–0.32	11.7–12.2	17.7–22.7
Larrea-Ambrosia-				
Grayia-Lycium	1000–1100	0.41–0.82	12.5–13.7	13.5–22.1
Larrea-Grayia-Lycium	1100–1325	0.34–0.48	14.1–17.6	9.3–23.1
Coleogyne-Larrea-				
Grayia-Lycium	1250–1350	0.42–0.51	16.7	26.0–29.1
Coleogyne-Grayia-Lycium	1350–1425	0.31–0.39	18.9–20.4	42.5–49.8
Coleogyne	>1425	0.37–0.41	22.8–24.1	45.1–50.0

slopes of Jackass Flats, *Menodora spinescens* is a codominant with *Coleogyne* (Site 14).

Southwest of the playa in Frenchman Flat (Site 38) occurs a special transition community that is dominated by *Lycium pallidum* (Fig. 4.5), a MDR species and the ecological equivalent of *L. andersonii*. This distinctive stand of *L. pallidum*, a deciduous species, is sharply defined on three sides by the evergreen *Larrea* and on the fourth, near the playa, *Lycium* is associated with *Atriplex canescens*, primarily a Great Basin species. Another transition association has *Lycium shockleyi*, an endemic of southern Nevada, and *Atriplex confertifolia*; this vegetation occurs on the southern boundary of NTS from 935 to 1200 m elevation, on the northwest face of Ranger Mountain, and in southeastern Frenchman Flat next to *Larrea-Atriplex* desert scrub (Sites 34–37).

Shrub composition of Mojave Desert sites

In 25 Mojave Desert sites that were monitored yearly by Beatley, she encountered a total of 24 species of shrubs. For these 929-m^2 plots, Beatley used the line transect method, which tends to oversample large shrubs, so her results are considered somewhat high. The mean number of shrub species per plot was 8, mean shrub height

sampled was 0.4 m, mean canopy cover was 17.4%, and mean density of living and dead shrubs was 1170 and 160 individuals ha^{-1}, respectively (Beatley 1979; Table 4.2). *Larrea* typically were the tallest shrubs, but averaged less than 0.9 m in height. For the 24 transition communities that were sampled, located mostly in Frenchman Flat, Yucca Flat, and Mid Valley, *Larrea* was often present, but the vegetation was largely characterized by *Coleogyne* and *Grayia*, and 33 species of shrubs were found. Mean number of shrub species per transition plot was 14, and mean shrub height was the same as in Mojave desert scrub; however, the vegetation was much denser because mean density of living and standing dead shrubs was 2060 and 270 individuals ha^{-1}, respectively, which produced the large mean canopy cover of 31.7% (Table 4.2). Highest values of average shrub height, averaging 1.15 m, were obtained for *Larrea* within *Larrea–Grayia–Lycium* transition associations.

Climatic determinants of vegetation types

Using 10-year environmental data that were obtained from the permanent study plots, Beatley (1974a, 1975, 1976a, b, 1979) concluded that temperature and precipitation parameters strongly

Table 4.2. *Woody plant diversity, density, height, and coverage in 68 representative permanent plots of 93 m^2 size established in the Nevada Test Site in 1963. Data from Beatley (1979).*

	No. sites	No. shrub species	Shrub density living	dead	Mean height, m	Mean coverage, %
Mojave Desert (938–1239 m elevation)						
Larrea-Ambrosia	6	18	1330	170	0.34	18.6
Larrea-Lycium-Grayia	7	15	900	80	0.51	14.9
Larrea-Grayia-Lycium	9	17	1300	140	0.47	18.9
Larrea-Atriplex	1	5	1450	760	0.27	21.3
Larrea-Psorothamnus	1	6	520	90	0.41	7.2
Menodora-Ephedra	1	8	1290	90	0.25	21.1
All sites	25	24	1170	160	0.41	17.4
Transition (939–1509 m elevation)						
Larrea-Grayia-Lycium	4	21	1510	390	0.47	28.2
Grayia-Lycium	4	15	2300	260	0.38	34.7
Coleogyne	3	9	3570	190	0.39	47.9
Coleogyne-Larrea-Grayia-Lycium	3	14	1780	300	0.46	28.9
Coleogyne-Grayia-Lycium	3	11	2810	300	0.39	43.3
Larrea-Atriplex-Coleogyne	1	10	1900	430	0.35	28.3
Coleogyne-Grayia-Artemisia	1	9	2300	590	0.47	40.5
Larrea-Lycium shockleyi-Atriplex	3	17	1250	80	0.22	15.0
Lycium shockleyi-Atriplex	1	4	1100	0	0.25	17.3
Lycium pallidum-Grayia	1	6	840	40	0.59	18.5
All sites	24	33	2060	270	0.40	31.7
Great Basin (940–1207 m elevation)						
Atriplex confertifolia	1	1	610	320	0.29	8.6
Atriplex-Kochia	1	2	1830	170	0.21	14.9
Atriplex-Ceratoides	1	6	1670	170	0.30	17.5
Atriplex canescens	1	3	440	40	0.36	5.4
All sites	4	8	1140	180	0.27	11.6
Great Basin (1677–1164 m elevation)						
Atriplex canescens	1	5	1000	60	0.44	16.0
Artemisia tridentata	1	5	1800	280	0.58	32.6
Artemisia tridentata-pinyon-juniper	1	5	1860	980	0.67	40.8
Artemisia nova	1	4	2840	730	0.35	37.3
Artemisia nova-pinyon-juniper	1	7	3880	450	0.30	37.0
Artemisia nova-pinyon-juniper	1	14	2050	230	0.74	34.8
All sites	6	21	2240	450	0.48	33.1

influence the species composition within each zone of vegetation and shrub cover and height of individual species. Beatley (1975) noted that open basins, such as Jackass Flats, at night have cold air movement downslope and away from the valley floor, but at night in closed basins, for example, Yucca Flat and Frenchman Flat, cold air accumulates over the playa, creating a temperature inversion in which the slopes are warmer than the lowland basin. In Frenchman Flat, at 938–943 m elevation extreme minimum temperature was −27.8 to −22.2°C, at 951–993 m was −22.2 to 20.0°C, and at 986–1131 m was 21.6 to −16.7°C. Likewise mean minimum extreme and mean minimum temperatures showed an increase with increase in elevation. At Yucca Flat, which has a

Table 4.3. *Qualitative analysis of desert scrub vegetation at the Nevada Test Site, classifying each on the basis of temperature and precipitation pattern, whether characteristic of Mojave or Great Basin Desert. Data from Beatley (1975).*

Scrub association	Daytime temperature regime	Nighttime tempera-ture regime	Mean precipitation
Larrea tridentata	Mojave	Mojave	Mojave
Lycium shockleyi	Mojave	Great Basin	Mojave
Lycium pallidum-Grayia spinosa	Mojave	Great Basin	Mojave
Grayia spinosa-Lycium spp.	Mojave	Great Basin	Intermediate
Coleogyne ramosissima	Mojave	Mojave	Great Basin
Atriplex confertifolia	Mojave or Great Basin	Mojave or Great Basin	Mojave, Intermediate, or Great Basin
Artemisia tridentata	Great Basin	Great Basin	Great Basin

higher playa (1200 m), extreme minimum temperatures ranged from a low of −26.1°C near the playa to −16.7°C at some sites above 1300 m.

The temperature inversion correlates closely with vegetational changes from Great Basin *Atriplex* communities next to the playa and Mojave Desert *Larrea* communities at the warmest sites. Beatley (1975) proposed that of the seven common desert communities at NTS, only *Larrea* communities must have Mojavean day/night temperature and precipitation regimes, whereas only *Artemisia* requires Great Basin day/night temperature and precipitation regimes. The other communities have mixed patterns of temperature and precipitation regime and hence they are classified as transitional (Table 4.3). According to her analysis, minimum temperatures may be crucial in the appearance of *Coleogyne* in the transitional desert zone. Beatley (1976b) also observed that percentage shrub cover for each of the principal desert vegetation associations increased with increases in the ratio of mean precipitation to mean temperature.

While focusing investigation on *Larrea* at NTS, Beatley (1974a) observed that total coverage of all shrub species increased linearly with increase in annual precipitation, but not for creosote bush in localities that received greater than 160 mm of rainfall (Fig. 4.6). For *Larrea* mean height of plants decreased sharply as plant density increased (Fig. 4.7), so that at its northern limit in Yucca Flat, where plant density is highest, mean plant height is less than 0.6 m. Mean height was not

strongly correlated with mean rainfall and even less well correlated with temperature parameters.

Shrubs that characterize the Mojave Desert lowlands occur at NTS where mean minimum temperatures are typically above 0°C, mean annual precipitation is less than 160 mm, and the ratio of mean annual precipitation to mean annual temperature is less than ten. Winter coldness

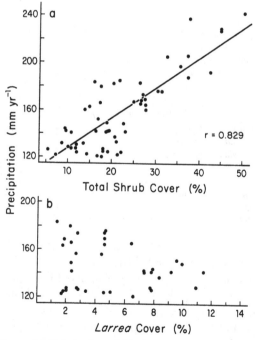

Figure 4.6 Relationship of Mojave Desert shrub cover to mean annual precipitation: (a) total shrub cover, (b) *Larrea tridentata* cover. Adapted from Beatley (1974a).

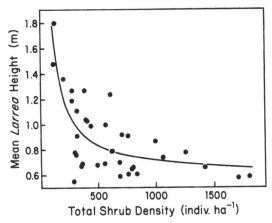

Figure 4.7 Height of *Larrea tridentata* of the Mojave Desert in relation to stand density (individuals ha⁻¹). Adapted from Beatley (1974a).

appears to be responsible for the absence of several common Mojave Desert shrub species, for example, *Ambrosia dumosa*, and Beatley (1976a) speculated that Mojave *Larrea* is intolerant of too much precipitation, being diminished in abundance and then absent from sites where annual rainfall exceeds 160 mm. The *Larrea–Coleogyne* transition occurs from 160–180 mm annual precipitation; blackbrush, which basically is a Mojavean species, survives better than *Larrea* above that zone where it has fairly warm nights during the coldest months. Plots in the transition sites had mean annual precipitation of 170–230 mm, as compared with 120–145 mm yr⁻¹ in the Mojave Desert sites, and this was probably the causal factor for higher canopy coverage in transition sites (Beatley 1979). As expected, transition associations with a majority of Mojave Desert perennials tended to have higher mean minimum temperature, whereas those dominated by *Lycium* or *Grayia* tend to be cold sites with relatively high precipitation. In her hypotheses on environmental controls for community occurrence, Beatley relied only on temperature and precipitation data. Doubtlessly, soil characteristics and other environmental variables are also important factors.

In the habitats on NTS where Great Basin vegetation prevails, the shrub communities of *Artemi-sia tridentata* do well where annual precipitation is about 200 mm and *P/T* ratios are relatively high (means up to 17.3), and these associations can tolerate and thrive at somewhat lower nighttime temperatures (means −3.4°C). Other Great Basin communities, which receive precipitation in the range of *Larrea* desert scrub, none the less tend to have cold mean minimum temperatures, ranging from −3.0 to −4.0°C.

None of the desert plant associations is apparently limited by maximum summer temperatures, which were roughly similar in all habitats that were sampled during Beatley's 10-year study. Maximum recorded temperatures ranged from 43°C in *Atriplex confertifolia* vegetation near the playa in Yucca Flat and in *A. canescens* communities near Forty-Nine Canyon to typical maxima of 44–48°C. One site in the Frenchman Flat lowlands, inhabited by *A. canescens*, had both the highest summer temperature (49.4°C) and lowest winter temperature ever recorded at NTS (−27.8°C).

Soil determinants of vegetation types

Ordination techniques were used by El-Ghonemy, Wallace, and Romney (1980b) to analyze the descriptive vegetation classification for 66 southern NTS localities. Their goals were to test the distinctiveness of the commonly cited vegetational associations and to determine which factors may control distributional patterns. For Mojave Desert and transitional desert plots, three main clusters were quantitatively delimited: *Atriplex canescens*, *A. confertifolia*, and then the remainder of the 58 plots. Using importance values calculated for shrub species from extensive measurements on relative density and relative cover, obtained from two 2 × 25 m quadrats per sample locality, El-Ghonemy and coworkers recognized similar vegetational associations as Beatley (1976a) but found that most were not very distinctive when subjected to cluster analysis. From principal component analysis they observed that individual species behaved relatively independently along the axes, suggesting that these communities are not discrete phytosociological units, but rather a changing association of individual species, each having its distribution affected by environmental gradients. Co-occurrence of species appears to be their mutual response to varying environments.

While searching for environmental correlations to vegetational grouping, El-Ghonemy, Wallace, and Romney (1980b, 1980c) determined that the most sharply defined groupings were attributed largely to soil properties. Species of *Atriplex* occupied sites that were poor in nitrogen, organic matter, and phosphorus, but rich in sodium, potassium, copper, and percent of clay. *Grayia spinosa* preferred sites that were rich in nitrogen and iron and poor in sodium, potassium, and clay content. *Acamptopappus shockleyi* and *Ambrosia dumosa* occupied the lower end of the gradient for exchangeable potassium. *Larrea* showed increasing abundance with an increase in exchangeable cations, although this correlation was thought to be due instead to the decrease in abundance of associated species. Other soil characteristics can negatively affect the growth and survival of *L. tridentata*; for example, it does not occur in dry lake areas that experience periodic flooding (Wallace & Romney 1972), is very sensitive to poor root aeration (Lunt et al. 1973), and generally is not present in pH neutral soils (Musick 1975).

PERENNIAL PLANTS OF ROCK VALLEY

Species composition and population structure

On the IBP validation site in Rock Valley (Fig. 2.15), located in a typical lowland zone of Mojave Desert, there were 6 vegetational associations and 21 species of perennial plants, including 16 shrubs and 5 herbaceous or suffrutescent perennials (Table 4.4). Approximately half of the site consisted of desert scrub dominated by *Larrea tridentata* and *Krameria erecta* and most of the remainder by *Ambrosia dumosa*. All were representative and common C_3 species of creosote bush associations, and the only succulent CAM species was a cholla (*Opuntia echinocarpa*), which

occurred at very low densities on the IBP validation site. The only species of C_4 shrubs, *Atriplex confertifolia*, fairly widespread in the MDR and many Great Basin plant communities (Caldwell 1985), was present in low numbers at Rock Valley. Also within Rock Valley, but not in the study areas, were two Great Basin associations, one at low elevations (940–1207 m) dominated by *Atriplex* spp. and the other at higher sites (1677–2264 m) dominated by sagebrush scrub and pinyon–juniper woodland (Beatley 1979).

Densities of perennials on the IBP validation site were sampled within 190 quadrats, each 2×25 m in size. The most abundant shrub species was *Ambrosia dumosa*, comprising nearly 30% of all shrubs with an estimated density in 1971 of 2531 individuals ha^{-1}, much higher than comparable densities sampled by Beatley (1976c). Density of *A. dumosa* was very constant for the seven plant associations on the validation site (Table 4.5). However, these shrubs are low, rarely exceeding 40 cm in height, and total biomass of white bursage was therefore low.

Six species of larger but less abundant shrubs actually contributed substantial biomass to the validation site. These were *Larrea tridentata*, *Lycium andersonii* and *L. pallidum*, *Krameria erecta*, *Grayia spinosa*, and *Ephedra nevadensis* (Mormon tea). *Krameria erecta* was second in relative abundance (16.2%, 1426 individuals ha^{-1}), followed by *G. spinosa* (12.2%, 1076 individuals ha^{-1}), but, unlike the other species, *Grayia* was notably more abundant in northern quadrats, where it constituted about one-fourth of all shrubs (Table 4.6). Creosote bush, the ecologically most important shrub, occurred throughout the validation site, with an aggregate relative density of 11% and 966 individuals ha^{-1}.

Similar relative densities were obtained in Rock Valley outside the validation site (Romney & Wallace 1980), but other published values of shrub density for Rock Valley were significantly higher, and may be in error (El-Ghonemy et al. 1980c). El-Ghonemy et al. reported that *Larrea* and *Ambrosia* were more abundant in Rock Valley than in nearby Mercury Valley, but earlier studies had reported higher bursage densities in Mercury

Table 4.4. *Perennial plant species occurring on the IBP validation site at Rock Valley. Growth habit, flowering periods, and elevational ranges (maximum) are those given by Beatley (1976) and apply only to distributions in central-southern Nevada.*

Species	Growth habit	Flowering period	Elevational range, m
Asteraceae			
Acamptopappus shockleyi	shrub	April–May	920–1770
Ambrosia dumosa	shrub	April–May	670–1380 (1530)
Encelia virginensis	shrub	April–July	1110–1250
Xylorhiza tortifolia	suffrutescent	April–June	700–1380
Cactaceae			
Opuntia echinocarpa	succulent shrub	May–June	700–1770
Chenopodiaceae			
Atriplex confertifolia	shrub	April–May	670–1900
Ceratoides lanata	shrub	April–June	950–2840
Grayia spinosa	shrub	April–May	920–1980
Ephedraceae			
Ephedra nevadensis	shrub	April–May	760–1830
Fabaceae			
Psorothamnus fremontii	shrub	April–May	700–1350
Krameriaceae			
Krameria erecta	shrub	April–June	700–1350
Lamiaceae			
Salazaria mexicana	shrub	April–June	700–1280
Malvaceae			
Sphaeralcea ambigua	herbaceous perennial	April–May	1010–1530
Nyctaginaceae			
Mirabilis pudica	herbaceous perennial	April–May	760–1990
Oleaceae			
Menodora spinescens	shrub	March–Apr	920–1830
Poaceae			
Erioneuron pulchellum	herbaceous perennial	April–May	980–1530 (1990)
Oryzopsis hymenoides	herbaceous perennial	April–June	730–2290
Rosaceae			
Coleogyne ramosissima	shrub	April–May	1220–1530 (1830)
Solanaceae			
Lycium andersonii	shrub	March–April	670–1530 (1820)
L. pallidum	shrub	March–April	700–1280
Zygophyllaceae			
Larrea tridentata	shrub	May–June	670–1380 (1590)

Valley, with up to 6400 individuals ha^{-1}, and even higher densities (9300) in Jackass Flats (Romney, Hale, Wallace, Lunt, Childress, Kaaz, Alexander, Kinnear, & Ackerman 1973). Total shrub densities greater than 18 000 ha^{-1} have been reported for Mojave desert scrub (Romney et al. 1973) as well as densities of individual species of 11 100 individuals ha^{-1} (*Acamptopappus shockleyi*). *Acamptopappus* tends to grow in groups within clumps of

other species and as individuals in the space between clumps (Wallace, Romney, & Kinnear 1980). Hunter (pers. comm.) in this area has measured bursage densities that commonly exceed 10 000 individuals ha^{-1}.

To estimate the initial population structure of scrub vegetation on the IBP validation site, frequency distributions of shrub size classes were determined for the ten most common species in

Table 4.5. *Mean densities (number of individuals ha⁻¹) of perennial plant species in vegetational subunits of the IBP primary study site at Rock Valley. Relative densities are given in parentheses. Data from El-Ghonemy et al. (1980d), with corrections.*

Species	Vegetational subunit						
	A	Ba	Bb	C	D	E	F
Ambrosia dumosa	2635 (30.0)	1098 (18.6)	2635 (36.4)	4026 (35.9)	2050 (15.9)	4758 (33.6)	3806 (25.2)
Grayia spinosa	2342 (26.7)	132 (2.2)	176 (2.4)	630 (5.6)	2196 (17.0)	3118 (22.0)	4978 (33.0)
Krameria erecta	878 (10.0)	1479 (25.1)	1318 (18.2)	878 (7.8)	1771 (13.7)	1098 (7.7)	1756 (11.7)
Larrea tridentata	732 (8.3)	1186 (20.1)	849 (11.7)	732 (6.5)	805 (6.2)	732 (5.2)	1610 (10.7)
Lycium pallidum	878 (10.0)	220 (3.7)	410 (5.7)	659 (5.9)	1244 (9.6)	878 (6.2)	1025 (6.8)
Ephedra nevadensis	439 (5.0)	659 (11.2)	805 (11.1)	1464 (13.1)	658 (5.1)	1171 (8.3)	644 (4.3)
Ceratoides lanata	439 (5.0)	15 (0.3)	88 (1.2)	1830 (16.3)	3806 (29.5)	1171 (8.3)	790 (5.2)
Lycium andersonii	293 (3.3)	952 (16.2)	732 (10.1)	322 (2.9)	366 (2.8)	483 (3.4)	205 (1.3)
Oryzopsis hymenoides	49 (0.6)	102 (1.7)	148 (2.0)	146 (1.3)	0 (0.0)	293 (2.1)	132 (0.9)
Xylorhiza tortifolia	0 (0.0)	29 (0.5)	37 (0.5)	37 (0.3)	0 (0.0)	293 (2.1)	76 (0.5)
Acamptopappus shockleyi	18 (0.2)	0 (0.0)	29 (0.4)	0 (0.0)	0 (0.0)	132 (0.9)	29 (0.2)
Psorothamnus fremontii	0 (0.0)	0 (0.0)	0 (0.0)	73 (0.7)	0 (0.0)	22 (0.1)	0 (0.0)
Coleogyne ramosissima	0 (0.0)	0 (0.0)	6 (0.1)	415 (3.7)	0 (0.0)	0 (0.0)	0 (0.0)
Salazaria mexicana	73 (0.8)	0 (0.0)	0 (0.0)	0 (0.0)	0 (0.0)	0 (0.0)	0 (0.0)
Opuntia echinocarpa	0 (0.0)	0 (0.0)	0 (0.0)	0 (0.0)	0 (0.0)	22 (0.1)	0 (0.0)
Encelia virginensis	0 (0.0)	17 (0.3)	0 (0.0)	0 (0.0)	0 (0.0)	0 (0.0)	0 (0.0)
Mirabilis pudica	0	0	3	0	0	0	0
Total	8776	5889	7236	11212	12896	14171	15051

Table 4.6. *Mean canopy areas of major shrub species in Rock Valley in 1971 and 1974 with associated estimates of total coverage by all perennial plants.*

Species	1971				1974				Cover 1971–4, %
	n	Mean canopy area, m²	Total cover, %	Relative cover, %	n	Mean area, m²	Total cover, %	Relative cover, %	
Ambrosia dumosa	1183	0.153	3.85	17.0	264	0.187	4.71	17.8	22.3
Atriplex confertifolia	38	0.300	0.34	1.5	–	–	0.34	1.3	–
Ceratoides lanata	234	0.059	0.25	1.1	60	0.093	0.41	1.6	45.4
Ephedra nevadensis	389	0.246	1.78	7.9	89	0.344	2.49	9.4	39.9
Grayia spinosa	589	0.116	1.25	5.5	128	0.202	2.18	8.2	74.4
Krameria erecta	731	0.278	3.95	17.5	172	0.261	3.72	14.1	−5.8
Larrea tridentata	518	0.409	3.99	17.6	145	0.470	4.59	17.3	15.0
Lycium andersonii	351	0.570	4.26	18.8	83	0.587	4.38	16.6	2.8
L. pallidum	227	0.517	2.76	12.2	64	0.646	3.45	13.0	25.0
Other species			0.18	1.8			0.18	0.7	–
Totals			22.61				26.45		17.0

1971 (El-Ghonemy, Wallace, & Romney 1980a). From the 190 sampling plots, 4282 individuals were randomly selected, identified, and measured for canopy width and height. Mean canopy dimensions were used in regression equations of dry weight on volume estimates (Wallace & Romney

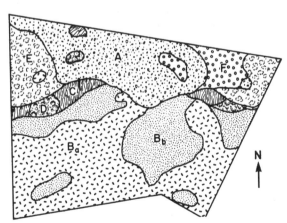

Figure 4.8 Distribution of six vegetation associations on the validation site in Rock Valley. A = *Ambrosia/Grayia*, B = *Ambrosia/Krameria* (Ba = *Krameria/Larrea*, Bb = *Ambrosia/Krameria*), C = *Ambrosia/Ceratoides*, D = *Ceratoides/Grayia/Ambrosia*, E = *Ambrosia/Grayia/ Ephedra*, F = *Grayia/Ambrosia*. Adapted from El-Ghonemy, Wallace, Romney, and Valentine (1980d).

1972) to calculate plant biomass size classes. In 1971, following good rains in 1969, each species appeared to be reproductively successful inasmuch as there were substantial numbers of young individuals, indicating recent recruitment (Fig. 4.8). When size-class frequency distributions were analyzed on the arithmetic basis, 7 species, including 5 of the most common ones, had a J-shaped distribution curve, with the greatest number of individuals appearing in the youngest size class. *Krameria erecta* and *Lycium andersonii* instead had unimodal, positively skewed distribution curves, whereas the fairly uncommon *Acamptopappus shockleyi* had an anomalous, discontinuous distribution curve (Fig. 4.9).

Canopy coverage and aboveground biomass calculations

Canopy coverage (m²) of Rock Valley perennials was calculated from density estimates and mean sizes of the plants that occurred in each of six vegetation zones at the Rock Valley validation site. Measurements of 18 species were made in winter, 1971, at the same time when plants were counted in quadrats. Mean cover of individuals within a species was estimated from two measurements of

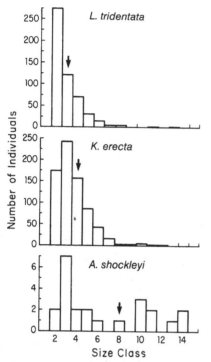

Figure 4.9 Frequency of individual shrubs by size class for *Larrea tridentata*, *Krameria erecta*, and *Acamptopappus shockleyi*. Size class intervals are based on the following biomass ranges: *L. tridentata*, 316 g., *K. erecta*, 64.8 g., and *A. shockleyi*, 11.2 g. From El-Ghonemy, Wallace, and Romney (1980a). Arrow indicates median.

shrub diameters (d_1 and d_2, m) along axes 90° apart:

$$A = \pi(d_1 d_2)/4$$

Total coverage was calculated for each zone, and site-wide coverage was estimated as the area-weighted mean of zone coverages. An overall mean size of perennials was computed from measurements of all plants, regardless of quadrat location.

In 1971 and 1972 shrub growth was modest, hence the 1972 and 1973 perennial coverages probably were not appreciably different from the 1971 estimates. However, during 1973 perennials experienced much greater growth than that observed during the two previous years. Consequently, the eight dominant shrubs were measured again in 1974 and mean areas were computed

as before. Mean sizes of the other perennials were not measured in 1974, but instead were conveniently assumed to equal those calculated in 1971, and for 1974 site-wide coverage was then estimated as previously described.

Mean canopy coverage for the validation site was estimated to be 22.6% for 1971 and 26.4% for 1974 (Table 4.6), for an increase of at least 17% due to the favorable growth conditions. The three largest canopies were those of *Lycium andersonii*, *L. pallidum*, and *Larrea tridentata*, and these shrubs accounted for approximately half of the coverage in 1971 and 1974, when just five species contributed about 80% of the total. Greatest increases in coverage from 1971 to 1974 were measured for the small shrubs, *Grayia spinosa*, *Ceratoides lanata*, and *Ephedra nevadensis*, having increases in mean cover of 74%, 45%, and 40%, respectively, and this improved their relative coverage in the plant communities. There were increases of about 25% in canopy coverage by *Ambrosia dumosa* and *Lycium pallidum*, a 15% increase by *Larrea tridentata*, but little change in canopy sizes of either *Lycium andersonii* or *Krameria erecta*. Under the favorable conditions of 1974, *A. dumosa* had the highest canopy coverage, barely surpassing *L. tridentata*; this was an expected result because bursage, a shoot-shedding perennial, forms vigorous new shoots following rains, whereas the evergreen creosote bush forms very short increments at the tips of existing shoots.

For the IBP analysis aboveground shrub dry biomass values (*W*, kg) were estimated from shrub volumes using equations given by Romney et al. (1973), but different equations were needed to make corrections in old data and to incorporate additional measurements. Here all regression equations have forced zero intercepts and are of the form $W = aV$. Shrubs collected for dimensional analysis were composed of both live and dead tissues, and the proportion of dead materials varied from species to species (Table 4.7); therefore, biomass was partitioned into live and dead components, and mean fractions of live tissue were computed to provide values for the regression equations. Values of *a*, the coefficient determined from regression analysis, are given in Table 4.8.

Table 4.7. *Coefficients for estimating total shrub biomass (living and dead) from shrub volumes, using an equation of W = aV, where W = dry biomass (kg), V = shrub volume (m^3), and a is the coefficient determined from regression analyses. n = shrub sample size.*

Species	n	Living/total biomass tissue	Coefficient (a)	r^2
Ambrosia dumosa	70	0.335	2.5	0.92
Atriplex confertifolia	33	0.451	4.7	0.90
Ceratoides lanata	66	0.337	3.2	0.88
Ephedra nevadensis	8	0.706	1.4	0.96
Grayia spinosa	54	0.521	2.3	0.92
Krameria erecta	43	0.672	2.0	0.90
Larrea tridentata	46	0.314	1.3	0.94
Lycium andersonii	62	0.78	2.2	0.92
L. pallidum	35	0.739	1.2	0.92

Table 4.8. *Estimated aboveground biomass of perennial species on the Rock Valley site in 1971 and 1974. Relative biomass percents are given in parentheses.*

Species	1971 Aboveground biomass, kg ha^{-1}		1974 Aboveground biomass, kg ha^{-1}	
	Total	Live	Total	Live
Ambrosia dumosa	306.6 (16.1)	102.7 (9.9)	410.8 (16.6)	137.6 (10.4)
Atriplex confertifolia	79.6 (4.2)	35.9 (3.4)	79.6[a] (3.2)	35.9[a] (2.7)
Ceratoides lanata	31.5 (1.7)	10.6 (1.0)	64.3 (2.6)	21.7 (1.6)
Ephedra nevadensis	126.0 (6.6)	88.9 (8.6)	195.3 (7.9)	137.9 (10.4)
Grayia spinosa	143.5 (7.6)	74.7 (7.2)	289.8 (11.8)	151.0 (11.4)
Krameria erecta	172.2 (9.1)	115.7 (11.1)	160.1 (6.5)	107.6 (8.1)
Larrea tridentata	318.9 (16.8)	100.1 (9.6)	417.8 (16.9)	131.2 (9.9)
Lycium andersonii	492.7 (25.9)	348.8 (33.6)	551.8 (22.4)	390.7 (29.5)
L. pallidum	199.1 (10.5)	147.1 (14.1)	266.8 (10.8)	197.2 (14.9)
Other species[b]	28.8 (1.5)	14.2 (1.4)	28.9 (1.2)	13.8 (1.0)
Totals	1899	1039	2465	1325

[a]Not measured in 1974; 1971 values were used.
[b]*Acamptopappus shockleyi, Coleogyne ramosissima, Menodora spinescens, Oryzopsis hymenoides, Psorothamnus fremontii, Xylorhiza tortifolia.*

Mean volumes (V, m^3) of shrubs were computed from measurements of canopy height (h, m) and two diameters (d_1 and d_2, m):

$$V = \pi h (d_1\, d_2)/4$$

In practice, mean shrub volumes were initially converted to mean biomass of aboveground tissues using the equation from Table 4.7. Living and dead biomass was calculated for each species from the mean ratio of these tissues. This procedure was done individually for plants from each of the six vegetation zones. Values were then multiplied by zone densities to yield aboveground biomass of living and dead tissues for each zone, as well as for the validation site as a whole. Only site-wide values are reported here.

To assess how much growth occurred following abundant fall–winter rains of 1973, mean shrub volumes were measured again at the beginning of the 1974 growing season. The same regression equations were used in 1971 and 1974 for estimating individual shrub biomass, and the values for ratios of dead/live tissues were assumed to have remained unchanged. Whereas dimensional measurements of shrub canopies were not made after the 1971 or 1972 growing seasons, there appeared to be little significant change in shrub volume or biomass over those two dry years. For early 1971 aboveground biomass of live shrub tissues was estimated to be 1039 kg ha^{-1}, and there was approximately 800 kg ha^{-1} of standing dry biomass (Table 4.8). *Lycium andersonii* was the most important species, having more than one-third of the living biomass, and *L. pallidum* was second in importance with 14%; these two species also had the lowest percentages of dead materials. Two-thirds of total aboveground biomass of *Ambrosia dumosa* and *Larrea tridentata* were dead stems, comprising about half of the dead wood pool.

In early 1974 the estimated living aboveground biomass of perennials at Rock Valley was 1325 kg ha^{-1}, an increase of 27.5% from the 1971 value (Table 4.8). Dead wood comprised an additional 1140 kg ha^{-1}, an increase of 32.5% in this biomass pool, presumably because the region had been fairly dry for several years. Overall, standing dead and living aboveground tissues increased by 30% between the two sample dates.

From 1971 quadrat counts of standing dead shrubs, the estimated density of the 15 species was about 994 ha^{-1}. The most abundant frequencies (individuals ha^{-1}) were those of *Ambrosia dumosa* (312), *Ephedra nevadensis* (222), *Grayia spinosa* (157), and *Larrea tridentata* (71), and the other species had fewer than 35 ha^{-1}. Relative abundance for dead specimens of *Ephedra* (0.22) was considerably greater than that of live plants (0.08) and an apparent anomaly among the species, although more recent observations have suggested that pocket gopher (*Thomomys umbrinus*) is an important cause of mortality in *E. nevadensis* (Hunter, pers. comm.). No dimensional analyses were made of dead shrubs; consequently, to approximate the mass of dead wood there one must assume that dead shrubs had the same mean biomass as live ones. Under that assumption, for

1971 the aggregate mass of all dead perennials was about 175 kg ha^{-1}. However, the mean biomass of dead plants probably was substantially lower than that of living ones, perhaps only 50%. Bamberg, Wallace, Romney, and Hunter (1980) estimated that standing dead biomass was about one-fourth the biomass of living tissue.

Root system architecture and belowground biomass calculations

At Rock Valley root system depth, as well as overall vegetation structure, is strongly influenced by soil structure, especially the presence of silica-lime and cemented hardpans that are commonly found in soil profiles at depths of 28–70 cm (Chapter 2; Romney et al. 1973). Unfortunately, in these soils root systems are very difficult to study. Instead Wallace and Romney (1972) excavated root systems from a sandywash area of Rock Valley, where relatively deep and uniform sediments were present. Most shrub species were characterized by a root system with deeply penetrating taproots, to 1–2 m, and relatively massive lateral roots that extended beyond the perimeter of the aboveground canopy. *Larrea*, *Ambrosia*, *Ephedra*, and *Lycium andersonii* share this pattern of root growth (Fig. 4.10a–d). Where caliche layers prevented any vertical penetration of roots, lateral root biomass increased. Wallace and Romney observed a pronounced absence of visible auxiliary roots on those shrubs. Root systems of *Grayia* and *Krameria* differed from those of typical shrubs in their tendency to develop a less massive taproot and a diminished lateral system but to have many fine roots and active root hairs (Fig. 4.10c).

The relationship between aboveground stem biomass and root biomass was examined for nine shrub species in Rock Valley in 1972 (Wallace, Bamberg, & Cha 1974; Wallace, Romney, & Cha 1980). Laborious techniques were used to excavate 1 to 3 m^3 of soil for inspection and removal of roots. Shrubs were separated into aboveground and belowground tissues, and all materials were oven-dried and weighed. Soil removed from around shrubs was not screened, but Wallace et

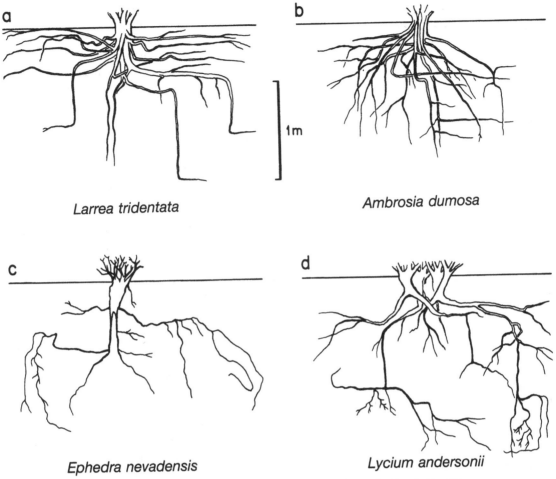

Figure 4.10 Rooting morphologies of Mojave Desert shrubs on the Nevada Test Site. a. *Larrea tridentata*. b. *Ambrosia dumosa*. c. *Ephedra nevadensis*. d. *Lycium andersonii*. From Wallace and Romney (1972).

al. (1974) increased all root dry mass by 15% to compensate for loss of fine roots before regressing adjusted root masses on stem masses. The resulting equations did not have zero intercepts, and so they are only applicable over ranges of stem masses that were analyzed by these authors. The regression equations were later revised when Wallace, Romney, and Cha (1980) estimated the very fine roots using a series of one-liter soil samples, collected in 1976 and floated with conventional salts. They added correction coefficients for fine roots in organic debris (1.32) and roots extending

into open spaces beyond the shrub zones samples (1.23). The aggregate correction coefficient therefore was 1.62 (not 1.73 as stated by Wallace et al. 1980).

Root/shoot ratios for the nine species ranged from 1.65 for *Lycium pallidum* to 0.44 for *Atriplex confertifolia*, with a mean specific value of about 0.9 (Table 4.9). Overall these values compare reasonably well with published data on root/shoot ratios from other warm desert communities (see review by Rundel & Nobel 1991). Root/shoot ratios between these studies differed considerably,

Table 4.9. *Estimated root/shoot ratios and community biomass of roots (kg ha^{-1}) in Rock Valley. See text for discussion of methods of calculations.*

Species	Root/shoot ratio	Biomass, kg ha^{-1}
Acamptopappus shockleyi	0.56	1
Ambrosia dumosa	1.15	118
Atriplex confertifolia	0.44	16
Ceratoides lanata	0.90[a]	10
Ephedra nevadensis	0.83	74
Grayia spinosa	0.72	65
Krameria erecta	0.79	70
Larrea tridentata	1.24	124
Lycium andersonii	0.83	252
L. pallidum	1.65	243
Total		963

[a]estimated value

especially for *Larrea tridentata*, suggesting possibilities of sampling bias.

Community-level estimates of root biomass were calculated using the corrected root/shoot ratios published for each species (Wallace, Romney, & Cha 1980) plus data on aboveground living biomass for 1971, as estimated in Table 4.9. For the Rock Valley validation site we calculated the estimated belowground biomass to be 963 kg ha^{-1} (Table 4.9). Because we corrected values of aboveground biomass, presented in this chapter, and used only living biomass data, this estimate of belowground biomass is approximately one-third the value given by Wallace et al. (1980).

Bamberg, Kaaz, Maza, and Turner (1974) reported on other root studies derived from soil samples taken to depths of 30 cm around the bases of the four species with greatest biomass, *Larrea*, *Lycium andersonii*, *Ambrosia*, and *Krameria*, as well as in spaces between shrubs. Large roots were removed from the soil by hand, whereas small roots were separated by screening and flotation in MgSO$_4$ solution. Beneath shrub canopies, greatest root biomass consistently occurred within the top 10 cm of the soil profile, and the smallest amounts of root materials were from the top 10 cm

of soil in spaces between shrubs. Estimations of root materials exceeded 5000 kg ha^{-1} (including an undetermined amount of dead roots) for just these four species – more than five times the estimate given in Table 4.9. Their higher estimate was attributed to presence of dead materials in samples and much increased collecting efficiency.

Net primary production

Field measurements of aboveground net primary production by shrubs were taken at Rock Valley for each year from 1971 to 1976. Six species (*Ambrosia dumosa*, *Ephedra nevadensis*, *Krameria erecta*, *Larrea tridentata*, *Lycium andersonii*, and *L. pallidum*), composing 86% of the total live aboveground shoot biomass on the validation site, were sampled during the spring of all six years. To access summer growth a second sampling was also conducted during early fall of 1971, 1972, and 1976. *Grayia spinosa* was sampled between 1971 and 1974, *Ceratoides lanata* between 1971 and 1973, and *Atriplex confertifolia* in 1972 and 1973. Earlier data on net primary production by shrubs in Rock Valley are available for 1966–8 (Wallace & Romney 1972).

Entire shrubs were harvested during peak growth in spring, 1971, and separated into old live stems, dead stems, new leaves, new stems, flowers, and fruits. All materials were oven-dried, and proportions of the various tissues were computed. From 1971 data it was concluded that the ratio of new plant material to old live stem mass was a suitable measure of production, and that it was not necessary to collect entire shrubs in order to estimate these proportions. Hence, in subsequent years workers sampled only whole branch portions of 15–20 shrubs of each species, and those samples were subdivided, dried, and weighed to estimate production ratios of new to old growth. Sampling was done twice each growing season, first at the time of peak leaf and flower development and second when fruit showed maximum development. Sampling dates differed from species to species in order to adjust to the phenology of each taxon. No measurement of secondary (radial) growth was made for old woody tissues, and thus

Table 4.10. *Estimated net production (kg ha⁻¹) by Rock Valley shrubs during the spring growing season. Relative production by each species is shown in parentheses.*

Species	1971	1972	1973	1974	1975	1976
Ambrosia dumosa	21 (11.5)	35 (17.0)	127 (18.6)	39 (17.7)	25 (11.9)	55 (14.5)
Atriplex confertifolia	10[a] (5.5)	11 (5.3)	38 (5.6)	11[a]	8[a]	17[a]
Ceratoides lanata	5 (2.7)	4 (1.9)	10 (1.5)	7[a]	6[a]	17[a]
Ephedra nevadensis	15 (8.2)	17 (8.3)	60 (8.3)	<1	51 (24.3)	31 (8.2)
Grayia spinosa	16 (8.7)	22 (10.7)	50 (7.3)	47 (21.3)	28[a]	66 (17.4)
Krameria erecta	52 (23.4)	52 (25.2)	105 (15.4)	34	32	37
Larrea tridentata	17 (5.3)	16 (7.8)	42 (6.2)	17 (7.7)	14[a] (6.7)	29 (7.6)
Lycium andersonii	29 (15.8)	31 (15.0)	138 (20.2)	33	23	66
L. pallidum	15 (8.2)	14 (6.8)	102 (15.0)	30	20	58
Other perennial species	3[a] (1.6)	4[a] (1.9)	10[a] (1.5)	2[a]	3[a]	4[a]
Totals	183	206	682	220	210	380

[a]Estimates based on indirect calculations.

no data were available to make direct estimates of this production component.

Other special problems were encountered from time to time. To estimate new leaf and stem production by the evergreen *L. tridentata*, workers had to discriminate in the field between old and new leaves. Inexperienced collectors were not always able to make this distinction, and in 1975 no reliable production coefficients were obtained for stem and leaf production.

From 1971 to 1973 species-specific spring net production was estimated by multiplying 1971 aboveground live biomass by ratios of new to old tissues of harvested samples. The same procedure was followed in 1974–6 except that aboveground live biomass estimates were based on 1974 measurements. New growth during the summer and fall of 1971, 1972, and 1976 was also estimated in the same manner. Production by species that were not sampled, for example, *Acamptopappus shockleyi, Coleogyne ramosissima,* and *Psorothamnus arborescens* was calculated from the arithmetic mean of the ratio of new to old tissues for shrub species that were sampled; however, overall contributions of such minor species to net primary production of the community was small. Missing data for *Atriplex confertifolia, Ceratoides lanata, Grayia spinosa,* and *Larrea tridentata*

were estimated using a log-linear model combining effects of growth allocation for those species, as determined from other years, and the year-to-year pattern of production for all species that were measured.

Estimated net primary production by shrubs at Rock Valley during spring growing seasons ranged from 682 kg ha⁻¹ to 183 kg ha⁻¹ (Table 4.10). During four relatively arid years at the site (1971, 1972, 1974, and 1975), shrub production was very similar, but the relative importance of productivity by individual shrub species was not constant from year to year. *Krameria erecta* was the most important producer in 1971 and 1972, having more than one quarter of the total new biomass. In the wet spring of 1973 its productivity doubled, but overall contribution to the community was diminished because other species were more productive. During the next two dry springs its production was very low, along with that of the other species in the community, but in the relatively favorable growing conditions of spring, 1976, *Krameria* retained low productivity while other species flourished. *Ephedra nevadensis* provided 24% of total shrub production in 1975, but less than 1% in the previous year. *Grayia spinosa* was a dominant shrub producer in 1974 and 1976. Although net productivity varied annually for

Table 4.11. *Estimated dry mass net production*
(kg ha⁻¹) by Rock Valley perennials in late
summer and early fall.

Species	1971	1972	1976
Ambrosia dumosa	12	6	39
Atriplex confertifolia	2	*	*
Ceratoides lanata	1	1	*
Krameria erecta	8	*	19
Larrea tridentata	14	1	9
Lycium andersonii	20	10	15
L. pallidum	11	5	17
Totals	68	23	99

*No measurements were made.

Ambrosia dumosa and *Larrea tridentata*, their
relative production was for the same period quite
constant for the community.

Shrubs may renew growth during late summer
and early fall, if they receive rains. During
summer drought the validation site received rainfall sufficient to stimulate growth in July and
August, 1968 (55 mm), August, 1971 (33 mm),
August and September, 1972 (28 mm), and September, 1976 (51 mm). Rainfall in July, 1974, measuring 26 mm did not stimulate observable production in any species. Shrub growth in early fall
was largely expressed in new leaves and some
young shoot elongation. Estimated net production
by perennials during the late summer and early

fall is given in Table 4.11. Neither *Grayia* nor
Ephedra showed any summer growth for those
years, but other species showed appreciable
amounts of vegetative growth. Late season production was 37% (1971), 11% (1972), and 30%
(1976) of estimated spring production.

Equations for predicting net annual aboveground primary production

Table 4.12 summarizes net annual aboveground
primary production (NAAP) for springtime 1966–
76 together with four temporal measures of precipitation. Production values were not significantly
correlated with rainfall in each calendar year ($r =$
0.44) or with rainfall within that year and past
calendar years ($r = 0.55$). Shrub production was
significantly correlated with hydrologic year
(September to August) precipitation ($r = 0.95$)
and with that occurring between September and
March ($r = 0.90$; Fig. 4.11). Earlier attempts to
model shrub production at Rock Valley did not
consider the importance of hydrologic year calculations (Webb, Lauenroth, Szarek, & Kinerson
1983). The equation for predicting NAAP (g m⁻²)
from September-to-August rainfall is:

$$NAAP = 0.31 (PPT) - 8.35$$

This equation extrapolates to a threshold of about
27 mm precipitation to support net production by
desert shrubs, and it is very close to the 25 mm

Table 4.12. *Estimated net annual aboveground production (NAAP) by perennials during the spring*
growing season in Rock Valley and associated rainfall totals (mm).

Year	NAAP, g m⁻¹ yr⁻¹	Calendar year pre-cipitation, mm	Calendar year + previous year precipitation, mm	Hydrologic year precipitation, mm	September–March precipitation, mm
1966	49.0	63	275	161	139
1967	31.0	182	245	148	47
1968	43.0	122	304	152	88
1971	18.3	104	226	100	47
1972	20.6	103	207	75	41
1973	68.2	208	311	250	218
1974	22.0	130	338	98	70
1975	21.0	63	193	113	93
1976	38.0	223	286	146	110

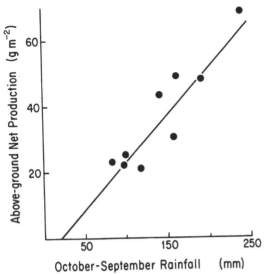

Figure 4.11 Aboveground net primary production of shrub species in Rock Valley in relation to hydrologic year precipitation from October to September. Data are for 1966–1968 and 1971–1976. From Turner and Randall (1989).

level postulated from empirical evidence by Beatley (1974b) for annual plant germination and growth.

Webb, Szarek, Lauenroth, Kinerson, & Smith (1978) used four years of data from Rock Valley and from two sites in Arizona to derive a simple linear model for NAAP in desert ecosystems, based on calculations from values of actual evapotranspiration (AET):

$$NAAP = 0.30 \, (AET) - 11.2$$

In this equation, AET for a given year, which unfortunately was not measured at Rock Valley, is assumed to be equal to precipitation. However, Le Houerou (1984) has observed that runoff in Algeria and Tunisia may be 2–10% of annual precipitation on areas of 1 to 100 km², and thus there is some error in equating AET and rainfall. This equation underestimates shrub production by 25 to 84% in 1966, 1968, 1973, and 1975, and overestimates it by 26 to 47% in 1967, 1974, and 1976. Webb et al. (1978) suggested that water use in both past and current years may provide the best correlation with NAAP for grasslands and desert

ecosytems. Our data (Fig. 4.10 and Table 4.12) clearly point out the strong correlation of NAAP with hydrologic year precipitation and the lack of significant correlation with two-year totals of precipitation.

Bamberg, Vollmer, Kleinkopf, and Ackerman (1976) estimated production for five species in 1972 and 1973 by contrasting data on NAAP using direct harvest methods with data obtained from summing CO_2 uptake of these shrubs on a monthly basis. Values used by those authors, except for *Krameria erecta*, differ from those in Table 4.10, given the revisions from our new analyses, but the recalculated values for NAAP do not affect the conclusions presented by Bamberg and coworkers. Aboveground net production by perennials for 1973, when there were excellent growing conditions, increased 350% relative to that in 1972. Measurements of CO_2 uptake in the two years exhibited roughly the same relationship, and net production estimate based on gas exchange analyses were 2–3 times greater than those obtained from harvest methods. These differences can be mostly ascribed to limitations in harvest procedures: no measurements of new root formation, no measurements of secondary stem and root growth, and simple harvest procedures done at only two times during the growing season.

The CO_2 uptake measurements by Bamberg et al. (1976) help to explain why the increased productivity between 1972 and 1973 was not uniformly manifested among all species. Table 4.10 shows that net production by *Ambrosia dumosa*, *Lycium andersonii*, and *L. pallidum* were, respectively, 3.6, 4.5, and 7.3 times greater in 1973 than in 1972; corresponding values for *K. erecta* and *Larrea tridentata* were 2.0 and 2.6. Cooler temperatures associated with increased rainfall in early 1973 may have been a primary factor for those differences, because the two species of *Lycium* and, to a lesser degree, *A. dumosa*, grew best in cool, moist conditions of early spring (Bamberg, Kleinkopf, Wallace, & Vollmer 1975). *Krameria* and *Larrea* are adapted to grow under hotter, drier conditions of summer and therefore were not as strongly affected as the other three species.

Table 4.13. *Estimated community rates of litterfall (kg ha⁻¹ yr⁻¹) for tissue components and total litter for six species of shrubs at Rock Valley. Estimates for 1975 pertain to data for March–December only. Shrub densities are revisions from Ackerman et al. (1975). Data from Strojan, Turner, and Castetter (1979).*

Species	Shrub density individuals ha⁻¹	Litterfall (kg ha⁻¹ yr⁻¹)					Total litterfall
		Year	Leaves	Stems	Flowers	Fruits	
Ambrosia dumosa	2498	1975	3.4	3.8	0.1	0.2	7.5
		1976	14.1	3.6	12.0	23.8	53.5
		1977	27.2	2.3	—0.4—		29.9
Ephedra nevadensis	734	1975	–	14.3	0	6.9	21.2
		1976	–	14.2	0	0.3	14.5
Krameria erecta	1404	1975	3.8	4.2	0.2	0	8.2
		1976	15.8	5.5	3.5	0.4	25.5
Larrea tridentata	953	1975	16.0	9.0	0.2	0.2	25.4
		1976	41.5	18.2	5.2	6.5	71.4
		1977	51.9	18.5	—3.4—		73.8
Lycium andersonii	729	1975	18.9	6.8	2.4	0.1	28.2
		1976	57.2	4.7	2.0	0.5	64.4
Lycium pallidum	511	1975	21.6	4.2	0.5	0.2	26.5
		1976	63.1	3.2	16.2	6.0	88.5
		1977	31.4	1.4	—0.1—		32.9
Total – six species		1975	63.7	42.3	3.4	7.6	1170
		1976	191.7	49.4	38.9	37.5	317.5
Estimated total all shrub species		1975					143
		1976					388

Litterfall

Whereas biomass and primary production processes were the major focus of shrub studies in Rock Valley, it is also important to understand the dynamics of litterfall. Quantitative studies of litterfall from six dominant shrub species were conducted between March, 1975 and December, 1977, as part of the IBP investigations (Strojan, Turner, & Castetter 1979). Litter traps, consisting of 0.91 m diameter cylinders with 6.4 mm mesh hardware cloth, were placed around individual shrubs and closed at the bottom with muslin cloth. Litter was collected at two- or four-week intervals in 1975 and 1976 from 58 individual shrubs of the six species, and in 1977 from 30 shrubs representing three species. Litter was oven-dried at 70°C and hand sorted into categories of leaves, stems, flowers, and fruits. Dry weights of litter collected between sample data were prorated appropriately to provide an estimate of litterfall for each calendar month.

Total litterfall and litterfall for each species varied greatly from year to year (Table 4.13). Seasonal patterns of litterfall were evident for all shrubs combined as well as among the six species that were studied (Strojan et al. 1979). In January and February, 1976, months not sampled in the previous year, litterfall was only 5% of the annual total, but a definite, small peak occurred at this time as old litter was dislodged from shrub canopies by heavy rains (February, 1976). Subsequent peaks of litterfall occurred in summer during drought and again in November–December at the end of fall. Similar but less distinctive peaks of litterfall were observed in 1977.

Lycium pallidum contributed the greatest amount to litterfall, accounting for 35% of all samples in 1975 and 44% in 1976, and *L. andersonii* contributed 25% and 22%, respectively

(Table 4.13). Although *Ambrosia dumosa* was the most abundant shrub species, its litterfall was only 2% of the total in 1975 and 5% in 1976, because canopy size was small and biomass was comparatively low.

Leaves comprised most litterfall, with 58% of the total in 1975 and 66% in 1976. Stems were second in importance, contributing 33% and 14%, respectively, whereas flowers and fruits combined were 9% and 20% in successive years.

Year-to-year variation in litter distribution among individual categories was evident in data for *A. dumosa* (Fig. 4.12). In this species, unusually good conditions for flowering and fruiting in 1976 produced litterfall rates for these organs 120 times those present in the previous year. Reproductive tissues formed 4.0% and 1.3% of litterfall for this species in 1975 and 1977, respectively, but 66.9% in 1976 (Table 4.13). For *Larrea tridentata*, annual variation was less pronounced; leaf tissues comprised 62.9, 58.1, and 70.3% of total litterfall in 1975, 1976, and 1977, respectively. *Larrea tridentata* and *Lycium pallidum* had a large relative proportion of reproductive tissues in 1976 litterfall as compared with the other two years.

Community rates of litterfall for Rock Valley have been estimated from data on individual shrubs in Table 4.13. These six shrub species comprised about 82% of the biomass of perennial plants at the site (Strojan et al. 1979). If the remaining species at the site produced litter at the same rate per shrub, an additional 26 kg ha^{-1} of litterfall would be added to 1975 totals and 70 kg ha^{-1} to 1976 totals. Thus, we estimate that total shrub litterfall from shrubs was 143 kg ha^{-1} in 1975 and 388 kg ha^{-1} in 1976.

Litterfall estimates for Rock Valley appear to follow rates of net primary production of aboveground tissues from the previous growing season. In 1975 net clipping weight production for the six shrub species discussed here was 145 kg ha^{-1} as compared with 117 kg ha^{-1} for litterfall (Strojan et al. 1979). For 1976 the comparable values were 318 kg ha^{-1} for litterfall. In a steady-state system, one expects such a balance, at least over multiple years. One source of loss in net primary production before reaching litterfall is herbivory. Chew and Chew (1970) estimated that small mammals (i.e., rodents and lagomorphs) consumed only about 2% of net primary production in the Sonoran Desert. Mispagel (1978) calculated that a major insect defoliator of *L. tridentata*, the grasshopper *Bootettix argentatus*, consumed 0.8–1.9% of annual leaf biomass. Combining all other herbivores on all shrub species at Rock Valley, it is doubtful that total herbivory can exceed 5% of net primary production. Thus, we estimate that greater than 95% of net primary production cycles into the litter compartment of the Rock Valley ecosystem.

SHRUB PHENOLOGY

General observations for the Nevada Test Site

Much qualitative phenological data were collected at localities on NTS of Mojave Desert shrubs. Wallace and Romney (1972) began studies (1968–70) at comparable stations in Frenchman Flat (975 m), Mercury Valley (1100 m), and Rock Valley (1020 m). Their work identified six phenophases:

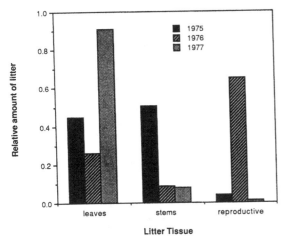

Figure 4.12 Relative distribution of litter categories by *Ambrosia dumosa* in Rock Valley. After Strojan, Turner, and Castetter (1979).

leaf buds, leaves, flowers, fruits, leaf fall, and dormancy, and they defined dates of initiation and the period of duration for these phases. Their data clearly identified the variable phenological responses of growth between years for the same species and between species, as shown for the most common perennials at Rock Valley (Figs. 4.13–4.14). *Atriplex confertifolia* and *Krameria erecta* were notable for having continued growth activity throughout the summer, when most shrubs were dormant in typical dry years, for example, 1969 and 1970. *Krameria* balanced its late growth with a start of vegetative growth one to two months later than that of other species.

Wallace and Romney also documented the effect that heavy July and August rains had in breaking shrub dormancy and thus allowing a second season of growth in 1968.

Ackerman, Romney, Wallace, and Kinnear (1980) provided a similar but broader analysis of phenological data for 1971–3 on NTS at the same three stations described above plus three in Yucca Flat (1200–1300 m) and a sagebrush station at Pahute Mesa (1720 m). Overall, this study provided phenological data on 16 species for the nine stations and interesting comparisons of growth response along the Mojave Desert–Great Basin elevational gradient. These data emphasized the

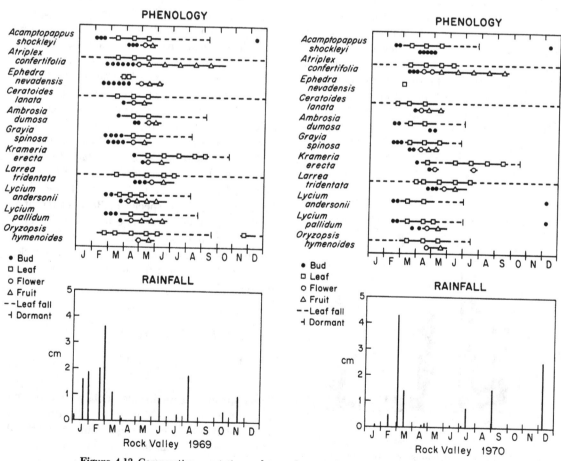

Figure 4.13 Comparative vegetative and reproductive phenology of shrubs and bunchgrasses in Rock Valley during 1969 and 1970. From Wallace and Romney (1972).

ROCK VALLEY

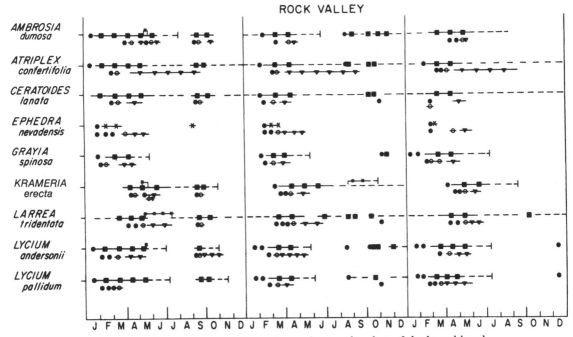

Figure 4.14 Comparative vegetative and reproductive phenology of shrubs and bunchgrasses in Rock Valley for 1970, 1971, and 1972 (left to right). ● = buds, ○ = flowers, □ = leaf production, △ = fruit, --- = leaf fall. From Ackerman, Romney, Wallace, and Kinnear (1980).

PAHUTE MESA

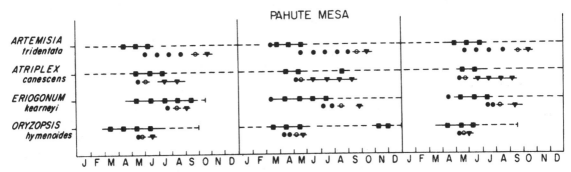

Figure 4.15 Comparative vegetative and reproductive phenology of shrubs and bunchgrasses at Pahute Mesa for 1970, 1971, and 1972 (left to right). ● = buds, ○ = flowers, □ = leaf production, △ = fruit, --- = leaf fall. From Ackerman, Romney, Wallace, and Kinnear (1980).

variations in phenological timing between species and between years for a single species. Using multiple sites also allowed Ackerman and co-workers to analyze the significance of temperature and precipitation gradients as influenced by site elevation. As in 1968, summer rains were sufficient in 1971 and 1972 to stimulate a second season of growth in Rock Valley (Fig. 4.13). Cooler temperatures and higher precipitation levels at Pahute Mesa permitted active growth of the perennial of sagebrush desert scrub into the summer months (Fig. 4.15).

Whereas the studies from 1968 to 1973 related the timing of seasonal changes in phenology to abiotic variables of rainfall, air temperature, and soil temperature, discussions of phenology were largely qualitative, and no attempt was made to develop a predictive formulation from abiotic variables. Likewise, Beatley's (1974b) model for explaining perennial associations included soil temperatures only in so far as they influenced germination of annual plants (Chapter 5). The particular phenological timing of shrubs was not part of that model, and no attempt was made to discriminate between phenologies of different species.

Data analysis for Rock Valley

Turner and Randall (1987) reviewed nine years (1968–76) of phenological data from Rock Valley to develop a more quantitative approach to the growth patterns of desert shrubs. They investigated annual variations in initiation dates of three important phenophases, appearance of leaves, appearance of flowers, and development of fruits, and restricted their analysis to responses of the seven species that comprised 95% of the shrub biomass at the IBP validation site. Methods of data collection paralleled those used in the earlier studies, and sets of 15–20 individuals of each species were marked and followed. Observations began in mid-January and continued weekly through the growing season. Initiation for each phenophase was noted as the first date for appearance of new leaves, flowers, or mature fruits among any of the individual plants in the study plots. They did not consider subsequent periods of leafing or flowering late in a season, nor did they include observations for onset of dormancy, both of which were discussed by Ackerman et al. (1980).

The variability in dates of first leaf initiation was notable between years for each species (Table 4.14). *Larrea tridentata* showed the least amount of variation, occurring each year in March, but leaf initiation of *Ambrosia dumosa* varied by eight weeks. *Krameria erecta* consistently formed new leaves 4–6 weeks after leaf initiation of the other species. The two species of *Lycium* were notable in having similar timing for leaf initiation every

year. Leaves were produced by all species in all years, although they were very ephemeral in the 'leafless' gymnosperm, *Ephedra nevadensis*.

Patterns of flower initiation showed the same type of variation between years as did leaf initiation (Table 4.14), although flowers were not produced for all species in all years. Most shrubs initiated flowers during a 4–6 week period, but *Ephedra* initiated its reproductive structures during a 10-week interval. Flowering of *Krameria* and *Larrea* was notably later than that for other species. Absence of flowers for three species in 1970 and *Ambrosia* in 1975 was attributed to occurrence of subfreezing temperatures in early spring of those years (Ackerman & Bamberg 1974; Turner & Randall 1987).

In some years the shrubs flowered but produced no fruits (Tables 4.15–4.16). Data on fruits showed similar patterns of variation as those discussed above for flower initiation.

Data presented in Tables 4.15–4.16 can be used for statistical analysis to determine mean dates of initiation for these three phenophases for the seven Rock Valley perennial species (Table 4.17). The mean dates are standard errors of the means based on Julian days, rounded to the nearest integer. The mean interval between first appearance of leaves and the first flowers was less than three weeks in *Grayia spinosa* and about four to six weeks in *Krameria* and *Lycium*. Average time between these events for *Ambrosia* and *Larrea* was eight weeks. In most species, leaf and flower production overlapped during the spring months (Figs. 4.13 and 4.14), and ripe fruit was first observed about two weeks after flower initiation for all species. Relationship between dates of first leafing and flowering for six species (excluding *Ephedra*) are shown in Figure 4.16. This is a graphical representation of phenological data that clearly separates the early-growth species, i.e., the two species of *Lycium* and *Grayia*, from the late-growth species, *Larrea* and *Krameria*. *Ambrosia dumosa* forms leaves early but flowers late.

At Rock Valley unusual variation in phenological timing has been accounted for by specific abiotic conditions. The cold temperatures in spring, 1969,

Table 4.14. *First dates of leafing by six species of shrubs in Rock Valley.*

Year	Ambrosia dumosa	Grayia spinosa	Krameria erecta	Larrea tridentata	Lycium andersonii	Lycium pallidum
1968	February 13	February 26	March 29	March 3	February 13	February 3
1969	March 27	March 20	April 20	March 22	February 14	February 14
1970	March 3	March 10	April 7	March 23	February 16	February 16
1971	February 1	February 2	March 29	March 6	January 25	January 25
1972	February 17	February 18	March 14	March 1	February 16	February 18
1973	February 28	February 15	April 18	March 12	February 26	February 20
1974	February 25	February 25	April 2	March 18	February 15	February 4
1975	February 27	no data	April 29	March 18	January 24	January 24
1976	February 18	March 4	April 9	March 26	March 4	March 4

Table 4.15. *First dates of flowering by seven species of shrubs in Rock Valley.*

Year	Ambrosia dumosa	Ephedra nevadensis	Grayia spinosa	Krameria erecta	Larrea tridentata	Lycium andersonii	Lycium pallidum
1968	April 26	February 9	March 1	April 29	May 6	March 8	March 18
1969	May 9	April 25	April 2	May 9	May 17	April 2	April 2
1970	none	none	March 31	May 11	May 18	none	April 13
1971	April 30	March 21	February 27	April 28	May 12	March 12	March 12
1972	April 16	March 18	March 4	April 13	April 19	March 9	March 7
1973	May 14	April 18	February 26	May 14	May 14	April 3	March 13
1974	May 2	April 25	March 18	May 10	May 10	April 2	March 26
1975	none	April 29	April 9	May 21	May 28	April 13	March 18
1976	May 7	April 9	March 26	May 21	May 7	April 9	April 2

Table 4.16. *First dates of fruiting by seven species of shrubs in Rock Valley.*

Year	Ambrosia dumosa	Ephedra nevadensis	Grayia spinosa	Krameria erecta	Larrea tridentata	Lycium andersonii	Lycium pallidum
1968	May 6	April 10	March 18	May 6	May 16	April 4	April 4
1969	May 19	May 2	April 14	May 19	May 26	April 18	April 18
1970	none	none	April 20	June 1	June 1	none	April 20
1971	May 10	April 15	March 22	May 27	May 20	March 31	none
1972	April 21	March 27	March 10	April 29	May 2	March 14	March 15
1973	May 21	May 7	April 3	May 29	May 21	April 18	March 22
1974	May 10	none	March 26	May 20	May 20	April 13	April 2
1975	none	May 21	April 18	none	June 6	April 18	April 2
1976	May 14	none	April 9	May 28	May 21	April 23	April 16

Table 4.17. *Mean dates (± one standard error of means) beginning three phenophases for seven species of shrubs in Rock Valley.*

Species	Leafing	Flowering	Fruiting
Ephedra nevadensis	–	April 5 ± 9.5	April 24 ± 8.1
Ambrosia dumosa	February 23 ± 4.9	May 2 ± 3.4	May 12 ± 4.2
Grayia spinosa	February 26 ± 5.1	March 17 ± 5.5	April 2 ± 4.8
Krameria erecta	April 7 ± 4.5	May 5 ± 4.6	May 22 ± 4.6
Larrea tridentata	March 15 ± 3.0	May 12 ± 3.4	May 22 ± 3.2
Lycium andersonii	February 13 ± 4.3	March 27 ± 5.1	April 10 ± 4.6
Lycium pallidum	February 11 ± 4.4	March 23 ± 4.0	April 5 ± 4.5

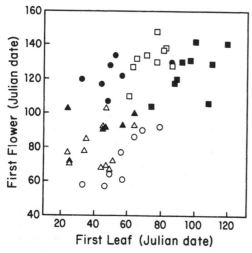

Figure 4.16 Relationship between mean dates of first leaves and first flowers, produced by six species of shrubs in Rock Valley. After Turner and Randall (1987).

accompanied by unusually heavy rainfall, delayed spring growth that year (Wallace & Romney 1972). Low temperatures in winter and spring, 1972 and 1973, were also suggested as the cause of late growth at Rock Valley (Ackerman et al. 1980).

Turner and Randall (1987) used environmental data, including measurements of rainfall and air temperatures at 2 m, for a quantitative approach to predict initial dates of leafing and flowering. These measurements were considered to be more reliable than soil moisture and soil temperature readings at the validation site. Their approach was regression analysis with dependent variables, expressed as Julian days. As shown in Tables 4.17–

4.18, initial dates of phenophases often ranged over periods of more than a month. Temperature and rainfall conditions in, for example, February of one year may bear importantly on leafing by one particular species. However, if this phenophase has already occurred by January of another year, then February conditions have no relevance. If observations were unlimited, one could then develop a set of sequential models, so that if, by a given date, a phenophase had not occurred, then a new model could be invoked, one that used more information than the first one. Unfortunately, this approach could not be supported by only nine years of observations, so instead for any particular phenophase the independent variables were limited to events occurring prior to the beginning of the latest month in which the phenophase was initiated.

Turner and Randall tested the following independent variables: (1) monthly mean air temperatures, with October–April inclusive and all combinations up to four contiguous months (when groups of months were used the independent variable was the arithmetic mean of the monthly means); (2) mean temperature of the coldest month over the same interval, i.e., December or January; (3) precipitation during single months (as in 1), and for all combinations up to four contiguous months; and (4) a weighted mean 'time of rainfall' in each interval involving three or more contiguous months.

In investigating yearly variations in phenology of annual plants, Slade, Horton, and Mooney (1975) used variables having values based on times

Table 4.18. *Regression equations relating initial dates of leafing and flowering of Rock Valley shrubs to air temperature and rainfall variables. TCM is mean temperature of coldest month. Other 'T' variables are mean temperatures of either individual months or inclusive combinations of months. 'P' variables are amounts of rainfall in single or groups of months. Code is explained in text.*

Species	Phenophase	Code	Equations predicting Julian dates	R^2
Ambrosia dumosa	leafing	4	$7.72\,(TND) + 0.47\,(PDJ) - 14.0$	0.755
Grayia spinosa		4	$6.57\,(TCM) + 0.42\,(PJ) + 20.1$	0.863
Krameria erecta		5	$-6.64\,(TJF) + 0.18\,(PF) + 136.7$	0.734
Larrea tridentata		2	$-0.58\,(PND) + 88.9$	0.783
Lycium andersonii		1	$3.43\,(TD) + 27.4$	0.327
L. pallidum		1	$5.33\,(TCM) + 17.7$	0.465
A. dumosa	flowering	5	$-2.53\,(TJM) + 0.08\,(PFM) + 139.1$	0.963
Ephedra nevadensis		5	$-27.58\,(TNF) + 1.02\,(PDJ) + 24.2$	0.961
G. spinosa		6	$8.86\,(TND) - 0.95\,(PN) + 26.2$	0.810
K. erecta		3	$-3.94\,(TA) - 0.69\,(PND) + 196.2$	0.748
L. tridentata		5	$-4.08\,(TMA) + 0.17\,(PDJ) + 176.4$	0.909
L. andersonii		3	$-5.79\,(TJM) - 0.76\,(PN) + 142.9$	0.771
L. pallidum		6	$6.53\,(TND) - 0.38\,(PND) + 45.8$	0.941

of occurrence for events of given intensities. In 1, 2, and 3 above, the temporal distribution of rainfall during various intervals was considered. When regression analyses or correlations were based on two independent variables, they restricted themselves to cases involving one temperature variable and one precipitation variable. No two-variable equations were adopted when F-tests gave insignificant values for entry of a second variable.

Regression equations that provided best fits to available phenological data for Rock Valley species are shown in Table 4.18. The codes given in this table indicate six different kinds of relationships between temperature or rainfall variables and timing of phenological events.

1 A positive correlation with temperature means that higher temperatures retard an event.

2 A negative correlation with rainfall means that higher rainfall accelerates an event.

3 Inverse relationships with both temperature and rainfall imply that higher temperatures and more rainfall both accelerate an event.

4 Positive relationships with both temperature and rainfall imply that higher temperatures and more rainfall both retard an event.

5 An inverse correlation with temperature and a positive correlation with rainfall imply that higher temperatures accelerate and higher rainfall retards an event.

6 An inverse correlation with rainfall and a positive correlation with temperature imply that higher temperatures retard and higher rainfall accelerates an event.

Rainfall and lower temperatures are almost always associated, as are periods of high temperatures and no rainfall. The question arises as to whether one of these variables predominates, or whether both are separately influential. This is particularly important for relationships 5 and 6, where both factors are acting in concert. In the foregoing analyses, partial correlations were computed to determine whether only a single variable was implicated. Equations in Table 4.18 include both temperature and rainfall variables, indicating that both factors played a significant role in explaining observed variations in phenologies for many species.

Returning to specific cases in Rock Valley, Turner and Randall (1987) concluded that leafing phenologies in *Ambrosia dumosa* and *Grayia spinosa* were retarded by both high winter temperatures and increased winter precipitation, whereas high winter temperatures alone slowed leafing in the two species of *Lycium*. *Krameria*, having a late growth cycle, showed a pattern where high late winter temperatures advanced the date of leafing. Higher levels of winter rainfall accelerated leafing in *Larrea*. Flowering in *Grayia* and *Lycium pallidum* were retarded by higher temperatures, whereas others were advanced by them. Increased rainfall was relatively neutral to flowering in *Larrea* and *Ambrosia*, retarded reproduction in *Ephedra nevadensis*, and stimulated flowering in the other four species. Overall, flowering phenologies were more highly correlated with the regressions in Table 4.18, suggesting that other, stronger environmental controls may be affecting this process.

SUMMARY

There are numerous forms of desert scrub vegetation at NTS, those of the Mojave Desert in lowest elevation in the south, the Great Basin Desert in the north, and transistion desert communities on the upland edges of Mojave desert scrub.

Dominant shrubs of the Mojave desert scrub on the Rock Valley validation site were creosote bush (*Larrea tridentata*), white bursage (*Ambrosia dumosa*), desert thorn (*Lycium andersonii*), box thorn (*L. pallidum*), and range ratany (*Krameria erecta*), which together contributed 80% relative shrub cover. Maximum mean canopy coverage was estimated to be 26%. White bursage comprised nearly 30% of all shrubs, yet this species had the same aboveground biomass as the ecologically important but less abundant creosote bush. Desert thorn had the largest aboveground biomass, totalling one-quarter of all shrubs and one-third of the

living biomass, and desert thorn and box thorn had the lowest percentages of dead stems. In contrast, two-thirds of total aboveground biomass of creosote bush and white bursage were dead stems. Root/shoot ratios of the five dominant species ranged from 0.79 to 1.65 and greater than 80% of community root biomass.

Estimated spring net primary production by shrubs ranged from 183 to 682 kg ha^{-1}, but the relative importance of productivity by individual species changed from year to year. Spring was the most productive season for shrub growth, but summer rains of at least 28 mm and rains in early fall stimulated growth in some of the shrub species. Net aboveground primary production values were not correlated with rainfall in a calendar year, but were to the September-to-August hydrologic year precipitation. Species did not increase productivity uniformly to precipitation, because temperature appears to be a primary factor affecting the growth response of each species. Variations have been carefully documented in phenological timing between species and between years for a single species. For example, in spring leaf initiation of range ratany was up to eight weeks later than the other species. Flowering in most species occurred during an interval of 4–6 weeks and was initiated about 4–6 weeks following leaf initiation. Analyses of Rock Valley data indicated that leafing phenologies of certain species were retarded by both high winter temperatures and increased winter precipitation, others by high temperatures alone, and still other species have an advanced date of leafing with high late winter temperatures. These data show how difficult it is to understand shrub growth and productivity at the community level and over years with highly variable conditions of rainfall and temperature.

5 Mojave Desert annuals

Herbaceous plants, displaying several differing growth forms, constitute a characteristic and diverse element of the flora for warm desert regions of North America (Chapter 3), and the Mojave Desert in particular. More than 85% of herbaceous species found on the Nevada Test Site and adjacent parts of central-southern Nevada are annuals, including those that complete their life cycles in the spring or in the summer, and herbaceous perennials and biennials (53%), which also may be either spring or summer active (Beatley 1976a; Mulroy & Rundel 1977). Whereas within a single week the standing biomass of herbaceous plants is generally small in comparison with that of the shrubs, net primary productivity by annuals during a favorable rain-year may approach 50% of the total for shrubs. Desert annuals tend to have relative high protein concentration in their tissues, but generally low levels of complex structural carbohydrates and toxic secondary compounds, so that these plants are important food resources for desert animals. Nitrogen buildup from annual plants may also serve as an important buffer for ecosystem nitrogen pools in these desert regions.

GENERAL ATTRIBUTES OF MOJAVE DESERT ANNUALS

Germination requirements

The physiological controls of germination in desert annuals have important ecological implications for their establishment and reproductive success. Many of the experimental scientific studies of desert annuals have been conducted on species of the Mojave Desert, beginning with early quantitative studies on germination requirements (Went 1948, 1949, 1955; Went & Westergaard 1949; Juhren, Went, & Phillips 1956). The research of Went and coworkers demonstrated that germination and establishment of Mojave Desert annuals is dependent not only on the total amount of rainfall at a site, but also on the month and temperature regime at the time of a heavy rain.

In Joshua Tree National Monument on the southern margin of the Mojave Desert, Went (1948) analyzed the effects of rainfall and temperature maxima and minima on the germination of desert plants. Summer annuals were those that germinated following the highly localized thunderstorms of July and continued growth if they received additional summer precipitation. Characteristic is *Pectis papposa* (Asteraceae), which grows at elevations in deserts of North America from 80 m to above 1400 m and often is very abundant. Other common summer-germinating and summer-flowering annuals were native and introduced species of *Amaranthus* (Amaranthaceae), annual grasses of *Aristida* and *Bouteloua* (Poaceae), and species of *Boerhaavia* (Nyctaginaceae), *Chamaesyce* (Euphorbiaceae), and *Mollugo* (Molluginaceae). *Bouteloua barbata* was an annual that below 1000 m elevation was able to germinate and produce ripe grains within a four-week period. For *Pectis*, germination occurred at 20°C and was notably high at 25–30°C after leaching fruits by running water for over 24 hours. Germination of *Amaranthus* was affected at similar high temperatures, but had no leaching requirements for its fruits.

A set of annual species germinated in late summer but flowered in late winter or spring. This set included usually the large spring annuals, *Abronia villosa* (sand verbena; Nyctaginaceae), the introduced *Erodium cicutarium* (storksbill; Geraniaceae), and *Oenothera deltoides* (evening primrose; Onagraceae).

A third and the largest set of annuals germinated only after significant autumn or winter rains; this included many of the notorious small 'ephemerals', what all treatments term winter annuals (Figs. 5.1–5.8), especially species of Boraginaceae, Brassicaceae, Asteraceae, Polemoniaceae, and Plantaginaceae. Went determined that the seeds (here used in a way to include small dry fruits that look like seeds) of winter annuals, as represented by *Lasthenia chrysostoma* (Asteraceae), germinated best at temperatures of 5–15°C, and sub-

Figures 5.1–5.4 Rosettes of winter annuals before the onset of rapid growth in early spring, Mojave desert scrub in California. **Figure 5.1.** (Top left) *Filago depressum* (Asteraceae), a minute plant (1 cm diameter) covered with long white silky hairs. **Figure 5.2** (Top right) *Malacothrix glabrata* (Asteraceae), with relatively thick, hairless leaves. **Figure 5.3.** (Bottom left) *Salvia columbariae* (Lamiaceae), chia, having distinctive cobblestone-like leaf surface. **Figure 5.4.** (Bottom right) *Camissonia boothii* (Onagraceae), with oddly pigmented, thin, smooth leaves.

Figures 5.5–5.8 Flowering winter annuals of Mojave desert scrub in California. **Figure 5.5.** *Geraea canescens* (Asteraceae). **Figure 5.6.** *Mohavea breviflora* (Scrophulariaceae). **Figure 5.7.** *Monoptilon bellioides* (Asteraceae). **Figure 5.8.** *Gilmania luteola* (Polygonaceae), an endemic of the Mojave Desert.

sequent experiments demonstrated best germination of winter annuals at 18/8–13°C day/night (Went 1949), but much interspecific variation was also observed (Juhren et al. 1956).

Went and Westergaard (1949) also analyzed the relationship of rainfall and temperature to the germination patterns of annuals in Death Valley, near the northern margin of the Mojave Desert and where mass spring flowerings occurred in 1935, 1940, and 1947. In field season 1945–6 the investigators found many localities of winter annuals, but

no summer annuals on the valley floor, and in salt flats and saline habitats no annuals were found. Success of winter annuals was attributed to one rainy episode from November 12 to 14, yielding 35 mm of precipitation, which was followed by minimum temperatures of 8–10°C. Laboratory studies indicated that characteristic species of winter annuals from Death Valley had highest percent germination under a temperature regime of 18/8°C day/night.

Rainfall seasonality, therefore, is critical for germination of desert annuals. For example, in 1941, the wettest year at Death Valley (107 mm), no mass flowering occurred that year or during the following spring, because significant rains were not received in November and December (Went 1955). He estimated that for autumn germination to occur a critical November rain must be at least 13 mm and preferably 2–4 times that amount. For *Pectis* in summer, he estimated that a 20-mm rain must occur before successful germination and establishment (Went 1949). Similar rains under warm summer conditions will not promote germination of winter annuals, and in the same manner, most summer annuals will not respond to fall or winter rains.

Beatley (1967, 1969a, b, 1974b) conducted extensive, long-term studies on desert ephemerals at NTS and tested and better quantified the germination patterns established by Went. Using permanent plots at 68 desert sites, which had weather stations and produced a total of 130 species of winter annuals, Beatley found that autumn rainstorms of 15–25 mm produced scattered germination of annuals, and little or no germination occurred when fall precipitation was less than 15 mm. Beatley (1974b) proposed a detailed phenological model linking annual plant growth and abundance to levels and seasonality of precipitation (Fig. 5.9). Whereas this model broadly fits observed patterns of annual plant development, it is a relatively simplistic construction and does not deal with other environmental factors, such as temperature, precipitation, periodicity, and herbivory, that may strongly impact annual plant establishment and development.

Rare, heavy spring rains may promote late germination of winter annuals. Such conditions occurred in 1964–5, when heavy rains in March and April, 1965, followed a fall with only 7 mm of precipitation, and there was a moderate germination of annuals (Beatley 1967). Total herb biomass, however, was relatively low. Likewise, the timing of fall rains may have strong differential impacts on individual species; although quantitative studies have not been carried out, late September and early October rains appear to be especially favorable to species of Polygonaceae, Hydrophyllaceae, Polemoniaceae, Fabaceae, and

Onagraceae (Mulroy & Rundel 1977). Heavy rains in December, when soil temperatures are near their lowest, tend to be favorable for mass germination of Boraginaceae and Brassicaceae.

Growth and maturation

Normally, winter annuals germinate in the autumn or early winter, but remain in a vegetative rosette or tuft (Figs. 5.1–5.4) until stem internodal elongation takes place, beginning in March or early April (Beatley 1969a; Mulroy & Rundel 1977). Once germinated, the development and survival of winter annuals is greatly dependent on a variety of environmental factors and local habitat conditions. Freezing nighttime temperatures or spring droughts may sharply reduce survivorship of annual plant rosettes.

Early studies of Mojave Desert annuals suggested that greater than half of the plants would survive to maturity once germinated (Went 1949; Went & Westergaard 1949; Juhren et al. 1956). This hypothesis, based on laboratory trials, was tested by Beatley (1966, 1967, 1969a, b) in population studies at NTS. Her survivorship studies were conducted in 1963–4 and 1964–5 at three sites, including Rock Valley. In 1963 annuals germinated in late September, following moderate rains, and completed their life cycle in late April and May, 1964, 7 to 8 months later. For 14 species average survivorship to maturity was 39% over this interval in Rock Valley (Table 5.1). Mortality of seedlings during that period was largely attributable to effects of drought (Beatley 1967). Highest rates of mortality occurred in the early spring, 1964, when rosettes began vegetative elongation. Precipitation in autumn and early winter, 1964, was only 7 mm, and thus inadequate to stimulate even scattered germination of annuals. However, as mentioned above, heavy spring rains triggered germination in late March and early April, and these plants completed their maturation within six to ten weeks. This time, mean survival of six species in Rock Valley was 83%, and drought-induced mortality was virtually absent. Beatley (1967) considered events in these two years to represent the extremes in growing seasons for the northern Mojave Desert Region. More recently, the dry

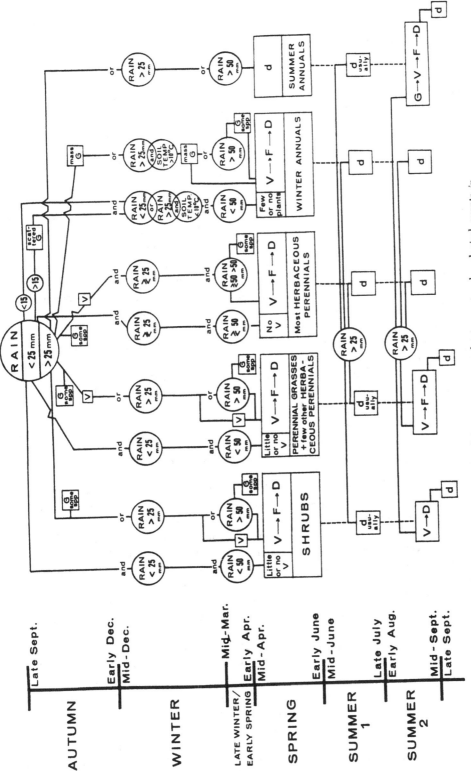

Figure 5.9 Flow diagram modeling environmental triggers to phenological events in Mojave Desert ecosystems. G = germination, V = vegetative growth, F = flowering and fruiting, D = senescence of seasonally deciduous organs, d = dormancy. After Beatley (1974b).

Table 5.1. *Survivorship for mixed species populations of winter annuals from germination to reproductive maturity at Rock Valley.*

Year	Interval studied	Number of species	Survivorship, %	Reference
1963–4	September–May	14	39	Beatley (1967)
1964–5	March–June	6	83	Beatley (1967)
1970–5	November–May	35	45	Turner (1972)
1971–2	December–May	30	19	Turner (1973)
1972–3	September–June	42	95	Ackerman, Bamberg, Hill, & Katz (1974)

years of 1988–9 and 1989–90 were extremely poor for annuals throughout MDR.

Facultative perennials

Several species that inhabit lowland desert may grow as winter annuals under typical desert conditions and as biennials or short-lived perennials whenever rainfall is unusually heavy (Beatley 1970, 1976a). By following the survivorship of seedlings that germinated from rains in spring of 1965, Beatley (1970) determined that typically annual plants of *Astragalus lentiginosus* var. *fremontii* were large, flowering plants in spring of 1966, a season that experienced very heavy rains, but at the lowest elevation many of the robust plants continued to grow and were still alive in spring, 1967. At higher elevations (above 1400 m), where rains are greater and more frequent, *A. lentiginosus* var. *fremontii* grows as a perennial. Beatley also observed that *Erioneuron pulchellum*, a low, tufted grass widespread in North American deserts, had populations at NTS with mixed reproductive strategies; in 1965 about 80% of the plants in a survey were winter annuals with life cycles of only several weeks, but up to 20% of the individuals survived and reproduced in 1966 and, for some, 1967. This facultative behavior of the life cycle is clearly related to having an abundance of soil moisture. At low elevations an individual of *Baileya pleniradiata* can grow either as an annual or biennial, and this also occurs in the winter annual *Xylorhiza canescens*, which is a perennial at moister, higher localities. Other common species of the Mojave Desert that probably have facultative life cycles are *Eriogonum inflatum*, *Mirabilis pudica*, and *Sphaeralcea* spp. (Beatley 1970).

Tidestromia oblongifolia, a C_4 species that is fairly common in lowland sandy deserts and washes, also has a facultative life cycle. This species typically is a perennial with a stout woody taproot, but it may flower in the first year and behave as a summer annual (Gulmon & Mooney 1977). In fact, in Death Valley Gulmon and Mooney discovered that nearly all individuals of *Tidestromia* died within two years. Even though this species has a relatively large taproot, it may be treated as a desert ephemeral whenever it uses the small amount of water stored in the uppermost horizon of the soil and cannot survive when soil water is depleted.

FIELD STUDIES AT THE ROCK VALLEY IBP SITE

Identifications of winter annuals were made during the spring, 1971–76, in 760 quadrats within the Rock Valley validation site. The 760 quadrats were in groups of four, with group members positioned 10 m in each cardinal direction from centers of the 190 quadrats used for sampling perennial plants. Quadrats were 0.2 × 0.5 m (0.1 m^2) in all years except 1972, when quadrats were 0.5 × 0.5 m (0.25 m^2). Densities of species were estimated in each of six vegetational zones within the validation site, and site-wide densities were computed as area-weighted means of zone densities. Estimates of aggregate densities of winter annuals in Rock Valley were also made during each of the previous eight years by Beatley (1974a), and these measurements were obtained in a plot near the southeastern corner of the validation site. Smaller numbers of plots in Rock Valley have been sampled annually since 1987 as part of ongoing studies of *Bromus rubens* invasions (Hunter 1990, 1991).

Species diversity at the Rock Valley study site

Seventy species of annual plants and four herbaceous perennials were collected in Rock Valley (Table 5.2), and of these 62 species were included

Table 5.2. *Annuals and herbaceous perennials from the Rock Valley site during the IBP study. Number of the six possible years of presence (1971–1976) is shown in parentheses, and the symbol '+' signifies that a species was observed at Rock Valley, but was not recorded in the study quadrat. Introduced species are noted with an asterisk.*

Amaryllidaceae	
Androstephium breviflorum	(+)
Asteraceae	
Calycoseris wrightii	(5)
Chaenactis carphoclinia	(6)
C. fremontii	(6)
C. macrantha	(3)
C. stevioides	(6)
Eriophyllum pringlei	(6)
Geraea canescens	(+)
Lygodesmia exigua	(6)
Malacothrix glabrata	(+)
Monoptilon bellidiforme	(2)
Rafinesquia neomexicana	(5)
Stylocline micropoides	(4)
Boraginaceae	
Amsinckia tessellata	(6)
Cryptantha angustifolia	(+)
C. circumscissa	(6)
C. dumetorum	(2)
C. micrantha	(6)
C. nevadensis	(6)
C. pterocarya	(6)
C. recurvata	(6)
Pectocarya heterocarpa	(6)
P. platycarpa	(6)
P. recurvata	(6)
Plagiobothrys jonesii	(3)
Brassicaceae	
Caulanthus cooperi	(6)
Descurainia pinnata	(6)
Guillenia lasiophylla	(6)
Lepidium flavum	(+)
L. lasiocarpum	(4)
Streptanthella longirostris	(6)

Campanulaceae	
Nemacladus glanduliferus	(1)
N. sigmoideus	(+)
Cuscutaceae	
Cuscuta nevadensis	(+)
Euphorbiaceae	
Chamaesyce micromera	(+)
C. setiloba	(+)
Fabaceae	
Astragalus acutirostris	(4)
A. didymocarpus	(4)
A. lentiginosus	(5)
Lupinus flavoculatus	(3)
L. shockleyi	(1)
Geraniaceae	
Erodium cicutarium	(+)
Hydrophyllaceae	
Eucrypta micrantha	(1)
Nama demissum	(2)
N. pusillum	(2)
Phacelia fremontii	(6)
P. vallis-mortae	(6)
Tricardia watsonii	(+)
Loasaceae	
Mentzelia obscura	(6)
Onagraceae	
Camissonia boothii	(4)
C. munzii	(6)
Oenothera primiveris	(+)
Plantaginaceae	
Plantago ovata	
Poaceae	
Bromus rubens	(6)
Stipa speciosa	(+)
Vulpia octoflora	(6)
Polemoniaceae	
Eriastrum eremicum	(1)
Gilia cana	(6)
G. transmontana	(6)
Ipomopsis polycladon	(6)
Langloisia schottii	(4)
L. setosissima	(6)
Linanthus bigelovii	(2)
L. demissus	(2)
Polygonaceae	
Chorizanthe brevicornu	(6)
C. rigida	(6)
C. thurberi	(1)
Eriogonum maculatum	(6)
E. nidularium	(6)
E. thomasii	(1)
E. trichopes	(6)
Oxytheca perfoliata	(5)
Ranunculaceae	
Delphinium parishii	(+)
Scrophulariaceae	
Antirrhinum filipes	(1)

Table 5.3. *Species richness of winter annuals recorded in 760 0.1 m² quadrats in Rock Valley and estimated site-wide mean densities and productivity between 1971 and 1976.*

Year	Number of species recorded	Mean density, individuals m⁻²	Net production, g m⁻²
1971	42	8	0.46
1972	42	11	0.30
1973	52	101	68.8
1974	46	88	1.75
1975	44	136	5.30
1976	56	454	14.7
Mean	47	133	15.2
Standard deviation	6	165	26.8

in the winter annual samples. Over half of the species in the vernal flora belonged to four families, Asteraceae (12 spp.), Boraginaceae (12 spp.), Polemoniaceae (8 spp.), and Polygonaceae (8 spp.). Several genera were important in terms of species diversity, especially *Cryptantha* (7 spp.), *Chaenactis* (4 spp.), *Eriogonum* (4 spp.), *Pectocarya* (3 spp.), *Astragalus* (3 spp.), and *Chorizanthe* (3 spp.).

Only two species of summer annuals were observed in Rock Valley, *Chamaesyce micromera* and *C. setiloba*, both present in 1971, and both are C₄ annuals (Mulroy & Rundel 1977). The paucity of summer precipitation has undoubtedly affected this low diversity of summer annuals. Other species of C₄ summer annuals, such as *Amaranthus* spp., *Aristida adscensionis*, *Bouteloua* spp., and *Allionia incarnata*, have been collected elsewhere at NTS (Beatley 1976a).

Only two introduced species occurred in the checklist of annual plants at Rock Valley site during the IBP study. The ecologically significant introduced grass *Bromus rubens* often occurred in large numbers within plots, but otherwise only *Erodium cicutarium* was present and only as a relatively uncommon species. In the last decade, however, the relative dominance and diversity of weedy introduced annuals has increased sharply at NTS, and Hunter (1990, 1991) has documented significant increases in densities of *B. tectorum* at Rock Valley and other Mojavean areas of NTS (Chapter 13).

Thirty-four species of winter annuals were pre-

sent in each of the six years of the IBP study, whereas at the other extreme 8 of the 62 species occurred in only a single year (Table 5.2). Maximum species diversity of winter annuals (56 spp.) was present in 1976, when there was the highest mean density of annuals, even though herb biomass was less than one-fourth that of 1973, a wetter year when 52 species were found (Table 5.3). Even during springs having low herb density and biomass, such as 1971 and 1972, a relatively high level of diversity (42 spp.) was observed.

Although precipitation regimes varied widely from year to year, dominant herb species remained remarkably constant during the IBP study. *Vulpia octoflora* (Poaceae) had the highest densities except in 1971, when it ranked third behind the introduced *Bromus rubens* (Poaceae) and *Chaenactis fremontii* (Asteraceae; Table 5.4), and *B. rubens* also ranked second in density in three years, emphasizing the ecological importance of those two grass species. *Chaenactis fremontii* and *Eriophyllum pringlei* were among the 10 most dense species in each year, *Pectocarya* spp. (four species of this genus could not be distinguished in the field before reproductive maturity), *C. carphoclinia*, and *B. rubens* in five years, and *Guillenia lasiophylla* and *Chorizanthe rigida* in four years.

Survivorship

Survivorship of mixed populations of annuals was measured in Rock Valley from 1971 to 1973 during

Table 5.4. *Survivorship for mixed species populations of winter annuals during four spring growing seasons from 1977 to 1981 at Jackass Flats, Nevada Test Site.*

Year	Interval studied	Number of species	Survivorship %	Reference
1977	February–April	31	71	Turner & Edney (1977)
1978	February–April	36	75	Turner (1979)
1979	March–May	26	69	Turner & Vollmer (1980)
1981	March–April	30	69	Turner & Vollmer (unpubl.)

the IBP study, and Beatley collected comparative data from Rock Valley in the 1963–4 and 1971–2 season. After IBP survival of annuals was also measured from Jackass Flats between 1976 and 1981 (Turner & Edney 1977; Turner et al. 1979; Turner & Vollmer 1980). Observations began at the initiation of germination of the first winter annuals and continued until plants were shedding fruits and seeds. Survival during that period was estimated from counts of plants in established quadrats.

Survivorship data of winter annuals of Rock Valley are summarized in Table 5.1. Only under the extraordinary conditions of 1972–3, when rainfall was abnormally high, did survivorship approach 100%, and it was highly variable in other years, dipping to 19% in 1971–2. Population survivorship of annuals in the IBP Rock Valley studies was directly proportional to total precipitation between October of the initiating year and April. It is interesting to note that heavy rains in October and November of 1972 stimulated widespread germination of annuals, but additonal rains from January to March caused little new germiantion. No annuals germinated in fall, 1973, when precipitation was largely absent, but a mid-January storm in 1974 produced a moderate amount of germination.

Counts of annual seedlings were made at Jackass Flats on October 24, 1976 and on February 7, April 6, and April 27, 1977. Between October, 1976, and February, 1977 (108 days) the survivorship of annual plants was 57%, but over February and March (59 days) 71% of the remaining population died, resulting in an aggregate survivorship of 40%. From April 6 to 27 there was a rapid increase in rates of mortality, with a mortality of 56%, and overall from October, 1976 through April, 1977 there was a survivorship of 18%.

Rates of spring seedling mortality at Jackass Flats were remarkably similar during four years of study (Table 5.4). Survivorship varied from a low of 69% in the spring of 1979 and 1981 to a high of 75% in the spring of 1978. Population survivorship in the early spring of 1978, when the site received 120 mm of rain, was only slightly better than that in 1977 (71%), when the same plots received about 2 mm of rain between February and April. These comparisons of spring mortality, ignoring fall and winter mortality, suggest that spring precipitation does not play a major role in influencing death of annuals. However, as just described, mortality rates increased sharply in April, 1977, as summerlike weather became dominant.

It is important to remember that all species in mixed populations of winter annuals do not experience the same survivorship. Beatley (1967) pointed out that survival of *Mentzelia albicaulis* and *Phacelia fremontii* was especially poor (0–7%) over the 1963–4 winter annual year. In Jackass Flats survival of *P. fremontii* was always the lowest or among the lowest among up to 36 species (Turner & Edney 1977; Turner et al. 1979; Turner & Vollmer 1980). Survival of *M. obscura* was better than its congener, but always less than the overall mean for survivorship. All studies consistently showed high survival rates of the introduced bromegrass, *B. rubens*, and species of Polygonaceae and Chenopodiaceae generally exhibited better than average survival.

Table 5.5. *Relative rank in absolute density of abundant winter annuals during spring, 1971–1976 at Rock Valley. NP = not present in that year.*

	1971	1972	1973	1974	1975	1976
Vulpia octoflora	3	1	1	1	1	1
Bromus rubens	1	7	–	2	2	2
Chorizanthe rigida	4	2	3	–	–	7
Chaenactis carphoclinia	8	6	4	5	–	4
Eriophyllum pringlei	6	5	5	10	7	5
Pectocarya spp.	10	4	2	–	3	6
Chaenactis fremontii	2	3	9	3	8	8
Streptanthella longirostris	–	–	8	–	–	–
Cryptantha circumscissa	8	–	–	7	10	–
Guillenia lasiophylla	–	–	6	8	4	3
Cryptantha recurvata	–	–	–	4	–	10
Chorizanthe brevicornu	7	9	–	–	–	–
Descurainia pinnata	–	–	7	–	5	–
Phacelia vallis-mortae	5	8	–	–	–	–
Cryptantha dumetorum	NP	NP	NP	6	NP	9
Gilia cana	–	–	10	–	–	–
Cryptantha nevadensis	–	–	–	9	–	–
Lepidium lasiocarpum	NP	–	–	NP	6	–
Caulanthus cooperi	–	–	–	–	9	–
Langloisia setosissima	–	10	–	–	–	–

Population densities

Population densities of winter annuals varied greatly from 1971 to 1976 at Rock Valley. These densities ranged from a mean high of 454 individuals m^{-2} in 1976 to a low of only 8 individuals m^{-2} in 1971 (Table 5.3). This is a range of densities of nearly 80x over just six years. Earlier studies by Beatley in Rock Valley between 1963 and 1970 found annual densities ranging from a low of 5 individuals m^{-2} in 1965 to a high of 124 individuals m^{-2} in 1969 (Beatley 1974b).

For 12 years of data there was no overall correlation between growing season precipitation (September through March) and annual plant density. Growing seasons with moderate but not unusually high levels of precipitation generally had higher densities of annuals than in years, such as 1973, when heavy rains occurred, probably due to thinning of high initial densities during good years as mean herb size increased.

Individual species of winter annuals showed quite distinctive patterns of change in relative density during the IBP study (Table 5.5). *Vulpia octoflora* achieved a density of more than 100 individuals m^{-2} in 1975. In 1971 and 1972, relatively dry years, *Vulpia* comprised only 9 and 19% of the individuals. Under the good growing condition of 1976, *Vulpia* was still the most abundant species, but its lower relative density of 17% may well have been an important factor in permitting high annual diversity in that spring.

Some of the tallest herbaceous species, such as species of Brassicaceae, had their highest densities in 1973 and 1976, when herb growth was excellent (Table 5.5). *Chaenactis* spp. overall, however, showed a decrease in relative density in 1973 when herb biomass was unusually high. Relative abundances of other taxa changed between years without a clear causality. Species of *Cryptantha*, commonly relatively low annuals, showed a high relative density under the poor growing conditions of 1974 but had low relative densities in a dry year (1972), a moderate year (1975), and a wet year (1973). *Bromus rubens* showed less variance between years, having 5–15% of total individuals throughout the sampling period, and *Pectocarya* also had relatively low variance with 2–9% of total individuals.

One means of comparing the relative structure of winter annual communities at Rock Valley is through the use of dominance-diversity curves, which depict the relative species abundances sequenced from highest to lowest relative density (Bowers 1987). In 1975, when *V. octoflora* had unusually high dominance, a distinctive pattern of dominance-diversity was present with a large number of relatively rare species (Fig. 5.10). Species diversity (H'), as calculated using a Shannon-Weiner index of relative densities, was a low 1.15. The two driest years, 1971 and 1972, had similar dominance-diversity curves, with high species diversities of 3.28 and 2.98, respectively (Fig. 5.10), but the two years having the best conditions for herb growth had different patterns. In 1973 there were fewer dominant species and more rare species than occurred in 1976, and species diversities were intermediate in these years.

Similarities between annual plant communities from year to year were also analyzed using plant compositional data with a Kendall rank-order correlation coefficient. This statistical technique uses ordinal ranking of species abundances for different years and tests for degree of similarity in their compositional makeup. Absolute concordance in the relative abundance of species would result in

a *Tau* value of 1.0 (Siegel 1956). Such analyses produce statistically significant correlations between all years, suggesting that there are overall patterns of significant correlation in relative community structure from year to year (Bowers 1987). Whereas annual densities fluctuate radically between years, the patterns of relative dominance remain largely unchanged.

Simple nonparametric correlation analyses have demonstrated that relative similarity of community structure between years is related to amounts of early-autumn precipitation in September and October. Years with more similarity in early-autumn precipitation are more alike in patterns of relative species abundances (Bowers 1987), whereas amounts of late-autumn precipitation (November and December) were not significantly correlated with such patterns. Patterns of relative species abundance from past years did not correlate statistically with patterns of abundance in any given year, indicating that biological variables of past productivity or reproductive output did not influence relative abundance for the years studied. Seed pools in the soil must therefore be sufficiently large to buffer out such influences.

Biomass and net productivity

Beatley (1969b) had conducted an extensive analysis of biomass of winter annual populations, sampled on the 68 permanent plots at the peak of reproduction in spring, 1964–6. She used this as a measure of net primary productivity by this group, realizing that some biomass had been consumed by vertebrate and invertebrate herbivores. In undisturbed communities of Mojave desert scrub, spring biomass varied from 1 to 442 kg ha^{-1} in 1964, 0 to 75 kg ha^{-1} in 1965, and 44 to 616 kg ha^{-1} in 1966. Variability at sites was high year to year and apparently unrelated to which shrub species was dominant.

To gather more detailed data on productivity, in 1971 and 1972 biomass of winter annuals was estimated during both April and May. These estimates were based on collections of plants in areas adjacent to the validation site. Plants were separ-

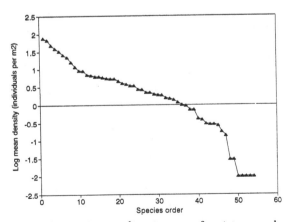

Figure 5.10 Dominance-diversity curves for winter annuals in Rock Valley from 1971 and 1972.

Table 5.6. *Estimated total annual production by important winter annuals from 1971 to 1976 at Rock Valley. All values are in kg ha^{-1} yr^{-1}. Insignificant production is indicated by a '+'.*

	Year					
	1971	1972	1973	1974	1975	1976
Bromus rubens	0.56	0.21	6.2	4.5	1.15	17.1
Chaenactis carphoclinia	0.63	0.22	25.0	1.5	0.07	22.8
C. fremontii	0.67	0.31	22.5	2.6	0.26	8.4
Chorizanthe rigida	0.27	0.88	25.3	0.4	+	37.2
Descurainia pinnata	0.04	0.01	38.0	0.1	0.28	0.3
Eriogonum trichopes	0.19	0.08	9.9	0.2	+	4.4
Guillenia lasiophylla	0.5	+	68.4	+	0.25	6.0
Oenothera munzii	0.08	0.11	12.7	0.3	+	6.2
Pectocarya spp.	0.04	0.12	14.1	0.2	1.70	4.5
Phacelia vallis-mortae	0.59	0.15	11.8	0.8	+	3.0
Strepthantella longirostris	0.21	0.16	136.6	+	+	+
Vulpia octoflora	0.07	0.11	100.7	2.8	9.83	3.4
12 species total	3.4	2.4	471	13.4	13.5	113
Total of all annuals	4.6	3.0	688	17.5	15.0	147

ated into leaves, stems, flowers, fruits, and roots, oven-dried, and then weighed. Average dry masses of various species were computed from these measurements. Mean masses of different species were combined with zone densities to estimate the biomass of various species in each zone for April and May. Zone values for each species were converted to site-wide standing crops in mid-April and died out before the May samples, whereas others reached their peaks later and were more important in May. Estimates of site-wide standing crops in mid-April and early May, for example, were 3.4 and 3.1 kg ha^{-1}, respectively. Total net production was estimated by combining April and May samples so as to reflect the maximum development of each species. This overall estimate (4.6 kg ha^{-1}) was from 38 to 52% higher than those based on April and May samples treated individually. Similarly, estimates of net production for April and May in 1972 were 2.0 and 2.3 kg ha^{-1}, respectively, while the overall estimate based on combined samples was 3.0 kg ha^{-1}. Again, the combined estimate was 30–33% higher than those based on separate months.

Annuals were so abundant and so large in 1973

that two discrete collections of plants were not made. However, species of annuals were collected throughout the spring at the apparent peak of reproductive effort. This procedure was followed in subsequent years of the analysis here, and estimated maximum standing crops were taken as measures of net primary production by winter annuals. Because winter annuals by and large have only a single season of growth, biomass and net production of individuals were treated as identical. On a community basis, having mixed species populations, of course the two measures are not the same given that all species do not reach maximum biomass at the same time.

During the six years of IBP studies, net production at Rock Valley ranged from lows of 4.6 and 3.0 kg ha^{-1} in 1971 and 1972, respectively, to a high of 688 kg ha^{-1} in 1973 (Table 5.6). This range of 230× between the high and low year is much greater than the 80× range in mean densities recorded over those same years. Beatley (1969b) had obtained values of 1.7, 3.5, and 1368 kg ha^{-1} in similar plots at Rock Valley, an even greater range. This disparity results from the large differences in mean biomass of individual herbs between years. Mean herb biomass of all species weighed individually each year ranged from a low of only 36.3 mg individual^{-1} in 1974 to a high of 1241 mg individual^{-1} in 1973, a 34× difference (Table 5.7).

Whereas it is difficult to use statistics to compare mean biomasses between species and between years for a single species, because sample sizes varied so widely, a comparison of measured values can give some idea of the impact produced during dramatically different years. During the excellent growing conditions of 1972–3, individual herbs became robust. Mean biomasses of individual plants exceeded 5000 mg in 3 species, *Amsinckia tessellata*, *Eriogonum trichopes*, and *Streptanthella longirostris*, and was above 1000 mg in 23 species and above 200 mg in 42 species of the 50 measured (Table 5.8). In contrast, no species of herb exceeded 145 mg in mean biomass in 1974.

In 1974 most species had a mean biomass of only 1–3% of their individual biomasses in 1973 (Table 5.8), despite being reproductively successful in both years. The very large variance in sizes of mature plants is a remarkable feature of desert annuals. For example, in *Caulanthus cooperi* and *Guillenia lasiophylla* average biomasses of flowering individuals were 16.1 and 13.5 g, respectively, but under the excellent growing conditions in spring, 1973, these same species had averages of 2556 and 2188 g, respectively, 160 times larger.

The primary production data for winter annuals at Rock Valley from Beatley (1969b; Wallace & Romney 1972) was used for comparisons with the IBP data by subtracting out the relative biomass allocation to root production during each IBP year. These values ranged from 5.4 to 8.1%. Net aboveground primary production for 11 years of data varied from a low of 0.02 g m^{-2} in 1965 to a high of 68.8 g m^{-2} in 1973 (Table 5.9). These values of aboveground net production (ANP) are significantly correlated with both September–March and September–August precipitation, but not with calendar year precipitation (Turner & Randall 1989). This pattern agrees well with the observations by Beatley (1974b) that a fall storm with at least 25 mm of precipitation is necessary for successful germination and growth of annual plants.

Turner and Randall (1989) have examined a number of predictive equations to estimate net primary production at Rock Valley. They found best results by incorporating Beatley's threshold effect of 25 mm precipitation into a log model that estimates ANP from September–March precipitation (SMP):

$$\log \text{ANP} = 1.976 \log (\text{SMP} - 26.2) - 2.746$$

This model is plotted in Figure 5.11 and fits the observed data from Table 5.9. More recent data on annual plant biomass from Rock Valley plots for 1987–91 also seem to fit this model relatively well, despite the increased dominance by *Bromus rubens* (Hunter 1991, pers. comm.).

Total annual production of a relatively few species provides a large proportion of the annual production by winter annuals. The estimated productivity of 12 important species comprised 68–90% of total herb production during the six years studied in Rock Valley (Table 5.8), while the relative importance of individual species in relation to biomass differed significantly. For example, *Vulpia octoflora* composed as much as 15% of herb biomass both in the dry 1971 and wet 1973 spring, but 65% of herb biomass in 1975 and only 2–4% in 1972 and 1976 (Table 5.8). *Streptanthella longirostris*, normally a minor species in the annual flora, was the biomass dominant in the wet 1973 and formed 20% of all herb biomass. *Chorizanthe rigida* was first in biomass important in both 1972 and 1976, comprising 29 and 25% of all herb biomass in those two years, whereas *Chaenactis fremontii* and *Guillenia lasiophylla* were the two most important species in 1971, followed closely by *Phacelia vallis-mortae* and *Bromus rubens*. In 1974 *B. rubens* was the dominant species, having 26% of all herb biomass (Table 5.8).

The relative allocation of annual plant biomass to root and shoot tissues remained relatively constant from 1971 to 1976 (Table 5.10). These values ranged from a low of 5.4% root tissue and a root/shoot ratio of 0.06 in 1974 to a high of 8.1% and a root/shoot ratio of 0.09 in 1975. From a study of eight winter annuals from southernmost Nevada, including two of the Rock Valley species, Bell, Hiatt, and Niles (1979) reported that roots com-

Table 5.7. *Mean dry masses (mg plant^{-1}) of winter annuals in Rock Valley.*

Year	Number of species weighed	Mean dry mass of all species	Standard deviation	Range species masses
1971	31	71.8	70	9–312
1972	35	43.4	68	6–410
1973	50	1241.0	1386	34–5974
1974	33	36.3	32	3–145
1975	37	169.9	216	11–1139
1976	36	59.4	83	4–415

Table 5.8. *Mean dry weight (g ha^{-1}) of important annual plant species at Rock Valley between 1971 and 1976. Sample sizes were highly variable, depending on the relative abundance of a species. Data for three species were lumped as* Pectocarya *spp.*

	1971	1972	1973	1974	1975	1976
Amsinckia tessellata	–	90.0	5253	68.0	1139	21.0
Astragalus lentiginosus	–	410.0	2557	22.5	192.2	259.7
Bromus rubens	49.1	39.2	810	44.0	109.8	24.1
Calycoseris wrightii	133.0	52.7	1637	–	151.6	–
Caulanthus cooperi	21.6	16.1	2556	–	207.7	33.2
Chaenactis carphoclinia	200.0	38.0	884	36.5	110.0	57.4
C. fremontii	65.0	41.7	1300	32.0	256.4	57.1
C. macrantha	52.5	–	–	46.6	74.4	
C. stevioides	33.4	24.9	1314	27.1	–	40.1
Chorizanthe rigida	38.2	47.5	249	64.2	45.1	172.8
C. brevicornu	19.9	19.6	254	44.1	125.0	61.9
Cryptantha circumscissa	40.0	7.6	186	14.0	10.8	11.2
C. micrantha	–	8.5	205	2.7	45.0	11.5
C. nevadensis	40.1	31.4	2863	22.8	74.5	30.8
C. pectocarya	18.0	–	1137	36.1	39.6	35.7
C. recurvata	17.7	28.5	541	27.4	79.7	17.6
Descurainia pinnata	22.1	10.5	1469	6.9	127.0	9.0
Eriogonum maculatum	–	21.7	191	25.7	43.6	–
E. nidularium	89.4	15.7	377	–	740.8	–
E. trichopes	111.8	59.5	5974	144.6	24.0	414.6
Eriophyllum pringlei	48.2	5.6	332	11.1	–	8.1
Gilia cana	29.0	30.0	541	15.0	47.0	31.5
G. transmontana	15.3	11.3	177	–	48.3	11.9
Guillenia lasiophylla	–	15.5	2188	14.7	108.8	11.9
Ipomopsis polycladon	21.0	15.2	592	–	26.9	26.2
Langloisia setosissima	91.3	26.8	1085	20.6	–	31.2
Lupinus flavoculatus	–	–	1903	64.6	187.4	–
Lygodesmia exigua	19.0	33.0	373	7.7	–	–
Malacothrix glabrata	18.7	61.0	–	37.0	207.8	46.1
Mentzelia obscura	133.0	62.3	1254	23.6	184.3	45.4
Oenothera munzii	83.7	68.8	1640	49.9	28.0	93.7
Pectocarya spp.	14.2	11.6	244	12.8	313.2	16.8
Phacelia fremontii	71.0	14.1	1445	10.7	113.5	13.8
P. vallis-mortae	161.3	52.2	2135	131.7	40.4	88.5
Rafinesquia neomexicana	185.0	96.5	847	65.0	72.4	–
Streptanthella longirostris	312.5	12.0	5908	–	–	–
Vulpia octoflora	9.1	9.0	198	6.9	95.3	4.3

Table 5.9. *Estimated net production at Rock Valley winter annuals and associated precipitation.*

Years	Net aboveground production, g m^{-2}	Calendar year precipitation, mm	September–March precipitation, mm	September–August precipitation, mm
1964	0.57	61	66	110
1965	0.024	213	30	107
1966	17.80	63	139	161
1967	4.50	182	47	148
1968	25.0	122	88	152
1971	0.46	104	47	100
1972	0.30	103	41	75
1973	68.8	208	218	250
1974	1.75	130	70	98
1975	1.50	63	93	113
1976	14.7	223	110	146

prised 12–22% of total biomass and root/shoot ratios of 0.14–0.28 until late in flowering, and then the contribution of roots at the termination of the study was 5–12% and root/shoot ratios were 0.05–0.14. A similar range of root/shoot ratios for Mojave and Sonoran Desert annuals has been

Table 5.10. *Percentages for biomass of winter anuals of Rock Valley allocated to various plant parts.*

Year	Leaves	Stems	Flowers	Fruits	Roots
1971	26.6	27.3	19.7	20.1	6.3
1972	27.9	19.8	25.5	19.8	7.0
1973	19.8	34.2	11.6	28.0	6.4
1974	36.4	20.3	28.7	19.2	5.4
1975	25.9	20.6	17.7	17.7	8.1
1976	22.3	24.5	19.0	27.5	6.7
Mean	24.8	24.5	22.0	22.1	6.6
Standard deviation	3.1	5.6	6.5	4.5	0.9

reported by Forseth and Ehleringer (1982). Reproductive allocation (flowers plus fruits) during the IBP survey ranged from about 40% in 1971 and 1973 to a high of 48% in 1974 (Table 5.10), with little interyear variance even though there were major differences in absolute biomass per year (Table 5.9). Bell et al. (1979) reported maximum reproductive allocations in eight species ranging from 14 to 55% for an average of 35%, and reproductive allocation was not correlated to plant size.

Microhabitat distribution

Data on overall herb density and biomass are useful for spatial projections of carbon and nutrient

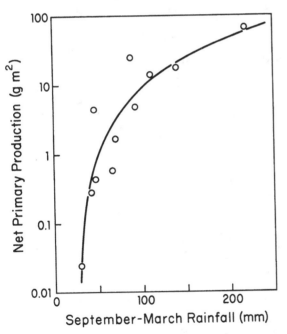

Figure 5.11 Modeling annual production in relation to precipitation for study areas at the Nevada Test Site. After Turner and Randall (1989).

Table 5.11. *Mean density (0.1 m⁻²) of winter annual plant species beneath shrub canopy (cover) and in the open, 1973–1976 in Rock Valley. Underlined values indicate significantly different densities in open and cover microhabitats (p < .05).*

Species	Year 1973 Cover	Open	1974 Cover	Open	1975 Cover	Open	1976 Cover	Open
Amsinckia tessellata	0.24	0.03	0.00	0.01	0.01	0.02	0.08	0.06
Astragalus acutirostris	0.01	0.00	0.01	0.00	0.01	0.00	0.03	0.02
A. didymocarpus	0.00	0.09	0.00	0.02	0.00	0.01	0.15	0.17
A. lentiginosus	0.00	0.01	0.02	0.02	0.03	0.01	0.02	0.01
Bromus rubens	0.45	0.01	2.41	0.74	1.34	0.67	20.66	2.59
Calycoseris wrightii	0.00	0.01	0.00	0.00	0.00	0.01	0.00	0.03
Camissonia boothii	0.02	0.02	0.05	0.02	0.00	0.00	0.15	0.18
C. munzii	0.27	0.01	0.05	0.05	0.00	0.01	0.83	0.31
Caulanthus cooperi	0.51	0.01	0.03	0.00	0.21	0.05	2.11	0.16
Chaenactis carphoclinia	0.49	0.42	0.06	0.62	0.01	0.11	1.94	3.96
C. fremontii	1.14	0.04	1.30	0.69	0.27	0.08	5.91	0.75
C. macrantha	0.00	0.00	0.01	0.01	0.00	0.00	0.00	0.01
C. stevioides	0.06	0.02	0.03	0.12	0.00	0.02	0.79	0.38
Chorizanthe brevicornu	0.20	0.07	0.00	0.06	0.07	0.01	1.28	0.33
C. rigida	0.24	0.53	0.02	0.07	0.01	0.02	1.32	2.25
Cryptantha circumscissa	0.04	0.09	0.06	0.46	0.04	0.14	0.34	0.78
C. micrantha	0.02	0.02	0.02	0.08	0.00	0.01	0.23	0.22
C. nevadensis	0.37	0.02	1.05	0.09	0.13	0.01	1.89	0.05
C. pterocarya	0.20	0.01	0.08	0.02	0.16	0.02	0.75	0.09
C. recurvata	0.98	0.04	0.56	0.43	0.20	0.08	3.42	0.51
Descurainia pinnata	1.18	0.07	0.02	0.19	0.10	0.16	0.52	0.22
Eriogonum maculatum	0.02	0.02	0.00	0.04	0.14	0.01	0.08	0.10
E. nidularium	0.00	0.01	0.00	0.01	0.01	0.02	0.02	0.04
E. trichopes	0.00	0.01	0.00	0.03	0.00	0.01	0.08	0.11
Eriophyllum pringlei	0.08	0.34	0.00	0.33	0.04	0.25	0.85	2.87
Gilia cana	0.18	0.17	0.00	0.00	0.01	0.03	0.45	0.48
G. transmontana	0.14	0.13	0.00	0.03	0.06	0.03	0.21	0.68
Glyptopleura marginata	0.00	0.01	0.00	0.00	0.00	0.00	0.00	0.01
Guillenia lasiophylla	1.37	0.05	0.17	0.01	0.69	0.13	8.42	3.30
Ipomopsis polycladon	0.02	0.20	0.00	0.03	0.00	0.05	0.19	0.88
Langloisia schottii	0.00	0.01	0.00	0.00	0.00	0.00	0.12	
L. setosissima	0.02	0.06	0.00	0.03	0.00	0.01	0.13	0.67
Lepidium lasiocarpum	0.04	0.00	0.00	0.00	0.00	0.22	0.04	0.10
Linanthus demissus	0.00	0.01	0.00	0.00	0.00	0.00	0.00	0.01
Lupinus flavoculatus	0.00	0.05	0.00	0.01	0.01	0.03	0.00	0.00
Lygodesmia exigua	0.08	0.02	0.02	0.03	0.04	0.01	0.28	0.12
Malacothrix glabrata	0.02	0.01	0.03	0.01	0.03	0.02	0.68	0.09
Mentzelia obscura	0.37	0.03	0.21	0.03	0.00	0.01	0.79	0.17
Monoptilon bellidiforme	0.00	0.00	0.00	0.00	0.00	0.00	0.01	
Nama demissum	0.00	0.04	0.00	0.00	0.00	0.00	0.00	0.10
N. pusillum	0.00	0.01	0.00	0.00	0.00	0.00	0.00	0.01
Nemacladus glanduliferus	0.00	0.01	0.00	0.00	0.00	0.00	0.00	0.00
Oxytheca perfoliata	0.00	0.00	0.00	0.01	0.00	0.00	0.17	0.09
Phacelia fremontii	0.16	0.03	0.02	0.01	0.00	0.01	0.13	0.13
P. vallis-mortae	0.47	0.00	0.38	0.01	0.04	0.01	1.92	0.05
Plagiobothrys jonesii	0.00	0.01	0.00	0.01	0.00	0.00	0.00	0.01
Rafinesquia neomexicana	0.10	0.00	0.00	0.01	0.00	0.00	0.09	0.01
Streptanthella longirostris	0.61	0.12	0.00	0.01	0.06	0.10	0.25	0.05
Stylocline micropoides	0.00	0.01	0.00	0.00	0.00	0.00	0.00	0.10
Vulpia octoflora	6.82	3.87	3.02	3.53	7.43	10.60	6.40	6.53

pools and fluxes for the Rock Valley ecosystem (Chapter 12); however, these values obscure the spatial variation that occurs in microhabitat selection by individual species and by the annual flora as a whole. Soil surface features as well as perennial plant canopies are important microhabitat variables affecting seed accumulation, water availability, and nutrient distribution (Shreve 1931; Beatley 1966). Indeed, desert biologists commonly find that species appear to have patterns of microhabitat distribution, with some species doing best beneath shaded shrub canopies while others grow well in full-sun openings between shrubs.

Using data on relative densities of 50 annual species, measured from 1973 to 1976 in the 760 quadrats, species distributions were related to their positions beneath shrub canopies and in open microhabitats. During this period 21 species were significantly associated with shrub canopies, five species were associated with open habitats, and the remainder showed no microhabitat preference (Table 5.11).

Clearly, the fertile islands of resources beneath shrubs provide favorable sites for annual plant seed accumulation, germination, and growth (Muller 1953). Favorable conditions of nutrient and water availability (Chapter 2) and partial shading from high solar irradiance during seedling establishment all favor annual plant growth (Mott & McComb 1974; Halvorson & Patten 1975; Patten 1978). Herb plots beneath shrubs at Rock Valley comprised only about 20% of the samples, but included 66–92% of the individual annuals sampled. Many of the species associated with shrub canopies were tall and erect in growth form, suggesting a strategy to harvest more PAR (Chapter 3).

SUMMARY

Desert annuals constitute the most diverse portion of the typical Mojave desert scrub community, even though in most seasons and years biomass of annuals is small in comparison with shrubs. The flora may have two distinct sets of annual species. One set includes the winter annuals, which germinate in cold months following heavy rains in fall, form vegetative rosettes through winter months, and then grow rapidly, reproduce, and die in the spring. Typical conditions for germination are 25 mm in a rainy episode and daytime temperature of about 18°C. Freezing temperatures and drought in the spring cause greatest mortality of winter annuals. The other set consists of species that germinate following the rare heavy rains of summer, grow rapidly, and complete the life cycle before temperatures decline sharply in the early fall.

During the IBP study in Rock Valley 70 species of annuals were present, of which 62 were winter annuals, approximately half of the species found in Mojave desert scrub of southern Nevada. Even during IBP years with low productivity there were 42 species, but only 34 species were found in all six years. Survivorship for mixed species populations of winter annuals varied from 19 to 95%, this remarkably in successive years, and each year the unique features of climate, with changes that influenced when species germinated and subsequently received more rain, strongly affected survivorship of individual species, dominance, and population densities.

In plots of winter annuals, spring biomass varied from 1368 kg ha^{-1} to nearly zero in Rock Valley over many years of careful data collection. In dry years estimated standing crop in mid-spring was about 3 kg ha^{-1} with a corresponding low net production. Mean herb biomass varied from 1241 mg individual^{-1} in spring, 1973, and only 36.3 mg individual^{-1} in 1974, a very dry year. In 1973 mean biomasses of several species exceeded 5000 mg, whereas in very dry years the maximum mean biomass was 142 mg. Net aboveground primary production for 11 years of data varied from 68.8 g m^{-2} to 0.02 g m^{-2}. Hence, these studies provided solid evidence to document the great year-to-year variance in the success of desert winter annuals. Analyses also showed that many species were significantly associated with shrub canopies, where there are more favorable sites for germination and nutrient and water availability for growth.

6   Adaptations of Mojave Desert animals

Desert environments produce extreme conditions for survival of animals, just as for plants (Chapter 3). High temperature and low water availability make thermoregulation an acute problem during the summer, whereas cold temperatures and low food availability during winter make aboveground activity energetically inefficient for most desert animals. Lack of water presents special problems for animals that produce liquid excreta, and these osmotic difficulties are magnified when diets contain materials that are high in solutes. Nutrient resources, be they plant tissues for herbivores or prey for carnivores, are often low in abundance and poor in quality, and individual food items tend to be available only for short intervals. Moreover, coexisting species must share those limited resources. Consequently, each resident species must have a set of adaptations to cope with stresses during its lifetime, and desert animals, like plants, utilize physiological, morphological, and phenological adaptations to either tolerate or avoid stresses. Unlike plants, however, animals have the ability to alter their environment through movement and thus possess intriguing suites of behavioral adaptive responses.

HEAT BALANCE AND THERMOREGULATION

Problems of heat in deserts

Strategies for maintaining reasonable thermal balance are of critical importance for ecological success of each animal species in a desert environment, especially the hot, dry conditions of summer months in the Mojave Desert, where midday temperatures commonly exceed 40°C and the ground surface may heat up to 70°C from intense solar radiation. For endotherms, typical core body temperatures of the eutherian (placental) mammals are 36–38°C and of birds are 39–41°C, and approximate lethal temperatures are 42–44°C and 46–47°C, respectively (Schmidt-Nielsen 1990); for ectothermic reptiles, temperature preferenda are below 39°C, and lethal temperatures generally are no higher than 46°C (Mayhew 1968; Bartholomew 1982b). Similar values are expected for ectothermic arthropods. When heat loads are extreme, high temperatures cause unfavorable changes in blood function, lipid and protein chemistry, and water status (Prosser 1973). Given that ambient midday (i.e., at about 1200–1400 hours) temperatures in summer often exceed optimal levels and may enter lethal limits for animals (Fig. 6.1), desert residents utilize a variety of physiological and especially behavioral strategies to maintain a reasonable daily thermal balance and thereby solve physiological problems of hyperthermia. These include relaxation of the limits within which homeostatic control is maintained, behavioral adjustment to avoid the problem of thermoregulation, and special physiological modifications and attributes (Bartholomew 1964, 1982b, c; Dawson 1967; Huey 1982; Congdon, Dunham, & Tinkle 1982).

Body temperature, heat balance, and thermoregulation are part of a larger picture of overall animal energetics and water balance, an adaptive system to allow animals to utilize food and water resources to survive and reproduce under desert

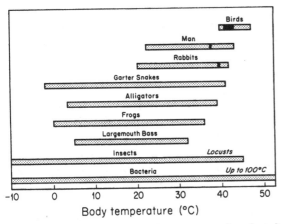

Figure 6.1 Survival limits for body temperature for selected animal groups and bacteria (from Gates 1980). The shaded area shows the normal operating range for endotherms.

conditions. Scarcity of water is probably the most important challenge of summer, because there is an unavoidable efflux of water vapor from the respiratory membranes when terrestrial animals are active (Schmidt-Nielsen 1964, 1972, 1990; Nagy & Peterson 1987, 1988). Daily and seasonal strategies of thermoregulation produce much energy savings, which in the desert setting is a crucial factor limiting productivity (Bartholomew 1982b; Nagy 1983, 1987).

A heat balance model

Every animal exchanges thermal radiation with its surroundings, with energy passing from a region of higher to one of lower temperature (Bartholomew 1982b; Tracy 1982). Ectotherms use this process to draw heat from the environment and thereby raise body temperature at essentially no energetic cost other than those calories used to orient the organism for purposes of thermoregulation. Once an ectotherm reaches its preferred temperature for activity, there are ways to use behavior and metabolism to more or less maintain body temperature near that point until environmental temperatures change so greatly that heat flows unavoidably back to the environment. Endotherms maintain a high and fairly constant internal

temperature, often at considerable energetic cost, because body heat is continually lost to the cooler environment from the animal's surface through conduction, convection, and radiation. When ambient temperature is greater than body temperature of an endotherm, physiological or behavioral adaptations are usually made to lower body temperature by the evaporation of water or the reduction of solar irradiance on the body surface.

For the body temperature of any animal to remain constant, the rate of metabolic heat production ($\dot{H}$, dT/dt) equals the rate of heat loss ($\dot{Q}$, dT/dt), and each is proportional to the expression:

$$\dot{H} = C \ (T_b - T_a) = \dot{Q}$$

where C is a conductance term, a measure of heat flow between the animal and its surroundings, T_b is the body temperature of the animal, and T_a is ambient temperature (Schmidt-Nielson 1990). To maintain steady state, when there is a change in either body or ambient temperature an adjustment must be made in the rate of metabolism or conductance. Similarly, if the rate of metabolism changes due to an increase in activity, then body temperature will increase and must be offset by loss of heat via the conductance term. For a desert animal experiencing hyperthermia while active on a summer day, it can retreat to a cooler microclimate, thereby reducing T_a and permitting body temperature to lose heat passively to its surroundings, or avoid hyperthermia by evaporative cooling, which can reduce body temperature.

The change in heat content of a resting animal (H_{tot}) is a function of several variables:

$$H_{tot} = H_c + H_r - H_e + H_s + H$$

where H_c is net conductive and convective heat exchange, H_r is net radiative heat exchange, H_e is net evaporative heat loss, H_s is net stored heat in the body, and H is metabolic heat production (Schmidt-Nielsen 1990). More simply stated, the energy generated by metabolism plus the energy taken up from the environment equals the energy stored and the energy released to the environment. On a hot day when body temperature equals its surroundings, heat conduction, convection, and radiation will approach zero, and metabolic heat production can only be dissipated by evaporative cooling. Conversely, on a cold day an active organism would lose heat to its surroundings and must provide the heat by oxidizing food or stored resources (for example, homeothermic endotherms), or it can thermoregulate in the sun (for example, ectotherms), or it can vastly reduce its rate of metabolism and experience a state of inactivity or dormancy with a reduction in body temperature (for example, ectotherms or heterothermic endotherms).

Relation of heat exchange to surface area and body mass

The strategy that is employed by an organism to control the direction of body temperature change is very much determined by, and is a function of, its surface area and body mass (Bartholomew 1982a). Rate of heat loss is proportional to surface area, so one expects that for organisms having the same body mass, increases or decreases in temperature are potentially more rapid for one with a high surface/volume ratio than for one with a low ratio. An organism having large mass is likely to experience much slower changes in temperature than one with a small mass, because its large volume allows heat to be stored and thereby buffers against rapid shifts in temperature. However, for an organism with large mass, it is also difficult to dissipate extra heat on a hot day except via a high rate of evaporative cooling. Thus, an animal of large size warms slowly, but is well protected from sudden temperature shifts or too frequent heating (Cowles 1979).

Thermoregulation for ectotherms

For terrestrial vertebrate ectotherms, for example, desert reptiles, activity is greatly governed by net gain of heat that is obtained from the sun. At low temperature reptiles are very sluggish – unable to move rapidly or at all – and consequently are vulnerable prey. By basking in the sun, body parts that touch the substrate are warmed via conduction (H_c), and the exposed surface, which is initially in energy equilibrium with ambient temperature, experiences additional heating via solar radiation (H_r). By choosing an appropriate substrate, position, and posture, as well as having good energy-absorbing properties (for example, dark coloration and high surface area), rate of body heating can be optimized and time of vulnerability can be reduced (Cowles & Bogert 1944; Norris & Lowe 1964; Norris 1967; Muth 1977a, b; Porter & Tracy 1983).

Once an ectotherm is active, energy produced by exercise adds metabolic heat. The animal continues to warm to its temperature preferenda, where physiological performance is at its peak; for desert species this is 35–39°C, several degrees higher than in nondesert, high-light ecosystems (Kay 1970; Minnich & Shoemaker 1970). At supraoptimal temperatures the animal operates well, but must use methods periodically to cool itself, staying away from lethal temperatures and if possible returning to the temperature preferenda. Thermoregulation for lowering body temperature during near lethal temperatures can be accomplished by using sites with cooler microclimates or less radiative heating, thereby lowering T_a or H_r, or by evaporative cooling, H_e.

Invertebrates (ectotherms) generally have much smaller masses, and being small has important effects on heat balance. These organisms are small enough that they tend to have no storage of body heat and rapidly assume the temperature of the surroundings. More importantly, rate of metabolism is exceedingly high, especially in flying species, and this generates heat for the organism (H). Flying insects can also generate significant amounts of body heat by preflight vibration of muscles (Heinrich & Bartholomew 1971; Cowles 1979). Terrestrial invertebrates of deserts have many different strategies to thermoregulate and cope with high ambient temperature (Hadley 1972; Crawford 1981), and colonial insects, such as ants and bees, can produce elevated nest temperatures via heat generated collectively by the workers.

Thermoregulation for endotherms

Homeothermic endotherms are those organisms for which the heat balance must stay in steady-state to maintain a constant, high body temperature, regardless of the external temperatures. Energy must be expended, often from stored materials (H_s), to maintain high body temperature during cold periods, and evaporative cooling (H_s) must be employed to alleviate hyperthermia. Heterothermic endotherms are fascinating cases wherein an animal has physiological adaptations to tolerate hyperthermia or hypothermia under certain environmental conditions in order to conserve energy or water reserves, thus the heat balance equation may have several different states for equilibrium. For example, an animal could experience torpor during summer (estivation) or winter (hibernation).

Activity patterns as related to thermoregulation

As in many other ecosystems, deserts have one set of organisms that forages aboveground during the day and another that becomes active at night. The majority of the ectothermic lizards have diurnal activity because they need solar radiation to heat their bodies (Bradshaw 1988a), but a consequence is that they are exposed to diurnal, predaceous mammals, birds, snakes, and other lizards. A noteworthy exception is *Xantusia vigilis* (desert night lizard), a very small, secretive, nocturnal animal with a low metabolic rate (Miller 1951; Mautz 1979). Resident granivorous and insectivorous birds are also diurnal desert animals.

The night landscape of the desert is dominated by activity of endothermic nocturnal rodents, particularly the predominantly granivorous Heteromyidae, and their predators, such as canids, owls, and snakes. Nocturnal organisms avoid the heat of the day to forage at night, when temperatures rarely exceed 30°C. A noteworthy exception here is *Ammospermophilus leucurus* (antelope ground squirrel), a rodent that is diurnally active and can be seen running at top speed during the hottest time of the summer day. Most species of snakes are nocturnal foragers, at least during hot weather, because these ectotherms are somewhat sensitive to high ground temperature (Cowles 1979). Moreover, after very hot days snakes may not emerge from shelter at night until after 10:00 p.m., to enable the ground to cool down. Many characteristic desert arthropods, such as scorpions, are also secretive nocturnal creatures (Cloudsley-Thompson 1988).

Small diurnal mammals are scarce in desert scrub because they have a limited capacity for heat storage and therefore would need high rates of evaporative cooling to dissipate heat uptake from the sun along with the heat generated by rapid metabolic turnover (Bartholomew 1982b). These animals have large surface/volume ratios, which means that they gain heat relatively rapidly. Such small mammals also lack the mobility to travel long distances to locate drinking water, and, in fact, some of these do not drink free water under natural conditions. Being nocturnal creatures, many desert rodents can evade the daytime heat, and they thereby uncouple thermoregulation from evaporative cooling. In contrast, small birds are mostly diurnal and can tolerate the heat because high body temperatures of 41°C are usually greater than that of the surroundings, so a bird typically loses heat passively by conduction, radiation, or convection.

In a winter-rainfall desert, such as the Mojave Desert, most activity of the resident animals occurs during the spring, when temperatures are moderate, winter annuals and shrubs have their highest rates of growth, and phytophagous insects emerge to eat the plants. As food resources deteriorate and animals have to deal with problems of heat balance, foraging becomes less frequent and for shorter duration (for example, Pianka & Parker 1972, 1975; Nagy 1973; Nagy & Medica 1977, 1986).

Uses of burrows

The burrow (Figs. 6.2–6.5) is a key adaptation for thermoregulation and energy balance in many desert vertebrates and invertebrates and it is where animals reside during periods of inactivity.

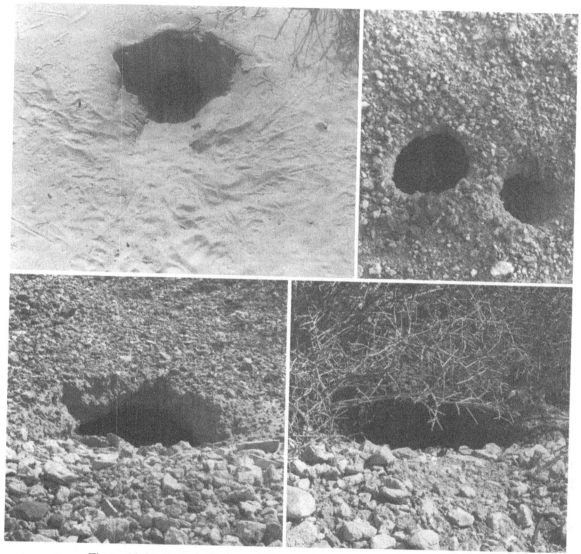

Figures 6.2–6.5 Burrows of desert animals. **Figures 6.2–6.3.** (Top) Characteristic burrows of heteromyid rodents in the Mojave Desert. **Figures 6.4–6.5.** (Bottom) Semicircular burrows of desert tortoise, *Gopherus agassizii.*

Typical burrows are fairly narrow tunnels fashioned by fossorial (burrowing) rodents, several common lizards, for example, *Dipsosaurus dorsalis* (desert iguana), *Cnemidophorus tigris* (whiptail lizard), and *Uta stansburiana* (side-blotched lizard), and a few snakes, for example, *Arizona elegans* (glossy snake), but wide, semicircular tunnels (Figs. 6.4–6.5) are excavated by the desert

tortoise (*Gopherus agassizii*). Excess and abandoned burrows are inhabited by other animals, including lizards and snakes as well as invertebrates. Some invertebrates build their own burrows (Edney 1977; Crawford 1981; Cloudsley-Thompson 1988).

Within a subterranean burrow, the microclimate varies much less than the macroclimate outside (Bradley, Miller, & Yousef 1974), and these are shelters with relatively stable temperatures (Fig. 6.6). In the summer heat, burrow temperature (−0.3 m) commonly is 33°C and varies only 1–2°C during a 24-hour period while temperature outside may range from over 40°C during the day to 25°C at night. During the coldest winter days the same burrow may maintain a steady temperature of 5°C while day/night temperature outside ranges from 10°C to below zero. A burrow has high humidity from the water efflux of the animal and for many species contains a food cache to nourish the animal during its belowground existence.

Nearly all small desert tetrapods use burrows as thermal refugia. Beginning with favorable weather in March and continuing to late October or early November, animals venture from their burrows and move abroad for the necessities of life, but they return to burrows for periods of rest and inactivity, to escape predators, and, for diurnal species, to avoid intense solar radiation. Overheated animals that enter a burrow during the heat of a day return to normal relatively quickly by lying on the relatively cooler floor and thus permitting conduction to remove excess body heat while the burrow walls remove back heat by radiation. For heteromyid rodents, foraging from burrows occurs every night, although some studies have shown that foraging decreases on those nights with a full moon (Price, Waser, & Bass 1984; Kotler 1984a, b, 1985). The nocturnal desert scorpions also tend to avoid moonlight (Cloudsley-Thompson 1988).

Thermoregulation of lizards

Many studies have described the daily and seasonal patterns of sunning and thermoregulation of lizards (Cowles & Bogert 1944; see references in Chapter 8). Lizards can precisely orient themselves to produce the optimal position for rapid warming up, including flattening their bodies and spreading ribs to present large surface area to gain heat (for example, *Phrynosoma*, horned lizards), or opposing measures to lose heat. Proper sunning methods are critical to enable lizards to forage throughout the daylight hours in spring, when food is usually abundant. A species with a low surface/volume ratio and large mass, such as the chuckwalla (*Sauromalus obesus*), needs a protected basking period because the animal heats up relatively slowly (Norris 1967).

After six weeks of very intense foraging in spring, activity often slows. In late spring, when midday heats up and food resources begin to dwindle, the daytime pattern of lizard activity is typically bimodal (for example, Pianka & Parker 1972, 1975; Parker & Pianka 1973; Pianka 1986), with early morning and late afternoon foraging and burrow habitation during the hottest hours. During summer months lizard foraging may decrease and be restricted to once a day or less, either in early morning or just before sunset, so burrows are heavily used. However, *D. dorsalis* is an example of a lizard that is active throughout the heat of the day during summer when sympatric species are resting within burrows or in shady microhabitats. Some species of snakes have daytime or crepuscular activity during the mild

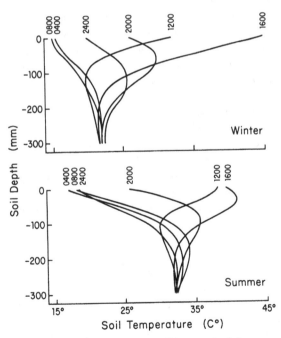

Figure 6.6 Soil temperature profiles in animal burrows, showing gradients of soil temperature with depth at six hours during the day in both winter and summer.

climate of early spring, shift to nocturnal activity in mid-spring and summer, and even have decreased activity July through early September (Cowles 1979). Within a species there may be differences in activity schedules of individuals, as in *G. agassizii*, in which juveniles that are less than 60 mm long are active at significantly lower temperatures than larger individuals (Berry & Turner 1986).

Aboveground activity often increases in early fall, but food resources then often are poor and daytime temperatures decrease markedly. By early November reptiles and many rodents enter burrows for winter hibernation, i.e., winter dormancy.

Nests of desert woodrats

Neotoma (woodrats and packrats) are rodents that construct large houses or nests consisting of piles of stones, stems, and other debris. These messy structures may measure two or more meters in diameter and a meter high, and they provide animals with a relatively stable thermal environment without exposure to the extreme high of summer days or the lows of winter nights. Desert woodrats are active year-round.

Use of shady microhabitats

Shade beneath shrub canopies is a common rest site for diurnally active lizards, to reduce solar radiation, and some species become arboreal when they climb into branches of shrubs, where they forage or rest in partial shade. One lizard, *P. platyrhinos* (desert horned lizard), almost totally buries itself in the loose soil mound of a shrub, thereby minimizing heating by exposing only the upper portion of the head and maximizing conduction with soil (Cowles 1979), Rabbits also seek shade beneath shrubs (Hinds 1973; Costa, Nagy, & Shoemaker 1976), and these animals shift to nighttime foraging during the hot summer.

Carnivores are generally the largest mammals of lowland desert scrub. Generally they use shade to limit exposure to direct sun during the hottest time of the day and experience cooling via evaporation. Kit fox (*Vulpes macrotis*) uses a den to

escape supraoptimal temperatures, and coyotes (*Canis latrans*) seek out dens, especially during the spring breeding season.

The few birds that are year-round residents in desert scrub tend to build nests in shrubs, shaded by plant shoots and far away from the intense heat at the ground. Moreover, egg laying and rearing of young are usually completed before the onset of harsh summer heat.

Sauromalus obesus lives in rocky outcrops and uses crevices between boulders as shade refugia. In the heat of summer, chuckwallas occasionally emerge from rock crevices to thermoregulate on rocks or to forage (Johnson 1965; Nagy 1973; Berry 1974). Crevices between rocks are also used by *Sceloporus* spp. (spiny lizards), many surface-active arthropods, and several small passerine birds (Smyth and Bartholomew 1966).

Some invertebrate microhabitats

One interesting invertebrate is the desert millipede (*Orthoporus ornatus*), which overwinters in a hibernaculum, but is trapped there because it is unable to dig its way to the surface through dry soil. After heavy rains the soil softens, which permits emergence (Crawford 1978). This sudden appearance after heavy summer rainfalls gives them the name stormworm.

It is difficult to generalize about patterns of activity and use of microhabitats for thermoregulation of invertebrates, but the subject relative to some species of the Mojave Desert has been reviewed (Crawford 1981). In general desert arthropods are unable to tolerate very high daytime summer temperatures, and therefore escape the heat by staying in burrows and having nocturnal activity.

Coloration

There are, of course, many examples of desert animals with light- or sand-colored skin, matching soil substrates and thereby reducing heat loads by reflecting some solar radiation, for example, the sidewinder (*Crotalus cerastes*) or zebra-tailed lizard (*Callisaurus draconoides*), which are effectively camouflaged against sand until they move. Hence, it is easy to appreciate that the color of an animal is adaptive for both thermoregulation and as protective coloration for predator–prey relations (Norris & Lowe 1964; Norris 1967).

Using spectroreflectometry Norris (1967) and Norris and Lowe (1964) quantified phenological and microhabitat differences in color of desert lizards. Small diurnal iguanid lizards show marked phenological changes; they have a dark phase when in burrows and become paler as they bask in the sun due to changes in melanin concentration in the skin. Within the thermal zone of surface activity and foraging, the typical lizard is color matched to the background, and in some cases difference in reflectance between the animal and substrate is within 1%. In two common species of the Mojave Desert, *D. dorsalis* and *C. draconoides*, the animal becomes 'superlight' and more reflective at very high temperatures, shining against the sand. Norris concluded that under cold conditions energy absorption with dark coloration is predominant, whereas at typical high temperatures background matching is predominant, especially as protective coloration. Superlight coloration can occur in some species when they are active under hyperthermic conditions because predators have retreated by then to shaded micro-

habitats. During extremely warm nights, diurnal species remain light in color. *Uta stansburiana*, which has aboveground activity year-round, is dark during the winter for facilitating solar heating. White belly scales on many species, for example, *P. platyrhinos*, probably help to minimize heat load on the ventral surface, although they also disrupt the shadow line, making that lizard less visible to predators.

Despite the apparently high thermal loading that would be expected to occur with a black color in a desert environment, a surprising number of desert species have black coloration, including many common invertebrates (Crawford 1981). Detailed energy balance studies of the polymorphic *D. dorsalis* have shown that a dark individual absorbs nearly 30% more radiation than a light-colored individual (Norris 1967). Some species have marked differences in coloration of juveniles and adults, especially in *S. obesus* (Fig. 6.7; Norris 1967). Such dark coloration would be advantageous if it could be utilized behaviorally to raise body temperatures to optimal levels more rapidly in cool morning and late afternoon hours. Other studies have postulated that dark pigmentation, especially black in many beetles, may be aposematic color to deter predators (Hamilton 1975; Cloudsley-Thompson 1977, 1979; Crawford

Figure 6.7 Adult and juvenile chuckwallas (*Sauromalus obesus*) basking on a rock.

1981). In certain deserts body color in beetles appears to have little effect on thermoregulation (Turner & Lombard 1990).

Black features of the raven (*Corvus corax*), a large diurnal bird, have been shown to reduce metabolic costs of thermoregulation (Marder 1973). Another fascinating case is the roadrunner (*Geococcyx californicus*), a large terrestrial cuckoo, which has brown mottled feathers, and under them a specialized, heavily pigmented, heat-absorbing black skin (Calder & Schmidt-Nielsen 1967; Ohmart & Lasiewski 1971). Body temperature of a roadrunner may fall by 4°C on a cold night, but at sunrise the bird warms up by basking at no extra metabolic cost by holding its back at right angles to incident solar rays and elevating its feathers and drooping its wings to expose the black skin. Conversely, other desert birds expose the thinly featured sides of the thorax by lifting wings from the body to facilitate heat loss when bird temperature is higher than air temperature (Bartholomew 1964).

Hibernation and estivation

During the cold season typical desert communities exhibit little aboveground animal activity because invertebrates are mostly in diapause (Crawford 1981) and terrestrial vertebrates often hibernate in burrows. Winter inactivity is not simply an attempt to thermoregulate, but instead an adaptive strategy to minimize energy expenditure (Bartholomew 1982b, c). Some of the smallest vertebrates, such as insectivorous desert shrews (*Notisorex*) and herbivorous pocket gophers (*Thomomys*), have high basal metabolic rates, and they remain active throughout the year.

Other small endotherms periodically abandon homeothermy to enable body temperature to drop and thereby reduce energetic costs during winter dormancy. In preparation for this, animals store fat within their bodies and may store large caches of foods in burrows. Ectothermic reptiles enter a cold-induced torpor, but some species may do so with metabolic rates that are significantly lowered from minimal low temperatures during summer months (Bennett & Dawson 1976; Gregory 1982).

Diapause in invertebrates, which is under hormonal control, is induced by changes in photoperiod or decreased temperature.

Two levels of hypothermia are recognized, relatively shallow hypothermia, wherein body temperature remains within 10°C of normothermia, and torpor or profound hypothermia, wherein body temperature falls to that of the surroundings or about 1°C above burrow temperature (Bartholomew 1972). Seasonal torpor, as during winter hibernation, is characterized by suspended respiration (often just one breath or less per minute), lowered heart rate, and low oxygen consumption (Bartholomew 1982b). To prevent body temperature from falling too low, i.e., below freezing, hibernating rodents have a hypothalamic regulator, which stimulates endogenous heat production. Every few weeks some species of hibernating rodents arouse, regain normothermia, eliminate accumulated metabolic wastes, perhaps feed on stored materials within the burrow, and then return to a dormant condition. If aboveground weather is warm, some burrow residents may venture abroad.

Perognathus longimembris, a nocturnal pocket mouse, is a species that exhibits little or no surface activity during parts of the year. This, the smallest of the heteromyid desert rodents (adult weight 6.5–10 g), does not appear in traps during the coldest winter months or during the hottest summer months. The explanation is that *P. longimembris* has a well-documented circannual rhythm whereby the animal can enter daily or seasonal torpor, but it is best expressed in winter (hibernation) and summer (estivation). Torpor occurs in their burrows, which have relatively stable temperature and may have low levels of oxygen (hypoxia) and high levels of carbon dioxide (hypercapnia).

Research on torpor of *P. longimembris* was conducted through laboratory programs, beginning with the seminal study by Bartholomew and Cade (1957), which documented induction of torpor at 9°C following 24 hours of starvation and through a wide range of temperatures. Winter dormancy of the little pocket mouse was documented in the field at Rock Valley by monitoring daily and seasonal movements of individuals exposed to ^{137}Cs from an aboveground source (French, Maza, & Aschwanden 1966). A new microdosimeter was used to show greater than 1.25 Roentgen d^{-1} summer surface activity (May through October), but almost no detectable radiation in January, meaning that there was high animal activity in summer and total inactivity on the surface in January. Absence of aboveground activity was directly confirmed in Owens Valley of eastern California by Kenagy (1973b), who none the less found that in the winter of 1969–70, during which surface food supply was high, *P. longimembris* was active all winter.

In the laboratory Lindberg and Hayden (1974) discovered that under summer temperatures pocket mice tended to occupy a location within the burrow system at about 20 cm depth, where temperatures were 26–30°C and where temperatures that aroused torpid subjects were achieved 2–3 h after sunset; locations closer to the surface would heat faster and thereby cause earlier arousal. In winter, locations of torpid subjects in burrows were at about 1 m soil depth, where winter soil temperatures were highest, but also below 31°C and often below the animal's lower critical temperature (French 1976, 1977). French determined that these pocket mice in the western Mojave Desert were dormant typically for about five months, mostly during the periods of coldest temperatures, but there was also evidence of daily estivation.

Unlike other rodents that experience hibernation, pocket mice do not store much fat, but rely on the burrow provision of seeds and fruits. About 130 g of cached seeds (millet) was sufficient to initiate torpor (French 1976), and the animal can utilize this cache any time without traveling to the surface. When the soil is heated sufficiently, either in spring or artificially in the laboratory, the animal returns to homeothermy at a rate of 0.6°C min^{-1} and begins foraging on the surface. French (1977) reported that males resumed foraging before females in springtime, and males also were most often captured in surface traps during a winter that followed a highly productive plant growing season.

The adaptive advantages of torpor for pocket mice are several. First, foraging is terminated during periods when little or no food is available on the surface (winter). Body energy is conserved by lowering basal metabolism and activity level (Bartholomew & Cade 1957; Bartholomew & Hudson 1961), and these tiny endotherms avoid high thermoregulatory costs during winter, which results from having a high surface/volume ratio. Water loss is minimized by spending much time within the burrow. Finally, by living in burrows much of the year, little pocket mice avoid competition with other heteromyid species, and simultaneously reduce the risk of being prey for community predators. On the last point, the early NTS study that used radioactive tracers produced evidence that noncaptive, tagged individuals lived at least 3–5 yr, a high longevity for a small rodent, suggesting that torpor may play a significant role by either slowing the aging process or reducing predation (French et al. 1966).

Estivation is also used as a physiological strategy by other desert organisms, including the ectotherms *Gopherus agassizii*, *Dipsosaurus dorsalis*, and *Uta stansburiana* and to lesser degrees by the endothermic *Ammospermophilus leucurus*, *Lepus californicus* (black-tailed jackrabbit), and *Peromyscus eremicus* (cactus mouse).

Burrow temperatures during hibernation are significantly lower than those during estivation, and some species employ group strategies to elevate burrow temperature. Antelope ground squirrel uses group huddling within the burrow, a strategy that provides higher winter burrow temperature than could be produced by a single individual (Karasov 1983). Group huddling is occasionally seen in desert tortoise and may be fairly common in snakes and lizards (Nagy 1988b), such as *Xantusia vigilis* in the Mojave Desert (Miller 1951), although huddling by ectotherms would not generate much heat and probably is a water-conserving adaptation by increasing relative humidity.

Facultative hyperthermia

A means of thermal adaptation, well documented for several desert endotherms, is the use of facultative hyperthermia, with which body temperature is allowed to exceed normothermia by several degrees during the hottest time of the day. This heat is temporarily stored, to be tolerated for a while. For some desert birds, controlled hyperthermia up to 46°C is tolerated presumably because this reduces the need to utilize evaporative cooling (Calder & King 1974). For other animals the excess heat is dissipated by conduction and radiation in the burrow or a shaded microhabitat.

One widely cited example of facultative hyperthermia is the antelope ground squirrel, which occurs commonly in Mojave desert scrub (Bartholomew & Hudson 1961; Hudson 1962; Bartholomew 1964, 1982b; Schmidt-Nielsen 1964; Hart 1971; Chappell & Bartholomew 1981; Karasov 1981). This species is seen even on the hottest days dashing at high speeds across the terrain in search of prey, although individuals also can be observed sitting quietly in shade, and it utilizes neither estivation nor hibernation. *Ammospermophilus* permits hyperthermia up to 43–44°C so that metabolic heat can be exchanged with its slightly cooler surroundings via radiation and conduction without resorting to evaporative cooling (Fig. 6.8). Once or twice per hour the hyperthermic animal must release enough heat to return to normothermia, and this is done by entering a burrow and unloading the heat via conduction and radiation with the soil of the burrow. After lowering its body temperature to 38°C, which only takes several minutes, the rodent resumes its aboveground activities.

Black-tailed jackrabbit also accepts a mild case of temporary hyperthermia, but it dissipates heat behaviorally by staying in shade and through evaporative cooling from respiratory surfaces (Shoemaker & Nagy 1976). In fact, jackrabbits must consume about 120 mg H_2 $kg^{-1} d^{-1}$ to employ that strategy for losing excessive body heat (Nagy, Shoemaker, & Costa 1976).

Hyperthermia is a term that applies strictly to

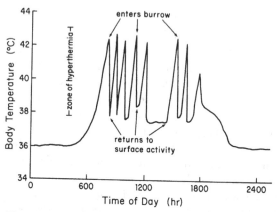

Figure 6.8 Diurnal changes in body temperatures of white-tailed antelope ground squirrel (*Ammospermophilus leucurus*). Points where high body temperatures force an animal in hyperthermia to re-enter a burrow and when lowered body temperature permits a return to the surface are shown by arrows. Adapted from Hudson (1962) and Chappell and Bartholomew (1981).

endotherms and not to ectotherms. All species of diurnal desert lizards of the Mojave Desert can tolerate body temperature in excess of 41°C, and *D. dorsalis* frequently operates at body temperatures of 45–47°C without permanent damage.

Countercurrent heat exchange systems

Vertebrates can prevent selected body parts from becoming too hot or too cold by making cardiovascular adjustments in the pathway of blood flow. Blood can be shunted near the surface of the skin to release excess heat to the environment or move through deeper vessels to preserve heat. Heath (1965, 1966) proposed a countercurrent heat exchange system whereby venous blood draining the head through the internal jugular vein loses heat to the cool arterial blood entering the head through the adjacent internal carotid artery. When a preferred body temperature is reached, contraction of the internal jugular constrictor muscle causes blood arriving from the body to be shunted into the external jugular vein, which has no countercurrent arrangement, and under these conditions the head receives cooler, body blood, allowing temperature differentials to occur between the brain and body. This mechanism may be important for keeping the brain cooler than body temperature and a reason for tolerance of short periods of hyperthermia.

Desert reptiles have received particular attention with respect to countercurrent heat exchange systems to prevent sensitive brain tissues from reaching lethal temperatures. An example is desert horned lizard (*P. platyrhinos*), but this probably applies to many other lizards and snakes and some birds. However, for small desert passerine birds, the actual difference in temperature may only be 0.2°C, a value too low to be physiologically significant. Such a system is important for large grazing endotherms (Taylor 1969; Taylor & Lyman 1972).

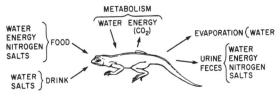

Figure 6.9 Diagram of typical influx and efflux parameters in water balance of a desert animal.

WATER BALANCE AND OSMOREGULATION

The heat balance of a desert organism is inseparable from the water budget because evaporative cooling is an important process to lower body temperature when it reaches a critically high level. Moreover, because the respiratory membranes of terrestrial animals are wet relative to the dry environment, some water vapor inadvertently is lost all the time. Two aspects of water relations of desert organisms make water loss a serious issue. First, during times of high temperature and low relative humidity, hence times when animal heat loads are high and evaporative losses would be maximal, there typically is no water to drink. Second, when the body loses water without replacing it, electrolytes in body fluids become more concentrated, thus putting the animal under physiological stress by departing from homeostasis. Residents of the desert that remain active during the summer must have a strategy to deal with a severe deficiency of water and an excess concentration of body salts.

Water flux rates

Water flux for an individual is defined as the rates of water gain and loss per day, which may result in either a net gain (if influx exceeds efflux) or a net loss (Fig. 6.9). Rarely do all gains and losses per day sum to zero because the processes involved are not directly linked. Water is lost by evaporation from the integument and respiratory tract and via excretion, i.e., feces, urine, and glands; water is gained from food, via drinking, or through the integuments (Table 6.1).

Relationship of water flux and body mass

Daily water efflux from free-ranging animals is strongly correlated with body mass, so that as body mass increases there is a proportional increase in total water flux (Fig. 6.10; Nagy & Peterson 1987, 1988). This is a linear relationship when presented as the log of water flux, expressed as milliliters per day, compared with log of body mass, expressed in grams. This relatively tight correlation can be expressed as a set of allometric equations, one for each category of vertebrates for free-ranging animals in the field:

Birds: $\log y = 0.137 + 0.694 \log x$
Eutherian mammals:
$$\log y = -0.487 + 0.818 \log x$$
Reptiles: $\log y = -1.185 + 0.726 \log x$

These equations were empirically derived by Nagy and Peterson (1988) from measurements obtained by numerous researchers, who have used tritiated or deuterated water to quantify the water influx of each individual. The slopes of the three lines (Fig. 6.10) do not differ significantly.

Figure 6.10 and the above equations help researchers to make initial predictions for water loss of terrestrial vertebrates. For example, a bird weighing 100 g should have a water flux that is

Table 6.1. *Sources of water gains and losses for terrestrial animals, including annotations to summarize water-conserving features that have been observed in typical warm deserts, adaptations that tend to minimize water losses.*

Gains	Losses	Water-conserving adaptations
Preformed (liquid) ingested food drink	Preformed excreted feces urine glands (incl. sweat, salivary, salt, etc.)	Feces with low moisture content; urine concentrated, specially as uric acid; urine retained as stored water (desert tortoise); sweating (in mammals) absent or used in economical ways for evaporative cooling; saliva not copious, not used for evaporative cooling (i.e., fur licking); salt glands (in some) concentrate electrolytes for excretion while resorbing water.
absorbed integument cloacal or oral membrane	evaporated integument respiratory tract	Skin or epidermis (exoskeleton) having low permeability to water and greatly reduced water loss (or gain); upper respiratory tract (of some) with narrow respiratory tract passage, designed to permit resorption of water within the nasal passage; (behaviorally) use of cooler microhabitats, which may be more humid, to lower body temperature and thereby decrease the water vapor concentration gradient between the animal and surrounding dry air (examples: burrows, rock crevices, and shade).
Metabolic oxidative water		

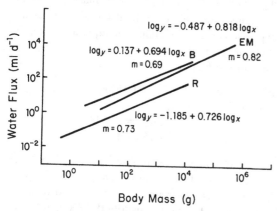

Figure 6.10 Allometric relations of log of water flux (ml d^{-1}) compared with log of body mass (g) for birds (B), eutherian mammals (EM), and reptiles (R). Redrawn from Nagy and Peterson (1988). Slope of line (m) is given.

approximately five times greater than a 10-g bird. It is also obvious from the figure that a 10-g bird would have a threefold higher water flux than a 10-g rodent, which in turn would be six times higher than a 10-g lizard. Such scaling relationships of structure or metabolism with body mass are well documented for a multitude of animal characters (Bartholomew 1982a; Schmidt-Nielsen 1984, 1990; Calder 1984).

Water economy index

Water economy index (WEI) is the ratio of daily water flux (m d^{-1}) to daily metabolic rate (kJ d^{-1}), an expression of water flux relative to energy metabolism (Nagy & Peterson 1988). Nagy and Peterson determined that free-living desert vertebrates tend to have a lower WEI than species living in nondesert habitats. The data for that conclusion are relatively few, but if correct the conclusion would be drawn that those species that can live in the desert are also capable of expressing tighter control over water loss, as some researchers have reported.

For arthropods Nagy and Peterson found that water flux is also scaled to body mass, but to the 0.943 power; however, the variance is extremely high for the species that have been sampled. None the less, desert, air-breathing species tend to have the lowest water flux rates as compared with insects from other habitats. For example, *O. ornatus*, the common desert millipede of North American deserts, loses water at a slower rate than other millipedes (Crawford 1972).

Evaporative cooling through the skin or epidermis

For desert animals, evaporative water loss through the skin (vertebrates) or epidermis (invertebrates), tends to be minimal or highly restricted. Under summer conditions, any highly evaporative surface area would be a potential pathway for an enormous loss of water, which is very scarce in the desert. Most vertebrates in the Mojave Desert do not sweat. Reptiles, birds, rodents, and carnivorous mammals do not sweat, but the introduced burro (*Equus asinus*) and secretive desert bighorn (*Ovis canadensis*) may, as they have sweat glands. Probably the best example of sweating in native mammals of North American deserts occurs in javelina (*Tayassu tajacu*), which tend to inhabit relatively lush desert regions where they can forage widely and consume juicy foods and drink water to replenish body fluids. Skin of terrestrial reptiles has low permeability for water, ions, and gases (Lillywhite & Maderson 1982).

Low cuticular conductance of water vapor characterizes typical invertebrates of the desert, and desert species tend to have lower evaporation rates than related species from mesic habitats (Edney 1977; Cloudsley-Thompson 1988). For example, the American desert scorpion (*Hadrurus arizonensis*) loses only 0.028% body weight per

hour in dry air at 30°C (Hadley 1974), and all scorpions have very low evaporation rates (Cloudsley-Thompson 1988). Chemistry of epicuticular lipids is considered to be particularly important for determining evaporation from the exoskeleton, and it is thought that desert species tend to have waxes with high melting points (Edney 1977; Hadley 1982).

Desert arthropods mostly are not able to tolerate water loss any better than related species in humid habitats (Edney 1974), but with their highly effective means of reducing epidermal evaporation, desert forms often need relatively little water. The small size and high surface/volume ratio of invertebrates rules out long-term evaporative cooling as a reasonable mechanism for lowering body temperature, because to maintain a given body temperature evaporation would have to be roughly proportional to the surface area (Schmidt-Nielsen 1964). Hence, large volumes of water would have to be lost per hour to accomplish effective cooling.

Evaporation from respiratory surfaces

The universal source of evaporative cooling is from the respiratory surfaces. To show how important this process can be on temperature regulation, cooling effect of evaporation was substantiated for *D. dorsalis* by Schmidt-Nielsen (1972), who observed that lizards placed in an environment of 30°C produced exhaled air that was 7°C lower than body temperature.

Mammals of Carnivora, including several top predators of the Mojave desert scrub, use a mechanism of panting to achieve thermal cooling (Crawford 1962; Schmidt-Nielsen 1964; 1972, 1990; Bartholomew 1982b). The nasal mucosa is the primary site for evaporation in this system, and the animal moves large volumes of air across this moist structure by sharp and rapid lung movement. Air movement may be unidirectional, inhaling through the nose and exhaling through the mouth, and this pattern maximizes evaporation. At an ambient temperature of 50°C, panting in canids can keep body temperature at 40°C.

Panting has also been documented as an important cooling mechanism in the Mojave lizards *D. dorsalis* (Templeton 1960) and *Crotaphytus collaris* (collared lizard; Dawson & Templeton 1963).

Because air temperature only occasionally exceeds the body temperature of birds (41°C), hyperthermia occurs relatively infrequently, but most birds pant when the condition arises. Panting is especially noticeable in the larger birds, such as the greater roadrunner (Calder & Schmidt-Nielsen 1967). Greater roadrunner can dissipate 78% of the metabolic heat by evaporation at an air temperature of 40°C and 137% at 44.5°C.

Several groups of middle-sized birds of deserts, e.g., mourning doves and owls, have a special cooling mechanism called gular flutter (Bartholomew, Lasiewski, & Crawford 1968; Schmidt-Nielsen 1972). The gular area is the floor of the mouth. During gular flutter, the gular area, which is highly vascularized, is moved rapidly with a relatively low expenditure of energy that does not involve the respiratory system. This releases much heat from the body. Doves, which perform gular flutter, are species that each morning fly, often over substantial distances, to springs, more mesic habitats, or human structures, to find drinking water to replenish what evaporated on the previous day (Bartholomew & Dawson 1954; Bartholomew & MacMillen 1960; MacMillen 1962).

Arthropod respiration involves spiracles, a set of minute pores through which water vapor diffuses and thereby affords some lowering of body temperature. Spiracular water loss increases with greater activity, which is in turn correlated with higher ambient temperature (Edney 1977). Coiling the body reduces evaporative water loss in *O. ornatus* (Crawford 1972).

Sources of water

For meeting their weekly water requirements, desert animals typically depend on water that is contained within, or physiologically extracted from, their food items (Schmidt-Nielsen 1964; Bartholomew 1982c; Minnich 1982). Insectivores and carnivores can obtain most or all of their water needs from the body fluids of their prey items, but water

stress sets in when hunting is poor and conditions of high evaporation occur. Among the favorite prey items of many insectivorous lizards are juicy larvae, which generally have 60–75% preformed water (Edney 1977). *Phrynosoma platyrhinos* consumes mostly harvester ants, which have higher water content than dry plants, but in order to do so they seem to possess a factor in their blood plasma that makes them resistant to ant venom (Schmidt, Sherbrooke, & Schmidt 1989). In general, adult terrestrial arthropods contain 62–77% water by weight, the percentage dependent mostly on the amount of stored fat (Edney 1977).

Especially herbivores have potentially large quantities of water that are available in fresh leaves, flowers, and fruits, which commonly consist of 80–90% water by weight. Therefore, during springtime, when winter annuals may be plentiful, a desert herbivore can extract considerable volume of water from fresh leaves while also gaining energy and nitrogen. The importance of water from fresh leaves has been well documented for a variety of Mojave Desert vertebrates, particularly *S. obesus* (Nagy 1972, 1973, 1975), *D. dorsalis* (Norris 1953; Minnich & Shoemaker 1970; Mautz & Nagy 1987), *G. agassizii* (Nagy & Medica 1977, 1986), *Lepus californicus* (Schmidt-Nielsen, Dawson, Hammel, Hinds, & Jackson 1965; Nagy, Shoemaker, & Costa 1976), and *Dipodomys microps* (Great Basin kangaroo rat; Kenagy 1972, 1973a). Larvae of butterflies and moths tend to have high rates of evaporation, but can afford to do so because they consume watery plant tissues; for example, larvae of *Manduca sexta* (hornworm) consume the large, juicy leaves of *Datura* (jimsonweed; Casey in Edney 1977).

Water stress from consuming dry foods

When favorite species of winter annuals die, desert animals are forced to forage increasingly on much drier plant materials and may instead show preference for watery tissues of perennial plants, such as the thickish leaves of *Ambrosia dumosa* (white bursage), flowers of *Larrea tridentata* (creosote bush), or occasionally even the water fruits of *Lycium pallidum* (box thorn). Typical summer food items of herbivores are dry and have very little extractable (preformed) water, but when desperate, herbivores will consume the juicy, mucilaginous fruits of cacti as a water source in midsummer. Some herbivores, particularly lagomorphs, which need much water to replenish that lost via evaporative efflux, in times of great water stress may also feed on cactus cladodes (*Opuntia*), while others supplement dry diets with ants, which contain over 50% water by weight (Edney 1977). Whenever the crop of winter annuals fails during a spring, populations of herbivores are placed at great risk.

Drinking water

Drinking water is of course a way that normally is not available to animals living exclusively within a desert scrub community during the summer. Large carnivores can range widely and therefore seek out water sources at higher elevations or, especially now, can visit human habitations, where standing water often can be found. However, under totally wild conditions *Canis latrans* is famous for the ability to dig around springs and within washes to reach water to a depth of at least 1 meter (Miller & Stebbins 1964).

Birds of any size also need daily or weekly drinks of water, and their ability to fly enables them to visit montane habitats, where natural springs may occur, or human facilities. Mourning doves (*Z. macroura*) make trips of several kilometers or more every morning to replenish water lost from body tissues in the preceding day. Bartholomew and Cade (1956) found that the house finch (*Carpodacus mexicanus*) drinks in all seasons of the year and obtains up to 44% of its water via drinking. Ground birds, such as the native desert quail (*Callipepla gambelii*) or introduced birds such as the chukar partridge (*Alectoris chukar*) from the Negev, also appear to require drinking water and therefore are found only where a source is locally available or standing water is provided through management procedures (Degen, Pinshow, & Shaw 1983; Pinshow, Degen, & Alkon 1983; Goldstein 1984; Goldstein & Nagy 1985). For the reason that desert birds for the most part

can fly and therefore leave desert scrub in search of water and nutrients, many biologists tend not to treat birds as true desert organisms. That notwithstanding, *Amphispiza bilineata* (black-throated sparrow), a common bird of Mojave desert scrub, is apparently more independent of drinking water than other passerines from North America (Smyth & Bartholomew 1966).

Reptiles will drink water in captivity, but were assumed to be nondrinkers in the field (Cowles 1979). However, for the desert tortoise, Medica, Bury, and Luckenbach (1980) documented a new phenomenon called nasal drinking. Although research on *G. agassizii* traditionally assumed that body water was derived solely from plant foods, Medica et al. observed that animals scraped out shallow depressions, 0.35–0.50 m in diameter and 30–50 mm deep, just before or during rainstorms, which served as water catchment basins and could hold a pool of water for up to 6 hours. These catchments were then used for nasal drinking, whereby the neck was outstretched and the snout was submerged. By drinking large quantities of standing rainwater, its large urinary bladder, which is used to retain urinary water, can be voided and refilled, via the kidneys, with rainwater. This was evidenced by observing uric acid and concentrated urine left at the drinking site. *Xantusia vigilis* (desert night lizard) and *Coleonyx variegatus* (banded gecko) have been observed to drink water droplets that condensed on the skin by lapping with the tongue (Lasiewski & Bartholomew 1969). No research has adequately evaluated how morning dew that condenses on plants may play a role in partial hydration of small animals of deserts, such as the Mojave Desert, where fog is not common.

Although the dromedary (*Camelus dromedarius*) is a common work animal of Old World deserts, during the last century these camels were introduced for trials to western North America. It was long assumed that the dromedary stored water in the hump or the rumen, giving it an extraordinary ability to travel for days in scorching desert without water, but massive water storage in these animals has no scientific basis (Schmidt-Nielsen 1964). Instead, a camel drinks at the end of a trek across the desert to restore water lost from its tissues; it tolerates an unusually high amount of dehydration (up to 27% of body weight) and at the same time has marvelous ways of reducing water loss (Wilson 1989). The thick fur coat acts as a barrier to heat, water loss is minimized from excreta, and sweating at the surface of the skin, hidden beneath the thick fur coat, can be greatly reduced when camels are dehydrated; they tolerate daytime hyperthermia and cool at night by radiation and conduction rather than by evaporation (Schmidt-Nielsen 1990).

Mojave Desert organisms do not have any unusual water-storing tissue or organs, with the possible exception of desert tortoises, which, as indicated above, have a large urinary bladder, and chuckwalla, which apparently has extracellular water storage in lateral lymph sacs (Minnich 1976, 1982).

Water requirements of desert amphibians

Amphibians are only seen in desert sites where standing water is available to successfully complete the egg and larval stages of the life cycle, for example, long-term oases and springs or freshwater summertime playas (Mayhew 1968; Shoemaker 1988). Amphibians characteristically have low upper lethal temperatures, highly permeable integument, and very dilute urine (Shoemaker 1988).

In the Mojave Desert, *Bufo punctatus* (red-spotted toad or desert toad) occupies those habitats in Death Valley and Joshua Tree National Monument, and in warmer deserts of North America *Scaphiopus* (spadefoot toad) is widespread. Spadefoot toads are seldom seen because each spends 10–11 months per year in a dormant stage within a burrow, which it makes with its characteristic shovel-like limbs. For this fossorial toad, enough water is extracted from the soil via osmosis through the highly permeable skin, especially the ventral epidermis, to avoid dehydration, and the waterholding capability of an amphibian burrow is enhanced by producing a mucus-like lining (Gordon 1982). *Scaphiopus* also has the ability to produce concentrated urine while in the burrow.

On the other hand, such amphibians are especially vulnerable to water stress outside the burrow because water is easily lost through that same porous structure.

Uptake of moisture by arthropods from fog

Arthropods have such a wide variety of food and water procurement strategies that for them it is difficult to generalize for desert life. An intriguing but very unusual mechanism of obtaining water occurs in fog deserts of Namibia, where at least two well-studied tenebrionid beetles, *Lepidochora argentogrisea* and *Onymacris unguicularis*, depend on fog moisture (Hamilton & Seely 1976; Seely & Hamilton 1976). These beetles stand along the crests of sand dunes assuming a stilted posture, which permits advective fog to condense as water droplets; water runs down to the mouth, where the beetle can drink freely.

In the California deserts, the desert cockroach (*Arenivaga investigata*) replenishes body fluids by absorbing water from unsaturated air within its underground burrow (Edney 1966; Edney, Haynes, & Gibo 1974). *Arenivaga* feeds on dead leaves of *Larrea*, *Atriplex*, and *Psorothamnus* by maneuvering from burrows just beneath the surface, but it must dig down into sand that has at least 82% relative humidity in order to extract water vapor via the salivary glands (Edney et al. 1974). Only the nymphs and wingless females of this desert cockroach perform this feat, and all forms drink by mouth.

Oxidative (metabolic) water

Granivores are an interesting subset of herbivores that rely on food items that are exceedingly dry, often having less than 3% preformed water content. Classical examples are kangaroo rats (for example, *Dipodomys merriami*), which do not drink water and survive almost entirely on seeds and other dry plant parts (Schmidt-Nielsen 1964, 1990). Such organisms obtain oxidative water (metabolic water), which is a product formed during the oxidation of carbohydrates,

$$C_6H_{12}O_6 + 6\ O_2 \rightarrow 6CO_2 + 6H_2O$$

$$(C_6H_{10}O_5)_n + n\ O_2 \rightarrow 6n\ CO_2 + 5n\ H_2O$$

as well as the more complex fats and proteins. Catabolism of 1 gram of glucose yields 0.60 g of water, and 1 gram of starch yields 0.56 g of water. Moreover, 1 gram of protein yields 0.50 g of oxidation water, and 1 gram of fat yields a remarkable 1.07 g of water. Particularly seeds are rich in carbohydrates or fats and frequently proteins, especially grasses (Poaceae).

Oxidative water certainly contributes to the water intake of all desert organisms, but its relative significance differs from species to species and group to group. Reptiles, for example, have much lower metabolic rates than endotherms, and as a result have lower rates of metabolic water production, but balancing that their rates of water loss by evaporation and excretion are also much lower than for endotherms. None the less, desert reptiles water loss tends to exceed gains from the production of oxidative water, therefore, without preformed water in the diet, these animals are in negative water balance (Minnich 1979, 1982).

Birds such as desert quail (*Callipepla gambelii*) consume a predominantly granivorous diet, but must still have a supplement of drinking water (Goldstein & Nagy 1985).

Respiratory resorption of water

Small mammals, such as heteromyid rodents of the Mojave Desert, and, to lesser degrees, small birds, for example, cactus wren (*Campylorhynchus brunneicapillum*), possess a useful water-conserving design of the respiratory tract wherein the nasal passages are extremely narrow (Schmidt-Nielsen, Hainsworth, & Murrish 1970; Gordon 1982; Schmidt-Nielsen 1990). When warm, moist air is exhaled, it exits through narrow nasal passages that have cool walls because water evaporates from that surface during inhalation. During exhalation, condensation thus occurs on the walls, and much water can there be recycled.

Osmoregulation

By and large the typical animal expends a great amount of energy to maintain a steady-state condition of the amounts of water and solutes in the body fluids and tissues (Gordon 1982). When the osmotic concentration of body fluids differs from the optimum, many physiological processes are negatively affected, and any can produce signs of stress on the organism. Terrestrial animals living under benign environmental conditions operate within a fairly narrow range of daily osmotic and solute concentrates around the optimum; when water content increases to the point where body fluids become diluted, excess liquid is released as urine; when water content decreases to the point of partial dehydration, water is obtained by feeding or drinking to replenish body tissues and water losses are reduced; when solute concentrations exceed desired levels, excess solutes are excreted. An animal living under stressful circumstances, for example, limited available water, must have behavioral and physiological methods to manage water loss and uptake with greater precision and generally the physiological capability to tolerate prolonged periods of dehydration and high solute concentration of body fluids.

Obviously, researchers of desert animals have made special efforts to investigate osmoregulation of these animals throughout the year, which will have times when the animal operates in a benign environment, with readily available water and solutes, and other times when animals are pushed to the limits of their tolerance. To limit water loss, the options available are (1) reducing water content of feces or urine, (2) preventing or reducing evaporation from the skin, (3) reclaiming respired water vapor via nasal countercurrent heat exchanger, and (4) salt glands, which are also a way to eliminate substantial quantities of dietary salt loads with little water loss (Schmidt-Nielsen 1964; Templeton 1964; Nagy 1988; Bradshaw 1988a). Many desert organisms have systems that tolerate osmotic concentrations in plasma and urine greater than the typical 320 mosM, for example, *Gopherus agassizii* (Nagy & Medica 1986) and *Camelus* (Schmidt-Nielsen 1964).

Concentrated urine and dry feces

Water conservation for desert animals favors mechanisms whereby urine is concentrated by resorption of water, and water is reclaimed before it exits the alimentary canal in fecal material. Whereas many organisms of mesic and aquatic habitats commonly excrete waste nitrogen as liquid urine containing dissolved urea, vertebrates and invertebrates of desert habitats show an overwhelming trend to excrete uric acid, which is crystalline and requires little water to be released, and fecal pellets can be relatively dry (Edney 1977; Minnich 1982; Gordon 1982; Schmidt-Nielsen 1990).

All birds excrete uric acid, but the maximum concentration of sodium or chloride is not very high relative to that of the blood because efficiency of the kidney to concentrate electrolytes is relatively poor (Dawson & Bartholomew 1968). For desert birds cloacal water loss typically exceeds 60% by weight of excreta, or at least 2% body weight, showing that birds have only moderate activities for reabsorbing water from excreta as compared with desert arthropods. None the less, water loss is decreased fairly significantly when birds are placed under water stress conditions, including their use of salt glands. However, at least in one species of the Mojave Desert, black-throated sparrow, renal extracts had extremely high concentrations of sodium chloride and high osmotic concentrations (Smyth & Bartholomew 1966), making this a species with verified physiological adaptations for summer drought conditions.

Mammals produce urea, but desert species tend to have concentrated urine consisting of much less water. Small rodents tend to produce very concentrated urea, as in the North American *Dipodomys* (kangaroo rat), which may have urine that exceeds 5000 mosM l^{-1}.

Salt glands

Nasal salt glands occur in many terrestrial lizards of deserts, including the majority of species in warm deserts of North America. These structures are designed to excrete potassium and sodium chloride from plasma without losing much water (Mayhew 1968; Minnich 1982). Salt can become encrusted around the nasal opening and on the snout and can be expelled by sneezing (Norris & Dawson 1964). Actually salt gland excretion is only pronounced in the herbivorous *Dipsosaurus dorsalis* and *Sauromalus obesus*, which take in very large quantities of potassium ions that must be eliminated efficiently by the body. Minnich (1970, 1976) determined that *D. dorsalis* is able to excrete 43% of the ingested potassium, 49% of the ingested sodium, and 93% of the ingested chloride by nasal salt glands, so that, even during summer drought, this animal shows no signs of electrolyte accumulation. Carnivorous lizards are less prone to accumulate high levels of electrolytes in the plasma, and hence show little salt gland activity.

Desert tortoises are herbivorous, but lack salt glands and accumulate high levels of electrolytes in the urinary bladder (Nagy & Medica 1986). Urine is retained as an internal water supply until it is evacuated when the animal drinks or when handled by humans. Terrestrial snakes and the venomous gila monster (*Heloderma suspectum*) of the Mojave Desert do not possess salt glands and therefore eliminate excess electrolytes mainly via the kidney–cloaca–bladder system (Minnich 1982).

Salt-excreting glands have also been reported for *Geococcyx californianus* (greater roadrunner) by Ohmart (1972), but they are not known in other Mojave Desert birds.

Levels of plasma electrolytes during drought

Physiological stress has been quantified in desert animals by measuring the levels of plasma electrolytes throughout the year. One finding of desert animals is that, by excreting potassium, sodium, and chloride ions via concentrated urine and salt glands, many species are able to retain homeostatic conditions while they are active aboveground. Carnivores consume watery, energy-rich foods and have few or no problems regulating plasma electrolytes, except when dehydration occurs, which must be remedied by locating drinking water or eating more food. Herbivorous species are the ones most subject to seasonal swings in electrolytes as the composition of the dietary quality changes.

An example of maintaining homeostasis of body fluids is *Sauromalus obesus*, a relatively large, herbivorous lizard (Nagy 1972, 1973, 1988a; Nagy & Shoemaker 1975). When it emerged from the winter hibernaculum, an individual consumed winter annuals containing abundant nutrients and energy-rich, watery tissues. Excess nitrogen from metabolism of amino acids, along with electrolytes that raise plasma osmolality were excreted with excess water as copious, dilute urine. As plants became desiccated in the approaching summer drought conditions, the diet consisted of drier materials and higher electrolyte concentrations, so urine became more concentrated and salt glands removed excess electrolytes from body fluids. None the less, during the drought the animal lost weight and entered daily estivation to avoid the most stressful times; as before, the diet was dry and contained abundant electrolytes, which were physiologically removed. During hibernation, weight loss and water flux were low (Nagy 1972).

Black-tailed jackrabbit is another herbivore that must consume lush vegetable materials and, if available, drink to supply its daily evaporative water loss (Nagy, Shoemaker, & Costa 1976; Shoemaker & Nagy 1976). Their urine is moderately concentrated, and feces is relatively dry (38% water). During the dry summer and fall months, preferred vegetable foods are not available, and jackrabbits are forced to consume dry and less palatable materials, woody stems and bark and, if available, watery stems of succulent plants; these items provide insufficient water, nutrients, and energy, so the animals are stressed and seek out more benign localities for survival.

Desert tortoise experiences the same transition from favorable herbaceous food resources in spring through dry foods in summer, but it deals with the changes in a different fashion. In spring, urine is stored in the urinary bladder, which becomes highly concentrated as the animal feeds on progressively drier plant materials that contain many ions (Nagy & Medica 1977, 1986; Minnich 1977). Plasma and blood urine have high osmotic concen-

trations (320–360 mosM; Danztler & Schmidt-Nielsen 1966) and therefore tortoises are under considerable physiological stress, and drink heavily and void the urinary bladder when they have a chance to drink rainwater, which drastically lowers electrolytes in the plasma but not to optimal levels (Medica et al. 1980). Thus, throughout the activity period of the year, desert tortoises are, for the most part, operating under physiological stress with elevated to very high levels of plasma electrolytes, but, by coping with this stress, the organism is capable of growing and surviving on an annual basis (Nagy & Medica 1986).

The species of bird most able to cope with dry Mojave Desert conditions is *Amphispiza bilineata*, which even under very dry conditions had a mean of 57% water in nonliquid excreta (Smyth & Bartholomew 1966), but it also was shown to concentrate urine quite well and rapidly. When vegetation is lush, black-throated sparrow obtains preformed water from its food and abandons drinking, and during dry periods it opportunistically obtains water by consuming termites, ants, and beetles.

Rodent granivores can maintain a very stable plasma level throughout the period of activity because they do not drink water and during most of the year do not eat food having abundant preformed water and high levels of salts. Urine is highly concentrated to excrete salts and conserve water, primarily obtained by oxidation of nutrients in its food. An exceptional species is *Dipodomys microps*, which feeds almost exclusively on fresh leaves of *Atriplex* (shadscale), which happen to have extremely high concentrations of salts, but this species of kangaroo rat has special teeth and behavior to trim away the salt-rich layers of the leaf to eat instead the more nutritious and less salty photosynthetic zone of the leaf (Kenagy 1972, 1973a,b). No studies have been done on osmoregulation of *D. microps*.

Adaptive designs of Mojave Desert vertebrates

The preceding pages described the range of adaptive approaches that animals employ in the winter-

Table 6.2. *Checklist of adaptations that aid common vertebrates of the Mojave Desert in maintaining water, salt, and energy balance.*

Adaptation	*Uta stansburiana*	*Dipsosaurus dorsalis*	*Gopherus agassizii*	*Crotalus cerastes*	*Dipodomys merriami*	*Perognathus longimembris*	*Ammospermophilus leucurus*	*Thomomys umbrinus*	*Lepus californicus*	*Amphispiza bilineata*	*Callipepla gambelii*	*Geococcyx californianus*
Osmoregulatory physiology												
Dry feces	+	++	+	?	++	++	++	?	++	+	+	?
Dry or concentrated urine	+	++		++	+++	++	+	?	+	++	+	+
Nasal salt glands	+	++					+	?	+	++	+	+
Storage of water			++	?							+	++
Low evaporative water loss	+	++	++	++	++	++	+	?	+	+++	++	?
Fluid volume regulation	?	++	?		++	++	?	?	+	+	+	+
Plasma ion regulation	+	++		+	++	++	?	?	+	+	+	+
Energy balance physiology												
Hyperthermia tolerance					+		++	+	+			+
Estivation or torpor	+	++	++	++	?	++	+	+	+			+
Hibernation		++	++	++	?	++						+
Fat storage	++	++	++	+	+	++	+		+	?	+	
Behavior												
Burrow use	++	+	++	++	++	++	++	++	+			
Nocturnal activity	++			++	++	++			+	++		
Drink water	?		++	?					?	+		
Group huddling		?	+		?	?	++			+	++	+
Dietary selection		+	+		++	++	++	++	+	++	++	++

rainfall Mojave Desert, and all of these features have been documented from other warm deserts of the world. No obvious classification emerges from these observations, no unique sets of species having particular adaptive designs for maintaining energy, water, and salt balance, as one finds in plants (Chapter 3). In lieu of a list of strategies, Table 6.2 summarizes the adaptations of 12 common vertebrate species, 10 of which are residents of Rock Valley. Much ecophysiological data is missing on even these widespread species, which emphasizes the amount of research that is still required to understand how the adaptations are integrated into the life history of these organisms.

ENERGY BUDGETS AND ECOLOGICAL ENERGETICS

A general energy budget

Heat budgets, discussed earlier in this chapter, can be used to analyze the sources of energy that permit animals to thermoregulate over a 24-hour period, which in turn determines part of the energy expenses of living per day. Chemical potential energy budgets are used to understand how ingested energy is partitioned into present-day maintenance, growth, and long-term survivorship, i.e., an integrated approach that considers impacts over longer periods of time. A simple chemical potential energy budget is described as an equation of rates:

$$I = F + U + M + P$$

where I is ingested energy, F is undigested energy that is excreted in feces, U is chemical energy that is excreted in the organic chemicals of urine, M is energy expended per unit time for metabolism, and P is energy used for production, i.e., for somatic growth, energy storage as lipids in the body, and reproductive structures, including eggs and embryos (Nagy 1988b). Hypothetically, by measuring these values over the course of a year, an annual energy budget can be constructed for each species and each population, showing which situations or seasons are energetically most important, most limiting, and most influential for controlling reproduction. This information can also be used to determine why certain species are more successful than others in a particular microhabitat.

As one can readily surmise, quantifying these parameters is not as simple as it seems. Belowground activity of animals is particularly difficult to quantify without perturbing the system. For ingestion, one can estimate caloric value of a daily food intake consisting of their typical foods using timed observations and analyses of stomach contents. Energy values for feces and urine are relatively easy to collect and analyze. Metabolism includes resting metabolism rate (RMR), typically measured during very short intervals on captive, starved animals, along with costs of movement and digestion, but as modified by body temperature (ectotherms) and thermoregulation (endotherms). Production values require whole-body sampling and biochemical methods that determine cost factors for making animal tissues from ingested nutrients.

Field metabolic rates

One important approach to studying metabolic rates and dietary intake in animals of diverse sizes and having different ecological requirements is in the measurement of field metabolic rates (FMR). FMR is the total energy cost of an animal in its native habitat for basal metabolism, thermoregulation, locomotion, feeding, predator avoidance, alertness, posture, digestion, growth, and reproduction (Nagy 1987).

One way to estimate FMR is called a time–energy budget. This is calculated by multiplying the duration of each type of the animal's activity (hr d^{-1}) by the approximate cost for conducting each activity (kJ hr^{-1}), then summing the products for all activities throughout the day sampled (Goldstein & Nagy 1985). This approach requires close surveillence of representative individuals during the course of sample days as well as measurement of each activity cost. Many researchers have recorded time budgets for a given population of birds, lizards, or mammals, but these are usually taken without knowing the energetic costs of each activity. Conversely, ecophysiologists often study the energetics of particular activities of animals, but have not recorded time invested in each per day per season under natural field conditions. In addition, some activity costs are difficult or impossible to sample under relatively natural conditions, for example, cost of reproduction, which does not occur as a discrete daily process, or integrated cost of thermoregulation for a 24-hour period. A time–energy budget for an animal with a simple activity schedule if derived this way may be fairly accurate, but accuracy depends very heavily on the correctness of the representative values used for the calculation.

More reliably, FMR can be measured using doubly labeled water. In this method, the washout rates of isotopes of hydrogen and oxygen are qualified after they are injected into field animals in the form of deuterated or tritiated water. The hydrogen isotope tracks primarily water loss, whereas the oxygen isotope washout rate, in equilibrium with oxygen in both water and respired CO_2, measures the summed loss of water and CO_2

(Nagy 1975, 1980). Many species of animals have now been studied using these techniques (Nagy 1988a, b; Nagy & Peterson 1988), providing an interesting database to assess FMR and body size in many groups of vertebrates from both desert and nondesert environments.

One clear picture from studies of FMR is that the energetic cost of living in endotherms is far greater than that of an ectotherm of the same size. During a spring or summer day, a mammal or bird of 250 g body mass spends about 310 kJ of energy per day for oxidative metabolism, while an iguanid lizard of the same mass uses only 19 kJ, a difference of 16× in energy required to survive a day. A portion of this difference is the result of higher mitochondrial densities, greater sodium transport, larger relative membrane surface area, and higher rates of enzyme and thyroid activity (Bennett 1972; Else & Hulbert 1981; Hulbert & Else 1981). However, a principal factor explaining the major difference in FMR is the metabolic response to daily thermal regimes. Endotherms must expend metabolic energy to maintain high body temperatures when ambient temperatures decrease at night, whereas ectotherms reduce both their body temperature and metabolic production of heat. Nagy (1987) calculates that the same 250 g endotherm has a RMR at 10°C ambient temperature that is about 200× higher than that of a lizard with similar mass at the same ambient temperature. These metabolic differences have profound implications for dietary needs of the two groups of organisms. In ecosystems where food resources are limited, ectotherms require far less food than do endotherms.

A number of authors have shown that climate and habitat type can affect the basal metabolic rate (BMR) of endotherms (Hulbert & Dawson 1974; McNab 1979; Weathers 1979; Dawson 1984). These studies suggested that desert birds and mammals have lower rates of BMR than those of endotherms from other habitats. Existing data on FMR in relation to body mass, synthesized by Nagy (1987), demonstrates that eutherian mammals from desert environments have FMRs that are 30% lower than eutherians (same body mass) in nondesert ecosystems. Similarly, desert birds

have FMRs that are less than half those calculated for nondesert species. For the two groups, slopes of FMR against body mass are not different for either mammals or birds, but the intercepts differ significantly. In contrast, when Nagy (1988b) compared FMR and other assimilation values of desert versus nondesert iguanid lizards, he found no significant differences.

Although the database is more limited, studies of FMR can tell biologists something about the metabolic implications of different dietary habits. Nagy (1987) showed that slopes of FMR versus body mass in eutherian mammals differed significantly between herbivores and carnivores. Small herbivores have somewhat higher FMR for a given body mass than do carnivores, with the relationship reversed for large body masses. However, for granivorous rodents there was no significant difference in FMR from that of small mammals having other diets.

Foraging efficiency

Foraging efficiency is calculated as the ratio of energy ingested while foraging to energy spent while foraging. Herbivorous lizards have significantly higher FMR than widely foraging insectivores and sit-and-wait insectivorous foragers (9.4–16.7, 1.4–2.5, and 1.1–1.6, respectively; Nagy 1988b), and the highest efficiency was recorded for S. obesus. Lizards also show clear trends in assimilation efficiency, i.e., $(I - F)/I$, with higher efficiencies associated with insect larvae and mammalian prey and lower values for plant materials that contain much inorganic and indigestible fibrous materials, although many terrestrial, herbivorous reptiles can digest and assimilate much of the cellulose (Iverson 1982; Karasov, & Diamond 1985).

For mammals, Nagy (1987) calculated that mean metabolizable energy contents for plant tissues is 10.3 kJ g^{-1}, of seeds 18.4 kJ g^{-1}, and of nectar 20.6 kJ g^{-1}. Insects provide 18.6 kJ g^{-1} for mammalian predators and 18.0 kJ g^{-1} for avian predators. Frugivorous and omnivorous diets provide a mean of 14 kJ g^{-1}.

Among the many interesting results from FMR

studies is a value on cost saving for reptilian estivation. For estivating desert tortoises, FMR were lowered by 50% (Nagy & Medica 1986), and for chuckwallas about 80% (Nagy & Shoemaker 1975). FMR of reptiles in winter hibernation was very low because body temperature is very low (Nagy 1988b).

Annual energy budgets

Using a series of regression equations developed by Nagy (1987), assuming a value for mean body mass for males and females, the mean daily FMR (kJ d^{-1}) and feeding rates (g d^{-1}) can be easily calculated. These values can be extended to an annual basis by multiplying by 365, and to a populational basis by multiplying by the number of animals per hectare. Annual estimates of food consumption should be increased by 1–3% to account for biomass allocation to growth and reproduction (McNeill & Lawton 1970; Turner 1970).

Annual energy budgets using seasonal means of FMR have been produced for only a few species that occur in the Mojave Desert. When body mass differences are normalized, desert tortoise and side-blotched lizard had similar assimilation and metabolic rates, but the lizard was three times more productive, and Nagy (1983) reported that energy allocated to reproduction in *U. stansburiana* was as high as 84% of total energy mobilized for single clutches of eggs. *Sauromalus obesus* had much lower rates of assimilation and metabolism and the lowest production rate.

Factors for initiating reproduction

Studies in all types of ecosystems have shown that females divert large quantities of energy and metabolites into reproduction. For example, in *U. stansburiana* 45–84% of energy flow, as determined by using doubly labeled water, was utilized for egg production (Nagy 1983). This species is extremely successful in desert habitats because it has rapid reproduction and may produce multiple clutches per year (up to seven; Turner, Medica, Bridges, & Jennrich 1982). Equally interesting is that a number of desert reptiles and endotherms

forego reproduction during any year when food resources are especially scarce (for example, in *S. obesus*; Nagy 1973, 1988a). Studies conducted on the kangaroo rat *Dipodomys merriami* have shown that there appears to be a linear dependence between ingestion of green vegetation and a reproductive response (Chapter 7).

SUMMARY

During hot, dry summer conditions, desert animals control body temperature in a variety of ways to avoid reaching lethal temperatures but typically by using water-conserving methods. For an animal experiencing hyperthermia, it can retreat to a cooler microclimate, such as a burrow, rock crevice, or shade of a shrub, where air temperature is lower and the body can lose heat passively to its surroundings. The largest animals lose heat very slowly and require a high rate of evaporative cooling to dissipate excess heat, this with water that must come preformed from its food or via drinking from sites outside the desert habitat. Many species opt to forage at night, when air temperature is significantly lower. However, many lizards and other ectotherms are diurnally active to utilize solar radiation for attaining preferred body temperature, and desert species often can tolerate a body temperature to 46°C. *Ammospermophilus leucurus*, a rodent, is a diurnal endotherm that tolerates near lethal hyperthermia before returning to its burrow to cool down. Small diurnal mammals are otherwise scarce because they have a limited capacity for heat storage and therefore would need to employ evaporative cooling, but small birds, which have higher body temperatures, are often warmer than the air and therefore lose heat passively by conduction, convection, or radiation. Some species, for example, *Perognathus longimembris*, can abandon homeothermy and enter estivation or torpor on a daily or seasonal basis, thereby also reducing energetic costs.

For free-ranging vertebrates daily water loss is tightly correlated to body mass and can be expressed as an allometric equation, but for the same sized animal a bird will lose 3 times more water than a rodent and 18 times more than a lizard. None the less, desert animals tend to have low water flux rates as compared with their closest nondesert relatives, in large part because evaporative loss from the surface is reduced and urine and feces tend to be highly concentrated or dry. Evaporation occurs largely from moist respiratory surfaces, although a few species have an adaptation to resorb nasal moisture during exhalation. Some birds actually increase water loss for evaporative cooling via gular flutter, and large mammals use panting. Most animals derive most of their water needs from the body fluids of their prey items or, for herbivores, from fresh plant leaves, given that drinking water only occurs following a heavy rain. The granivores consume items having less than 3% water, but can gain their water via oxidative (metabolic) conversion of carbohydrates, fats, and proteins, for example, in *Dipodomys merriami*, which never needs water to drink and can subsist on dry grain. When under water stress, there may be a buildup of electrolytes in body fluids, resulting in physiological stress, although excretion of uric acid or, less commonly, excretion of salt via nasal glands helps to minimize potentially toxic levels of those electrolytes.

Field metabolic rate is one measure of quantifying energetic costs of animals, and this is calculated by determining approximate cost per each activity during the year from combined time budget analyses and physiological measurements. Recently this has also been quantified reliably using doubly labeled water. Eutherian desert mammals have field metabolic rates that are 30% lower than related species of the same size, and herbivorous lizards have higher rates than insectivores. Thus, preliminary data on desert animals strongly suggest that their energy budgets are more efficient than those of nondesert habitats.

7 Mammals

In Mojave desert scrub during the hot daytime mammals are infrequently seen, but at night the desert landscape becomes a very active place when they emerge from their sites of safety to forage (Chapter 6). Chief among these are the small fossorial rodents, particularly Heteromyidae, which constitute a community of several different species and size classes busily harvesting dry fruits and seeds to cache within the burrows. Population densities of heteromyids characteristically show marked seasonal and yearly fluctuations at any site as well as between sites, and density changes have been monitored relatively easily because heteromyid rodents can be efficiently captured and released without injury. Therefore, these rodent communities have been heavily used in studies of patterns and processes in community ecology and population biology (Reichman 1991), particularly at long-term sampling sites, such as in Rock Valley.

MAMMALS OF THE NEVADA TEST SITE

The mammal fauna of the Nevada Test Site consists of 47 species (Allred & Beck 1963; Allred, Beck, & Jorgensen 1963; Jorgensen & Hayward 1963, 1965; O'Farrell & Emery 1976), of which half are rodents (Table 7.1). Many rodent species are members of the families Cricetidae (7 spp.) and Heteromyidae (9 spp.); these groups are common aridland species throughout western North America, including other Mojave Desert areas (Miller & Stebbins 1964). Other mammals occurring on NTS are wideranging species: 4 species of insectivorous bats, 3 species of shrews and 3 of lagomorphs, 7 species of carnivores, and 6 species of large grazers, including occasional wild horses (*Equus caballus*), domestic cattle (*Bos taurus*), and wild burros (*E. asinus*).

Clear biogeographic patterns can be seen in rodent distributions between the Mojave Desert communities in the southern portion of NTS and cooler Great Basin communities to the north. Typical of the Mojave Desert and Sonoran Desert are Merriam's kangaroo rat (*Dipodomys merriami*; Fig. 7.1) and desert kangaroo rat (*D. deserti*), desert wood rat (*Neotoma lepida*), cactus mouse (*Peromyscus eremicus*), little pocket mouse (*Perognathus longimembris*; Fig. 7.2), southern grasshopper mouse (*Onychomys torridus*), and round-tailed ground squirrel (*Spermophilus tereticaudus*). Characteristic of higher elevation Great Basin communities are Ord's kangaroo rat (*D. ordii*) and Great Basin kangaroo rat (*D. microps*), Great Basin pocket mouse (*Perognathus parvus*), sagebrush vole (*Lagurus curtatus*), and Townsend's ground squirrel (*Spermophilus townsendii*). In certain desert scrub localities, such as Frenchman Flat and Yucca Flat, small mammals from the warm and cold deserts are sympatric, thus re-emphasizing that the habitats there are transitional between the northern and southern types. Some rodent species, such as white-tailed antelope ground squirrel (*Ammospermophilus leucurus*; Fig. 7.3) and southern pocket gopher (*Thomomys umbrinus*, formerly *T. bottae*; Fig. 7.4), are widespread and occur throughout NTS. Fifteen species of rodents have been captured in *Larrea–Ambrosia* desert scrub (Allred et al. 1963),

Table 7.1. *Mammals of the Nevada Test Site (Jorgensen and Hayward 1965; Moor and Bradley 1987; Medica 1990). Species observed in Rock Valley (Turner 1973; Medica 1990) are indicated by an asterisk.*

Family	Species	Common Name
Arvicolidae	*Lagurus curtatus*	sagebrush vole
Bovidae	*Antilocapra americana*	pronghorn antelope
	Bos taurus	cow (domestic)
	Ovis canadensis	bighorn sheep
Canidae	*Canis latrans*	coyote
	Urocyon cinereoargenteus	gray fox
	Vulpes macrotis	kit fox
Cervidae	*Odocoileus hemionus*	mule deer
Cricetidae	*Neotoma lepida*	desert woodrat
	Onychomys torridus	southern grasshopper mouse
	Peromyscus crinitus	canyon mouse
	P. eremicus	cactus mouse
	P. maniculatus	deer mouse
	P. truei	pinyon mouse
	Reithrodontomys megalotis	western harvest mouse
Equidae	*Equus asinus*	burro (feral)
	E. caballus	horse (feral)
Erethizontidae	*Erethizon dorsatum*	porcupine
Felidae	*Felis concolor*	mountain lion
	Lynx rufus	bobcat
Geomyidae	*Thomomys umbrinus*	southern pocket gopher
Heteromyidae	*Dipodomys deserti*	desert kangaroo rat
	D. merriami	Merriam's kangaroo rat
	D. microps	Great Basin kangaroo rat
	D. ordii	Ord's kangaroo rat
	Microdipodops megacephalus	dark kangaroo mouse
	M. pallidus	pale kangaroo mouse
	Perognathus formosus	long-tailed pocket mouse
	P. longimembris	little pocket mouse
	P. parvus	Great Basin pocket mouse
Leporidae	*Lepus californicus*	black-tailed jackrabbit
	Sylvilagus audubonii	desert cottontail
	S. nuttallii	Nuttall's cottontail
Mustelidae	*Mustela frenata*	long-tailed weasel
	Spilogale putorius	wester spotted skunk
	Taxidea taxus	badger
Procyonidae	*Bassariscus astutus*	ringtail
Sciuridae	*Ammospermophilus leucurus*	white-tailed antelope ground squirrel
	Spermophilus tereticaudus	round-tailed ground squirrel
	S. townsendii	Townsend's ground squirrel
	S. variegatus	rock squirrel
	Tamias dorsalis	cliff chipmunk
Soricidae	*Notiosorex crawfordi*	desert shrew
	Sorex merriami	Merriam's shrew
	S. tenellus	Inyo shrew
Vespertilionidae	*Antrozous pallidus*	pallid bat
	Myotis californicus	California myotis
	Pipistrellus hesperus	western pipistrelle
	Plecotus townsendii	Townsend's big-eared bat

Figure 7.1 Merriam's kangaroo rat, *Dipodomys merriami.*

Figure 7.2 Little pocket mouse, *Perognathus longimembris.*

Figure 7.3 White-tailed antelope ground squirrel, *Ammospermophilus leucurus.*

Figure 7.4 Southern pocket gopher, *Thomomys umbrinus*.

Figure 7.5 Black-tailed jackrabbit, *Lepus californicus*.

whereas each of the other major NTS vegetation types may be inhabited by ten or more species of rodents.

The black-tailed jackrabbit, *Lepus californicus* (Fig. 7.5), typically is very conspicuous within communities on playas, bajadas, and middle-elevation woodlands. This species is most active in late winter and early spring and often shows dramatic declines in population size during late summer and fall, when food availability is extremely low, and then jackrabbits are restricted to scattered local sites with favorable conditions. Desert cottontail (*Sylvilagus audubonii*) is widespread in the major vegetation types, whereas Nuttall's cottontail (*S. nuttallii*) is restricted to Great Basin communities.

Of the seven free-ranging carnivores coyote on NTS (*Canis latrans*) and kit fox (*Vulpes macrotis*) are the most widespread, conspicuous, and numerous, particularly in *Larrea–Ambrosia* desert scrub of valley bottoms and bajadas. Other species are more typical of upland communities. Mountain lions (*Felis concolor*) are rare and secretive, usually restricted to the northern mesas, but a few sightings have been made as far south as Mercury (O'Farrell & Emery 1976).

Large grazers also are restricted to the higher elevations of NTS, where they can obtain adequate food and water. Mule deer (*Odocoileus hemionus*) occur on the high mesas in sagebrush desert scrub and pinyon–juniper woodland except during winter months (Jorgensen & Hayward 1965). Pronghorn antelope (*Antilocapra americana*) and desert bighorn sheep (*Ovis canadensis*) have been rarely observed at NTS.

To date the pallid bat (*Antrozous pallidus*), California myotis (*Myotis californicus*), western pipistrelle (*Pipistrellus hesperus*), and Townsend's big-eared bat (*Plecotus townsendii*) have been collected at NTS, but careful studies of bat populations need to be conducted to determine whether other species are present.

Water requirements and heat avoidance of Rock Valley terrestrial mammals

As reviewed in Chapter 6, there have been many ecophysiological studies of desert mammals that inhabit the Mojave Desert from sites having no sources of freestanding water to drink. Summer-active rodents and lagomorphs require either physiological or behavioral adaptations to tolerate extreme abiotic stresses encountered in desert habitats (Schmidt-Nielsen 1964, 1972; Nagy 1988a). Excellent pertinent research has been published from locations within the Mojave or Sonoran Desert on the two species of lagomorphs, which are herbivores, and most desert rodents, which are fossorial herbivores, granivores, or insectivores, that occur in Rock Valley.

Desert cottontail (*Sylvilagus audubonii*), studied near Tucson, Arizona, uses burrows year-round, but must forage aboveground every day. Hinds (1973) determined water loss curves for these rabbits and demonstrated high evaporative water vapor loss at high ambient temperatures. Adaptations were behavioral ones that maximized resting in shade and thus minimized activity during the most stressful daylight hours.

In a Mojave Desert site near Barstow, California, water relations and energy budgets were studied in detail for the black-tailed jackrabbit (*Lepus californicus*), which year-round became active within 30 minutes of sunset and reposed under shrubs or retreated to self-constructed burrows throughout the day (Costa, Nagy, & Shoemaker 1976). Jackrabbits had high water requirements, despite behaviorial adaptations to rest during daytime heat and the physiological capability to produce fairly concentrated urine (Nagy & Shoemaker 1976; Nagy 1988a). They consumed as much plant material as possible to obtain vital water during stressful summer conditions, but still could not maintain body mass and lost more water than they gained (Shoemaker, Nagy, & Costa 1976).

Ecophysiology of valley pocket gophers (*Thomomys umbrinus*) has been studied in desert habitats. These fossorial herbivores feed on roots and young plants found in the process of digging tunnels of their extensive burrow system and never forage beyond the edge of a burrow opening (Vleck 1979, 1981). Gettinger (1984) has shown in a southern California montane population that individuals of *T. umbrinus* were active day and night, regardless of season.

Rodents characteristically are active year-round and at Rock Valley were especially visible during spring and summer. However, *Perognathus longimembris* typically lives in the burrow in a state of torpor and hypothermia during the hottest part of the year and when food supply is exhausted (Bartholomew & Hudson 1961; Tucker 1965, 1966; French 1976, 1977).

Most desert rodents are granivores and nocturnal foragers, during hot days residing belowground in burrows, where temperature 0.5 m beneath the ground rarely exceeds 35°C (Schmidt-Nielson 1975). Southern grasshopper mouse and desert wood rat utilize preformed water, and their daytime water losses are minimized because burrows are nearly always saturated with water vapor (MacMillen 1972). Grasshopper mice are carnivorous, often eating orthopterans, and they can produce high urine concentrations of urea (MacMillen & Lee 1967). Heteromyid rodents, and in particular kangaroo rats (*Dipodomys*), are renowned for their ability to obtain oxidation ('metabolic') water from starch and fat metabolism of dry fruits and seeds (Schmidt-Nielsen 1964, 1990; MacMillen & Christopher 1976; Grubbs 1980). The most common diurnal rodent is antelope ground squirrel, which obtains most of its water from insects, supplemented by oxidation water, but it has a unique tolerance of hyperothermia and behavioral adaptations for using the burrow periodically to cool down (Chapter 6).

Carnivores obtain much water from their prey, but range widely enough to encounter or search for water during long drought periods.

Rodent species and body sizes in Mojave desert scrub communities

Rodents are key components of these desert mammal communities, being the most numerous consumers of plant materials and, at the same time, chief prey items of the top carnivores. A typical rodent community in the Mojave Desert may consist of nocturnal, carnivorous grasshopper mice, diurnal, ground squirrels, fossorial herbivorous gophers, and nocturnal granivores – often a guild of three or more species – that daily consume dry fruits and seeds, but seasonally forage on water-containing green leaves. Up to eight seed-eating species, some common and others infrequent or rare, may occur on a bajada or sand dune, whereas habitats with low productivity often have a smaller rodent fauna.

Much ecological research has been conducted on desert rodent communities to determine why and how members of the granivorous guild are able to coexist and be fairly diverse even though they appear to utilize similar resources. Brown (1973) was intrigued that each desert rodent community has species having different mean body weights. He live trapped rodents on 18 semistabilized sand dunes in western U.S. deserts, about half that were located in the Mojave Desert, and found that each dune could have been colonized by 13 granivorous species, but characteristically had fewer species. Moreover, common species were those having major differences in adult body weight. On each dune the common coexisting species differed by a body weight ratio (larger/ smaller) of at least 1.5, and rare species also tended to have body weights not represented by other species in that community.

The pattern of size classes of body weights can be seen in the Mojave Desert and western Great Basin Desert sites of Nevada and eastern California (Fig. 7.6), which resemble rodent communities found at NTS. For example, sand dune 4 in Pershing County, Nevada had *Dipodomys deserti* (*ca.* 100 g), *D. microps* (uncommon, *ca.* 60 g), *D. merriami* (*ca.* 38 g), *Peromyscus maniculatus* (18 g), and *Perognathus longimembris* (7 g). Brown hypothesized that the parameter of body size is involved in how the animals utilize the resources, and having different body sizes would suggest that they are somehow partitioning the resources to minimize interspecific competition (Brown 1973, 1975; Brown & Lieberman 1973; Brown, Reichman, & Davidson 1979).

Although many animal ecologists agreed with

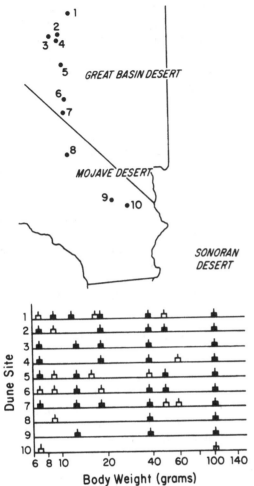

Figure 7.6 Distribution of body sizes for desert granivorous rodent communities in desert regions of the western United States. After Brown (1973).

desert communities of Sonoran and Great Basin deserts had nonrandom patterns and fit the model of a community partitioned by species having non-overlapping body weights. However, for granivores in the Mojave Desert the null hypothesis was not rejected, in part, the authors suggested, because the number of sampled sites was too small. By and large desert rodent communities generally have predictable categories of body size, having minimum ratios for species pairs of approximately two.

Another test of the competition model was made by Larsen (1986) in analyzing competitive release of species when other species of the guild are not present. According to the niche compression hypothesis of competition, if two or more species are using similar and overlapping resources, then breadth of microhabitat use should decrease when the number of species increases and conversely should increase when species are not present. According to another competition hypothesis, breadth of microhabitat use should expand when there is an increase in conspecifics using that resource. Larsen tested these hypotheses at desert sites in Nevada, Utah, and eastern California, and he found that breadth of microhabitat use decreased with an increase in number of potential granivorous competitors. *Dipodomys merriami* used a broader range of microhabitats when *D. ordii* was absent, suggesting that they are potential competitors and thus explaining why they are not commonly trapped at the same site. Merriam's kangaroo rat also expanded usage when its own population density was high, i.e., showing intra-specific competition. In contrast, *P. longimembris* exhibited no change in microhabitat utilization when occurring either with or without *Dipodomys*. This indicates that kangaroo rats and pocket mice generally are not competitors within a guild. However, earlier experiments, reviewed by Larsen, provided some data showing that within the genus *Perognathus*, just as within *Dipodomys*, interspecific competition, especially interference competition, would occur between coexisting species having similar body size.

Brown that adult body weight appears to be important in partitioning desert rodent communities, methods used to draw this conclusion were not judged to be rigorous enough to test this hypothesis against a stochastic model, one that would not invoke competition (Connor & Simberloff 1979; Strong, Szyska, & Simberloff 1979). Using a much larger data set Bowers and Brown (1982) made tests to evaluate the null hypothesis that these desert communities are random assemblages with respect to body size. They found that

POPULATION DENSITIES OF MAMMALS IN ROCK VALLEY

Records of mammals in Rock Valley

Surveys (Table 7.1) of Rock Valley identified 21 species of mammals, including rodents (11), carnivores (5), bats (3), and lagomorphs (2). Although workers expected to find shrews, particularly the desert shrew (*Notiosorex crawfordi*; Jorgensen & Hayward 1963), multiple years of pitfall trap studies produced no specimens. None of the bats known from NTS roosted on the validation site.

Over the years coyote, kit fox, bobcat (*Lynx rufus*), badger (*Taxidea taxus*), and western spotted skunk (*Spilogale putorius*) were the carnivores observed in Rock Valley, but especially during IBP studies no effort was made to capture them. Coyotes and kit foxes, with carnivorous birds, were the major top carnivores of small mammal and reptile prey of this desert site. Field observations made in 1971 estimated that one to four individuals of coyotes and also of kit foxes occupied the primary 50-ha site, but they were infrequently seen (Turner 1972).

Both the jackrabbit and desert cottontail were commonly observed but not trapped on the validation site. Jackrabbit density was studied on five occasions, April of 1971–3, September, 1971, and July, 1972, by conducting drives across the study site (Balph 1971). Few sightings were made in July and April (*n* = 3–7), but 17 jackrabbits were observed in September, 1971 (Maza 1974). No studies were made to record autumnal differences for dry versus wet years for either jackrabbits or carnivores.

Survey methods for censusing heteromyid rodents

Over the years several different designs were used at Rock Valley to live-trap nocturnal rodents. The earliest inventories (Allred et al. 1963; Jorgensen 1963) used double line transects and 6.24 ha grids. Subsequent studies of rodent activities by Jorgensen and Hayward (1965) used two simple transect techniques. One was a single U-shaped transect containing 200 Young-type small mammal live-traps placed at 9.2 m intervals. Their second design was a grid having six linear transects, each with 12 traps spaced 22.9 m apart. For the later ecological studies in Rock Valley (French et al. 1966; Maza, French, & Aschwanden 1973; French et al. 1974), live traps were placed 15 m apart in a 165 m × 165 m (12 × 12 traps) rectilinear grid pattern covering 2.72 ha. Each trapping station could have one or two set traps, and the entire grid could be shifted to a slightly different position by moving traps diagonally half the distance to the next row of traps. Obviously, estimates of rodent activity, population density, and home range sizes could be and were very different, depending on the trapping protocol, and the 2.72-ha rectilinear grid was eventually adopted for research within the enclosures before and during the IBP project.

Densities of heteromyid rodents within enclosures

Five species of nocturnal heteromyid rodents were live-trapped during the many studies at the Rock Valley, *P. formosus*, *P. longimembris*, *D. deserti*, *D. merriami*, and *D. microps*. Most published papers and unpublished reports of NTS desert communities have found that the most abundant rodents were pocket mice (Beatley 1976b), especially *P. formosus* or *P. longimembris*. In Rock Valley plots A, B, and C (Fig. 2.15), which were locations for rodent studies conducted in the 1960s, *P. formosus* was abundant and had population densities up to 62 individuals per hectare, whereas the other species was infrequent. A short distance away on the IBP plots 11 and 12 *P. longimembris* were especially common and *P. formosus* had lower densities. A simple explanation for that pattern is that *P. formosus* prefers soils that are sandy and loose in texture, whereas *P. longimembris* inhabits coarse, firmly packed and rocky soils (Turner 1972; Bowers 1986).

In Rock Valley, the three larger heteromyid species, all kangaroo rats, were typically less abundant than pocket mice in the study plots, and *D. deserti*, which prefers loose, sand dune-like habi-

Table 7.2. *Densities of rodents trapped on the Rock Valley validation site during spring 1971–1975. All values are for combined numbers of adults and young of the same year.*

Species	1971 May	1972 April	1973 April	1974 April	1975 April
Perognathus formosus	0.93	0.32	0.66	1.96	0.15
Perognathus longimembris	9.81	5.73	7.29	20.46	4.12
Dipodomys merriami	0.86	0.55	0.15	1.13	0.99
Dipodomys microps	0.50	0.44	1.84	3.33	0.58
Ammospermophilus leucurus	0.16	0.00	0.00	0.39	0.34
Onychomys torridus	0.13	0.22	0.23	0.26	0.13
Peromyscus crinitus	0.09	0.02	0.01	0.00	0.04
Other	0.00	0.00	1.11	2.96	1.11
Total	12.48	7.28	11.29	30.49	7.46

tats, was rarely trapped – only 12 from 1963 to 1971 – hence, densities were never determined.

Table 7.2 contains data on changes in population densities of the four common heteromyid species at Rock Valley as studied for the IBP project. *Perognathus longimembris* was the most abundant species and constituted over two-thirds of total spring and summer densities, and with *P. formosus* pocket mice constituted about 80% of the heteromyid populations throughout the study. Moreover, these species commonly showed great variance in seasonal and yearly densities. For example, at Rock Valley densities of *P. longimembris* ranged from 28.5 individuals ha^{-1} in summer, 1973, to lows of 1.1 individuals ha^{-1} in summer, 1971, and spring, 1975, whereas the other ranged from 10.7 individuals ha^{-1} in summer, 1973, to a low of 0.2 individuals ha^{-1} in summer, 1971, and spring, 1975. The coincidence of high and low values most likely correlated with high survivorship under conditions of favorable food resources in summer, 1973, and extremely low survivorship during two stressful seasons when either food resources or predation kept population densities low, respectively. Maza et al. (1973) observed large changes in population densities of *P. formosus* during a seven-year study (1962–8) in plot C.

Densities of kangaroo rats were, on average, similar to those observed between 1963 and 1968

in studies that were conducted within plots A and C (French et al. 1974). For *D. merriami*, densities ranged from 0.2–1.1 individuals ha^{-1} in spring and 0.4–1.1 individuals ha^{-1} in summer, whereas for *D. microps* estimates were 0.4–3.3 individuals ha^{-1} in spring and 0.4–2.5 individuals ha^{-1} in summer. Highest densities occurred for *D. merriami* in 1974 and 1975 and for the other in 1973 and 1974. Both species were fairly evenly distributed within the plots.

Population densities of heteromyid rodents, especially pocket mice, are known to fluctuate dramatically at desert locations (Whitford 1976), even though no major change has occurred in the perennial vegetation. Reynolds (1958) hypothesized from Arizona data that high rodent densities followed seasons of abundant winter annuals. To test that hypothesis from 1963 to 1968 Beatley (1969a, 1976b) collected data on autumn precipitation, early spring densities of annuals, and postreproductive, summer rodent densities at 15 permanent study plots within Jackass Flats, 8 in the south, central, and east (Site Group A) and 7 in the north and west (Site Group B). For Site Group A, which had a fairly good crop of winter annuals in 1964, subsequent rodent reproduction was moderately successful (Fig. 7.7). The next year rodent populations were extremely low when Jackass Flats received almost no autumnal rain and the crop of annuals mostly failed, but in 1966, after heavy

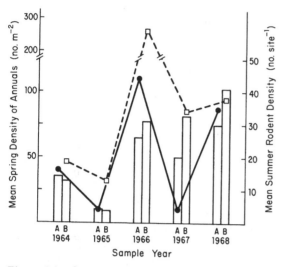

Figure 7.7 Relationship of rodent population size and production of winter annuals at two sites on the Nevada Test Site. After Beatley (1969a). The annual plant densities are shown for sites A (solid line) and site B (dashed line), while histograms show rodent densities.

rains and extraordinary annual production, rodent populations strongly rebounded. In 1967 another poor crop of annuals preceded a slight depression in rodent densities, but in the final year of study heteromyid populations increased by 50% to their highest level following another burst of annuals. For Site Group B a similar pattern prevailed, although from 1966 to 1968, years with abundant annuals at those plots, rodent populations remained high.

Analyses of fluctuations in rodent densities

Wide year-by-year fluctuations in heteromyid rodent populations in Rock Valley were analyzed for data from 1963 to 1968 and the IBP studies for 1971 to 1975. The status of spring populations is largely influenced by numbers of young weaned during the previous year as well as survivorship of adults during the fall and winter months. For Rock Valley sites, French et al. (1974) characterized 1963, 1964, and 1967 as years of poor reproduction, 1965 and 1968 as good, and 1966 as very good. Reproduction in 1971, prior to the IBP study, was poor; it was adequate to good in 1972,

remarkably good in 1973, poor in 1974, and adequate to good in 1975. French's team concluded that variability in reproduction resulted principally from changes in litter sizes and incidence of pregnancy, two variables which themselves are positively correlated. Under very favorable conditions, even the young-of-the-year reproduce, yielding maximal rates of population increase, as in 1965 during the study by French and again in 1973 in a study by Chew (1975).

To date, most observations have been generally consistent with Beatley's conclusions that winter annual productivity, in one of several possible ways, strongly affects rodent reproduction. Shoots of herbs was found to comprise 30–36% of the diet of *D. merriami* in Nevada populations during the reproductive season, but was minimal in non-reproductive months (Bradley & Mauer 1971).

Relationship of rodent reproduction to green vegetation

Studies conducted at sites in the Sonoran Desert provided some evidence that successful reproduction in *Dipodomys* follows the ingestion of green vegetation (Van de Graaff & Balda 1973; Reichman & Van de Graaff 1975). By quantifying reproductive status of males and females in field sites where annual vegetation was either abundant or virtually absent, these investigators found a significant linear dependence between ingestion of green vegetation and a reproductive response.

From research conducted at plots on NTS, Beatley (1976b) strongly held the same view, noting that for *D. merriami* some exogenous water, obtained by consuming winter annuals, appeared to be necessary under field conditions in desert habitats to initiate reproductive activity but not necessarily in moister environments. However, closely examining *D. merriami* from the Mojave Desert, Soholt (1977) found no significant correlation between onset of reproduction in a sample period and nature of the diet in the preceding period, for example, in females for the occurrence of estrus or events during gestation or in males for changes in their reproductive status. With laboratory experiments Soholt demonstrated that fresh vegetable matter mainly provided daily free water, which is critical for lactation.

It is possible still that rodents may be obtaining physiologically stimulating substances, such as estrogenic factors, from the vegetative materials, thereby indirectly controlling heteromyid rodent reproduction (Chew & Butterworth 1964; French et al. 1974).

Some field data obtained in Rock Valley still raise questions about the relationship of winter annuals to spring reproduction. During at least two years rodent reproduction was not apparently correlated with either rainfall or primary productivity in the expected way. Rainfall between October, 1964 and March, 1965 was low, approximately 30 mm, yielding extremely sparse populations of winter annuals, and yet rodent reproduction during spring, 1965 was good (French et al. 1974). In one enclosure densities of *P. formosus* increased from about 5 to over 50 individuals ha^{-1} between the spring and mid-summer sampling dates of 1965. Another year, 1972, had low winter rainfall but good rodent reproduction in spring. Moreover, poor reproduction of rodents in 1974 seemed aberrant following a season with good winter rains (about 70 mm) and a dense crop of annuals. Other studies in Rock Valley on factors influencing reproduction have also been made.

Further insights into the factors influencing these rodent population sizes emerged from experimental studies with *P. formosus* in plots A and C (Chew, Turner, August, Maza, & Nelson 1973; Chew 1975). A fence was constructed to bisect plot C into north and south halves of about 4.4 ha each. In 1972 densities of pocket mice were then adjusted so that spring densities began in the ratio of 10 : 4 : 1 in the three sample areas. Chew et al. determined that the number of young weaned per successful pregnancy, incidence of sexual activity, duration of sexual activity, and longevity or survivorship were all inversely related to initial population density.

If increasing interactions between pocket mice negatively affects reproduction and survival, then the effect could be alleviated by artificially reducing such stresses. In 1973, therefore, Chew ran experiments in plot C to manipulate rodent densi-

ties and observe the consequences. Previous sheet metal barriers were arranged in an orderly pattern in C south, *Perognathus* were trapped outside the enclosure and added to both halves of plot C, so that by late June 225 pocket mice occupied C south and 199 were in C north, and reproduction led to subsequent increases in both populations (Chew & Turner 1974). The study predicted that pocket mice in C north, because animals were not confined to patches, would (1) show less intense sexual activity and of shorter duration, (2) wean fewer offspring per pregnancy, (3) exhibit higher mortality, and (4) tend to move more frequently through the barrier fence. Of these predictions, number 2 was rejected, numbers 1 and 3 were partially supported by observations, and number 4, the least interesting one, was supported.

Another result of the 1973 experiment by Chew et al. was the contradiction to 1972 observations, when reproduction and survival were inversely related to density. Because pocket mice were added in large numbers to plot C and spring reproduction was very successful that year, artificially high densities of pocket mice were reached, almost 100 individuals ha^{-1} during August, whereas French et al. (1974) obtained highest summer densities only half that value. Although the high 1973 pocket mouse densities were artificially induced, they were sustained without unusual mortality through October, so that density-dependent effects, observed in 1972, were not manifested.

Chew (1975) attempted to integrate his experimental results using *P. formosus* with earlier findings by French et al. (1974). His approach was to compute an index of pocket mouse reproduction (RP) by dividing the number of adult females present at the start of the breeding season by the number of young registered later in that year. These RP values were based on data supplied by French (1963–8) and from the experiments conducted in plot C between 1971 and 1973. Chew computed an average RP for each year, based on various plot-specific values available, including winter annual production estimates for 1963–5 by Beatley (1969a, b) and for 1966–8 reported by Wallace and Romney (1972). Herbage biomass values for 1972 were those reported for Zone 20 of the validation site (Turner 1973, 1975; Turner & McBrayer 1974) or measured in plot C (Nelson & Chew 1977). The question was posed whether RP values were significantly correlated with the spring density of females, herb biomass, or both. Chew (1975) used nonparametric rank correlation tests because of apparently nonnormal dispersions of variables (even following transformations to logarithms). He concluded that RP was not correlated with herb biomass but was inversely correlated with spring densities of females.

These analyses were repeated using several modifications and corrections. First, three of the female density values used in the first analysis were incorrect, being roughly twice the correct values. Second, plot-specific data were used rather than means from all observations in the same year. Plot-specific herb biomass estimates were not always available, in which cases the single available value for all plots was used. This approach increased the number of cases to 25.

The reanalysis, using the above corrections, showed that the rank correlation test (Snedecor 1956) was nonparametric and tested the hypothesis that there was no correlation between rankings. The statistic r_s may range from -1.0 (complete discord) to $+1.0$ (complete accord), and its statistical significance depends on the sample size. With 25 cases r_s must be >0.38 or <-0.38 (5% level) or >0.49 or <-0.49 (1% level) to reject the null hypothesis. Rank correlation tests involving first RP and herb biomass and secondly RP and female rodent density gave r_s values of -0.04 and -0.48, respectively. Hence, in the first instance there was no significant relationship, but the second test strongly rejected the null hypothesis of no correlation, suggesting a significant correlation between reproductive index and density of female pocket mice. Likewise, tests using multiple regression analysis of similar data sets demonstrated that female density entered as the first variate with a multiple R of 0.46 at the 5% significance level, whereas herb biomass did not even enter.

Chew (1975) reviewed the original studies by Beatley at NTS along with those of Van de Graaff

and Balda (1973) and Reichman and Van de Graaff (1975), but he could generate little support for a direct, quantitative relationship between net primary production by winter annuals and subsequent rodent populations. He commented that 'there is no basis for deciding whether the vegetation is a trigger to reproduction, or for what it is necessary'. This statement needs to be tempered. First, there certainly are data indicating an effect of winter rainfall, which in some way influences growth and quality of vegetation. Chew's rank correlation tests of Beatley's 5-year sample strongly rejected the null hypothesis of no correlation between densities of winter annuals and numbers of rodents trapped between July and September. Several years of the Rock Valley data can also be interpreted in the same manner. None the less, Chew (1975) interpreted the 1973 data to mean that when food resources are sufficiently high, density-dependent constraints on heteromyid rodent reproduction can be overridden.

In analyzing these studies retrospectively it is difficult to say exactly what conclusions are correct. The data were instructional in identifying at least two processes affecting reproduction in *Perognathus*, food productivity by annuals and animal population density, factors that apparently do not interact as in a two-factor regression model, but possibly, under extreme circumstances, in a mutually exclusive manner. Rock Valley observations were also made without knowledge about patterns of predation on rodents within the plots, and this ultimately needs to be added in generation of a more robust model for natural processes of population biology.

Natural densities of other rodents in Rock Valley

Population studies at the Rock Valley validation site were also carried out with three species of nocturnal cricetid rodents, *Onychomys torridus*, *Neotoma lepida*, and *Peromyscus crinitus*. Whereas several individuals of *Onychomys* were captured per night, this amounted to spring densities ranging from 0.13 individuals ha^{-1} (1975) to 0.26 ha^{-1} (1974). Summer densities ranged from 0.1 (1971) to 0.4 individuals ha^{-1} (1975). Relatively higher numbers in 1974 apparently followed from the favorable conditions during the 1973 breeding season. The other two species were rarely trapped and therefore their true abundance was unknown. *Neotoma* occurred mainly among rocky outcrops at the base of the mountains (i.e., south of the validation site) or occasionally within a clump of *Yucca schidigera*. Because yuccas were uncommon, no reliable density estimates were obtained. *Peromyscus crinitus* only inhabited rocky areas and along drainage channels. As far as is known, none of these species affected in any significant way the models for heteromyid rodent populations.

The diurnal *Ammospermophilus leucurus* and *Spermophilus tereticaudus* were present in all plots at Rock Valley, and especially *Ammospermophilus* were most efficiently trapped with $5 \times 5 \times 16$ inch Tomahawk live-traps set and inspected regularly during the day. Estimated spring densities ranged from zero (1972 and 1973) to 0.4 ha^{-1}. Antelope ground squirrels were easily observed at all times during the study, hence the zero values reflected insufficient trapping effort, possibly also because trapping was done during periods of low abundance. Elsewhere Bradley (1968) observed a home range of 36.8 ha with a daily range of 10 ha and that individuals were seldom trapped in the same half of their ranges, cautioning one not to draw any conclusions for the validation site. Few data have been published on the population biology of *S. tereticaudus* (Neal 1965; Hawbecker 1975).

Burrowing *Thomomys umbrinus* were present in low numbers within all enclosures, and a $26\ \mathrm{m} \times 48\ \mathrm{m}$ grid was laid down 35 m southwest of plot C to study their population biology. These animals were live-trapped using either a metal box with a spring-loaded door or a plastic tube with a metal closure, each baited with fresh apple or sprayed with apple juice. Burrows were excavated to determine the relationship between gopher density and extent of burrow systems (Dingman & Bandoli 1973; Dingman & Byers 1974). Animal abundance was estimated as 3.8–4.7 ha^{-1} during the course of the study. Those densities were

rather low for pocket gophers, which can have populations at least ten times that amount (Howard & Childs 1959), but Rock Valley has low herbaceous productivity and relatively poor soils for gophers. These studies reported that substantial amounts of two species of *Ephedra*, but especially *E. nevadensis*, were consumed by *T. umbrinus*.

FORAGING ACTIVITIES OF HETEROMYID RODENT COMMUNITIES

Interesting surveys of rodent distribution with respect to vegetation types have been carried out at NTS and other sites within the Mojave Desert, beginning with a short-term study in 1959 and 1960 when Allred and Beck (1963) trapped rodents in five distinct vegetation zones in Yucca Flat and Frenchman Flat to show relative abundances of the rodent species in relation to plant cover. This was followed in the same locations with extensive trapping and further analyses of seasonal and diurnal activity of rodents (Jorgensen & Hayward 1965). In 1962 Beatley began her data collections to determine relationships between plant productivity and rodent populations, a study which resulted in a very sound regional analysis and synthesis of kangaroo rat occurrence with respect to plant community structure (Beatley 1976b, c).

Preferred habitats of heteromyid rodents

Beatley concluded, after many years of analyzing plant associations and trapping records, that there is a nearly perfect correlation of *Larrea* Mojave desert scrub with the occurrence of large populations of *D. merriami* (Fig. 7.8). Preferred habitats of *D. microps* are the transitional *Grayia–Lycium* and *Coleogyne* communities and Great Basin *Artemisia* desert scrub, and highest animal densities occur where mean annual precipitation is about 200 mm and where shrub cover is high (23–51%). Beatley (1976c) concluded that the '*D. microps* environment', where density of that species is highest, has greater shrub cover than sites for the other species of *Dipodomys*. *Dipodomys*

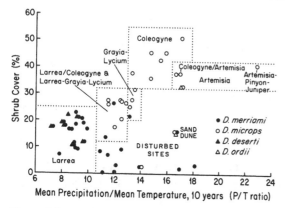

Figure 7.8 Distribution of *Dipodomys* species relative to vegetation type at the Nevada Test Site. Adapted from Beatley (1976c).

Modes of foraging relative to shrub cover

Shrubs are extremely important for foraging heteromyid rodents, not only as resources for food, but also apparently for protection, and there have been intensive studies of shrub and open space utilization by the different species. Working in Mojave desert scrub near Joshua Tree National Monument, Thompson (1982a) found that *D. merriami*, *D. deserti*, and *P. longimembris* spent about 85% of their foraging time in shrub microhabitats and, conversely, they purposely avoided open areas. None the less, all individuals at some time had to cross open areas to forage under different shrubs.

An individual of either *D. deserti* or *D. merriami* typically visited eight or more shrubs per foraging bout (Fig. 7.9) and had a slight tendency to forage at the periphery of large shrubs, where it

deserti mostly occurs in low *Larrea–Ambrosia* desert scrub in very loose sand habitats with low rainfall. Occurrence of all three species within a given plot, such as the IBP validation site, generally signifies that within the plot were microhabitats that each of the species preferred. Heteromyid rodents occur at low densities in communities dominated by shadscale, *Atriplex confertifolia*, but here there are small populations of *D. merriami* with few individuals of *D microps*.

As described in Chapter 6, *D. microps* differs from other kangaroo rats because individuals cannot survive exclusively on a granivorous diet, and normally the leaves of *Atriplex confertifolia* are harvested every night as a source of free water (Kenagy 1972, 1973a, b). *Dipodomys microps* apparently forages on shadscale leaves year-round, and therefore one would expect that occurrence of this kangaroo rat should be coincident with the shrub on which it depends, and generally this is seen. None the less, Beatley reported that thriving populations of *D. microps* occur at NTS where no shadscale is present, suggesting that alternative leaves are utilized in the diet. Moreover, *D. microps* occurs in all-deciduous *Grayia–Lycium* communities on Yucca Flat, where no fresh leaves are available for several months each year.

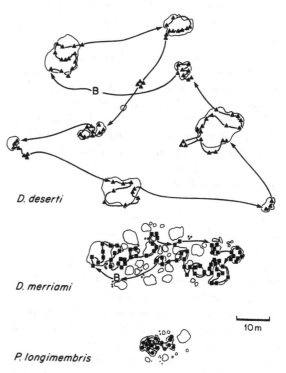

Figure 7.9 Foraging pattern of *Dipodomys* and *Perognathus* within a desert community of heteromyid rodents in Rock Valley. B = burrow. Adapted from Thompson (1982a).

Table 7.3. *Mean seed densities (number m^{-2}) under the canopies of three shrub species and in intershrub spaces at Rock Valley at two dates in 1972. Data from Nelson and Chew (1977).*

| Month | Under canopy | | | Intershrub spaces |
	Ambrosia dumosa	*Larrea tridentata*	*Lycium andersonii*	
June	4586	8028	7397	570
October	1948	3989	3166	231

stopped about six times per minute to dig and glean seeds and dry fruits (Thompson 1982a). While foraging kangaroo rats moved slowly, but during transit between shrubs they traversed open space very quickly (Thompson 1985, 1987). Transit speeds in excess of 26 km h^{-1} have been recorded for *D. deserti*, which inhabits sites where vegetation is sparse; however, this very large species seems to have the greatest preference for shrub microhabitats. Interestingly, when the tiny *P. longimembris* coexists with *D. deserti*, the largest kangaroo rat, foraging by kangaroo rats increases in open areas, suggesting that competition is acting to limit habitat overlap. *Dipodomys deserti* is also the species that spends the most time observing rather than foraging, probably to help detect predators in its fairly unprotected habitat. None the less, *D. deserti* is commonly seen foraging in open areas, actually in three times as many open-area sites as conspecifics, where it digs and gleans frequently. Hence, *D. deserti* appears to use open-area resources more than the other species, probably because its rapid bipedal locomotion offsets its risk of predation, but its mean trip distance is 22.1 m, only half that of *D. merriami*, which is foraging mostly under shrubs (Thompson 1982a).

Individuals of *P. longimembris* tend to stay close to the burrow entrance and forage slowly and in a course-grained manner under nearby shrubs. The little pocket mouse, smallest of the granivorous guild (7 g), moves very slowly and appears to dig and glean almost continuously while sorting and sifting the uppermost 1–2 mm of soil in search of food, harvesting scattered, individual seeds. An individual tends to visit only one or two shrubs per foraging bout and have a maximum transit speed up to only 2 km h^{-1}, typically across very short distances (Fig. 7.9). Hence, *P. longimembris* clearly avoids open areas and does not appear to utilize open areas for much of its foraging.

Pattern of seed distribution for foraging granivores

Nelson and Chew (1977) investigated the nature of seed reserves in open areas (EX, exposed areas) and under shrubs of suitable size (UC, under canopy) within plots A and C, where population studies were being conducted from 1972 to 1974 on *P. formosus*. Dry fruits and seeds were carefully screened from soil, including analyses of the top 2 centimeters of soil, in which pocket mice forage. Undershrub seed densities were at least five times greater than in exposed areas of plot C with high rodent densities, having means in 1972, a year of low herb production, of 2435–7682 m^{-2} for UC and 231–570 m^{-2} for EX (Table 7.3). In 1973 seed reserves were at least ten times greater because spring herbage had very high production, and even the foraging of artificially high densities of *Perognathus* in plot C (100 individuals ha^{-1}) did not seem to markedly decrease the seed reserve. In 1974, large seed reserves were able to accommodate high rodent density in plot C even though herb production that spring was average. It is unlikely that the seed reserve in most moderately productive desert scrub community is limiting or is a factor in controlling the diversity of species found there.

Many researchers have suggested that species of kangaroo rats prefer to harvest dense clumps of seeds, whereas pocket mice tend to harvest scattered, individual seeds. For example, Bowers

(1982) found that *D. merriami* foraged for a low diversity of seeds, whereas *P. longimembris* and *P. formosus* collected a wide range of seed types. These data support the hypothesis that kangaroo rats forage for rich clumps of very common species while pocket mice harvest the seeds encountered as they sift through soil. However, during foraging experiments in laboratory arenas, individuals of *P. longimembris* harvested mostly clumped seeds when tested with a homospecific opponent or individuals of *D. merriami*, whereas they harvested predominantly scattered seeds, and fewer of them, when opposed by much larger individuals of *D. deserti* (Thombulak & Kenagy 1980). These three species were collected on the same desert sand dune in Nevada and commonly coexist there, hence the foraging interactions of these three species may be more complex than hitherto appreciated.

Foraging in open areas versus beneath shrubs

Although direct observations demonstrate that kangaroo rats and pocket mice spend approximately equal foraging times under shrubs, *Dipodomys* are caught more often in open-area livetraps as they cross from shrub to shrub. This sample bias casts suspicion on the accuracy of live trapping data for reconstructing nocturnal patterns of heteromyid rodent activity (Thompson 1982b).

Numerous recent studies, especially some conducted in Nevada, have demonstrated that foraging of heteromyid rodents tends to be reduced during nights when the moon is full or whenever plots are illuminated artificially (Kotler 1984a, b, 1985; Price, Waser & Bass 1984; Longland & Jenkins 1987; Bowers 1988). For example, Kotler (1984a) and Bowers (1988) found that feed trays placed in open areas were used at a slower rate when sites were illuminated at night. Thompson (1982b), working in sites near Joshua Tree National Monument, noted that moonlight avoidance is practiced by *D. merriami* and *D. microps*, but not *D. deserti*. Kotler (1985) also showed that the quadrupedal species of *Perognathus* were captured at higher frequencies by owls and therefore were at

greater risk than faster-moving, bipedal kangaroo rats. This perhaps explains why pocket mice prefer to forage under shrubs, i.e., for greater safety. However, Jorgensen and Hayward (1965) detected no correlation between nighttime activity and presence of a full moon at NTS sites, and Kenagy (1976) in Owens Valley found no apparent influence of moonlight levels on nighttime activity of kangaroo rats, which showed activity in light levels two to four orders of magnitude brighter than that at full moon.

Heteromyid rodent home ranges

Beginning in 1959 research on desert scrub communities at NTS identified numerous interesting activity patterns of rodents (White & Allred 1961; Allred & Beck 1963; Jorgensen & Hayward 1965), and data from those early studies, conducted at Yucca Flat and Frenchman Flat, stimulated detailed studies on activity patterns of Mojave Desert heteromyid rodents and the dynamics of their home ranges.

It has long been known that several species of *Perognathus* are seldom caught in traps during winter months, and in desert communities *P. formosus* is rarely seen after October and *P. longimembris* after September at NTS (Jorgensen & Hayward 1965; Nelson & Chew 1977; French et al. 1974). This is especially true when surface food supplies dwindle. As discussed in Chapter 6, the large species of pocket mouse has been shown to utilize torpor (Bartholomew & Cade 1957), indicating that torpor is physiologically and ecologically important in this species. Bartholomew and Cade speculated that using torpor may be a major reason why pocket mice have high longevity relative to other heteromyid rodents (French et al. 1966; Maza, French & Aschwanden 1973), and this is a way that species can use similar resources for very different life-history strategies.

Kenagy (1976) performed a fine-tuned study of daily and seasonal activities of *D. merriami* and *D. microps* at a Mojave desert field site in Owens Valley in eastern California. First, he determined that the two species did not differ in time when nighttime surface activity began, and over 80% of the animals were active on the surface within 20 minutes of activity onset. Second, Kenagy could find no diminution of surface activity by kangaroo rats with relationship to moon phase or brightness, i.e., their emergence from burrows may not be a direct response to ambient light intensity.

Perhaps the most exhaustive rodent study done at NTS was a 7-year monthly census of marked individuals of *P. formosus* in plot A and some comparisons with rodents in plots B, C, and D to elucidate home range dynamics of enclosed populations (Maza et al. 1973). Data were collected during the 1960s, but computer analyses had to be developed before the complex data set could be properly analyzed. Analyses were eventually based on monthly movements of 340 individuals that had been captured at three or more different locations within the enclosure and at least on ten different occasions. Principal focus of the analysis was to determine shapes and sizes of home ranges. Data obtained by Maza and co-workers clearly showed that home ranges of these pocket mice were essentially circular (Fig. 7.10). Average size of a home range, in which the individual spent 39% of the time, had a radius of

15.5 m, a radius of 31.0 m for 86% of the time, and of 46.5 m for roughly 100% of the time. When an animal shifted its home range, it did so usually with one major long-distance excursion, and then it set up a new circular home range. Sexually active males were those that showed about half of the long-distance excursions, and these mostly during spring reproductive months, and male home ranges typically were larger than those of females. When population density was high, home range appeared to be lower. Mean radius of circular home ranges of kangaroo rats was greater for *D. merriami* (r = 37.3 m) and significantly smaller for the larger species *D. microps* (r = 26.5 m). Surprisingly, the smallest heteromyid, *P. longimembris*, had a home range (r = 19.0 m) somewhat larger than *P. formosus*, even though the reverse would be expected based on body size considerations.

Assembly rules of heteromyid rodent guilds

Obviously, the dynamics of granivorous heteromyid guilds in desert communities are more complex than we currently understand. Although each community tends to have distinct size classes (body weights) for adults of the constitutive species (Brown 1973), still it is unclear which general ecological factors could produce this pattern for each assemblage, because unambiguous examples of interspecific competition are few. Large and small individuals appear to forage in the same microhabitats, utilizing similar seed and dry fruit resources although harvesting in somewhat different ways, and they often construct their burrows side by side. Seed resources are known to be highly variable in distribution, and their density may itself be a component of microhabitat structure (Price & Reichman 1987; Chapter 11). All species cache seeds in large quantities, more than they ever use. The species have different modes of locomotion and ways of using shelter, thereby presenting different predation risks, but carnivorous mammals and birds hunt them all. Moreover, each species has a different mean home range, but range of an individual by and large overlaps those of other species as well as conspecific individuals. Cer-

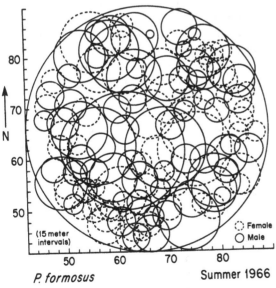

Figure 7.10 Foraging pattern of *Perognathus formosus* in plot A, Rock Valley. Adapted from Thompson (1982a).

Table 7.4. *Mean dry body weight of small rodents at Rock Valley. Data reported here are calculated fresh weights of adult male and female animals collected in April and July, 1972.*

| Species | Male | | Female | |
	Mean (range)	n	Mean (range)	n
Perognathus longimembris	2.6 (1.8–3.6)	110	2.5 (2.1–3.6)	142
Peromyscus crinitus	5.4	1	4.8	1
Perognathus formosus	6.4 (6.0–7.8)	10	6.2 (6.0–6.6)	3
Onychomys torridus	6.4 (5.4–7.5)	9	8.0 (6.6–11.4)	8
Dipodomys merriami	13.2 (11.1–15.3)	22	13.4 (11.1–16.2)	13
Dipodomys microps	19.7 (17.1–22.2)	13	17.3 (13.2–20.7)	11
Thomomys umbrinus	22.3	10	–	–
Ammospermophilus leucurus	23.0 (20.0–25.5)	5	22.4 (21.0–24.0)	5

tainly the species are not partitioning only one desert resource, or else, if they are, important parameters have not yet been identified.

In lieu of a single explanation, ecologists continue to cite lists of apparent ecological differences between species as ways that granivores coexist. To date studies have not been designed specifically to study minute-to-minute social interactions between species and conspecifics either at low population densities, when food resources and space are not limiting, or at high densities, when food resources and space are in short supply. None the less, studies in enclosures at Rock Valley informed researchers that even the highest population densities observed did not cause species to disppear from the guild. Studies are needed throughout the colonization phase of these guilds, to learn whether new arrivals encounter competition from established species, and if so along which ecological or behavioral parameters.

SECONDARY PRODUCTION BY MAMMALS

Mean dry body weights of adult rodents at the Rock valley study site are given in Table 7.4. Males and females were not significantly different in weight for any of the eight species. Dry body weights of the very small adults of *Perognathus* and *Onychomys* were 2–6 g, whereas those of *Dipodomys, Ammospermophilus,* and the infrequent individuals of *Spermophilus* and *Neotoma*

(data not in Table 7.4) were 120–130 g and 120–180 g, respectively. Seasonal changes in body weight of *P. formosus* were observed in 1966, when values were highest in late spring to early summer and declined during the hot and dry conditions of late summer.

SUMMARY

Terrestrial mammals of Rock Valley are the typical native species found in the northern Mojave Desert, including 11 species of rodents, 2 species of lagomorphs, and 5 species of carnivores. Fossorial rodents, especially the granivorous Heteromyidae, are key components of this *Larrea–Ambrosia* desert scrub community. Kangaroo rats and pocket mice nightly harvest large volumes of seeds and dry fruits and cache them in burrows for later use.

The great majority of informative mammal studies done at Rock Valley were conducted on the heteromyid rodents within fenced plots. Population densities of common species fluctuated dramatically from season to season and year to year. High survivorship usually occurred under conditions of favorable food resources and low survivorship during stressful seasons. Research data from Rock Valley suggest that variability in reproduction in kangaroo rats resulted mostly from changes in litter sizes and on inception of the reproductive cycle, both strongly affected by winter annual productivity because green vegetation is a prerequisite for reproductive success of common species. For *Perognathus* certain analyses of Rock Valley data suggest a significant correlation between reproductive index and density of females, indicating that density-dependent constraints and food resources must be used together for a complex model for these rodent populations.

Much field data have been collected on foraging activities of heteromyid rodents, in an attempt to determine how coexisting species can use the same food items and habitat. *Dipodomys deserti* appears to use open area resources more than other species, perhaps because its rapid bipedal locomotion offsets risk of predation, while other species forage mostly hidden beneath shrubs. Foraging experiments have demonstrated that individuals of a species harvest mostly clumped seeds when placed under certain conditions and predominantly scattered seeds under other ones, suggesting that foraging interactions within multispecies heteromyid rodent communities are complex. Large and small individuals of different species appear to forage in the same microhabitats, utilize similar food resources, and construct their burrows side by side.

8 Reptiles

The herpetofauna of the Nevada Test Site (Table 8.1; Allred, Beck, & Jorgensen 1963; Tanner & Jorgensen 1963; Tanner 1969; O'Farrell & Emery 1976; BECAMP 1991a) consists of 15 species of lizards (suborder Sauria), 17 species of snakes (suborder Serpentes), and desert tortoise (*Gopherus agassizii*, order Testudines). The reptiles are ecologically diverse, occupying a full range of habitats from lowland desert scrub, playa margins, and washes to the highest pinyon–juniper (*Pinus–Juniperus*) woodland. Several characteristic species of the Mojave Desert have not been observed at NTS, but otherwise this preserve possesses an excellent profile of the regional reptiles and has been a prime location for studying their ecological characteristics.

REPTILES OF THE NEVADA TEST SITE AND ROCK VALLEY

Most species of reptiles that occur in Mojave Desert habitats at NTS (Table 8.1) are those also found nearby at Death Valley National Monument as well as in the southernmost localities at Joshua Tree National Monument (Miller & Stebbins 1964; Rowlands, Johnson, Ritter, & Endo 1982), although at each preserve some slightly distinctive geographical races may occur. All three preserves have populations of desert tortoises (*G. agassizii*). The three snake faunas are roughly comparable, but JTNM has more species because it has habitats supporting six species of rattlesnakes (*Crotalus*). None of the three preserves has reported the venomous *Heloderma suspectum* (banded gila monster), which is more common in the Colorado

subdivision of the Sonoran Desert (Stebbins 1985); however, gila monsters have been sited near NTS on the north side of the Spring Mountain Range (Medica, pers. comm.).

Uta stansburiana, side-blotched lizard (Fig. 8.1), is the most widely distributed species of lizard at NTS, observed in every major habitat type. Also very common are *Cnemidophorus tigris* (western whiptail lizard), *Callisaurus draconoides* (zebra-tailed lizard; Fig. 8.2), *Gambelia wislizenii* (long-nosed leopard lizard; Fig. 8.3), *Phrynosoma platyrhinos* (desert horned lizard; Figs. 8.4–8.5), and *Sceloporus magister* (desert spiny lizard; Fig. 8.6), which occur in Mojave Desert and transitional desert communities.

In Mojave Desert sites Allred et al. (1963) reported 11 species of lizards in *Larrea–Ambrosia* communities, 7 species in *Grayia–Lycium* communities, and 6 in disturbance sites dominated by *Salsola* (Russian thistle). Transitional desert sites with *Coleogyne* (blackbrush) and *Atriplex–Kochia* contained 8 and 6 species, respectively. Low species diversity of lizards occurred in pinyon–juniper woodland, where only three species have been found. These include three species that are not seen in Mojave desert scrub, *Sceloporus occidentalis* (western fence lizard), *S. graciosus* (sagebrush lizard), and *Eumeces skiltonianus* (western skink), as well as the nearly ubiquitous *U. stansburiana*. Four species are largely restricted to the Mojave Desert Region: *Coleonyx variegatus* (banded gecko; Fig. 8.7), *Dipsosaurus dorsalis* (desert iguana; Fig. 8.8), *Sauromalus obesus* (chuckwalla; Fig. 8.9), and *Xantusia vigilis* (desert night lizard; Fig. 8.10).

Table 8.1. *Reptiles of the Nevada Test Site (Allred et al. 1963; Tanner & Jorgensen 1963; Tanner 1969; O'Farrell & Emery 1976; BECAMP 1991a). Species observed in Rock Valley are indicated by an asterisk.*

Family	Species	Common Name
Lizards:		
Gekkonidae	*Coleonyx variegatus	western banded gecko
Iguanidae	*Callisaurus draconoides	zebra-tailed lizard
	*Crotaphytus collaris	collared lizard
	Dipsosaurus dorsalis	desert iguana
	*Gambelia wislizenii	long-nosed leopard lizard
	*Phrynosoma platyrhinos	desert horned lizard
	*Sauromalus obesus	chuckwalla
	Sceloporus graciosus	sagebrush lizard
	*S. magister	desert spiny lizard
	S. occidentalis	western fence lizard
	*Uta stansburiana	side-blotched lizard
Scincidae	Eumeces gilberti	Gilbert's skink
	E. skiltonianus	western skink
Teiidae	*Cnemidophorus tigris	western whiptail
Xantusiidae	*Xantusia vigilis	desert night lizard
Snakes:		
Colubridae	*Arizona elegans	glossy snake
	*Chionactis occipitalis	western shovel-nosed snake
	Diadophis punctatus	ring-necked snake
	*Hypsiglena torquata	night snake
	*Lampropeltis getulus	common kingsnake
	*Masticophis flagellum	coachwhip
	M. taeniatus	striped whipsnake
	*Phyllorhynchus decurtatus	spotted leaf-nosed snake
	*Pituophis melanoleucus	gopher snake
	*Rhinocheilus lecontei	long-nosed snake
	*Salvadora hexalepis	western patch-nosed snake
	*Sonora semiannulata	ground snake
	*Tantilla planiceps	California black-headed snake
	Trimorphodon biscutatus	western lyre snake
Leptotyphlopidae	*Leptotyphlops humilis	western slender blind snake
Viperidae	*Crotalus cerastes	sidewinder
	*C. mitchellii	speckled rattlesnake
Turtles:		
Testudinidae	*Gopherus agassizii	desert tortoise

The desert tortoise, *Gopherus agassizii* (Fig. 8.11), has a fairly broad distribution in lowland Mojave and Sonoran deserts, including southeastern California, southern Nevada, the extreme southwestern corner of Utah, the western half of Arizona west of the Mogollon Rim and also east of Tucson, and most of lowland Sonora, Mexico, but fossil records document that it occurred at least 10 000–12 000 years ago in Chihuahuan Desert of southernmost New Mexico (Van Devender, Moodie, & Harris 1976). Within its current range they inhabit sites with sandy loam, light gravel, or heavy gravel soils, not those of loose sand or alkali hardpan, because the soil must have good denning potential, i.e., be firm enough to resist cave-ins of the long, wide, semicircular burrows (Figs. 6.4–6.5; Woodbury & Hardy 1940, 1948; Luckenbach 1982; Schamberger & Turner 1986). Desert tor-

Figure 8.1 Side-blotched lizard, *Uta stansburiana*.

Figure 8.2 Zebra-tailed lizard, *Callisaurus draconoides*.

Figure 8.3 Long-nosed leopard lizard, *Gambelia wislizenii*.

Figure 8.4 Desert horned lizard, *Phrynosoma platyrhinos*.

Figure 8.5 Desert horned lizard, *Phrynosoma platyrhinos*.

Figure 8.6 Desert spiny lizard, *Sceloporus magister*.

Figure 8.7 Banded gecko, *Coleonyx variegatus*.

Figure 8.8 Desert iguana, *Dipsosaurus dorsalis*.

Figure 8.9 Chuckwalla, *Sauromalus obesus*.

Figure 8.10 Desert night lizard, *Xantusia vigilis*.

Figure 8.11 Desert tortoise, *Gopherus agassizii*, photographed in Rock Valley.

toises live in desert scrub where shrubs, in addition to burrows, provide cover, and a majority of the burrows and pallets, especially of juveniles, are protected by canopies or basal branches of shrubs (Burge 1978; Berry & Turner 1986).

Most species of snakes are seen infrequently in NTS desert habitats because they tend to be active at only certain times of the day (many are nocturnal), mostly have camouflaging color patterns, are present typically in low densities, and tend to avoid contact with humans. Consequently, there have been relatively few studies on snakes at NTS.

Chionactis occipitalis (western shovel-nosed snake) is the most commonly seen snake, particularly in *Larrea–Ambrosia* and *Grayia–Lycium*

communities on bajadas, where these snakes are active from May to August (Allred et al. 1963). Two species of rattlesnakes are present, *Crotalus cerastes* (sidewinder) and *C. mitchellii* (speckled rattlesnake), most notably in *Larrea–Ambrosia* desert scrub. The speckled rattlesnake occurs relatively frequently in Rock Valley. In addition to the rattlesnakes, three other species of small, nocturnal snakes are venomous, but are not considered dangerous: *Trimorphodon biscutatus* (western or Sonora lyre snake), *Tantilla planiceps* (blackheaded snake), and *Hypsiglena torquata* (night snake).

HABITATS AND NICHES OF ROCK VALLEY LIZARDS

In Rock Valley alone about 70% of the reptile species at NTS have been observed. On bajada slopes with *Larrea–Ambrosia* desert scrub and along the washes researchers have found 7 species of lizards, 14 species of snakes, and a thriving population of desert tortoise (Table 8.1). Three other species of lizards are present in the rocky hillsides surrounding Rock Valley: *Crotaphytus collaris, Sauromalus obesus* (Fig. 8.9), and *Xantusia vigilis* (Fig. 8.10).

The natural history and ecology of desert lizards in the western United States have been popular research topics for the past 50 years, resulting in over a thousand scientific articles. A significant number of those studies were conducted on populations within the Mojave Desert, mostly at, or adjacent to, Death Valley, Joshua Tree National Monument, and NTS, especially at Rock Valley. As a general rule, the habitat choice, overall dietary preferences, life history, and behavior of wideranging lizards are similar from site to site, and overviews of lizard natural history can be obtained from several references (Stebbins 1954, 1985; Miller & Stebbins 1964; Mayhew 1968; Pianka 1986).

Characteristics of the common species

Uta stansburiana is an extremely common and widespread sit-and-wait forager of western North America, active nearly year-round, although at typical sites being seen mostly from March through November and showing little or no midday activity during clear, hot days from early May to mid-September (Tinkle 1967; Alexander & Whitford 1968; Kay 1970; Parker & Pianka 1975). In Rock Valley *Uta* is generally not found on the surface from mid-December to mid-January (Medica, pers. comm.). The side-blotched lizard searches from elevated perches, such as rocks or logs, even small mounds that build up at bases of shrubs, to increase surveillance and thereby enlarge the foraging area. An individual typically makes a short dash for its prey, for example, grasshoppers, crickets, beetles, ants, termites, and insect larvae, and then returns to a perch. In fall and spring individuals have been observed consuming cotyledons of newly germinated annuals (Medica & Nagy, pers. comm.).

Cnemidophorus tigris is a wide-foraging carnivore that is active mostly at moderate temperatures and relatively early in the daytime. Preferred body temperature is about 39°C (Medica 1967), although they can tolerate temperatures up to 47°C (Kay 1970; Asplund 1974; Vitt & Ohmart 1977b). When summer days are clear and hot, western whiptails stay in shade, where they have been observed to congregate in the coolest microhabitats under shrubs and grasses. These lizards utilize shade throughout their aboveground activity, even basking in only filtered light, and they move rapidly across open areas. The diet consists primarily of a variety of invertebrates, especially termites.

The fast-running *Callisaurus draconoides* is a diurnal, sit-and-wait insectivorous forager, which lies motionless on the substrate until a prey item is observed (Norris & Lowe 1964; Kay, Miller, & Miller 1970; Pianka & Parker 1972; Tanner & Krogh 1975; Vitt & Ohmart 1977a; Smith, Medica, & Sanborn 1987). This animal prefers sandy terrain with sparse vegetation, where it forages in open spaces between shrubs. Its prey is varied, depending on seasonal abundance, ranging from larvae in early spring, before many insect groups have emerged, to orthopterans, hymenopterans, homopterans, and coleopterans in the summer (Vitt & Ohmart 1977a; Smith et al. 1987), even occasional plant materials (Kay et al. 1970; Smith et al. 1987). *Callisaurus* maintains daily activity throughout the hot summer days, using clever forms of behavioral thermoregulation (Muth 1977a, b) and tolerating body temperatures as high as 46°C under natural conditions (Packard & Packard 1970).

Sceloporus magister is a widespread diurnal insectivore of lowland habitats, but rarely is common at any site (Parker & Pianka 1973; Vitt & Ohmart 1974). Desert spiny lizards tend to be arboreal, and within the Mojave Desert they often are collected on *Yucca brevifolia*, the only prominent arborescent species of Mojave desert vegetation, and *Y. schidigera* (Mojave yucca), a fru-tescent species. The diet changes seasonally but consists mostly of adult beetles, ants, and insect larvae. Large lizards often specialize on ants.

The only gecko in the fauna is *Coleonyx varie-gatus*, a small, nocturnal, slow-moving insectivore (Parker 1972b; Parker & Pianka 1974). Adults are active April through October, but immatures are also intermittently active November through March. In springtime insect larvae are a large component of the diet, whereas in summer ter-mites, beetles, and orthopterans are taken. This lizard forages cautiously and raises its disruptively colored tail as a distraction to prey and predators (Vitt, Congdon, & Dickson 1977).

Tanner and Krogh (1973b) described the natural history and ecology of the desert horned lizard based mostly on NTS studies from 1965 to 1971, using data collected by various researchers at Yucca Flat, Frenchman Flat, and Mercury Valley. Desert horned lizards are diurnal insectivores, and a large portion of the diet consists of ants, especially *Pogonomyrmex californicus*, solitary-foraging seed-harvester ants (Rissing 1981; Collins 1988). *Phrynosoma platyrhinos* consumes numer-ous types of terrestrial insects as they are encoun-tered, but on occasion this species also uses the reddish-orange fleshy fruits of *Lycium andersonii*, which they may harvest by climbing into the shrub (Banta 1961; Tanner & krogh 1975). This lizard is not territorial, and individuals may travel hun-dreds of meters per day while foraging and search-ing for appropriate cover beneath desert shrubs. They tend to be active relatively early and late each day (Pianka & Parker 1975). As with other lizards, they are relatively inactive during midday in summer, instead finding cover under shrubs or in loose soil or plant debris (Cowles 1979), although on occasion they use burrows of other animals.

Gambelia wislizenii is a voracious diurnal pred-ator that has a very wide distribution in the arid and semiarid flatland communities of western North America (Parker & Pianka 1976). The long-nosed leopard lizard tends to have a narrow activity time and during the summer appears in early morning before many other lizards in the community. In the warm deserts its diet consists largely of other lizards, including *Uta*, *Cnemi-dophorus*, *Callisaurus*, *Phrynosoma*, and possibly *Sceloporus*, which were also the most common species in Rock Valley. Small rodents, for example, *Perognathus longimembris*, are occasion-ally consumed.

Characteristics of other common Mojave desert species

A very unusual reptile that does not use burrows is *Xantusia vigilis*, a nocturnal, intensely secret-ive, and very small lizard (mean adult body weight of 1.1–1.3 g) that seeks shelter in the shady piles of broken limbs among the leaf bases of species of *Yucca* or occasionally between loose slabs of rock (Miller 1951; Zweifel & Lowe 1966; Mautz 1979). This tiny lizard has no hibernation period and at least in the southern part of its range is almost always active year-round, hunting at night for ter-mites, small beetles, crickets, moths, and other arthropods. *Xantusia* avoids inhabiting or hunting within nests of *Neotoma* (packrats), which are predators (Miller & Stebbins 1964). Lethal tem-perature for the desert night lizard is 37–39°C, so these animals are normally active at lower tem-peratures than sympatric species (Regal 1966; Kour & Hutchison 1970).

The chuckwalla is a remarkable, large, diurnally active iguanid lizard that inhabits rock outcrops, which it uses for basking and among which it can seek shelter from the heat and predators in rock crevices (Johnson 1965; Nagy 1973; Berry 1974). Home ranges are about 0.2 ha in spring and summer. While basking on a rock, *Sauromalus* may show threatening behavior to an intruder or chase it from the rock, but in general the animal does not defend a territory. The herbaceous diet of chuckwallas changes monthly, but tends to be fairly selective (Sanborn 1972; Nagy 1973). For six to eight weeks after emerging from winter hiber-nation, body weight of *Sauromalus* increases sharply as it consumes fresh plant materials (Nagy 1973). For example, in one population its spring diet mostly included fresh leaves of *Camissonia claviformis*, *Ambrosia dumosa*, and *Mirabilis bige-*

lovii and flowers of *Stephanomeria exigua* (in April). Flowers (in May) and then fruits (June through July) of *Larrea tridentata* plus fresh or drying leaves of *A. dumosa* dominated the diet into summer, when chuckwallas ate very little, and through August and September dry leaves of *Ambrosia, Larrea,* and *Mirabilis* were harvested. From July through September few individuals become active, and each probably emerges from rock crevices only once every two or three days. In Rock Valley a favored food item was *Sphaeralcea* (Sanborn 1972).

Crotaphytus collaris is saxicolous, i.e., it tends to inhabit steep slopes with exposed rock outcroppings (Banta 1960; Tanner & Krogh 1974b; Vitt & Ohmart 1977c). The collared lizard is a carnivore that feeds on invertebrates and other lizards.

Dipsosaurus dorsalis is a common lizard of the Mojave Desert, but has never been observed in Rock Valley, although it may be expected there. The desert iguana is a sandy-white, actively foraging herbivore that has a very wide range of thermal tolerances and the highest maximum body temperature of these lizards (Norris 1953; Kay 1970; Weathers 1970; Minnich & Shoemaker 1970; Pianka 1971). It will voluntarily tolerate body temperatures up to 47°C. *Dipsosaurus* uses shade or retreats to burrows at very hot times of the day and aggressively challenges intruders, using its tail as an organ of aggression. This lizard, like *Callisaurus* can move very quickly across the desert landscape (mean running speed in nature, 7.28 m s^{-1}) to avoid predators (Belkin 1961). However, unlike *Callisaurus*, *D. dorsalis* is an herbivore that relies very heavily on flowers of *Larrea*, which are harvested from the shrub that at the same time also provides some shade (Cowles 1946). So successful is this lizard that population densities of 350–450 individuals ha^{-1} have been reported (Krekorian 1983, 1984). Typical home ranges of males and females are about 0.15 ha, and there is much overlap of home ranges and a much different territorial social structure than insectivorous iguanids (Krekorian 1976).

POPULATION BIOLOGY AND ECOLOGY OF LIZARDS IN ROCK VALLEY

A major emphasis of research programs at NTS has been studying the population biology of Mojave Desert lizards. In the 1960s Tanner and associates from Brigham Young University (Tanner & Jorgensen 1963; Jorgensen & Tanner; Tanner & Hopkin 1972; Tanner & Krogh 1973a, b, 1974a, b, 1975) began preliminary population studies at Yucca Flat on the lizard community, but collection of data was shifted mostly to Frenchman Flat, Mercury Valley, and Rainier Mesa and concentrated on the population biology of *Sceloporus magister, S. occidentalis, Uta stansburiana, Phyrnosoma platyrhinos, Gambelia wislizenii,* and *Callisaurus draconoides.* Another group of researchers, headed by Turner from UCLA, began marking lizards in 1963 at the 9-ha plots in Rock Valley. Rock Valley studies, which were concentrated on *U. stansburiana, Cnemidophorus tigris, P. platyrhinos,* and *G. wislizenii,* were initiated not only to determine the factors that control year-to-year variations in population density, but also to determine effects that gamma irradiation could have on natural populations of animals and their reproduction (Chapter 13). In fact, the basic studies on undisturbed, nonirradiated populations were needed in order to compare treated results from irradiated plots (plot B). Combined the two research groups obtained very detailed data for constructing life tables of common lowland species in the northern Mojave Desert and at Rock Valley in particular.

Sampling techniques

Throughout the course of Rock Valley studies attempts were made to mark all lizards within and around the plots by toe clipping, and during active months lizards were captured on a regular basis to keep track of growth parameters, survivorship, and reproductive condition. For example, *U. stansburiana* was captured (noosing or by hand) by a team of two or three researchers one week per month within each study plot. Once a lizard was registered, it was marked with quick-drying paint, which lasted two weeks, to avoid recapturing that lizard during the same month (Turner, Hoddenbach, Medica & Lannom 1970). These data sets served as baseline data for sampling during the IBP study at Rock Valley, and, in fact, the IBP studies were conducted by the same researchers. Lizards were collected off plots for autopsies to determine egg-laying potentials of the species, so as to leave the populations in plots unaffected by sampling.

Methods for determining lizard densities and biomass

Densities of the five principal lizard species inhabiting the Rock Valley site were measured between 1971 and 1974, and estimates dating from 1964 were available for some of these species. During the IBP, densities of *U. stansburiana* were based on number of different marked individuals captured on the plots or near the validation side between mid-March and mid-April. Densities of *G. wislizenii* and *P. platyrhinos* were based on numbers obtained in one of the 9-ha enclosed plots adjacent to the IBP validation site. Densities of *C. tigris* and *C. draconoides* were estimated in May and June by extrapolating periodic counts of these lizards along transects across the validation site and multiplying by a scaling factor to adjust for lizards at least 18 months of age that were not seen on any given day. For *Callisaurus* a scaling factor of 2.0 was used, this based on earlier observations on this species (Tanner & Krogh 1975).

For *Cnemidophorus*, researchers discovered that yearlings (6–10 months of age) were not counted along transects with the same facility as were the older individuals, so that transect counts were often very incomplete for this age class. For example, in 1972 yearling whiptail lizards comprised 31.3% of 150 lizards of known age registered in two fenced plots adjoining the validation site. Therefore, to compensate for undercounts a scaling factor of 3.66 was used for 1971 and of 3.33 for 1972–4, this multiplied times the number of adults that were 18 months of age or older found along the transect. Estimates of all species densi-

ties pertained only to spring populations and did not include hatchlings (hatched in the same year).

Dry weights were estimated as 30% of live biomass (Turner et al. 1976). Additional details on methods were given by Medica et al. (1971) and Medica and Smith (1974).

Activity patterns

Observations of seasonal lizard activity at Rock valley were compiled from 1968 to 1972. *U. stansburiana* was active year-round throughout the study, and *P. platyrhinos* was generally the first of the other species to become active in late winter, having a date of first appearance from mid-February to mid-March. The latest species to appear was *C. tigris*, two to four weeks later than *Phrynosoma*.

Population densities within study enclosures

Uta stansburiana was the most abundant lizard in Rock Valley, with mean spring densities of about 50 individuals ha^{-1} (Table 8.2). Data collected from 1966 to 1972 showed that densities for this species varied from 23.5 to 80.3 individuals ha^{-1}. *Uta* also exhibited a pronounced microgeographical variation in abundance because individual plots within Rock Valley had a twofold difference in density (Turner, Medica, Bridges, & Jenrich 1982). Relative abundances of *U. stansburiana* were high in areas with denser shrub cover, for example, *Lycium–Grayia* communities, whereas *C. tigris* was more abundant in areas with sparser shrubs, for example, *Larrea–Ambrosia* communities.

Mean spring densities of *Cnemidophorus* did not differ significantly between 1964 and 1970 and the IBP measurements for 1971–4 (Table 8.2); nevertheless, densities varied widely during the IBP study from a record low of 8.6 individuals ha^{-1} in 1971 to a record high of 38.2 individuals ha^{-1} in 1974 (Table 8.2). However, IBP data for 1971–4 must be viewed with caution because the transect count techniques required use of a scaling factor based upon calibration used from 1967 to 1970 data.

Table 8.2. *Mean densities and dry biomass (± one standard error) of lizards in Rock Valley. Ranges of observations are given in parentheses.*

Species	1971–1974		1964–1970	
	Mean density, individuals ha^{-1}	Mean biomass, g ha^{-1}	Mean density, individuals ha^{-1}	Mean biomass, g ha^{-1}
Uta stansburiana	46.5 ± 10.9 (32.5–78.8)	42.8 ± 5.6 (31.8–57.2)	51.4 ± 8.1[a] (34.6–80.3)	40.7 ± 3.8[b] (33.6–46.8)
Cnemidophorus tigris	19.1 ± 6.8 (8.6–38.2)	61.0 ± 15.5 (32.3–96.5)	15.5 ± 1.5[c] (9.9–21.6)	50.2 ± 6.8[d] (32.5–69.5)
Gambelia wislizenii			2.2 ± 0.8[e] (1.6–2.9)	21.5 ± 2.0[f] (18.0–27.4)
Phrynosoma platyrhinos	1.7 ± 0.3 (0.9–2.3)	8.9 ± 2.7 (1.1–12.6)	2.4 ± 0.4[g] (1.2–4.7)	69.2 ± 15.2[h] (54.0–84.4)
Callisaurus draconoides	2.6 ± 0.4 (1.8–3.6)	11.4 ± 1.5 (7.8–15.0)		

[a] 4 plots, 1966–1970 (Turner et al. 1982)
[b] 4 plots, 1965–1967 (Turner et al. 1976)
[c] 3 plots, 1964–1970 (Turner 1977)
[d] 3 plots, 1964–1967 (Turner et al. 1969a)
[e] 2 plots, 1964–1970 (Turner et al. 1969b; Turner et al. 1982)
[f] 3 plots, 1964–1970 (Turner et al. 1969b)
[g] 2 plots, 1964–1970 (Medica et al. 1973)
[h] 2 plots, 1969–1970 (Turner and Chew 1981)

Estimates of density for *C. draconoides* were also based on scaling up counts (×2.0) along transects, although this scaling factor had no experimental basis. Tanner and Krogh (1974b) reported that *Callisaurus* was always the most abundant lizard species in other portions of NTS, and for Frenchman Flat they estimated population densities of 12.5–15.5 individuals ha^{-1}, with biomasses of 156 g ha^{-1} in 1968 and 175 g ha^{-1} in 1969. Estimated densities from the IBP study of 1.8–3.6 individuals ha^{-1} for Rock Valley populations may have been too low.

Turner and Chew (1981) found that densities of *P. platyrhinos* within small, 0.44-ha enclosures in Mercury Valley were 10–32 individuals ha^{-1}, substantially higher than those observed in Rock Valley (Table 8.2) and reported from Frenchman Flat (5 individuals ha^{-1}) by Tanner and Krogh (1973b). Turner and Chew concluded that fences may have enforced local increases in density and

may also have had some favorable influence on survival.

No data from 1971 to 1974 are presented for *G. wislizenii*. Values used in IBP site reports may be invalid because they were based on registries in one of the fenced plots in which this species was declining to abnormally low numbers. Densities of *S. magister* and *Coleonyx variegatus* were not obtained in Rock Valley. Geckos were caught commonly in pitfall traps that operated in Rock Valley before 1971, and based on this experience Medica (unpubl. data) has suggested that its density may have been as high as 30 individuals ha^{-1} (see also Parker 1972a). Tanner and Krogh (1973a) reported estimated densities for *S. magister* ranging from a few to 50 individuals ha^{-1}, the difference probably being influenced by habitat quality. In Rock Valley this species occurred infrequently, although it was often observed on the hardware cloth fences that surrounded the plots.

Comparative size relationships of Rock Valley lizards

At Rock Valley the largest common lizard was *Gambelia wislizenii*, a predator on other lizards and even *Perognathus*. For adult female leopard lizards average minimum SVL (snout-to-ventral length) was 101 mm with a dry weight of 13.1 g, and adult males SVL was 94 mm SVL with a dry weight of 9.2 g. *Callisaurus draconoides*, *Cnemidophorus tigris*, and *Phrynosoma platyrhinos* were about 70 mm SVL, but the latter were significantly heavier. The smallest adult lizards were *Uta stansburiana*, having minima of 46 mm SVL for adult females and 48 mm for adult males, and *Coleonyx variegatus*, having a minimum of 50 mm SVL and a dry weight of 1 g. Yearlings of *Uta* were almost the size of adults, but those of *Cnemidophorus* were about 75% SVL and 60% of adult weight.

Dry biomasses for lizards of Rock Valley are estimated in Table 8.2. Highest mean biomass was calculated for *C. tigris*, which included an estimate of 96.5 g ha^{-1} for 1974, which far exceeded observations in Rock Valley between 1964 and 1967 (Turner, Medica, Lannom, & Hoddenbach 1969a, b). From 1964 and 1970 densities of *Cnemidophorus* were based on total registries accumulated in fenced areas, and rosters of this nature averaged about 90% of capture–recapture estimates (Turner et al. 1969; Medica, Hoddenbach, & Lannom 1971), so the IBP densities of 61.0 g ha^{-1} were probably higher than actually occurred. Biomass of *Uta* for 1971–4 was 42.8 g ha^{-1}, very close to those values calculated in earlier studies; however, values for *Phrynosoma* were substantially lower than those calculated from the abnormally high densities in Mercury Valley obtained by Turner and Chew (1981). Biomass estimate for *Callisaurus* was 11.4 g ha^{-1}, but needs to be challenged because estimates of densities had low confidence. Based on assumed densities, combined dry biomass of *S. magister* and *Coleonyx variegatus* could have been 15–25 g ha^{-1}, but no estimate has been made for *Gambelia*. Consequently, a site-wide estimate of lizard biomass is uncertain, but probably was about 150 g ha^{-1} dry weight for yearlings and adults combined.

Relationships of egg production to environment

Long-term ecological studies at Rock Valley provided a useful body of information on the reproductive biology of dominant species observed over multiple years of study. These investigations confirmed postulated relationships between winter rainfall – as it effects community productivity – and egg production by some lizard species (see Mayhew 1966a, b; Zweifel & Lowe 1966), especially that the principal variable in egg production is frequency of clutches. Clutch frequency may be extremely labile in desert species, and intensive field effort is required to evaluate this topic accurately.

Quite remarkable are females of *U. stansburiana*, which laid one to six clutches annually, depending in part on age, but principally on the relative abundance of winter rains (Hoddenbach & Turner 1968; Medica & Turner 1976; Turner, Medica, & Smith 1974; Turner, Medica, Bridges, & Jennrich 1982). Clutch deposition begins in early- to mid-April in very favorable years, and mean intervals between clutches were reported as 21.5 days (range 14–35), being shortest in the middle of the activity season (Turner, Hoddenbach, Medica, & Lannom 1970). Clutch size also varies between years (Hoddenbach & Turner 1968; Medica & Turner 1976), but contributes less to annual variations in egg production than does variation in clutch frequency. Clutch size was positively correlated with female body size. Body masses of females were also inversely correlated with population densities, so that egg production by females in dense populations was slightly reduced, whereas it was enhanced in sparse populations. This was one of two regulating mechanisms identified in the Rock Valley research on dynamics of *Uta* populations (Turner et al. 1982). A density-dependent effect on egg production has been described in this species, but it does not operate in the potent manner that has been described for desert pocket mouse, *Perognathus formosus* (Chew 1975).

Females of *Cnemidophorus* lay one or two clutches of eggs per year (Turner, Medica, Lannom, & Hoddenbach 1969a), and those of *Phrynosoma* and *Gambelia* usually lay one clutch, but may lay another (for example, 1965 and 1969) or none (1964 and 1970; Turner et al. 1969; Medica, Turner, & Smith 1973a). Nagy (1973) found that *Sauromalus* typically produced one clutch, but did not lay eggs in 1970 at a site in the southern Mojave Desert. Gravid female chuckwallas have occasionally been observed in the vicinity of fenced Rock Valley plots from late May to early June, when they venture downslope from the Specter Range to deposit eggs in the more friable soil in lower Rock Valley (Medica, pers. comm.); for these clutch size is 6–13 eggs with a mean of 7.8. Largest clutches of the species inhabiting lowland deserts have been found in *S. magister* (Parker & Pianka 1973; Vitt & Ohmart 1977c).

Survivorship

Survivorship of lizards in Rock valley was estimated in two ways. Western whiptail and leopard lizards were monitored in the enclosures over a period of years, and from those observations minimal survival rates were inferred from comparisons of those rosters (Turner, Lannom, Medica, & Hoddenbach 1969; Turner, Medica, Lannom, & Hoddenbach 1969b). Survival of *Uta stansburiana* was estimated in the same manner (Turner et al. 1982) and by extrapolating data from frequencies of recaptures for marked individuals in cohorts of known size (Turner, Hoddenbach, Medica, & Lannom 1970; Turner 1975). Turner (1977) estimated adult annual survival rates of 22% for *U. stansburiana*, 54% for *G. wislizenii*, and 40% for *C. tigris*, by fitting reported survival rates to an exponential model. This approach supported the idea of age-constant adult mortality, but provided no information on year-to-year differences in death rates. Since then, Medica and Turner (1984) documented observed maximum life spans of 118 months for *Gambelia*, 94 months for *Cnemidoph-*

orus, 82 months for *Phrynosoma*, and 58 months for *Uta*.

Survival of *Uta* was analyzed in detail to develop a population model for this species (Turner et al. 1982). Experiments within enclosures showed a definite density-dependent effect on annual survival of lizards that were older than eight months of age as well as improved survival in the absence of predatory *Gambelia*. Hence, survival rates differed between years owing to changes in relative abundances of the two species. However, the analysis could not factor out the possible effects of using the enclosure as well as several abiotic factors.

Effects of rainfall on lizard productivity

The postulated relationship between female egg production and amount of winter rainfall was apparently seen for *U. stansburiana* using field experiments in which plots were artificially irrigated to receive the equivalent of 50 mm of additional precipitation during fall. Body masses and egg production of females increased in treated plots relative to those in nonirrigated plots (Turner et al. 1982). However, hatchling abundance (following egg laying) was never directly assessed in Rock Valley in the same year, so that the numerical response of lizard populations was only first measured in the ensuing spring. This means that any effects of winter rain (for example, in fall–winter of 1966–7) were not observed until the spring of the next year (spring, 1968).

As described in Chapters 4 and 5, Beatley (1974b) emphasized the importance of individual rain events of at least 25 mm between late September and early December in promoting germination and growth of winter annuals during the ensuing spring. However, total plant production variables (perennials and annuals) are more highly correlated with total rainfall between October and March than with fall rainfall. October-to-March rainfall was also the variable found to be most effective in the *Uta* model (Turner et al. 1982), presumably because higher plant productivity resulted in greater production of invertebrate prey items for that lizard species.

Table 8.3. *Estimated spring densities (individuals ha⁻¹) of lizards in Rock Valley and relevant winter rainfall data (October–March).*

Year	Rainfall, mm	*Uta stansburiana*	*Cnemidophorus tigris*	*Callisaurus draconoides*	*Phrynosoma platyrhinos*	*Gambelia wislizenii*
1965	31.2		9.9		1.4	
1966	30.5	56.8	18.1		1.2	
1967	139.1	80.3	15.2		2.2	
1968	47.0	42.5	15.7		2.4	
1969	78.0	34.6	16.9		2.9	
1970	209.3	42.8	21.6		4.7	
1971	83.1	40.0	9.5	3.6	2.1	1.0
1972	46.8	32.5	10.4	2.5	1.7	1.0
1973	41.4	34.5	19.2	1.8	2.3	1.0
1974	209.3	78.8	38.2	2.6	2.6	1.0

Estimated spring densities (year n) and relevant rainfall between October (year $n - 2$) and March ($n - 1$) are presented for five lizard species in Table 8.3. Only densities of *Phrynosoma* were statistically correlated with winter rains. Rank correlation tests gave values of r_s equal to 0.42 and 0.49 for *Cnemidophorus* and *Uta*, respectively, falling well below critical values (at 5% level) of 0.63 and 0.67. For *Phrynosoma*, $r_s = 0.80$, exceeding the critical value of 0.77 at the 1% level of significance. Similarly, when common logarithms of *Uta* and *Cnemidophorus* densities were regressed on logarithms of October–March rainfall, the F-tests of nonzero slope were insignificant. Corresponding analysis of *Phrynosoma* densities gave a statistically significant F-value of 11.9 ($F_{0.01} = 10.6$).

The model for explaining population dynamics of *U. stansburiana* incorporated several other variables, including March air temperatures, abundance of predators, and density and composition of the *Uta* population itself (Turner et al. 1982). Hence, the general absence of an obvious relationship between winter rains (year $n - 2$, $n - 1$) and spring lizard densities (year n) was not totally surprising. This suggests that closer analysis is needed of recruitment *per se* and its relationship to winter rainfall. Table 8.4 gives estimated densities of juveniles just after the first hibernation for the three dominant species. Analyses of these juveniles gave results similar to those using total

Table 8.4. *Estimated spring densities (individuals ha⁻¹) of yearling (juvenile) lizards in Rock Valley.*

Year	*Uta stansburiana*	*Cnemidoph- orus tigris*	*Phrynosoma platyrhinos*
1965	–	2.70	0.22
1966	42.0	9.33	0.61
1967	67.5	3.00	1.28
1968	27.6	1.88	0.83
1969	27.0	4.52	1.33
1970	39.4	7.52	2.89
1971	28.8	1.44	0
1972	23.4	2.61	0
1973	29.0	5.28	1.22
1974	70.1	4.78	2.60

densities. Rank correlation tests and regression analyses of abundances of juvenile *Uta* and *Cnemidophorus* were insignificant, whereas tests of juvenile *Phrynosoma* showed spring densities to be positively and significantly correlated with amount of winter rainfall.

Residuals of regression analyses of densities and rainfall showed that two years, 1966 and 1971, were distinctly aberrant, at least in terms of a rainfall-numbers hypothesis. Rainfall relevant to 1966 observations (October, 1964–March, 1965) was only 30.5 mm, the lowest recorded between 1965 and 1974, yet numbers of juvenile *Uta* and *Cnemidophorus* in spring, 1966, indicated good recruit-

Table 8.5. *Energy flow and production estimates for lizards in Rock Valley.*

Species	Year	Energy flow, kcal ha^{-1} yr^{-1}	Production	Reference
Uta stansburiana	1965–1966	1831	336	Turner et al. (1976)
	1966–1967	2794	535	Turner et al. (1976)
	1967–1968	2115	536	Turner et al. (1976)
	1966–1967	3309	442	Nagy (1983)
Cnemidophorus tigris	1965–1966	–	185	Turner & Chew (1981)
Phrynosoma platyrhinos	1969–1970	–	305	Turner & Chew (1981)

ment in the previous summer. Unusually good reproductive success for summer, 1965, was also observed in *Gambelia wislizenii* (Turner et al. 1969). Hoddenbach and Turner (1968) described events during spring, 1965, viz., late rains (in March and early April), delayed germination by annuals, exceptional vegetative growth by shrubs, and an unusual trend in mean clutch size of *Uta* to produce larger clutches toward the end of the egg-laying season. How these events may have combined to produce the observed flush of reproduction in *Uta* was not clear, but if *Uta* juvenile data are analyzed from 1967 to 1974, there is a significant correlation with winter rainfall ($r = 0.79$, $F = 9.76$). Omitting 1966 from the analysis had little effect on the re-evaluation of *Cnemidophorus* juvenile data. Rainfall events relevant to spring, 1971, were those between October, 1969 and March, 1970, when at least 83 mm of rain was recorded, but about 70% of that was in February and March. Thus, most winter rain was too late to be beneficial for germination of annuals and lizard reproduction, and numbers of juveniles of the three dominant lizard species were low (Table 8.4). No other years provided evidence that February–March rains had beneficial effects on lizard reproduction.

Mortality

Except for *U. stansburiana*, there are few data on how death rates of hatchling lizards in Rock Valley varied between 1965 and 1974, and nothing on

what factors influenced this process. In New Mexico, Whitford and Creusere (1977) attributed a pronounced increase in the abundance of *Cnemidophorus tigris* to better than average overwinter survival of hatchlings, but in Rock Valley there were no data on early mortality, either by predation, which may be a very important driving variable, or abiotic factors.

Secondary production

Secondary production by *U. stansburiana* in Rock Valley was estimated from IBP data (Turner et al. 1976; Nagy 1983), and other production values for *C. tigris* and *P. platyrhinos* were developed by Turner and Chew (1981). These estimates are summarized in Table 8.5. To standardize these estimates, production for *Uta* was examined in terms of population densities and biomass using the methods employed for the analysis of the rodent *Perognathus formosus* (Turner & Chew 1981).

One finding was that production by *Uta* was not statistically correlated with density, as it was with *Perognathus*. This probably is because the time of highest lizard density, summer, is not the time of highest biomass, fall. Growth of hatchling lizards until they enter hibernation is proportionately large, with high growth rates (Tanner & Krogh 1973a, b, 1974a, 1975), but yields only a very small, gradual increase in overall production. Several production models were derived, and not surprisingly all emphasized the importance of

Table 8.6. *Minimum snout-to-vent length (SVL) and dry weight (g) of adult snakes in Rock Valley. Species are arranged by weight from highest to lowest.*

Species	Sample size	Minimum SVL, mm	Mean body weight, g	Weight range, g
Pituophis melanoleucus	6	975	85.6	66.3–130.1
Crotalus mitchellii	3	640	76.7	62.8–94.0
Lampropeltis getulus	3	700	43.5	30.8–62.1
Crotalus cerastes	17	400	32.8	13.2–55.8
Masticophis flagellum	19	590	29.9	14.8–55.6
Salvadora hexalepis	2	630	21.5	16.1–27.0
Rhinocheilus lecontei	7	569	20.3	10.0–30.2
Phyllorhynchus decurtatus	2	289	4.7	4.1–5.3
Sonora semiannulata	6	250	2.7	2.1–3.7
Chionactis occipitalis	77	200	2.6	1.4–3.7

summer reproduction and ensuing growth of the hatchlings (Turner & Chew 1981). Although spring density was inferred to act in a regulating manner in the *Uta* population model (Turner et al. 1982), such an effect was not clearly expressed in the production models.

SNAKES OF ROCK VALLEY

Absolute abundances of snakes inhabiting Rock Valley remain unknown, although field work in the three 9-ha circular enclosures (plots A–C) between 1962 and 1971 provided some useful values. Research included the operation of hundreds of pitfall traps, which provided some data on the relative abundances of the smaller snakes.

Chionactis occipitalis (western shovel-nosed snake) is a small nocturnal snake and the most common one at NTS (Tanner & Jorgensen 1963) and in Rock Valley, where densities were estimated as 5–10 individuals ha^{-1} (Medica 1973). This species feeds on centipedes, insects, and scorpions. Less common small snakes (having mean dry body mass of 2.5–5 g) that occurred in Rock Valley were *Sonora semiannulata* (ground snake) and the nocturnal *Phyllorhynchus decurtatus* (spotted leaf-nosed snake).

Intermediate-sized snakes (mean dry body mass of 20–30 g) included the nocturnal *Rhinocheilus lecontei* (long-nosed snake), the mostly nocturnal *Crotalus cerastes* (sidewinder), and the diurnal *Salvadora hexalepis* (western patch-nosed snake) and *Masticophis flagellum* (coachwhip). *Salvadora*

hexalepis has a very broad thermal tolerance (37°C), which enables this species to emerge in the early morning hours when temperatures are otherwise too cool for most lizards and snakes; it basks and then is ready to hunt for lizards when they appear (Jacobson & Whitford 1971). The coachwhip is a common, very fast-moving snake that prowls throughout the day in search of large arthropods, lizards, other snakes, and small rodents, but, contrary to a widespread belief, it does not whip prey to death. The sidewinder, which can be abroad during the day, prefers habitats with wind-blown sand (Norris & Kavanau 1966) and consequently occurred throughout Rock Valley at lower densities, perhaps 0.2–0.5 individuals ha^{-1}, which is the likely density of the long-nosed snake and western patch-nosed snake (Medica 1973).

The largest snakes (mean dry body mass around 80 g) of Rock Valley were diurnally active *Crotalus mitchellii* (speckled rattlesnake), *Pituophis melanoleucus* (gopher snake), and the rarely seen *Lampropeltis getulus* (common kingsnake). The speckled rattlesnake shifts from a diurnal to a nocturnal activity pattern during hot weather, beginning in May and continuing through September (Moore 1978).

The greatest contributors to snake dry biomass (Table 8.6) were *Chionactis occipitalis* and *M. flagellum*, both about 20 g ha^{-1}, and *Crotalus cerastes*, estimated as being 12 g ha^{-1}. Assuming densities of 0.05–0.1 ha^{-1} for the rarely observed species produces a total dry biomass of 70 g ha^{-1} for all snakes. Even allowing for serious underestimates of abundance of certain species by three to four times, total snake biomass could not exceed 250 g ha^{-1}.

Virtually nothing is known about secondary productivity or population dynamics of snakes in the Mojave Desert. Their highly seasonal activity patterns, secretive ways, and nocturnality have made it difficult for researchers to study them over long periods of time.

DESERT TORTOISE

To maintain a successful population of desert tortoises, there must be sufficient herbaceous or succulent plant materials, which provide energy, water, and minerals (Chapter 6). Best localities have at least 20 g m^{-2} dry weight of spring annual production (Schamberger & Turner 1986). Hence, within the boundaries of overall distribution, *Gopherus agassizii* has many types of populations, ranging from those with high densities of up to 75 km^{-2} to large areas where there are few or no individuals. Even at a given site, a decrease in population density may be substantial whenever the crop of spring annuals fails (Turner, Medica, & Lyons 1984). The population in Rock Valley, which has been studied for three decades, is a site with low population density year after year.

Surface activity pattern

The typical pattern of surface activity of *G. agassizii* has been described (Miller 1932, 1955; Woodbury & Hardy 1948), but research at Rock Valley, begun in 1963, has contributed greatly to our modern understanding of its physiological ecology and life history parameters (Chapter 6).

The key to unraveling the physiological ecology of desert tortoises of Rock Valley was a 16-month study conducted in plot C from February, 1976, to capture individuals as they emerged from winter hiberation, to June, 1977, when tortoises were in the second season of drought-induced estivation (Nagy & Medica 1977, 1986). Eleven tortoises, ten of which were probably prereproductive subadults (body weight 500–1420 g) and one large male (body weight 2200 g), probably sexually mature, were subjects of the study.

During early spring, when the vegetative rosettes of winter annuals were stimulated to elongate, tortoises emerged from winter hibernation during a seven-week period, beginning in late March. A typical individual left the burrow one time every four days, and for the first week or so did not feed. A spring feeding foray, once in three days, lasted about three hours, and tortoises foraged primarily on three annuals, *Camissonia munzii*, *Langloisia setosissima*, and some *Bromus rubens*, consuming about 63 g fresh mass of herbs per fresh weight kilogram of body mass. This food provided abundant water, which accumulated in their large urinary bladders, but salts were not excreted and accumulated in the blood plasma and urine. Body weight increased due to intake of much water, but dry weight of each organism decreased because tortoises could not consume enough dry weight of plant materials to satisfy energy expenditures.

In late spring and early summer winter annuals and other herbs ceased growing and experienced desiccation, as summer drought commenced. Surface activity of tortoises was limited to one 1-hour foray every six days, and the remainder of the time was spent estivating in shallow summer burrows. Tortoises consumed more grass, especially *Oryzopsis hymenoides*. During this period each tortoise lost body water, with a 5.5% decrease (mid-May through June), and lost body mass, but the individual had a positive energy balance because more dry matter was consumed.

Foraging was resumed in mid-July with arrival of a summer thundershower, and immediately after body weight of two individuals increased 9.2%. That sudden increase in body weight was difficult to explain in terms of eating plants. Instead, Medica, Bury, and Luckenbach (1980) documented a new phenomenon, called nasal drinking. Although workers on desert tortoises traditionally assumed that body water was derived solely from plant foods, Medica et al. (1980) observed that animals scraped out shallow depressions, 0.35–0.50 m in diameter and 30–50 mm deep, just before or during rainstorms to serve as water catchment basins, which could hold a pool of water for up to six hours. These catchments were then used for nasal drinking, whereby the neck is outstretched and the snout is submerged. By drinking large quantities of standing rainwater, its large urinary bladder, which is otherwise used to retain dilute urinary water, can be voided and refilled, via the kidneys, with rainwater, as evidenced by uric acid and concentrated urine left at the drinking site.

After a tortoise is fully rehydrated, it may also consume dry plant matter. Following rain events tortoises fed during 3-hour foraging bouts every two or three days on dry grasses and herbs, not on the abundant shrub materials (Nagy & Medica 1986). Late July and August were dry, and body weight decreased 7.8% and urine became concentrated, but tortoises experienced a period of positive energy balance as they consumed dry plant materials. In early September another heavy rain occurred and through drinking they increased body weight immediately. In late summer through early fall the animals consumed rosettes of spring annuals, mostly *Camissonia* and *Bromus*. During this phase body dry matter content increased by more than 25%.

Although preferred body temperature of active desert tortoise is 28–34°C, individuals are opportunist and will leave their burrows at suboptimal ambient temperatures to take advantage of rainstorms and available food, even by coming out at night, if it is raining, to drink. Ambient temperatures declined in the fall at Rock Valley, yet tortoises continued to eat herbs, but at a reduced rate. Body weight remained fairly constant and water content remained high, but plasma and urine became highly concentrated during November.

In late November, cool temperatures forced tortoises into their burrows for winter hibernation. During winter no tortoises were seen abroad, and they probably did not eat. Judging from individuals dug up during winter and those examined at the end of hibernation, animals lost almost no weight during hibernation. They stayed in hibernation until mid-March, 1977, although some individuals were observed near burrow entrances.

In early April, 1977, tortoises were actively feeding in Rock Valley in a manner similar to that of the previous year. Following spring rain in 1977 there was a gain in body weight of 14.1%.

During the course of a full year an individual tortoise at Rock Valley spent 98.3% of its time in the burrow or in a pallet and only 0.3% – approximately 29 hr yr^{-1} – feeding. Many burrows caved in or were destroyed during winter, and collectively tortoises constructed new burrows, especially many shallow 'summer' ones.

Studies at Ivanpah Valley, California, produced estimates of home ranges for unenclosed populations at 22 ha (range, 3–89 ha; Turner et al. 1981) and 19 ha (range, 2–73 ha; Medica et al. 1982). Recently Barrett (1990), using radio transmitters attached to free-ranging tortoises, estimated home ranges of 19.07 ha (S.D. 4.63 ha; range 3–53) on a steep desert bajada in central-southern Arizona (Sonoran Desert). Home ranges for males and females were not significantly different, and tortoises occupied territory on steeper slopes during winter months, which were probably warmer microhabitats than at the bottom of the bajada.

Feeding habits

Feeding habits of desert tortoises were not studied as part of the IBP in Rock Valley, but have been studied in other parts of the Mojave Desert. At Ivanpah Valley, California, food consumption and population studies were conducted from 1980 to 1982 (Turner, Medica, & Lyons 1981, 1984; Medica, Lyons, & Turner 1982). In spring, 1981, 43% of the diet of desert tortoise was a variety of desert annuals and grasses, and this total frequency of herbaceous tissues was maintained through summer by increasing grass consumption while annuals disappeared (Table 8.7). Under extremely dry conditions of 1981, annuals at Ivanpah Valley had very low productivity and accounted for 5–6% in diets in spring and summer, 1981; during the same year grasses, which comprised 25% of the diet in spring, decreased to less than 4% in summer. Consumption of seeds also decreased sharply from 1980 to 1981. To compensate for those losses in herbs and seeds, tortoises

Table 8.7. *Relative frequencies of food items in desert tortoise feces from Ivanpah Valley, California in 1980 and 1981. The category 'other' includes arthropod parts and unidentified plant material.*

| | 1980 | | 1981 | |
	Spring	Summer	Spring	Summer
Annuals	18.6	9.2	6.1	5.0
Grasses	24.7	33.0	25.1	3.6
Cacti	37.0	48.7	57.9	86.9
Perennial plants	2.3	1.4	2.7	0.8
Seeds	16.0	7.1	5.6	3.3
Other	1.4	0.6	2.6	0.4

consumed larger amounts of cactus tissue, especially fruits of *Opuntia*, increasing that from 37% of tortoise diet in spring, 1980, to 49% in that summer, to 53% in spring and 87% in summer, 1981 (Table 8.7). In both years parts of shrubs, including *Ephedra, Larrea, Grayia, Lycium,* and *Krameria*, comprised a very small portion of the diet (Turner et al. 1984), and consumption of arthropods was also a very minor component of the tortoise diet. Interestingly, grasses and desert annuals constituted over 80% of the tortoise diet in northern localities of the Sonoran Desert.

Population samples in Rock Valley

Gopherus agassizii has been the subject of many recent research projects, especially to understand how to protect natural populations of this species from human interactions (Turner 1986; Berry 1986a, b). But even before this flurry of recent research began the longest sustained study of a natural population had been initiated in Rock Valley.

Between 1963 and 1967, young individuals, estimated to have been hatched since 1959, were tagged and body measurements were obtained within the 9-ha enclosed circular plots A, B, and C. These individuals subsequently were followed over the years in order to develop a detailed description of their behavior, ecology, and life history parameters. No new individuals were observed between 1968 and 1973, and of course there was no emigration or immigration for the enclosed plots. In 1972 and 1973 18 individuals

were seen in those plots, thereby inferring a density of about 0.7 ha^{-1}. However, this density is artificial due to the presence of fences and is a relatively low density in comparison with many sites in California (Berry & Nicholson 1984). Fenced enclosures certainly restricted movements of Rock Valley resident tortoises, but probably also interfered to some extent with normal predation by canids, therefore possibly enhancing survivorship.

Body size and weight

In 1972 minimum plastron (shell) length of male tortoises at Rock Valley was 180 mm, and mean dry body weight was 544 g (range, 283–820 g). Females (including immatures) were considerably smaller, having a minimum plastron length of 126 mm and mean dry body weight of 203 g (range, 103–820 g). However, this does not imply that the species has any sexual dimorphism but means instead that the age classes of males were somewhat older. If the Rock Valley data are used at face value, estimated dry biomass of desert tortoise during the IBP study was around 200 g ha^{-1} (Medica & Smith 1974).

Growth rates

Medica, Bury, and Turner (1975) determined that plastron length had an average increase of 9.1 mm yr^{-1} between 1963 and 1973. Yearly growth increments were estimated for a 5-year period (1969 through 1973) and ranged from a low of 1.8 mm in 1972 to a high of 12.3 mm in 1969. Highest growth rates seemed to be present following winters of high rainfall. However, the study continued and was reanalyzed for 1963–85, for which size increases were followed for 15 of the original individuals (Turner, Medica, & Bury 1987). There was a relatively linear increase in plastron throughout the interval, with a mean of 7.6 mm yr^{-1}. Turner et al. observed a slightly elevated growth rate during years 5–8, but did not find a strong correlation of shell growth to availability of forage. Germano (1988) made similar measurements in 1985 and found that yearly plastron growth rate from 1959 to 1973 was 10.9 mm versus 9.1 mm obtained by Medica et al. (1975), and that for 1973–84 was 7.6 mm, identical to the findings of Turner et al. (1987). He showed that growth rate decreased from year 5 to adulthood, with peaks that often occur following peaks of winter rainfall, but correlations between growth rate and rainfall could not be found and there were no such peaks as tortoises reach adulthood and sexual maturity after year 17.

Germano (1988) also tested the scute annuli, concentric rings of hard epidermal layers on the scutes (polygonal sections) of the shell (Fig. 8.11), as an estimator of age. In this natural population he found that total number of annuli per scute was typically one or two less than age of the individual. The best way to age a tortoise is to count the number of bone rings, one for every year, but this is a destructive even though very reliable method of aging; nevertheless, the correlation of scute annuli, a nondestructive technique, to number of bone rings was extremely tight ($R^2 = 0.94$), signifying that the nondestructive technique was highly accurate. Each scute annulus probably signifies one year of growth, and differences in old animals of one or two annuli may be that they reached adulthood, when annuli are no longer formed, one or two years before his study. If this procedure is highly predictive of ages from juveniles and young adults, this means that an investigator can determine growth rate and ages with only one handling of an individual.

Death rates in free-ranging animals

Many aspects of population dynamics of desert tortoises could not be studied under realistic conditions in Rock Valley enclosures. However, in the free-ranging population at Ivanpah Valley from 1980 to 1981 observed mortality was 4%, but 18% of the 1981 population did not survive to 1982 (Turner, et al. 1984). There was no significant difference in mortality between males and females. Increased death rates seem to be correlated with dry conditions and low food availability during summer, 1981. In addition to problems of food availability, predation on desert tortoises by coyotes (*Canis latrans*) and kit foxes (*Vulpes macrotis*) has been observed to increase in dry years when rodent and rabbit populations diminish in number (Berry 1984). At Goffs, California, Turner and Berry (1984) estimated annual death rates of 2% for desert tortoises greater than 180 mm in length.

In the late 1980s Charles Peterson, a researcher at UCLA, discovered death in native populations from an illness, simply called upper respiratory disease syndrome (URDS). This typically fatal disease was introduced via diseased captive individuals that, with good intentions, were released back into the wild. URDS has assumed epidemic levels in some populations of the Mojave Desert, while microbiological cause of this illness is currently unidentified. Diseased individuals show external symptoms of nasal discharge and wheezing and possibly lethargy, and internal examinations reveal degeneration of upper respiratory epithelium and differences in blood and urine parameters.

Egg production

Female tortoises in the Mojave Desert most commonly lay two clutches of eggs per year, from early May to mid-July (Turner et al. 1984; Turner & Berry 1984; Turner, Hayden, Burge, & Roberson 1986). One clutch per year or no egg production occurs in a small number of females in typical years and in a majority of females during dry years. Also, females that have recently achieved sexual maturity tend to have only one clutch. Very rarely three clutches per year have been observed (Turner & Berry 1984). By weighing females every one to two weeks, estimates of mean clutch frequency at Ivanpah Valley were 1.60 in 1980 and 1.10 in 1981 and at Goffs were 1.89 in 1983, 1.57 in 1984, and 1.75 in 1985. Females were also x-rayed to determine clutch sizes, which were positively correlated with body weight. At Goffs mean clutch size (4.50) was lower than other workers had predicted.

Egg production data and techniques from California studies were used to interpret events at Rock Valley. Estimated ages of reproductive maturity of Rock Valley females is 17–18 years, and three females that were 23 or 24 years old carried a clutch of eggs in mid-May (Turner, Medica, & Bury 1987). Using scute annuli to estimate age, Germano (1988) identified gravid females by year 19.

SUMMARY

Rock Valley has the characteristic reptiles that occur on bajadas of the northern Mojave Desert. Within plots and at the IBP validation site were seven species of lizards, *Uta stansburiana*, *Cnemidophorus tigris*, *Phrynosoma platyrhinos*, *Callisaurus draconoides*, *Gambelia wislizenii*, *Coleonyx variegatus*, and *Sceloporus magister*. Highest densities (max. 80.3 individuals ha^{-1}) were recorded for *U. stansburiana*. One reason for high abundance of *Uta stansburiana* was that females may lay up to six egg clutches annually; clutch number tended to be high after abundant winter rains, and clutch size was positively correlated with female body size. This species had the lowest adult annual survival rate (22%) and shortest maximum life span (58 months). *Cnemidophorus tigris* had the highest mean dry biomass, an estimated adult annual survival rate of 40%, and a maximum life span of 94 months. But even having extensive data from long-term demographic studies was not sufficient to determine which factors controlled population densities of these two species. For *P. platyrhinos*, which has a specialized diet of ants, spring densities were positively and significantly correlated with amount of winter rainfall.

Whereas little is known about the population biology of snakes from Rock Valley, the locality has been used since 1959 for long-term studies on the desert tortoise (*Gopherus agassizii*). This herbivore forages during the spring on vegetative rosettes of winter annuals and stores preformed water in the large urinary bladder. As summer approaches, desert tortoise mostly stays in its burrow and must tolerate high levels of salts in blood plasma, loss of body mass, and water stress. During a rare summer rainstorm this animal can scrape out depressions for water collection, where they relieve water stress by nasal drinking. In Rock Valley researchers have found that one annulus is added per scute yr^{-1} and that plastron growth rate was 10.9 mm yr^{-1} from 1959 to 1973 and 7.6 mm yr^{-1} from 1973 to 1984. Females become sexually mature at 17–18 years and normally carry one or two clutches of eggs per year. Success of desert tortoise populations, their energy and water budgets as well as reproductive capacity, depends very strongly on the success of desert annuals.

9 Birds

Desert scrub habitats, which have low annual precipitation, simple vegetational structure, and low primary production, support few resident bird populations unless permanent water is available (Serventy 1971). For each continent there is a very short list of species that are truly characteristic of thriving under desert conditions. For example, despite biogeographic dominance of desert and semi-desert environments in Australia, which comprise the central 70% of that continent, Keast (1959) identified only 17 bird species as characteristic of the region, just 3% of total breeding avifauna. Likewise, in North America only 31 species of birds are listed as desert inhabitants (MacMahon 1979). Nevertheless, in spring and fall transient populations of many bird species migrate through dryland habitats on the Nevada Test Site, and some bird species are seasonal or year-round residents, having some adaptations for desert life (Dawson & Bartholomew 1968; Dawson 1984).

BIRDS OF THE NEVADA TEST SITE

The avifauna of the Nevada Test Site consists of 220 species (Table 9.1; data as of 1991) of which at least 160 were classified as transients (Allred et al. 1963; Hayward et al. 1963; O'Farrell & Emery 1976; Castetter & Hill 1979; BECAMP 1991b). Many transients were sighted during biannual mass bird migrations as part of the western North America flyway, and birds rest and feed at NTS during cool months, especially when precipitation occurs (Chapter 2) and creates temporary standing water. On-site breeding has only been confirmed at NTS for 31 species, largely those living at the higher, cooler, less arid habitats. Sightings of numerous waterfowl show that any bird is a candidate to be present in desert environments, because birds in general possess physiological characteristics and behavioral traits preadapted either to tolerate or avoid drought and high-temperature problems (Bartholomew & Cade 1956), but the desert avifauna largely reflects range extensions into the desert from less xeric, surrounding regions.

Of 192 species documented in the first study by Allred et al. (1963), greatest species diversity was observed in pinyon-juniper woodland (92 spp.) and *Larrea–Ambrosia* desert scrub (83 spp.), whereas lowest diversity was found on playas with *Atriplex–Kochia* scrub (38 spp.) and post-disturbance *Salsola* stands (37 spp.), which had low plant diversity and ephemeral surface water. Spring and seepage areas, montane communities, and sites located by reservoirs were the richest bird habitats, having 167 species, then 88% of the known avifauna, indicative that daily water availability is critical in deserts to support high bird diversity.

Small passerine birds, including such native species as sage sparrow (*Amphispiza belli*), mountain bluebird (*Sialia currucoides*), and white-crowned sparrow (*Zonotrichia leucophrys*) and the introduced starling (*Sturnus vulgaris*) use NTS as an important winter feeding ground. Horned lark (*Eremophila alpestris*) and house finch (*Carpodacus mexicanus*) are year-round residents in mesic mixed communities, but are restricted to fall and winter months in arid Mojave Desert sites;

Table 9.1. *Birds of the Nevada Test Site (Hayward, Killpack, & Richards 1963; Moor & Bradley 1974; Castetter & Hill 1979; unpublished list by P. D. Greger from BECAMP 1991b). Species observed at the Rock Valley validation site during the IBP study (Hill & Burr 1974) are indicated by an asterisk.*

Family	Species	Common name
Order Podicipediforme		
Podicipedidae	*Aechmophorus occidentalis*	western grebe
	Podiceps nigricollis	eared grebe
	Podilymbus podiceps	pied-billed grebe
Order Pelecaniformes		
Pelecanidae	*Pelecanus erythrorhynchos*	American white pelican
Phalacrocoracidae	*Phalacrocorax auritus*	double-crested cormorant
Order Ciconiformes		
Ardeidae	*Ardea herodias*	great blue heron
	Botaurus lentiginosus	American bittern
	Bubulcus ibis	cattle egret
	Butorides striatus	green-backed heron
	Casmerodius albus	great egret
	Egretta thula	snowy egret
	Ixobrychus exilis	least bittern
	Nycticorax nycticorax	black-crowned night-heron
Threskiornithidae	*Ajaia ajaja*	roseate spoonbill
	Plegadis chihi	white-faced ibis
Order Anseriformes		
Anatidae	*Aix sponsa*	wood duck
	Anas acuta	northern pintail
	A. americana	American wigeon
	A. clypeata	northern shoveler
	A. crecca	green-winged teal
	A. cyanoptera	cinnamon teal
	A. discors	blue-winged teal
	A. platyrhynchos	mallard
	A. strepera	gadwall
	Aythya affinis	lesser scaup
	A. americana	redhead
	A. collaris	ring-necked duck
	A. valisineria	canvasback
	Branta canadensis	Canada goose
	Bucephala albeola	bufflehead
	Chen caerulesces	snow goose
	Cygnus columbianus	tundra swan
	Melanitta perspicillata	surf scoter
	Mergus merganser	common merganser
	M. serrator	red-breasted merganser
	Oxyura jamaicensis	ruddy duck
Order Falconiformes		
Accipitridae	*Accipiter cooperii*	Cooper's hawk
	A. gentilis	northern goshawk
	A. striatus	sharp-shinned hawk
	*Aquila chrysaetos	golden eagle

Table 9.1 (*cont.*)

Family	Species	Common name
Order Podicipediforme		
	Buteo jamaicensis	red-tailed hawk
	B. lagopus	rough-legged hawk
	B. regalis	ferruginous hawk
	B. swainsoni	Swainson's hawk
	*Circus cyaneus	northern harrier
	*Haliaeetus leucocephalus	bald eagle
	Pandion haliaetus	osprey
Cathartidae	**Cathartes aura*	turkey vulture
Falconidae	**Falco mexicanus*	prairie falcon
	F. peregrinus	peregrine falcon
	*F. sparverius	American kestrel
Order Galliformes		
Phasianidae	*Alectoris chukar*	chukar
	Callipepla gambelii	gambel's quail
Order Gruiformes		
Rallidae	*Fulica americana*	American coot
	Gallinula chloropus	common moorhen
Order Charadriiformes		
Charadriidae	*Charadrius alexandrinus*	snowy plover
	C. montanus	mountain plover
	C. semipalmatus	semipalmated plover
	C. vociferus	killdeer
	Pulvialis dominica	lesser golden-plover
	P. squatarola	black-bellied plover
Laridae	*Chlidonias niger*	black tern
	Larus californicus	California gull
	L. delawarensis	ring-billed gull
	L. philadelphia	Bonaparte's gull
	L. pipixcan	Franklin's gull
	Sterna caspia	Caspian tern
	S. forsteri	Forster's tern
Recurvirostridae	*Himantopus mexicanus*	black-necked stilt
	Recurvirostra americana	American avocet
Scolopacidae	*Actitis macularia*	spotted sandpiper
	Calidris alpina	dunlin
	C. bairdii	Baird's sandpiper
	C. himantopus	stilt sandpiper
	C. mauri	western sandpiper
	C. melanotos	pectoral sandpiper
	C. minutilla	least sandpiper
	Catoptrophorus semipalmatus	willet
	Gallinago gallinago	common snipe
	Limnodromus scolopaceus	long-billed dowitcher
	Limosa fedoa	marbled godwit
	Numenius americanus	long-billed curlew
	Phalaropus lobatus	red-necked phalarope
	P. tricolor	Wilson's phalarope
	Tringa flavipes	lesser yellowlegs
	T. melanoleuca	greater yellowlegs
	T. solitaria	solitary sandpiper

Table 9.1 (*cont.*)

Family	Species	Common name
Order Columbiformes		
Columbidae	*Columba livia*	rock dove
	*Zenaida macroura	mourning dove
Order Cuculiformes		
Cuculidae	*Coccyzus americanus*	yellow-billed cuckoo
	Geococcyx californianus	greater roadrunner
Order Strigiformes		
Strigidae	**Asio flammeus*	short-eared owl
	A. otus	long-eared owl
	*Athene cunicularia	burrowing owl
	Bubo virginianus	great horned owl
Tytonidae	*Tyto alba*	common barn-owl
Order Caprimulgiformes		
Caprimulgidae	*Chordeiles acutipennis*	lesser nighthawk
	C. minor	common nighthawk
	*Phalaenoptilus nuttallii	common poorwill
Order Apodiformes		
Apodidae	**Aeronautes saxatalis*	white-throated swift
Trochilidae	*Calypte costae*	Costa's hummingbird
	Selasphorus platycercus	broad-tailed hummingbird
	S. rufus	rufous hummingbird
Order Coraciiformes		
Alcedinidae	*Ceryle alcyon*	belted kingfisher
Order Piciformes		
Picidae	*Colaptes auratus*	northern flicker
	Melanerpes lewisi	Lewis' woodpecker
	Picoides scalaris	ladder-backed woodpecker
	P. villosus	hairy woodpecker
	Sphyrapicus varius	red-naped sapsucker
Order Passeriformes		
Aegithalidae	*Psaltriparus minimus*	bushtit
Alaudidae	**Eremophila alpestris*	horned lark
Bombycillidae	*Bombycilla cedrorum*	cedar waxwing
Corvidae	**Aphelocoma coerulescens*	scrub jay
	Corvus brachyrhynchos	American crow
	*C. corax	common raven
	Cyanocitta stelleri	Steller's jay
	Gymnorhinus cyanocephalus	pinyon jay
	Nucifraga columbiana	Clark's nutcracker
	*Pica pica	black-billed magpie
Emberizidae	*Agelaius phoeniceus*	red-winged blackbird
	*Amphispiza belli	sage sparrow
	*A. bilineata	black-throated sparrow
	Calcarius lapponicus	Lapland longspur
	*Chondestes grammacus	lark sparrow
	*Dendroica coronata	yellow-rumped warbler

Table 9.1 (*cont.*)

Family	Species	Common name
	*D. nigrescens	black-throated gray warbler
	*D. petechia	yellow warbler
	D. townsendi	Townsend's warbler
	*Euphagus cyanocephalus	Brewer's blackbird
	Geothlypis trichas	common yellowthroat
	Guiraca caerulea	blue grosbeak
	Icteria virens	yellow-breasted chat
	Icterus galbula	northern oriole
	*I. parisorum	Scott's oriole
	*Junco hyemalis	dark-eyed junco
	Melospiza lincolnii	Lincoln's sparrow
	M. melodia	song sparrow
	*Molothrus ater	brown-headed cowbird
	Oporornis tolmiei	MacGillivray's warbler
	*Passerculus sandwichensis	savannah sparrow
	Passerina amoena	Lazuli bunting
	P. cyanea	indigo bunting
	Pipilo chlorurus	green-tailed towhee
	P. erythrophthalmus	rufous-sided towhee
	Pheucticus melanocephalus	black-headed grosbeak
	Piranga ludoviciana	western tanager
	Pooecetes gramineus	vesper sparrow
	Quiscalus mexicanus	great-tailed grackle
	Q. quiscula	common grackle
	Setophaga ruticilla	American redstart
	Spizella atrogularis	black-chinned sparrow
	*S. breweri	Brewer's sparrow
	*S. passerina	chipping sparrow
	*Sturnella neglecta	western meadowlark
	Vermivora celata	orange-crowned warbler
	V. ruficapilla	Nashville warbler
	V. virginiae	Virginia's warbler
	*Wilsonia pusilla	Wilson's warbler
	*Xanthocephalus xanthocephalus	yellow-headed blackbird
	Zonotrichia atricapilla	golden-crowned sparrow
	*Z. leucophrys	white-crowned sparrow
Fringillidae	*Carduelis pinus	pine siskin
	C. psaltria	lesser goldfinch
	C. tristis	American goldfinch
	Carpodacus cassinii	Cassin's finch
	*C. mexicanus	house finch
	C. purpureus	purple finch
	Coccothraustes vespertinus	evening grosbeak
	Loxia curvirostra	red crossbill
Hirundinidae	*Hirundo pyrrhonota	cliff swallow
	*H. rustica	barn swallow
	Riparia riparia	bank swallow
	*Stelgidopteryx serripennis	northern rough-winged swallow
	*Tachycineta bicolor	tree swallow
	*T. thalassina	violet-green swallow
Laniidae	*Lanius ludovicianus	loggerhead shrike

Table 9.1 (*cont.*)

Family	Species	Common name
Mimidae	*Dumetella carolinensis*	gray catbird
	Mimus polyglottos	northern mockingbird
	*Oreoscoptes montanus	sage thrasher
	Toxostoma dorsale	crissal thrasher
	*T. lecontei	Le Conte's thrasher
	T. rufum	brown thrasher
Motacillidae	**Anthus rubescens*	American pipit
Muscicapidae	*Catharus guttatus*	hermit thrush
	C. ustulatus	Swainson's thrush
	Myadestes townsendi	Townsend's solitaire
	*Polioptila caerulea	blue-gray gnatcatcher
	P. melanura	black-tailed gnatcatcher
	*Regulus calendula	ruby-crowned kinglet
	Sialia currucoides	mountain bluebird
	S. mexicana	western bluebird
	Turdus migratorius	American robin
Paridae	*Parus gambeli*	mountain chickadee
	P. inornatus	plain titmouse
Passeridae	*Passer domesticus*	house sparrow
Ptilogonatidae	**Phainopepla nitens*	phainopepla
Sittidae	*Sitta canadensis*	red-breasted nuthatch
	S. carolinensis	white-breasted nuthatch
Sturnidae	**Sturnus vulgaris*	European starling
Troglodytidae	*Campylorhynchus brunneicapillus*	cactus wren
	Catherpes mexicanus	canyon wren
	Cistothorus palustris	marsh wren
	*Salpinctes obsoletus	rock wren
	Thryomanes bewickii	Bewick's wren
	Troglodytes aedon	house wren
Tyrannidae	*Contopus borealis*	olive-sided flycatcher
	C. sordidulus	western wood-pewee
	Empidonax difficilis	western flycatcher
	E. oberholseri	dusky flycatcher
	*E. wrightii	gray flycatcher
	*Myiarchus cinerascens	ash-throated flycatcher
	*Pyrocephalus rubinus	vermilion flycatcher
	Sayornis nigricans	black phoebe
	*S. saya	Say's phoebe
	Tyrannus forficatus	scissor-tailed flycatcher
	*T. verticalis	western kingbird
	*T. vociferans	Cassin's kingbird
Vireonidae	*Vireo gilvus*	warbling vireo
	V. solitarius	solitary vireo
	V. vicinior	gray vireo

both species experience significant population increases during winter, and at that time may be seen in large flocks consisting of thousands of birds (Richards 1962). Two granivorous species, black-throated sparrow (*Amphispiza bilineata*; Fig. 9.1) and mourning dove (*Zenaida macroura*; Fig. 9.2), are largely late spring and summer residents on NTS.

Figure 9.1 Black-throated sparrow, *Amphispiza bilineata.*

Species on the IBP validation site in Rock Valley

During the IBP studies 53 species of birds were observed at the validation site in Rock Valley (Table 9.1). Included were six species of swallows (Hirundinidae) and swifts (Apodidae), aerial insectivores, which are widely foraging, strong fliers, and several large, diurnally active falconiform predators that prey on small mammals and lizards and which have wide home ranges. Forty-three species of passerine birds were observed on the IBP validation site, many of which were transients. Notably absent in Rock Valley were any large, diurnal, granivorous ground birds, such as Gambel's quail (*Callipepla gambelii*; Fig. 9.3) and the introduced chukar (*Alectoris chukar*), which occurred elsewhere on NTS.

NATURAL HISTORY OF COMMON MOJAVE DESERT BIRDS

There have been comparatively fewer scientific studies of Mojave Desert birds than on sympatric rodents (Chapter 7) and lizards (Chapter 8), prob-

ably because for flying organisms there are problems in quantifying many ecological and physiological parameters. The research, which has been periodically reviewed (Bartholomew 1964, 1972; Calder & King 1974; Dawson 1984; Wiens 1991), has shown that most desert birds avoid heat stress by seeking cooler microhabitats, for example, resting among rocks (Smyth & Bartholomew 1966), if they cannot fly to sources of water, but some also can tolerate hyperthermia to 46°C for short periods (Calder & King 1974). By tolerating hyperthermia an individual reduces its need to utilize evaporative cooling. As described in Chapter 6, a few species of desert birds also can cool themselves using gular flutter (Bartholomew, Lasiewski, & Crawford 1968).

Gambel's quail

Gambel's quail (*Callipepla gambelii*; Fig. 9.3), also called desert quail, is a widespread and successful species in the Mojave and Sonoran deserts. Unlike other granivorous desert animals of this region, it typically is active during daylight hours, and this presents interesting questions concerning the

Figure 9.2 Mourning dove, *Zenaida macroura*.

Figure 9.3 Gambel's quail, *Callipepla gambelii*.

means by which these ground-dwelling birds are able to obtain adequate food and water while avoiding potentially lethal heat stress.

As discussed in Chapter 6, diurnally active species frequently utilize behavioral traits to minimize summer heat stress. Studies of Gambel's quail in the Sonoran Desert of California have shown that these animals tend to remain inactive in the shade during the hottest part of the day (Goldstein 1984). A time budget established that a bird spent 26% of its time and 48% of daylight hours in that inactive condition (Fig. 9.4). Foraging activities summed to 6.8 hr d^{-1}, half in the morning and half in the evening. Rapid running was used almost entirely to dash between shrubs during the hottest part of the day, and flight activity was virtually absent except for moving down steep banks or arroyo wells. Running and flying combined amounted to less than 2 min d^{-1}.

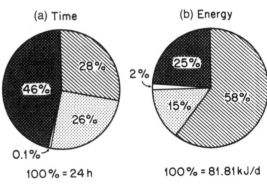

Figure 9.4 Time budget for Gambel's quail (*Callipepla gambelii*) at a site in the Sonoran Desert of California. From Goldstein and Nagy (1985).

Despite their extreme thermal environment, Gambel's quail have negligible costs associated with thermoregulation, and their metabolic thermoneutral zone extends from 34 to 44°C, allowing them to avoid costs associated with cooling the body simply with behavioral traits, i.e., by utilizing shade microhabitats (Goldstein & Nagy 1985). Foraging activities accounted for 58% of the 82 kJ daily energy use, calculated from time-activity budgets (Fig. 9.4). Although there was virtually no energy expended for flight or running, foraging cost was not notably low because they may travel up to 3.5 km d^{-1} (Goldstein & Nagy 1985), which is more than five times farther than typical distances that are traveled by small mammalian granivores of similar mass (Garland 1983).

In the Deep Canyon area near Palm Desert, California, seeds and dry fruits comprised 94% of the dry weight of food in the crop of desert quail, ants comprised 5%, and leaf material was the remainder (Goldstein & Nagy 1985). A similar diet has been reported for the same species in Arizona (Hungerford 1962).

Analyses of the daily feeding rates and metabolic water relations for Gambel's quail have revealed that free-living individuals require about 5 ml of water per day when they are fully acclimated to summer conditions. Oxidative water that is metabolized from their food (8.1 g d^{-1} dry weight) would provide about 55% of this water requirement, leaving a remainder of 2.4 ml d^{-1} to come from preformed water in foods or from drinking. If these birds do not drink, they would therefore need to average 23% water content in their food to survive. Seeds, however, average less than 3% water. Thus, the two options are consuming juicy arthropods, i.e., ants, or consuming watery vegetation. Both options appear to be utilized during the peak of water stress, when cactus fruits and ants are included in the diet. However, when free water is available, the proportion of arthropods in the diet does not change markedly, suggesting that ants may be a nutritional requirement for nitrogen. Absence from a place like Rock Valley is probably an indication that *C. gambelii* needs drinking water as well as some dense vegetational cover (Phillips, Marshall, & Monson 1964; Small

Table 9.2. *Species of birds in Rock Valley (1971–3) that were year-round (Y) or seasonal residents (winter, W; summer, S), but some of the seasonal residents were transients (T) during poor years. Elsewhere on NTS these same species breed (B), probably breed (P), or do not breed (from Greger, unpublished list, 1991). Species are grouped according to food preferences (based on Peterson 1990).*

	Residency in Rock Valley	Breeding status on NTS
Carnivore		
Aquila chrysaetos	Y	P
Athene cunicularia	Y	P
Circus cyaneus	W	N
Falco mexicanus	Y	B
F. sparverius	Y	B
Lanius ludovicianus	Y	B
Omnivore		
Corvus corax	Y	B
Pica pica	W	N
Sturnus vulgaris	W	P
Insectivore		
Anthus rubescens	W (1972)	N
Myiarchus cinerascens	S, T	P
Salpinctes obsoletus	W	B
Insectivore/frugivore		
Mimus polyglottos	S, T/B (1973)	B
Toxostoma lecontei	Y/B	B
Granivore (primarily)/insectivore/frugivore (rarely)		
Amphispiza belli	S, T/B (1973)	B
A. bilineata	W, S/B	B
Carpodacus mexicanus	Y	B
Eremophila alpestris	W, T	B
Junco hyemalis	W	P
Spizella breweri	S, T/B (1973)	B
Zenaida macroura	S	B

1974; Degen, Pinshow, & Aikon 1982; Degen, Pinshow, & Shaw 1983; Goldstein & Nagy 1985; Nagy 1988a).

For desert quail living under conditions of drought, there is a shutdown within the kidney of reptilian-type nephrons by an antidiuretic hormone, and this produces a substantial reduction in urine production (Braun & Dantzler 1972, 1974).

Mourning doves

Mourning dove (Fig. 9.2) is a common granivore of desert scrub habitats, but, as discussed in Chapter 6, this species must fly daily to the nearest source to drink water (Bartholomew & Dawson 1954; Bartholomew & MacMillen 1960; Mac-Millen 1962; Dawson 1984).

Passerine birds

Small passerines that are granivorous also consume arthropods and may likewise include juicy fruits (Table 9.2). Ecophysiological studies of these species are limited to several studies on water relations, which indicate that *Amphispiza bilineata* (Fig. 9.1) is the most tolerant species and can live in desert habitats without drinking water and is considered to be the best adapted seed-eating bird of North American deserts (Smyth & Bartholomew 1966; Bartholomew 1972). Hence, black-throated sparrow occurs on barren desert slopes far from surface water. Ecophysiological research on California populations of black-throated sparrow determined that these birds can live for months without drinking on diets consisting exclusively of dried seeds, from which oxidative water was obtained.

The house finch, *Carpodacus mexicanus*, can also tolerate some drought, and, although it likes to drink, the animal can retain its body weight during a drought by eating juicy fruits (Bartholomew & Cade 1956; Miller & Stebbins 1964; Calder 1981).

Insectivorous passerines are buffered from water stress because a number of them use juicy fruits to obtain preformed water during drought (Table 9.2). Very few physiological measurements have been made on these desert species (Sheppard 1970).

Greater roadrunner

Absent from Rock Valley was the great roadrunner (*Geococcyx californianus*; Fig. 9.5), which is more typical of communities of warm desert habitats with higher biomass and abundant prey items. This is a diurnally active bird that preys primarily on lizards and small snakes. Its abilities to tolerate voluntary hypothermia and utilize early morning sun for thermoregulation are noteworthy adaptations for conserving energy (Chapter 6; Calder & Schmidt-Nielsen 1967; Ohmart & Lasiewski 1971). The great roadrunner also has salt-excreting glands for water economy (Ohmart 1972).

Le Conte's thrasher

Le Conte's thrasher (*Toxostoma lecontei*) is a common bird of flat, open creosote bush desert scrub and along washes, especially where soil is sandy (Miller & Stebbins 1964). These birds can move rapidly across open ground while foraging for insects, particularly larvae, which are often obtained by digging and extracting the prey with its long, curved bill. There are no published studies on the ecophysiology of this species.

BIRD STUDIES OF ROCK VALLEY
Sampling methods

Selective ecological studies of bird populations were carried out at Rock Valley for the IBP program during two dry years (1971–2) and the following wet year (1973). Sampling was done within three plots between February and June and in November and December, 1971; between February and June and in November, 1972; and twice weekly between February and November, 1973. Study areas were circular plots A and C adjacent to the validation site, and the other (25 ha) was located within the site boundaries and contained portions of the representative vegetation zones of the site.

Densities of breeding birds were estimated to the nearest 0.25 territory in each year using the Williams spot mapping census technique (Williams 1936; Kendeigh 1944), which has been demonstrated as a very reliable but time-consuming method (Christman 1984). Beginning in February, five to eight censuses were made

Figure 9.5 Great roadrunner, *Geococcyx californianus*.

each month during the breeding season. Nests were located by observing movements of mating pairs and augmented by a nest-searching program for those species exhibiting weak territorial behavior. In 1973 live body mass was measured for individual breeding birds that were captured in mist nets in the vicinity of their nests.

Residency

Of the 53 species of birds on or near the primary study site in Rock Valley from 1971 to 1973, 21 species were residents but only 8 of these were year-round residents (Table 9.2). Five of the permanent residents were carnivores: golden eagle

(*Aquila chrysaetos*), prairie falcon (*Falco mexicanus*), American kestrel (*Falco sparverius*, which also eats grasshoppers; Fig. 9.6), burrowing owl (*Athene cunicularia*; Fig. 9.7), and loggerhead shrike (*Lanius ludovicianus*). The other three species were the common raven (*Corvus corax*), which is an omnivore that will eat carrion (nowadays specializing on road kills), house finch, primarily a granivore, and Le Conte's thrasher, primarily an insectivore. Several of the seasonal residents in Rock Valley were widely reported from warm desert habitats, including ash-throated flycatcher (*Myiarchus cinerascens*), rock wren (*Salpinctes obsoletus*; Fig. 9.8), mourning dove (summer, 1973), horned lark (*Eremophila*

Figure 9.6 American kestrel, *Falco sparverius*.

alpestris), and black-throated sparrow, and from Great Basin habitats appeared sage sparrow (*Amphispiza belli*) and Brewer's sparrow (*Spizella breweri*). American pipit (*Anthus rubescens*), a bird that breeds in tundra and at high elevations, was a winter resident in 1972. The nonresident white-crowned sparrow (*Zonotrichia leucophrys*) was commonly observed in all three years, yellow-rumped warbler (*Dendroica coronata*) in 1971, and chipping sparrow (*Spizella passerina*) in 1973.

Breeding in Rock Valley

Two species, Le Conte's thrasher and black-throated sparrow, nested on the study plots in Rock Valley during 1971–3. In 1973, following a highly favorable period of winter rainfall, Brewer's

sparrow, sage sparrow, and mockingbird (*Mimus polyglottos*) also maintained breeding territories within the plots.

Brewer's and sage sparrows normally nest at higher elevations in cool *Artemisia* desert scrub and woodland communities within southern Nevada (Hayward et al. 1963; Bradley & Deacon 1967; Austin & Bradley 1971), and these species are the two most common and characteristic breeding birds of *Artemisia* communities throughout the western United States (Wiens & Rotenberry 1981; Rotenberry & Wiens 1989). Other seasonal residents need special features for successful breeding. For example, *Myiarchus cinerascens* nests in cavities of tree trunks (Miller & Stebbins 1964) and *Mimus polyglottos* typically does not nest in sparse desert scrub, but shows preference

Figure 9.7 Burrowing owl, *Athene cunicularia.*

for taller vegetation for both foraging and nesting (Hill 1980).

For Le Conte's thrasher initiation of egg laying varied by only four days between 1971 and 1973, and Hill (1980) suggested this as evidence that photoperiodic factors control the onset of breeding. Thrashers always attempted two or three broods. Breeding of black-throated sparrow began in May for 1971, March for 1972, and April for 1973. Coincidentally, shrubs flowered and fruited late in 1971, corresponding to the latest start of breeding for the sparrows. These observations suggest that time of breeding in this species may be controlled by the phenologies of certain plants or by the availability of the phytophagous insects, each of which responds to temperature cues. Sparrows raised only one brood each season, with peak nesting in May of all three years.

Habitat partitioning

The two regular breeding species appeared to partition the habitat effectively. Although stomach contents were not analyzed, observations indicated that the thrashers fed primarily on ground-dwelling arthropods, while black-throated sparrows consumed phytophagous insects and unspecified plant materials. Le Conte's thrasher began nesting in February by constructing its nest in the center of shrubs, typically about 0.40 m from the ground. These birds almost always nested in species of *Lycium*, because their nests required strong support. Black-throated sparrow nested in the lowest parts of shrubs as close as 0.20 m from the ground. *Lycium andersonii* was the most frequently used, but where that species was uncommon *Grayia* and *Atriplex confertifolia* were also utilized. No data were obtained on feeding and nest-site preferences of the three species that bred

Figure 9.8 Rock wren, *Salpinctes obsoletus.*

on the plots in 1973, although Petersen and Best (1985) have demonstrated that in Idaho sage sparrows, using *Artemisia tridentata*, specifically built nests in the center of shrubs at a mean distance of 0.34 m from the ground and avoided locations on the southwestern side of the plant, where winds and afternoon solar radiation were greatest.

Production and population densities

Estimated breeding densities of birds at the Rock Valley plots are shown in Table 9.3. Black-throated sparrow comprised more than 90% of the breeding pairs in 1971 and 1972 but dropped to 75% during 1973, even though number of breeding pairs increased significantly, because three additional species were added to the breeding population. Le Conte's thrasher had identical and

low breeding densities for the three years of the study. Dry biomass values for breeding pairs were estimated for small samples of the population.

Turner and Chew (1981) estimated production by black-throated sparrow and Le Conte's thrasher in Rock Valley for 1973, when breeding conditions there were unusually favorable. Production was estimated to be 29 kcal ha^{-1} for black-throated sparrow and 10.5 kcal ha^{-1} for Le Conte's thrasher. Their calculations required a number of assumptions as well as egg weight estimates extrapolated from linear dimensions and the use of selected observations on thrashers in Arizona (Russell, Gould, & Smith 1973).

Hill (1980) collected quantitative data on nesting characters for four of the breeding bird species observed in the 1971–3 IBP study (Table 9.4). Mean clutch size increased in 1973 both for Le

Table 9.3. *Estimated breeding densities and dry biomass of birds in Rock Valley, 1971–1973.*

Species	Breeding pairs per 100 ha			Dry biomass, g ha^{-1}		
	1971	1972	1973	1971	1972	1973
Toxostoma lecontei	3.0	3.0	3.0	1.4	1.4	1.4
Amphispiza bilineata	43.1	52.4	60.6	4.1	5.0	5.8
A. belli	0	0	2.3	0	0	0.3
Spizella breweri	0	0	11.7	0	0	0.9
Mimus polyglottos	0	0	2.3	0	0	0.8
Total	46.1	55.4	79.9	5.5	6.4	9.2

Table 9.4. *Nesting by four species of birds in Rock Valley, 1971–1973. Numbers of nests under observation are given in parentheses.*

Species	Year	Mean clutch size	Eggs hatched, %	Mean number fledged per nest
Toxostoma lecontei	1971	3.3 (3)	57.1 (2)	not observed
	1972	3.3 (3)	85.7 (2)	not observed
	1973	3.8 (4)	73.3 (4)	2.0 (4)
Amphispiza bilineata	1971	3.0 (6)	not observed	1.5 (2)
	1972	3.3 (6)	84.6 (4)	0.3 (4)
	1973	4.0 (9)	78.1 (8)	2.6 (14)
A. belli	1973	4.0 (1)	75.0 (1)	3.0 (1)
Spizella breweri	1973	3.0 (2)	100.0 (2)	2.5 (2)

Conte's thrasher and black-throated sparrow, apparently reflecting increased food resource availability. Percent of eggs hatched did not differ significantly between a dry year (1972) and a wet one (1973), but the mean number of young fledged per nest was much greater in 1973 for black-throated sparrows.

Few data on Brewer's and sage sparrows were obtained from the study site, which supported small populations of these birds in comparison with stands of *Artemisia* elsewhere that support several hundred individuals per 100 ha (Wiens & Rotenberry 1981). Throughout the species range, including the 1973 plots in Rock Valley, mean clutch size of both sparrow species was three eggs (Table 9.4; Rotenberry & Wiens 1989).

Densities of breeding bird pairs in Rock Valley (46–80 pairs per 100 ha) were low in comparison with typical North American desert communities (Table 9.5). Densities exceeding 90 pairs per 100 ha were available in the literature from Sonoran and Mojave desert scrub locations (Hensley 1954; Sheppard 1968). During IBP surveys (Table 9.5) Rock Valley ranked third of the four desert sites for breeding bird densities and had the fewest species of breeding birds. Highest breeding densities and dry weight biomass values were obtained at the Silverbell site (Russell et al. 1973). For Chihuahuan Desert earlier studies by Dixon (1959) reported densities of 23–37 pairs per 100 ha, and Raitt and Maze (1968) estimated densities of 21–44 pairs per 100 ha. On the other hand, Hill (1980)

Table 9.5. *Breeding bird densities and biomasses at four desert sites that were studied in North America during the IBP program. Data are from Russell et al. (1973) for the Sonoran Desert, Raitt (1972) for the Chihuahuan Desert, Balph et al. (1971) for the Great Basin Desert, and Hill (1980) for the Mojave Desert.*

Site	Mean annual precipitation, mm	Year	Nesting species	Pairs 100 ha^{-1}	Biomass, g ha^{-1}
Sonoran (Silverbell)	280	1972	22	390	215
Chihuahuan (Jornada Range)	230	1971	7	40	11.5
Great Basin (Curlew Valley)	350	1972	6	120	17.7
Mojave (Rock Valley)	130	1971–3	2–5	46–80	5.5–9.2

reported that breeding densities of Le Conte's thrasher (3 pairs per 100 ha) and black-throated sparrow (43–61 pairs per 100 ha) were among the highest reported for those species (cf. Sheppard 1970; Russell et al. 1973). However, biomass of breeding birds in Rock Valley (5–9 g ha^{-1}) was significantly lower than that observed in Arizona (Russell et al. 1973) and roughly 30–50 % of values from localities in Utah (Balph 1973) and New Mexico (Raitt 1972).

SUMMARY

There are relatively few species of resident birds during periods of drought in Mojave Desert scrub. Ecophysiological studies have been carried out on several of the common species, and observers have concluded that desert birds tend to seek the coolest microhabitats and may be able to tolerate hyperthermia for short periods. Excluding mourn-

ing dove, which typically flies every day to drinking sites, typical desert birds rely on preformed and metabolic water from their foods.

The scanty evidence on bird ecology from Rock Valley suggests that the small resident avifauna is a reflection of the relative scarcity of water and foods because higher densities and more bird breeding occur in wet years and within more mesic portions of NTS. Le Conte's thrasher and black-throated sparrow were the two species that nested at Rock Valley during the two dry years, but production and densities of breeding bird pairs were generally low in comparison with similar data from other warm desert sites in North America. During the wet year of 1973, several additional species assumed residency in Rock Valley and laid clutches of eggs. Hence, yearly variability in desert bird populations are highly dependent on climatic factors.

10 Arthropods

Invertebrates comprise the largest and most diverse segment of faunal biomass in deserts, and in particular, desert arthropods play fundamental roles in ecosystem processes of carbon and nutrient cycling and plant reproductive biology, especially as pollinators. Included also are important herbivores, which limit net primary production by vascular plants, both of fresh shoots and reproductive structures above ground and roots and stems below ground. Many arthropods, for example, microarthropods, ants, and beetles, play important roles in decomposition processes. Still other species are predators, consumers of other arthropods. To complete the food web, one finds that especially the large, palatable forms are preferred, juicy prey items for insectivorous vertebrates, such as lizards, cricetid rodents, and predaceous birds, and most vertebrates consume arthropods whenever water and typical food items become scarce.

TROPHIC SPECIALIZATIONS

At least seven categories of trophic specialization for desert invertebrates have been identified (Crawford 1981). Above ground there are foliage (leaf and stem) herbivores, pollinators, and granivores. Operating to regulate each of these groups are diverse assemblages of carnivorous invertebrates, in addition to the vertebrate predators. Within each trophic guild feeding habits for individual species vary from generalist to highly specialist. Belowground consumers include root herbivores, which feed on living tissues, and coprovores, necrovores, and detritivores, which help to recycle dead organic matter. Root herbiv-

ores, particularly cicada, beetle, and lepidopteran larvae, are important members of the soil fauna beneath shrubs. Desert coprovores and detritivores, along with associated carnivores, often live on soil surface among litter. Dung beetles and isopods are perhaps the best known of detritivores and coprovores, but many other invertebrate groups are represented (Crawford 1979).

For arthropods in the aboveground environment, seed plants provide the primary source of food. Food may be in the form of living foliage for defoliators, phloem exudate for sap-feeders, nectar for pollinators and nectar robbers, and ripening fruit to dried, dispersed seeds for granivores. These types of resources are highly variable in both spatial and temporal distribution, and thus provide special problems for life history adaptations. Leaves of perennial plants may be present for six to twelve months of the year, whereas ephemeral species may be available only for a few weeks in a single year and not necessarily every year. Models of coevolved strategies for herbivory and defense between defoliators and plants have been described in theoretical models (Rhoades & Cates 1976; Cates, & Rhoades 1977). Flowering may be greatly reduced or absent in years without rain, eliminating resources for pollinators and some granivores, and this in turn affects many of those highly coevolved insect–plant interactions (Strong, Lawton, & Southwood 1984).

Types of soil and litter habitats

Soil habitats show species abundance and diversity patterns that are highly nonrandom in horizontal space. Resource availability and reduced abiotic stress combine to promote highest frequencies and abundances beneath shrub canopies. Beneath the canopy resources of water and nutrients are often more available than in open areas (Chapter 2), and seasonal cycles of temperature are also buffered. For soil arthropods, soil and litter environments are more stable in resource availability than aboveground habitats, and food is readily available for such animals as Orthoptera (crickets and grasshoppers), microarthropods, and scarab and tenebrionid beetles (Coleoptera).

Three broad types of soil and litter habitats have been identified: (1) crevice-type habits beneath and between objects at the soil surface, (2) the soil matrix and associated burrows, and (3) the open surface itself (Crawford 1981). Tenebrionid beetles (darkling beetles), which are scavengers of decaying organic matter and fungi, and many arthropod predators inhabit soil crevices during the day and become active on the soil surface at night. Most desert scorpions and many Orthoptera and Coleoptera utilize crevice habitats. Studies of Old World deserts have shown that relative abundance and distribution of many crevice-dwelling arthropods may be strongly influenced by presence of surface stones, which help to provide appropriate microhabitats (Crawford 1981). Hymenoptera (ants, wasps, and bees) may be crevice-dwellers in spaces beneath rocks or other surface objects, but the majority utilize soil burrows for their colonies. Such colonies may be relatively permanent, as for ants, or shorter-lived, as for termites, when plant tissues serving as the food resource and shelter is used up. A classic group of soil predators in desert regions are ant lions (Neuroptera), which trap prey in conical pits built in sandy soil (Strange 1970; Tuculescu, Topoff, & Wolfe 1975).

Life-history strategies

The ecological success of desert invertebrates, as well as desert vertebrates, is highly dependent on their ability to couple reproductive activity to periods of resource availability in an environment of highly irregular intensity of abiotic and biotic stress. Thus life-history strategies become important components for adaptation of desert arthropods and can serve as another way to categorize species.

Numerous species are multivoltine, whereby a species synchronizes a short life span with multiple generations, with each year being timed to the availability of ephemeral resources. Among desert arthropods thrips (Thysanoptera), mirid bugs and aphids (Homoptera), many moths and butterflies (Lepidoptera) and flies (Diptera), and taxonomically diverse coprovores and necrovores generally fit this category. Taylor (1980) has modeled the theoretical population dynamics of short-lived multivoltine species to suggest how cueing of activity, rapid breeding, and patterns of entry into dormancy or dispersal stages have evolved in these species.

Short-lived invertebrates with a single generation per year, a univoltine life-history strategy, are notably abundant and diverse in desert environments. Prominent among these are isopods (Isopoda), sunspiders (Solpugida), and many spiders (Araneida), beetles, and orthopterans. Among insect orders with multivoltine strategies; however, there may also be species with univoltine strategies.

Arthropod herbivores and detritivores that utilize more or less perennial food items have relatively long lives. For example, detritus, in the form of organic matter, is present in the soil throughout the year. Seasonal activity patterns of detritivores are determined by a complex of factors, and litter moisture content, microorganism activity, and predation pressures may all influence population size. Perennial plant tissues, for example, those of the evergreen shrub *Larrea tridentata* (creosote bush), form another constant food resource. Populations of detritivores and herbivores that are active most of the time provide food resources for arthropod carnivores. None the less, overall activity of the arthropod community is low during cold months and accelerates in spring as warm temperatures and moisture stimulate biomass production of food plants.

SURVEYS OF ARTHROPODS ON THE NEVADA TEST SITE

O'Farrell and Emery (1976) have already published a cumulative but still incomplete list of arthropod species that were collected on the Nevada Test Site by biologists from Brigham Young University mainly 1959–66, and briefly reviewed published papers from which the fauna was compiled. Certainly the list of over 1000 species is preliminary, given that even among the spiders researchers recorded 17 taxa that probably were new species (Allred & Beck 1967). Of insects, more than 20% belonged to family Miridae (159 species), which typically are sap-feeders. Other families well represented at NTS were the plant-eating weevils in Curculionidae (56 spp.), detritivorous beetles in Tenebrionidae (59 spp.), pollinating beeflies in Bombylliidae (61 spp.), and aboveground foraging ants in Formicidae (57 spp.), and these same families are common throughout Mojave desert scrub.

FIELD IBP STUDY METHODS

Detailed arthropod studies were conducted in Rock Valley from 1971 through 1976 as part of the IBP studies, and a working list of species was compiled (Tables 10.1–10.2; see Turner 1973, for more complete listing). Two separate sets of sampling procedures were used. Soil organisms were collected using extraction from soil samples and with buried pitfall traps, while canopy-dwelling arthropods were sampled mostly using vacuum procedures.

Soil samples

In 1974 soil arthropods were sampled in the vicinity of four species of perennials, *Ambrosia dumosa* (white bursage), *Krameria erecta* (Pima rhatany), *L. tridentata*, and *Lycium andersonii* (desert thorn), which combined contributed about 66% of perennial cover on the validation site (Chapter 4). One-liter soil samples were taken from each of three depths (0–0.10 m, 0.10–0.20 m, and 0.20–0.30 m) from each of three positions around a

shrub, shrub base, canopy margin, and at three mean shrub radii from shrub base. Shrubs were chosen randomly, but none was sampled more than once. For every week in 1974 the sample program yielded nine samples from each of the representative four shrubs. Arthropod were extracted from soil by a modified Tullgren method (Newell 1955; Edney, McBrayer, & Franco 1974).

Pitfall traps

Pitfall captures of arthropods were presumed to be a reasonable but not perfect measure of relative abundance when comparisons are based on equal trapping efforts. However, because so little was known about life histories of soil arthropods in desert regions, one cannot infer unequivocally that changes in makeup of samples reflected changes in population states and sizes of individual species. Instead, observed population increases may have been due to accelerated rates of productive activity or to recovery of dormant arthropods (Franco, Edney, & McBrayer 1979).

Species dwelling on the soil surface were sampled in Rock Valley from 1971 to 1976 using four grids with 300 pitfall traps, placed 15 m apart, and from 1974 to 1976 additional traps were used at the edge of the validation site. Franco et al. (1979) have reviewed sampling methods for obtaining samples during all periods of insect activity.

Analysis methods of soil arthropods

In order to compare yearly changes in apparent densities of trapped arthropods, we computed indices of relative abundance (I) based on numbers of individuals collected for each taxon as a function of trapping effort:

$$I_i = \frac{100C_i}{T_i}$$

where C is total number of individuals captured for each taxon and T is trapping effort expended (number of trap-nights) during the week that species i was active.

Table 10.1. *Taxonomic position of non-insect arthropods collected in Rock Valley during 1971 and 1972, or reported from this site in other studies. Methods of capture are indicated at pitfall traps (P), D-vac (V), or other (O).*

Chelicerata		Homalonychidae	
Class Arachnida		*Homolonychus theologus*	(P)
Order Acarina		Linyphiidae	(P, V)
Argasidae	(P)	Loxoscelidae	
Calyptostomidae	(O)	*Loxosceles unicolor*	(P)
Erythraeidae	(V)	Lycosidae	
Neophyllobiidae	(O)	*Geolycosa rafaelana*	(P, O)
Oribatidae	(O)	*Pardosa ramulosa*	(P)
Tetranychidae	(P, V)	*Schizocosa* sp.	(P)
Trombiculidae	(O)	*Tarentula kochi*	(P)
Order Pseudoscopionida	(O)	Oxyopidae	
Order Phalangida		*Oxyopes tridens*	(P)
Phalangiidae		Pholcidae	
Eurybunus riversi	(P)	*Psilochorus utahensis*	(P, V)
Order Solpugida		Plectreuridae	
Ammotrechidae	(P)	*Kibramoa paiuta*	(P)
Eremobatidae	(P, V)	*Plectreurys tristis*	(P)
Order Scorpionida		Salticidae	
Vejovidae		*Habronattus agilis*	(P, V)
Anuroctonus phaeodactylus	(P)	*Marpissa californica*	(P, V)
Hadrurus arizonensis	(P)	*Metacyrba* sp.	(P)
H. spadix	(P)	*Peckhamia* sp.	(V)
Vejovis becki	(P)	*Pellenes* sp.	(P, V)
V. boreus	(P)	*Phidippus* sp.	(P, V)
V. confusus	(P, V)	Theraphosidae	
Order Araneida		*Aphonopelma steindachneri*	(P, O)
Agelenidae		*Aphonoplema* sp.	(P, O)
Agelenopsis aperta	(P)	Theridiidae	
Araneidae		*Achaearania* sp.	(P, V)
Metepeira gosoga	(V)	*Latrodectus hesperus*	(P)
Caponiidae		Thomisidae	
Orthronops gertschi	(P)	*Apollophanes texanus*	(P, V)
Clubionidae		*Ebo* sp.	(P, V)
Micaria gosiuta	(P)	*Misumenops* sp.	(P, V)
Syspira eclectica	(P, V)	*Xyaicus* sp.	(P)
Ctenizidae		Mandibulata	
Aptostichus stanfordianus	(P)	Class Crustacea	
Dictynidae		Order Isopoda	
Dictyna personata	(P)	Armadillidae	
Diguetidae		*Venezillo arizonicus*	(P, V)
Diguetia cantities	(P)	Class Chilopoda	
Filistatidae		Order Scolopendromorpha	
Filistata utahana	(P)	Scolopendridae	
Gnaphosidae		*Scolopendra michelbacheri*	(P)
Callilepis sp.	(P)	Class Diplopoda	
Cesonia classica	(P)	Order Spirobolida	
Haplodrassus eunis	(P)	Atopetholidae	(P)
Herphllus hesperolus	(P)		

Table 10.2. *Major taxa collected in can traps and from shrubs in Rock Valley, 1971–1975.*

Taxon	General feeding habits
Diplopoda	Saprovores
Insecta	
Collembola (3 families)	Fungivores
Coleoptera (35 families)	
Chrysomelidae (7 genera)	Herbivores
Curculionidae (16 species)	Herbivores
Melyridae (6 genera)	Predators
Scarabaeidae (8 species)	Largely herbivores
Tenebrionidae (33 species)	Omnivores
Diptera (39 families)	
Embioptera (1 family)	Fungivores
Hemiptera (15 families)	
Lygaeidae (4 species)	Largely herbivores
Miridae (7 species)	Herbivores
Pentatomidae (6 species)	Herbivores or predators
Homoptera (19 families)	
Cicadellidae (12 species)	Herbivores
Membracidae (3 species)	Herbivores
Pseudococcidae	Herbivores
Hymenoptera (37 families)	Herbivores
Formicidae (22 species)	Omnivores
Isoptera	
Termitidae	Saprovores
Lepidoptera (32 families)	Herbivores
Neuroptera (5 families)	Predators
Orthoptera (8 families)	
Acrididae (4 species)	Herbivores
Gryllacrididae (2 species)	Saprovores
Mantidae (2 species)	Herbivores
Tettigoniidae (7 species)	Herbivores
Psocoptera (3 families)	Omnivores
Thysanoptera (3 families)	Predators or herbivores
Thysanura (2 families)	Saprovores

Because year-to-year differences in abundance could not be directly related to absolute density, attempts were made to estimate actual densities for five kinds of tenebrionid beetles by marking individuals and using capture-recapture analysis (Thomas & Sleeper 1977). Density of tenebrionids was estimated by employing an earlier 'geometric'

model (Edwards & Eberhardt 1967; Overton & Davis 1969):

$$N = n/[1 - (n/t)]$$

where N is total number of individuals present, n is total number of different individuals captured, and t is total number of captures (both marked and unmarked beetles). Densities were estimated by dividing N by size of the area that was sampled. However, very few marked individuals were recaptured, and, with extremely large populations, the model was susceptible to large errors and produced mostly unsatifying estimates (Thomas & Sleeper 1977).

Vacuum procedures for shrub-dwelling arthropods

Insects, mites, and spiders were sampled on shrubs at Rock Valley from March through September, 1971–5. Principal shrubs were sampled: *Ambrosia dumosa, Grayia spinosa* (hopsage), *Krameria erecta, Larrea tridentata, Lycium andersonii,* and *L. pallidum* (desert thorn). Numbers of each species sampled were based on relative densities within that desert scrub community (see Bamberg, Wallace, Kleinkopf, & Vollmer 1974), and more than 3700 shrubs were sampled. Approximately half of these were equal numbers of *A. dumosa* and *Larrea tridentata*, and 18% was taken from *K. erecta*.

Each shrub was rapidly covered by a 1-m³ tent to prevent the escape of flying arthropods and then vacuumed continuously for 5 minutes with a D-Vac Model 1 (Mispagel 1978). Mining and gall-dwelling species could not be collected in this manner. All but the smallest specimens were manually separated from plant materials in shadow boxes that were maintained in the field. Plant debris was placed in Tullgren funnels for 24 hours to remove remaining microarthropods. All specimens were preserved and archived at California State University, Long Beach.

Analysis methods of aboveground arthropods

Densities of various species or species groups were estimated by combining average numbers of individuals taken per shrub or shrubs with densities of host plants. Densities were calculated for two-week periods, an interval selected to maximize sample sizes and to permit detection of major changes in size and composition of arthropod populations. Density estimates reflect collections made in two plots (1974 and 1975) or all four plots (1971–3). Data from different plots were combined to derive a common density estimate (d):

$$d = N\Sigma \; a_i b_i \; /(\Sigma \; b_i)^{\alpha}$$

with N being total number of individuals collected, a_i density of the host shrub within the ith plot, and b_i number of shrubs sampled from the ith plot.

The efficiency with which various species were taken from shrubs depended on the types of arthropods and time of collection. Certain groups were particularly refractory to collect by D-vac, for example, weevils (Curculionidae) and some other beetles. Sleeper (pers. comm.) found that numbers of vacuumed grasshoppers (*Bootettix argentatus*) from *Larrea* were 15–27% less than actual populations. These errors depended on life stages, for example, earlier instars were more susceptible to capture, and time of day. At night vacuum collection of *Bootettix* was highly efficient (90%), but only 70–75% were obtained in daytime collections (Mispagel 1978). None the less, estimates of numbers of all arthropods were based on actual numbers collected without making *ad hoc* adjustments to compensate for unequal efficiencies of capture.

Quantifying arthropod biomass

As just described, biweekly estimates of arthropod densities were combined with measurements of mean dry weights of individual taxa to provide a measure of monthly standing biomass. Mean dry weight of adults and immature stages of many species or species groups were measured as well as water content as a percent of live weight.

To estimate standing stocks accurately one needs to know size or age composition of various populations during each month of the activity season. Although relative numbers of immature and adult individuals could be inferred from composition samples, proportions of different nymphal stages or larval instars were not evaluated. Nearly all species or groups that were analyzed had a univoltine life-history strategy; hence, we assumed that mean masses of immature individuals increased linearly from the earliest stage collected to the final immature stage. Adults were assumed to have a constant mass. Mean dry weights of arthropod species generally did not differ greatly between years, except for the grasshopper *B. argentatus*.

For many groups of small phytophagous or sap-feeding arthropods, all representatives of a single family or related groups of families were combined, for example, all Acrididae (grasshoppers and crickets). Because confamilial shrub dwellers were usually of about the same size in any single month, this source of error was modest. For more complex populations, immatures and adults were treated separately and estimated standing stocks were combined to give total biomass estimates.

For the following analyses, weekly sampling data from traps within the 12 fenced enclosures were pooled to provide monthly values for biomass, and monthly values were also generated for population density per unit area.

GROUND-DWELLING ARTHROPODS IN ROCK VALLEY

For convenience in analyzing voluminous data, we assigned soil arthropods as detritivores, predators, or phytophages and separated data on ants, which fit numerous ecological niches for special analysis.

In 1976 more than 35 000 individuals of ground-dwelling arthropods were found in pitfall traps in the 12 100-m^3 enclosures (Table 10.3). Overall abundance of individuals rose rapidly from March to May, reaching a peak population level in September and October before falling sharply in November, when only about 25% of peak numbers

Table 10.3. *Seasonal abundance of arthropods (excluding ants) captured in pitfall traps on 12 100-m²* *fenced plots in Rock Valley (1976). These data are modifications of original IBP values (Turner 1977).*

Organisms	Feb.–March	Apr.–May	June–July	Aug.–Sept.	Oct.–Nov.	Total
Detritivores						
Tenebrionids	456	2 445	2 510	4 192	527	10 130
Others[a]	46	235	427	498	134	1 340
Predators						
Spiders	239	1 104	3 162	1 498	281	6 284
Others[b]	114	677	1 226	693	228	2 938
Phytophages						
Weevils	743	1 440	2 227	4 237	5 983	14 630
Total	1 598	5 901	9 552	11 118	7 153	35 322

[a]isopods, millipedes, orthopterans
[b]centipedes, scorpions, solpugids

were obtained. February populations were less than 5% peak levels. Phytophagous arthropods comprised the most abundant group, having 40% of all individuals for the year. However, more than half of this total was obtained in September and October. In October phytophagous soil arthropods comprised 83% of all sample individuals. Predatory soil arthropods, representing 23% of overall captures, were the most frequent guild in samples of June and July. Detritivores comprised 29% of sampled individuals with peak numbers from July through September.

Aggregate monthly density of soil arthropods in Rock Valley ranged from a minimum of 1200 individuals m^{-2} in July to more than 12 000 individuals m^{-2} in December (year not specified). This contrasts with estimates for other types of ecosystems (Franco et al. 1979); densities in more mesic systems were significantly higher, rising to as many as 834 500 individuals m^{-2}. However, densities at Rock Valley study plots were comparable to those reported for the Australia desert (Wood 1971), the California desert (Wallwork 1972), and the Chihuahuan Desert (Santos, DePree, & Whitford 1978). Edney, Franco, & McBrayer (1976) estimated that all soil arthropods in Rock Valley collectively metabolized about 1577 J m^{-2} yr^{-2}.

Detritivores

Beetles

Darkling beetles (Tenebrionidae, Coleoptera), including about 50 species, comprised the ecologically most dominant group of detritivores at NTS, and species with wide ecological distributions, such as *Eleodes obscura*, tended to be more abundant than were species having relatively narrow ecological distributions (Tanner & Packham 1965). Tenebrionidae were most diverse in *Larrea–Ambrosia* and *Grayia–Lycium* communities (Allred et al. 1963).

Field studies in Rock Valley have yielded 30 taxa of tenebrionid beetles (Table 10.4), and they often exhibited habitat preference for specific soil types. On the IBP study site *Centrioptera muricata*, *Cryptoglossa verrucosa*, *Triorophus laevis*, and *Trogloderus costatus* were almost exclusively restricted to fine-textured sandy soils, whereas *Anepsius brunneus*, *Edrotes robus*, and *Eusattus dubius* were collected on coarser, gravelly soils. *Araeoschizus sulcicollis*, the most abundant species, was found associated with ant nests. Although some tenebrionids were active essentially year-round, those of Rock Valley were first observed in March and disappeared from the plots by mid-October. The majority were most active in late

Table 10.4. *Tenebrionoid beetles collected at Rock Valley in 1971 and 1972.*

Species	Pitfall traps	D-vac	Other
Alaephus nevadensis	+	+	
Anepsius brunneus	+	+	
Araeoschizus sulcicollis	+		
Asidinia semilaevis	+		
Auchmobius subboreus	+		
Blapstinus pubescens	+		
Bothrotes sp.			+
Centrioptera muricata	+		
Chilometopon abnorme	+		
Conibiosoma elongatum	+	+	
Coniontis nevadensis	+		
Cryptoglossa verrucosa	+		
Edrotes robus	+		
Eleodes armata	+		
E. dissimilis	+		
E. extricata	+		
E. grandicollis	+		
Eupsophulus castaneus	+	+	+
Eusattus agnatus	+		+
E. dubius	+	+	
Euschides luctata	+		
Helops attenuatus			+
Helops sp.			+
Metoponium convexicolle	+	+	
Notibius substriatus	+		
Philolithus pantes	+		
Sphaeiontis dilatata	+		
Trichiasida acerba	+		+
Triorophus laevis	+	+	
Trogloderus costatus	+		

summer or early fall, and seasonal activity peaked in August and September (Table 10.3). Activity patterns of individual species differed significantly. For example, *Eusattus dubius* was an uncommon species with peak activity in late April, *Eleodes armata* and *Trogloderus costatus* had population peaks in September, and *A. sulcicollis* reached its peak in May or June (Table 10.5).

In 1976 estimated minimal active density for *A. sulcicollis* was 47 000 individuals ha^{-1}, whereas nine other species had estimated densities of 1000 or more individuals ha^{-1} (Table 10.6). Like other common tenebrionids, *A. sulcicollis* showed a sharp decline in abundance in the dry year 1974 (Table 10.7).

In comparison with other beetles, Tenebrionidae are relatively large. For Rock Valley *Eleodes grandicollis* averaged 660 mg dry weight and *E. armata* and *Cryptoglossa verrucosa* were 242 and 301 mg, respectively (Table 10.8). However, the abundant *Araeoschizus sulcicollis* were tiny and averaged only 1 mg dry weight.

Using data on abundance and dry weight on 15 species, total estimated biomass of Tenebrionidae was 623 g ha^{-1}, 75% of the total for all detritivores. The greatest biomass contribution came from *Eusattus dubius*, which was the second most abundant species (Table 10.9) and had a mean dry weight of 11.8 g individual^{-1} (Table 10.8), resulting in an estimated value of 206 g ha^{-1} (Table 10.10), or 19% of the total for all detritivorous arthropods.

Scarab beetles (Scarabaeidae) are fairly common detritivores at NTS (Allred & Beck 1965), and at least eight species were observed during Rock Valley studies. Of these *Diplotaxis moerens* was included in biomass calculations and had a mean dry weight of 32.7 mg (Table 10.8).

Orthoptera

Orthopterans are best known as phytophagous herbivores, but may be ecologically important as detritivores at Rock Valley. Detritivorous species included the camel cricket (*Ceuthophilus fossor*) and the cockroach *Arenivaga erratica*. Peak numbers of *Ceuthophilus* were always observed early in the year; trapping between October, 1973 and February, 1974 yielded large numbers of first and second instar nymphs during October and progressively later stages (through fourth instar) into February. However, this was not the actual time of maximal abundance (Table 10.5), because Allred, Beck, and Jorgensen (1963) reported peak numbers of adults during weeks 40–43 (October 1–28). During the four years of sampling (1971–4), both species had peak abundances in 1971, but low population densities of *Ceuthophilus* occurred in 1973 and 1974 and of *Arenivaga* in 1972 and 1973 (Table 10.7).

Total community biomass of detritivorous orthopterans was 91 g ha^{-1} (Table 10.10), more

Table 10.5. *Annual numbers of soil arthropods captured in pitfall traps between 1971 and 1974 at Rock Valley in calendar weeks active and mean week of peak activity.*

	Numbers trapped				Calendar weeks active	Mean week of peak activity
	1971	1972	1973	1974		
Solpugida						
Eremobatidae	30	52	115	83		
Scorpionida						
Hadrurus arizonensis	35	36	46	17	15–37	31
Vejovis spp.	59	32	23	8	10–38	32
Araneida						
Apollophanes texanus	10	37	18	25	10–33	19
Herpyllus hesperolus	25	37	38	31		
Marpissa californica	4	35	19	11	10–32	23
Psilochorus utahensis	204	320	216	51	10–39	30
Syspira eclectica	46	47	66	49	14–38	22
Isopoda						
Venezillo arizonicus	218	221	225	35	11–39	32
Chilopoda						
Scolopendra michelbacheri	23	29	19	21	10–38	32
Orthoptera						
Arenivaga erratica	28	26	17	27	11–38	31
Ceuthophilus fossor	37	40	30	33	10–38	12
Hemiptera						
Reduvius sonoraensis	50	31	16	5	10–38	23
Coleoptra, Curculionidae						
Eucyllus vagans	6	10	69	14		
Miloderes mercuryensis	10	21	28	14		
Ophryastes varius	3	17	24	3		
Coleoptera, Tenebrionidae						
Araeoschizus sulcicollis	122	183	196	59	10–39	22
Asidina semilaevis	46	21	23	21	32–38	33
Centrioptera muricata	81	84	266	22	14–38	25
Conibiosoma elongatum	11	24	21	19	10–39	30
Cryptoglossa verrucosa	109	95	82	69	11–37	25
Edrotes ventricosus	32	33	20	17	18–39	30
Eleodes armata	34	29	30	18	18–38	38
Eusattus dubius	3	9	24	8	11–27	16
Metoponium spp.	11	22	9	8	14–36	26
Philolithus pantex	15	19	19	15		
Triorophus laevis	201	66	30	16	11–36	24
Trogloderus costatus	87	73	93	16	14–39	35

than 10% of total detritivore biomass and generally greater than that of phytophagous species.

Other arthropods

Insects are not the only important detrivores in Rock Valley or desert ecosystems in general, and terrestrial isopods (Isopoda) and millipedes (Diplopoda) are notably important (Cloudsley-Thompson & Chadwick 1964).

Two species of isopods are known from NTS (Allred & Mulaik 1965). Most widespread of these is *Venezillo arizonicus* (Armadillidae), which occurs in *Larrea-Ambrosia* and *Grayia-Lycium*

Table 10.6. *Estimated densities of arthropods collected in pitfall traps on 100 m² enclosures at Rock Valley in 1976. Actual removals and standard errors of density estimates are also given. See Mispagel and Sleeper (1983) for more details.*

	Removals	Minimal active density, individuals ha^{-1}	Standard error
Solpugida	2 367	19 808	1 241
Scorpionida			
Hadrurus arizonensis	21	175	354
Vejovis spp.	399	3 325	356
Araneida			
Apollophanes texanus	436	3 633	258
Gnaphosidae	977	9 183	529
Theraphosidae	24	200	28
Isopoda			
Venezillo arizonicus	436	3 633	623
Chilopoda			
Scolopendra michelbacheri	151	1 258	166
Diplopoda	50	415	190
Orthoptera	854	7 158	960
Coleoptera, Curculionidae			
Eucyllus unicolor	1 211	10 133	2 210
E. vagans	10 130	85 208	20 700
Miloderes sp.	1 580	13 208	2 510
Neocercopedius sp.	598	7 275	2 040
Ophryastes spp.	1 000	8 417	990
Coleoptera, Tenebrionidae			
Anepsius brunneus	602	5 017	806
Araeoschizus sulcicollis	5 627	46 975	4 990
Asidina semilaevis	31	258	46
Auchmobius subboreus	214	1 783	316
Centrioptera muricata	19	158	65
Conibiosoma elongatum	661	5 508	839
Craniotus blaisdelli	120	1 000	215
Cryptoglossa verrucosa	2	17	13
Edrotes ventricosus	148	1 317	301
Eleodes armata	10	83	42
E. dissimilis	38	317	84
Eusattus dubius	930	10 883	1 620
Metoponium spp.	873	7 483	425
Philolithus pantex	24	200	61
Triorophus laevis	558	4 692	386
Trogloderus costatus	276	2 383	497
Coleoptera, Chrysomelidae			
Larvae	213	1 768	

desert scrub communities as well as mixed mesic communities (Allred et al. 1963), and this sowbug was a significant detritivore among soil arthropods at Rock Valley. *Venezillo* generally was active from March through October (females) or December (males) and reached its peak abundance in August of most years (Table 10.5). Individuals averaged 29 mg in dry weight (Table 10.8). In 1976 minimal active densities for this species were estimated to be about 3600 individuals ha^{-1} in enclosed plots

Table 10.7. *Estimated indices of relative abundance (I) for soil arthropods sampled in pitfall traps at Rock Valley between 1971 and 1974. See text for discussion.*

Taxon	Feeding guild	Index of relative abundance			
		1971	1972	1973	1974
Solpugida	predator	0.65	1.00	2.05	1.73
Scorpionida					
Hadrurus arizonensis	predator	0.92	1.05	1.05	0.71
Araneida					
Apollophanes texanus	predator	–	0.80	0.39	0.89
Psilochorus utahensis	predator	4.43	5.71	3.60	1.70
Syspira eclectica	predator	1.22	1.02	1.83	1.02
Isopoda					
Venezillo arizonicus	detritivore	4.74	4.09	3.88	0.92
Chilopoda					
Scolopendra michelbacheri	predator	0.52	0.50	0.36	0.48
Orthoptera					
Arenivaga erratica	detritivore	0.78	0.46	0.47	0.75
Ceuthophilus fossor	facultative detritivore	0.84	0.74	0.58	0.59
Coleoptera–Tenebrionidae					
Araeoschizus sulcicollis	detritivore	2.77	3.15	3.92	1.09
Asidina semilaevis	detritivore	3.29	2.10	2.40	1.75
Centrioptera muricata	detritivore	2.25	2.33	5.78	0.69
Cryptoglossa verrucosa	detritivore	2.73	2.26	1.86	1.58
Conibiosoma elongatum	detritivore	0.23	0.48	0.44	0.40
Edrotes ventricosus	detritivore	1.07	0.97	0.45	0.57
Eleodes armata	detritivore	0.77	0.50	0.56	0.39
Eusattus dubius	detritivore	0.14	0.38	0.92	0.27
Metoponium spp.	detritivore	0.29	0.48	0.23	0.40
Triorophus laevis	detritivore	5.28	1.65	0.58	0.40
Trogloderus costatus	detritivore	1.89	1.83	1.79	0.62

at Rock Valley (Tables 10.6, 10.9). However, population size was observed to vary significantly between years; in 1974, a dry year, populations of *V. arizonicus* were less than 20% of those that were measured during the three previous years (Table 10.5). In 1972 local densities varied by 100% between local sites having similar vegetation.

Desert millipedes are large detritivores that may be active diurnally and nocturnally, but, unlike centipedes, activity patterns of desert millipedes result from complex interactions between precipitation and temperature, which affect soil organic matter (Wooten, Crawford, & Riddle 1975; Crawford 1981). Two of the six species at

NTS (Chamberlin 1962, 1963, 1965) were reported in Mojave Desert habitats (Allred et al. 1963); *Arinolus nevadae* and *Orthichelus michelbacheri* were most apparent during winter months and thus provide a very different activity pattern than most other detritivorous arthropods in Rock Valley (Fig. 10.1, Table 10.5).

In typical years at Rock Valley, rainfall in October resulted in emergence of millipedes. In 1976, when careful biomass measurements were made (Table 10.10), millipedes accounted for 20–60% of trapped biomass of detritivorous arthropods, and they had a minimal active density of 1000 individuals within a 9-ha enclosure.

Table 10.8. *Mean dry weight (mg) and water content (% live weight) of adult arthropods at Rock Valley. Standard error of mean weight is given in parentheses.*

Taxon	n	Mean dry weight, mg	Mean water content, % live weight
Phalangida			
Eurybunus riversi	3	92.9	74
Solpugida			
Eremobatidae			70
Scorpionida			
Anuroctorus phaeodactylus	7	393 (44)	68
Hadrurus arizonensis	9	1453 (110)	69
Vejovis becki	14	59.9 (8.4)	69
Araneida			
Loxosceles unicolor	3	8.8 (0.7)	73
Psilochorus utahensis	14	1.3 (0.2)	72
Isopoda			
Venezillo arizonicus	12	28.8 (1.3)	54
Chilopoda			
Scolopendra michelbacheri	4	80.9	77
Coleoptera, Chrysomelidae			
Pachybrachis sp.	3	2.1	64
Coleoptera, Curculionidae			
Eucyllus vagans	13	2.1 (0.3)	
Miloderes sp.	1	1.8	40
Ophryastes varius	2	0.5 (0.07)	45
Smicronyx imbricatus	12	0.23 (0.01)	
Coleoptera, Scarabaeidae			
Diplotaxis moerens	18	32.7 (2.2)	61
Phyllophaga sociatus	4	59.4 (3.2)	64
Coleoptera, Tenebrionidae			
Anepsius brunneus	2	0.5 (0.07)	
Araeoschizus sulcicollis	5	1.0 (0.2)	45
Asidina semilaevis	37	178 (7)	72
Auchmobius subboreus	3	15.7 (1.2)	61
Centrioptera muricata	6	134 (26)	50
Coniontis nevadensis	10	18.6	67
Cryptoglossa verrucosa	8	301 (15)	51
Edrotes ventricosus	43	25.2	67
Eleodes armata	21	242 (5)	58
Eleodes grandicolllis	17	660 (27)	64
Eusattus dubius	2	11.8 (3.7)	
Metoponium convexicolle	13	4.9 (0.3)	44
Philolithus pantex	11	117	68
Triorophus laevis	2	7.0 (0.3)	60
Trogloderus costatus	19	52.1 (3.3)	45
Orthoptera			
Bootettix argentatus	69	40.1	70
male	14	27.3 (0.9)	
female	24	66.8 (2.1)	
Capobotes fulginosus	4	837	60
Ceuthophilus fossor	1	41.9	71
Cibloacris parviceps	12	104	65
Insara covilleae	1	33.1	68
Ligurotettix coquilletti	1	33.1	68
Tanaocerus koebelei	2	65.9	73

Taxon	n	Mean dry weight, mg	Mean water content, % live weight
Hemiptera			
Nysius ericae			
male	89	0.26 (0.01)	25
female	38	0.44 (0.02)	
Phytocoris nigripubescens			
male	8	0.14 (0.01)	
Centrodontus atlas			
male	38	0.60 (0.01)	42
female	21	1.0 (0.02)	
Chlorochora sayi	10	92.7	39
Dendrocoris contaminatus	2	15.9	57
Multareis cornutus			
female	29	1.1 (0.007)	
Multareiodes bifurcatus			
male	17	0.76 (0.05)	48
female	20	1.0 (0.05)	
Scaphitopius nicricollis			
male	4	0.25 (0.005)	
female	4	0.62 (0.01)	
Lepidoptera (larvae)			
Anacampsis sp.	8	4.0	69
Hyles lineata	9	603	84
Semiothisa larreana	17	12.3	70
Vanessa cardui	14	10.8	86
Hymenoptera			
Crematogaster mutans	11	0.16 (0.03)	
Iridomyrex pruinosum	12	0.06 (0.003)	
Pheidole desertorum	2	0.10	
Veromessor pergandei	22	1.4 (0.4)	69
majors	15	1.9	69
minors	7	0.5	

Biomass of detritivores of Rock Valley

On the basis of soil biomass sampled, detritivores formed the most important arthropod guild, with 42% of total biomass abundance, having peaks in May and August. Large numbers and biomass of *E. dubius* (Tenebrionidae) were responsible for the spring peak, contributing 51, 64, and 46% of detritivore biomass in March, April, and May, respectively. Many species of tenebrionid beetles contributed to a late summer peak but especially *Asidina semilaevis*, *Philolithus pantex*, *Trogloderus costatus*, *Araeoschizus sulcicollis*, and *Auchmobius* *subboreus* (Table 10.10). Such beetles comprised the largest components of detritivore biomass in every month sampled except February, when millipedes comprised more than 60% of the biomass. Millipedes were important detritivores only during winter months, when other groups were less active, and did not occur in samples from late spring through early fall, when overall activity by detritivorous arthropods was highest. Orthopteran detritivores were sampled throughout the year, but they reached peak abundance in late spring and summer (Table 10.10). Maximum relative importance of orthopterans for detritivore biomass occurred in

Table 10.9. Monthly and total removals of detritivorous arthropods from 12 fenced 100-m² plots in Rock Valley during 1976.

Species and species groups	Month										Total captures	Total removals ha^{-1}
	Feb	Mar	Apr	May	Jun	Jul	Aug	Sep	Oct	Nov		
Tenebrionid beetles												
Centrioptera muricata	–	1	1	10	5	1	1	–	–	–	19	157
Trogloderus costatus	2	46	27	39	27	12	32	71	20	–	276	2 290
Eleodes armata	–	–	–	–	–	–	2	2	6	–	10	83
E. dissimilis	–	–	–	2	–	4	11	19	2	–	38	315
Araeoschizus sulcicollis	2	82	542	438	338	949	1 667	1 379	202	28	5 627	46 704
Conibiosoma elongatum	–	3	93	175	62	96	138	80	14	–	661	5 486
Edrotes ventricosus	2	16	1	1	7	8	10	49	48	6	148	1 228
Eusattus dubius	69	218	256	306	71	3	–	4	7	2	936	7 768
Asidina semilaevis	–	–	–	–	–	3	27	1	–	–	31	257
Philolithus pantex	–	–	–	–	–	–	20	4	–	–	24	199
Craniotus blaisdelli	–	–	–	–	–	–	–	–	78	42	120	996
Trioropus laevis	1	5	47	171	104	122	91	5	12	–	558	4 631
Metoponium spp.	5	13	20	229	177	189	154	53	29	4	873	7 245
Anepsius brunneus	–	–	–	85	131	128	148	89	16	5	602	4 996
Auchmobius subborues	–	–	–	–	9	64	87	48	6	–	214	1 776
Orthoptera	1	22	15	181	163	98	104	103	33	6	726	6 026
Isopoda	–	–	6	29	66	89	103	93	47	3	436	3 618
Diplopoda	14	9	4	–	–	–	–	4	19	–	50	415
Miscellaneous[a]	–	–	–	–	–	–	–	–	–	–	916	7 603
Total	96	415	1 012	1 666	1 160	1 766	2 595	2 004	539	96	12 265	101 800

[a]includes additional tenebrionid beetles

Table 10.10. *Monthly and total dry weight biomass (g ha⁻¹) of detritivorous arthropods removed from 12 fenced 100-m² plots in Rock Valley during 1976.*

Species and species groups	Feb	Mar	Apr	May	Jun	Jul	Aug	Sep	Oct	Nov	Total estimated biomass, g ha⁻¹
Tenebrionid beetles											
Centrioptera muricata	–	1.4	1.4	14.0	7.0	1.4	1.4	–	–	–	27
Trogloderus costatus	0.9	19.9	11.7	16.9	11.7	5.2	13.8	30.7	8.6	–	119
Eleodes armata	–	–	–	–	–	–	6.0	6.0	18.0	–	30
E. dissimilis	–	–	–	0.5	–	1.0	2.7	4.7	0.5	–	9
Araeoschizus sulcicollis	–	0.6	4.3	3.5	2.7	7.5	13.1	10.9	1.6	0.2	44
Conibiosoma elongatum	–	–	0.5	0.9	0.3	0.5	0.7	0.4	1.0	–	3
Edrotes ventricosus	0.4	3.1	0.2	0.2	1.4	1.6	2.0	9.6	9.4	1.2	29
Eusattus dubius	15.2	47.9	56.3	67.3	15.6	0.7	–	0.9	1.5	0.4	206
Asidina semilaevis	–	–	–	–	–	2.3	21.1	0.8	–	–	24
Philolithus pantex	–	–	–	–	–	–	15.6	3.1	–	–	19
Craniotus blaisdelli	–	–	–	–	–	–	–	–	12.0	6.5	18
Triorophus laevis	0.1	0.3	2.7	9.9	6.0	7.1	5.3	0.3	0.7	–	32
Metoponium spp.	0.2	0.5	0.8	9.3	7.2	7.7	6.3	2.2	1.2	0.2	36
Anepsius brunneus	–	–	–	0.4	0.6	0.6	0.7	0.4	0.1	–	3
Auchmobius subborues	–	–	–	–	1.1	8.0	10.9	6.0	0.7	–	27
Orthoptera	0.1	2.7	1.9	22.5	20.3	12.2	12.9	12.8	4.1	0.7	91
Isopoda	–	–	0.3	1.6	3.6	4.9	5.6	5.1	2.6	0.2	24
Diplopoda	25.9	16.7	7.4	–	–	–	–	7.4	35.2	–	93
Total	42.8	93.1	87.5	147.0	77.5	60.7	118.1	101.3	96.3	9.4	834

Table 10.11 *Species diversity of arachnids (scorpions, solpugids, and spiders) in North American warm deserts. Adapted from Chew (1961) and Polis and Yamashita (1991).*

	Number of species		
	Scorpions	Solpugids	Spiders
Sonoran Desert	4	11	>54
Coachella Valley, CA (sand)			
(Polis and McCormick 1986)			
Deep Canyon Reserve, CA	6	9	71
(Frommer 1986)			
Mojave Desert			
Larrea-Ambrosia scrub	6	9	25
Grayia-Lycium scrub	7	18	34
(Allred et al. 1963)			
Chihuahuan Desert			
White Sands, NM	1	4–5	23–31
(Muma 1975)			
Portal, AZ	8	10	50

June, amounting to 26% of the total. Isopods also peaked at the same time, but they were absent or virtually absent during winter (Table 10.10) and never formed more than 8% of detritivore biomass during any single month.

Predatory soil arthropods

Predatory soil arthropods of the Mojave Desert included representatives of many divergent taxa, especially Scorpionida, Solpugida, Araneida, Phalangida, and Pseudoscorpionida. The ecological importance of these groups in desert environments has been reviewed in detail by Polis and Yamoshita (1991).

Scorpions

Scorpions are nocturnal, solitary predators that use a poisonous stinger to kill spiders, large insects, and even small vertebrates (Polis 1990). In general, scorpions are 'sit-and-wait' predators with generalist feeding habits (Hadley & Williams 1968), and they have numerous adaptations to life in desert environments (Hadley 1979, 1990; Polis 1988) but have evolved to favor low reproductive

potential and density-dependent regulation of population size (Polis & Farley 1980).

Nine species have been reported at NTS, and these are primarily restricted to warmer Mojave Desert communities (Gertsch & Allred 1965). Six species of scorpions were present in Rock Valley (Table 10.1), a number comparable to that present in many other sites of American warm deserts (Table 10.11).

The largest scorpion in Rock Valley was *Hadrurus arizonensis* (giant desert hairy scorpion), which averaged nearly 15 g in dry weight (Table 10.8) and at maturity reached a length of 140 mm. Both *H. arizonensis* and the smaller *H. spadix* were common on IBP plots, the former more abundant on fine-textured, sandy soils and the other more on coarse-textured, gravelly soils. Overall the most abundant scorpions throughout the validation site were species of *Vejovis*, particularly *V. becki*, a small scorpion averaging only 60 mg in dry weight. *Anuroctonus phaeodactylus*, an intermediate-sized scorpion with a mean dry weight of 393 mg, was mostly observed on fine-textured soils.

Scorpion activity was greatest during late summer and early fall (Table 10.5). At Rock Valley many scorpions were trapped in August and September, and populations declined sharply during October (Table 10.12). Approximately 95% of captures were species of *Vejovis* (Table 10.12), and these accounted for approximately 70% of scorpion biomass (Table 10.13). Overall, scorpions comprised only about 4% of individual predatory arthropods sampled, but more than 40% of that biomass. In September, 1976, scorpions at their peak biomass comprised nearly 85% of predatory arthropod biomass (Table 10.13); this was when densities of *Vejovis* were estimated to be nearly 400 individuals ha^{-1} as compared with 21 individuals ha^{-1} for *H. arizonensis*.

Solpugida

Solpugida, sunspiders and windscorpions, are well represented in warm desert regions of the world, except in Australia (Crawford 1981), and most of the 120 species known from North America are found in arid and semiarid regions (Muma & Muma 1988). They are active predators, feeding on insects, other arthropods, and small vertebrates, especially lizards, and they hunt primarily at night (Polis & McCormick 1986). Field studies at NTS have encountered 29 species of solpugids (Muma 1963; O'Farrell & Emery 1976), with greatest density in *Grayia–Lycium* communities (Allred et al. 1963). For Rock Valley specimens identifications of solpugids for the IBP were restricted to the family level because most individuals were juveniles. Ammotrechidae and Eremobatidae were present.

During 1971 and 1972 considerable spatial variability in solpugid activity was observed on the Rock Valley study plots (Turner 1973). As a group, solpugids were most abundant in late spring and summer (Table 10.12) and during that period formed a significant component of noninsect predatory arthropods in Rock Valley. They comprised about one-third of estimated biomass of this group overall and nearly 50% of the active biomass of predatory arthropods during late spring and early summer (Table 10.11). Solpugids also formed about 22% of all predatory arthropods that were trapped, but their population densities were high in 1973, a very productive year, and relatively low in 1971, a dry year. Population estimates for 1976 suggested that there were approximately 20 000 individuals ha^{-1} (Table 10.12).

Spiders

Many spiders (Araneida) are well adapted to desert conditions throughout the world and are important predatory carnivores (Gertsch 1949; Chew 1961; Cloudsley-Thompson & Chadwick 1964; Polis & McCormick 1986). Five guilds of desert spiders are distinctive enough to be recognized on the basis of their foraging habits: nocturnal hunters, runners, ambushers, agile hunters, and web-builders (Hatley & MacMahon 1980). At NTS they form a diverse group with 94 species from 65 genera and 22 families, but undoubtedly many more species occur there (Allred & Beck 1967).

Greatest abundance of spiders was reported from *Larrea–Ambrosia* communities, but greatest diversity occurred in transitional desert communities with *Coleogyne*. Spider diversity within individual Mojave Desert communities at NTS was relatively low as compared with sites in the Sonoran Desert (Table 10.11), but overall abundance for communities was rich. From 1959 to 1965 seasonal abundance of spiders at NTS peaked for females in July and for males in May, whereas maximum species diversity peaked in July for females, but in June for males (Allred & Beck 1967). Populations and diversities of immature spiders reached a maximum in July.

Psilochorus utahensis was the most abundant spider throughout NTS as well as at Rock Valley. Other widespread and abundant species were *Haplodrassus eunis*, *Syspira eclectica*, and *Tarantula kochi*. *Psilochorus* and *Syspira* were primarily active during late spring and summer, whereas *Haplodrassus eunis* was observed mostly in the coolest months from late fall to early spring (Table 10.5). During the IBP studies at Rock Valley 38 species of spiders were collected mostly in pitfall traps, with some also in D-vac samples, but data

Table 10.12. *Monthly and total removals of predaceous arthropods from 12 fenced 100-m² plots in Rock Valley during 1976.*

Species and species groups	Feb	Mar	Apr	May	Jun	Jul	Aug	Sep	Oct	Nov	Total removals	Estimated total removals ha⁻¹
Chilopoda	–	14	10	15	24	31	20	12	24	1	151	1 253
Scolopendra michelbacheri	2	89	138	467	570	493	360	62	103	83	2 367	19 646
Solpugida												
Scorpionida											392	3 254
Vejovis spp.	–	9	10	30	40	61	119	111	9	3	392	3 254
Hadrurus spp.	–	–	4	4	3	2	5	1	–	–	19	158
Araneida												
Psilochorus utahensis	2	31	39	603	1 266	1 526	1 108	183	50	39	4 847	40 230
Apollophanes texanus	–	45	55	86	109	83	38	13	2	5	436	3 618
Theraphosidae	–	–	–	–	–	1	10	–	13	–	24	199
Gnaphosidae	25	136	188	133	49	128	98	48	90	82	977	8 109
Miscellaneous											1 516	12 580
Total	29	324	444	1 338	2 061	2 325	1 758	430	291	213	10 729	89 051

Table 10.13. *Monthly and total dry weight biomass (g ha⁻¹) of predatory arthropods removed from 12 fenced 100-m² plots in Rock Valley during 1976.*

Species and species groups	Feb	Mar	Apr	May	Jun	Jul	Aug	Sep	Oct	Nov	Total estimated biomass, g ha⁻¹
Solpugida	0.2	8.5	13.2	44.8	54.7	47.3	34.5	5.9	9.9	8.0	227
Chilopoda											
Scolopendra michelbacheri	–	6.8	4.8	7.2	11.6	15.0	9.7	5.8	11.6	0.5	73
Scorpionida											
Vejovis spp.	–	4.5	5.0	14.9	19.9	30.4	59.3	55.3	4.5	1.5	201
Hadrurus spp.	–	–	17.2	17.2	12.9	12.9	8.6	21.5	4.3	–	89
Araneida											
Psilochorus utahensis	–	0.3	0.4	6.0	12.6	15.2	11.0	1.8	0.5	0.4	48
Apollophanes texanus	–	0.2	0.3	0.4	0.6	0.4	0.2	0.1	–	–	2
Theraphosidae	–	–	–	–	–	1.6	16.3	–	21.2	–	39
Gnaphosidae	0.2	2.1	2.9	2.0	0.8	2.0	1.5	0.7	1.4	1.3	15
Total	0.4	22.4	43.8	92.5	113.1	124.8	141.1	91.1	53.4	11.7	694

presented here on spiders is based solely on the pitfall samples.

The very abundant *P. utahensis* is a relatively small spider and in Rock Valley had a dry weight of only about 1 mg (Table 10.8). This species reached peak abundance in 1972, as did *Apollophanes texanus* and *Marpissa californica*, and declined sharply in numbers during 1974, an arid year (Table 10.5). *Psilochorus* tended to have its population peak in July, and in July, 1976, they comprised 88% of the individual spiders (Table 10.12) and 6% of total predaceous arthropods that were collected in Rock Valley (Table 10.13). For 1976 this species comprised 77% of all spiders trapped and 45% of all predaceous arthropods; it also comprised 46% of total spider biomass and 7% of predaceous arthropod biomass. Population densities for *P. utahensis* were estimated from pitfall measurements to be 40 000 individuals ha^{-1}, but this was probably an overestimate of actual density because pitfall traps may attract such spiders, providing ideal habitats to augment naturally available homesites.

Theraphosid spiders as a group, including at least two species of tarantulas, formed the second largest component of spider biomass (38%) during 1976, with large peaks of abundance at Rock Valley in August and October (Table 10.13). Annual patterns of activity in desert tarantulas has been described in detail. In Rock Valley they comprised about 15% of individual spiders over the year, and *Appolophanes texanus*, third in abundance of Rock Valley spiders, had 7% of individual spiders collected, but only 2% of spider biomass. Gnaphosid spiders, most notably *Herpyllus hesperolus*, represented only 0.4% of collected samples in 1976 although their relatively large size yielded 15% of total spider biomass.

Limited population data from Rock Valley were also obtained for three additional species, *Loxosceles unicolor*, *Marpissa californica*, and *Syspira eclectica*, which were collected monthly from 1971 to 1974. Data on relative population sizes on *M. californica* and *S. eclectica* are presented in Table 10.5 (see also Turner 1972, 1973; Turner & McBrayer 1974).

Phalangida

Phalangids (harvestman or daddy-long-legs) resemble spiders, but differ in a number of respects. Most species are nocturnal in activity, feeding on small arthropods, snails, and earthworms. Other species are scavengers, utilizing decaying organic matter, dead invertebrates, and even bird droppings. A few species are fungivorous or phytophagous. Three species of phalangids have been reported from NTS, all of which are relatively widespread (Allred 1965). A single species, *Eurybunus riversi*, was collected in Rock Valley during the IBP studies, and peak abundance of this phalangid occurred in late winter and early spring.

Centipedes

Desert centipedes (Chilopoda) are highly mobile, nocturnal carnivores on the soil surface (Crawford 1981). Six species within five families were found at NTS (Allred et al. 1963), and greatest diversity was present in *Grayia-Lycium* communities, where four species were collected. However, greatest abundance of centipedes occurred in cooler upland habitats.

Only a single species, the ecologically widespread *Scolopendra michelbacheri*, was found in *Larrea-Ambrosia* and *Grayia-Lycium* communities in Rock Valley. Despite its relatively large size, with 81 mg as a mean dry weight (Table 10.8), this centipede never reached the abundance of other large arthropod predators at Rock Valley. On the study area *S. michelbacheri* was active from March through October and reached peak abundance in July (Table 10.12). In 1976 minimal active density for this species was estimated to be 1258 individuals ha^{-1} (Table 10.6), and from 1971 to 1974 population densities did not vary greatly between years.

Mites

The predominating prostigmate mites are probably predaceous, although some of the most numerous mites of this group (nanorchestids and pachygnathids) are not. Tydeid mites probably feed on nematodes, and their late season increase in numbers coincided with an increase in sampled abundance of nematodes (Franco et al. 1979).

Biomass of predatory arthropods of Rock Valley

Predatory arthropods were surprisingly the second largest guild of soil arthropods, with 35% of total biomass. This high relative biomass, in comparison with that of other guilds, suggests that their food supply must significantly extend to foliage arthropods and small vertebrates. Scorpions and solpugids formed the most important groups of predatory arthropods that were collected in soil traps in 1976, with 33 and 42% of total biomass, respectively (Table 10.13). Species of *Vejovis*, the smaller scorpions, comprised nearly 70% of scorpion biomass and exhibited peak numbers in August and September, and smaller numbers of the larger *Hadrurus* were present from April to October. Maximum numbers and biomass of solpugids were present May to August, but some individuals were active during winter months, when most other predators were absent. A secondary peak of individual solpugids was recorded in October, 1976 (Table 10.12), representing young animals that emerged from eggs laid before gravid females were removed. Spiders collectively comprised 15% of total predator biomass, but nearly 60% of the collected individuals (Table 10.13). The single most important spider was *Psilochorus utahensis*, with peak activity from June through August. Almost half of all individuals of predatory arthropods sampled were this species, and it was the most commonly collected predator in previous years at Rock Valley. Species of tarantulas (Theraphosidae) were important in August and October.

Phytophagous soil arthropods

Phytophagous arthropods accounted for 6–9% of aggregate standing stocks throughout 1974, as compared with predatory forms, which comprised from 8% (April–June) to 18% (July–December) of total biomass. Collembolans, the most abundant

soil insects at Rock Valley, are probably fungivores.

Weevils

Phytophagous arthropods formed the largest number of individuals of soil arthropods that were sampled in 1976, due in particular to small weevils (Curculionidae), such as *Eucyllus vagans*, *E. unicolor*, *Miloderes mercuryensis*, and *Ophryastes varius*. However, this group was the smallest guild for biomass, with only 23% of total soil arthropods. *Eucyllus vagans* comprised more than half of that biomass, in particular because very large numbers were present in September and October (Tables 10.14, 10.15). Species of *Ophryastes* were the second largest component of biomass, with 37% of total phytophagous arthropod biomass, and they had a spring peak activity in May. Biomass of individual weevils was quite small. In Rock Valley an adult of *E. vagans* averaged 35 mg dry weight, whereas that of *M. mercuryensis* was less than 2 mg (Table 10.8). Quantitative data were collected on five species of defoliating broad-nosed weevils at Rock Valley. These species were significant defoliators only as adults, because larval stages fed on root tissues.

Species of *Eucyllus*, *Miloderes*, and *Ophryastes* showed peak abundances under the favorable growing conditions of 1973 and low population densities in the more environmental stressful years of 1971 and 1974 (Table 10.5). In 1976 *E. vagans* was estimated to be the most abundant phytophagous weevil, having about 85 000 individuals ha^{-1}, as sampled from soil pitfall traps (Table 10.3). In contrast, lepidopteran larvae, stick insects (Phasmatidae), and miscellaneous arthropods formed only a small fraction of phytophagous soil arthropods that were sampled in 1976, contributing only 5% of individuals and less than 1% of biomass for the phytophagous group (Tables 10.14, 10.15).

Table 10.14. *Monthly and total removals of phytophagous arthropods from 12 fenced 100-m² plots in Rock Valley during 1976.*

Species and species groups	Feb	Mar	Apr	May	Jun	Jul	Aug	Sep	Oct	Nov	Total	Estimated total removals ha⁻¹
Coleoptera												
Curculionidae												
Eucyllus vagans	19	122	97	135	729	280	207	3 120	4 573	848	10 130	84 079
E. unicolor	1	8	4	4	22	35	230	488	394	25	1 211	10 051
Mildoreres mercuryensis	1	141	236	284	195	688	40	12	4	–	1 580	13 122
Neocercopedius sp.	55	355	169	18	1	–	–	–	–	–	598	4 963
Ophryastes spp.	2	39	130	363	200	77	86	54	43	6	1 000	8 042
Chrysomelidae (larvae)	2	7	31	152	9	1	7	1	2	1	213	1 768
Lepidoptera (larvae)	1	75	461	134	15	1	1	4	29	32	753	6 250
Orthoptera												
Phasmatidae	–	3	1	8	3	1	5	15	3	1	40	332
Miscellaneous											89	739
Total	80	750	1 129	1 098	1 174	1 063	576	3 694	5 048	913	15 614	129 605

Table 10.15. *Monthly and total dry weight biomass (g ha⁻¹) of phytophagous arthropods removed from 12 fenced 100-m² plots in Rock Valley during 1976.*

Species and species groups	Feb	Mar	Apr	May	Jun	Jul	Aug	Sep	Oct	Nov	Total estimated biomass, g ha⁻¹
Coleoptera											
Curculionidae											
Eucyllus vagans	0.5	3.0	2.4	3.3	17.9	6.9	5.1	76.7	112.3	20.8	249
E. unicolor	–	0.1	–	–	0.2	0.3	1.7	3.6	2.9	0.2	9
Mildoreres mercuryensis	–	2.6	4.3	5.1	3.5	12.1	0.7	0.2	0.1	–	29
Neocercopedius sp.	0.2	1.5	0.7	0.1	–	–	–	–	–	–	3
Ophryastes spp.	0.3	6.7	22.2	61.9	34.1	13.1	14.7	9.2	7.3	1.0	171
Chrysomelidae (larvae)											
Lepidoptera (larvae)	–	0.1	0.6	0.2	–	–	–	–	–	–	1
Orthoptera											
Phasmatidae	–	0.1	–	0.3	0.1	–	0.2	0.6	0.1	–	1
Totals	1.0	14.1	30.2	70.9	55.8	32.4	22.4	90.3	122.7	22.0	463

Spatial patterning of soil arthropods in Rock Valley

At the validation site in Rock Valley spatial patterns of abundance by soil arthropods were studied using another set of measurements, which investigated variation of arthropod density and biomass in relation to shrub canopies and soil depth. Studies utilized funnel extractions of soil volumes to provide data on mites and other microarthropods that were not adequately sampled by pitfall traps.

Preliminary analyses by IBP researchers showed that neither shrub species nor its size significantly affected abundances of soil arthropods; hence, data from all shrubs were combined for analysis. Details of this procedure were given by Franco et al. (1979), who recognized distance from shrubs and sample depth as the two major sources of variation in arthropod abundance. Another examined covariate, soil moisture, was also important. In general, abundance of arthropods was greatest at bases of shrubs and near the soil surface. Numbers decreased with distance from shrubs and with depth. The basic pattern was also influenced by season; between January and March and November and December more than 80% of arthropods in samples were found within the top 0.10 m of soil, whereas during June and September, when soil temperatures were higher and water content was lower, proportions of arthropods in the upper 0.10 m fell to 30 and 20%, respectively. Perhaps these changes were related to migration phases rather than to changes in numerical states of populations (Edney, McBrayer, & Franco, 1974; Edney, Franco, & McBrayer 1976).

Concentration of arthropods around the bases of shrubs may reflect effects of plants on soil moisture and temperature, but the overriding factor was probably amount of organic matter in soil. Soil carbon was measured in 20 samples, and arthropod abundance was highly correlated ($r = 0.89$) with amounts of soil carbon.

Franco et al. (1979) used their sampling data to create a model that was capable of estimating densities and biomass of soil arthropods on a landscape scale. One step in this procedure was to estimate area of influence by shrubs on arthropod abundance, and this was found to be about 1.5 shrub radii. Beyond that distance shrubs had no apparent effect on arthropod numbers. This study also recognized the importance of a density gradient within the 1.5 radius distance and considered total coverage by all perennial species. When viewed in this manner, two prominent peaks in aggregate density were present, one in March, with about 5000 individuals m^{-2}, and another in November–December, with about 12 000 individuals m^{-2}. A smaller peak in August (about 3000 individuals m^{-2}) may have reflected increased sampling susceptibility, as stimulated by late July rainfall, rather than real changes in population states. Biomass tracked changes in numbers fairly closely except in May, when a conspicuous increase in numbers of beetle larvae resulted in an estimated aggregate biomass of 37 mg m^{-2}, or about twice that estimated for February and April. Estimated biomass in November–December was 26–31 mg m^{-2}. July was the low point for 1974, both in terms of number (about 1650 individuals m^{-2}) and biomass (8.7 mg m^{-2}). Mites were always more numerous than insects and as much as 17 times more abundant in November. Insects usually sustained somewhat higher standing stocks than mites because insect values included the relatively large masses of beetle larvae. In March, for example, estimated insect biomass was about five times that of all mites. However, in November and December, 1974, tydeid and cryptostigmate mites were so dense (over 10 000 individuals m^{-2}) that aggregate mite biomass was from 8.4 (in November) to 2.1 times that of insect biomass.

FOLIAGE ARTHROPODS

Foliage arthropods that were regularly vacuumed from plants in Rock Valley from 30 shrubs permitted assessment of diversity and population sizes of foliage herbivores that are associated with each species as well as overall estimates of biomass and productivity.

Numbers of arthropod species sampled did not vary greatly between years. The highest number of species identified was 197 (1972), whereas the low number was 159 (1975; Fig. 10.1), but individual shrub species supported very different densities of foliage arthropods (Table 10.16). Highest densities of arthropods were found on *L. tridentata* and *A. dumosa*, with 5-year mean numbers of arthropods collected of 13.9 and 13.7 individuals per sample, respectively. For *A. dumosa*, these high densities resulted from an abundance of thrips in 1972 and 1973, with almost 30 per sample on average in 1973 (Table 10.17). Numbers of arthropods sampled for the four other dominant shrubs in Rock Valley averaged about 4–5 individuals per sample. Year-to-year variations in foliage arthropod densities showed evidence of pattern. In 1973, an unusually favorable year for shrub growth (Chapter 4), highest measured densities of arthropods occurred in four of six species (Table 10.16). Poor conditions for shrub growth in 1974 resulted in low densities of foliage arthropods, particularly on *Grayia spinosa*, *Krameria erecta*, and *Lycium andersonii*; all species of foliage arthropods were generally less abundant in this year, but the relative effect was particularly notable for thrips.

Aggregate shrub arthropod index

Possible relationships between net primary production by shrubs and abundance of foliage arthro-

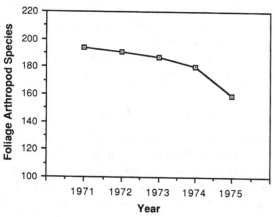

Figure 10.1 Number of arthropod species collected with D-vac samples at the IBP validation site in Rock Valley, 1971–1975.

pods was tested by developing an aggregate shrub arthropod index for the community, derived from the data in Table 10.17 when weighted by proportional importance of each shrub species. This index was 5.0 in 1971, 9.3 in 1972, 15.8 in 1973, 3.2 in 1974, and 8.0 in 1975. Net primary production and shrub arthropod index were generally related, but the results were not statistically significant. However, as shown in Figure 10.2, shrub arthropod index was significantly correlated with fall precipitation for the five-year Rock Valley sample. Statistical correlations of shrub arthropod index and arthropod densities for individual shrub species with monthly temperatures were generally not significant. Numbers of total foliage arthropods from *Lycium* and defoliating insects from *Krameria* were inversely correlated with mean air temperature from the preceding December.

Types of foliage arthropods

Foliage arthropods in Rock Valley were divided into categories of feeding specialization: defoliators, sap-feeders, thrips, fungivores, predators, parasitoids, and others (not classified). Phytophagous species (defoliators, sap-feeders, and thrips) formed the great majority of sampled arthropods (Table 10.17). Excluding thrips from consideration, the number of collected phytophagous arthropods during five years (13 375) was 4.6 times the number of carnivores (predators and parasitoids, 3159). This ratio varied for individual years from a low of 3.2 in 1972 to a high of 8.6 in 1974. If highly variable populations of thrips were included, the overall ratio was about 8.3. By comparison, herbivore/carnivore ratios for other North American deserts have been reported as 2.9 for sagebrush (*Artemisia*) scrub and 4.5 for shadscale (*Atriplex confertifolia*) scrub in Curlew Valley (Osborne 1975) and 2.2 for creosote bush (*Larrea*) scrub at Jornada Range in the Chihuahuan Desert (Johnson, Johnson, & Bellows 1975).

Defoliating insects

At Rock Valley the guild of defoliating insects was dominated by three orders, Lepidoptera, Coleop-

Table 10.16. *Numbers of arthropods collected in D-vac samples from shrubs in Rock Valley, 1971–1975.*

Species	Year	Number of samples	Number of arthropods collected	Mean number per sample	Standard error of mean
Larrea tridentata	1971	220	1939	8.8	0.54
	1972	281	3290	11.7	1.07
	1973	210	3148	15.0	2.26
	1974	175	1894	10.8	1.68
	1975	98	2176	22.2	3.38
Ambrosia dumosa	1971	173	541	3.1	0.36
	1972	238	4352	18.3	6.61
	1973	201	8390	41.7	10.45
	1974	173	454	2.6	1.71
	1975	236	949	4.0	0.53
Lycium andersonii	1971	130	622	4.8	0.66
	1972	185	724	3.9	0.34
	1973	149	821	5.5	0.91
	1974	170	116	0.7	0.10
	1975	72	373	5.2	1.63
L. pallidum	1971	30	69	2.3	0.50
	1972	30	155	5.2	1.24
	1973	26	229	8.8	3.09
	1975	72	263	3.7	0.50
Grayia spinosa	1971	30	69	2.3	0.34
	1972	34	303	8.9	2.93
	1973	85	546	6.4	2.47
	1974	158	102	0.6	0.09
	1975	113	400	3.5	0.33
Krameria erecta	1971	129	799	6.2	1.03
	1972	203	712	3.5	0.31
	1973	171	1796	10.5	2.42
	1974	171	215	1.3	0.13
	1975	121	607	5.0	0.59

tera, and Orthoptera. Small numbers of Diptera and Hymenoptera were also recorded as defoliators. Defoliators often were relatively large individuals, but numerically much less abundant than the smaller sap-feeders.

Lepidoptera

Overall, lepidopteran larvae formed the most important defoliators at Rock Valley and NTS.

Limited field surveys at NTS had previously identified 72 species of butterflies and moths in 13 families (Allred 1969b), but during the IBP program at Rock Valley special attention was given to lepidopteran larvae because of their significance as herbivores on vascular plants. Pitfall and D-vac samples permitted proper identification of 26 families of Lepidoptera (Table 10.18).

Lepidopteran larvae as a group were the most ecologically important defoliators in Rock Valley.

Table 10.17. *Mean numbers of various trophic groups of arthropods collected in D-vac samples from shrubs in Rock Valley, 1971–1975.*

Species	Year	Defoliators	Sap-feeders	Thrips	Fungivores	Predators	Parasitoids	Others
Larrea tridentata	1971	1.32	4.21	0.30	0.14	0.88	0.75	1.22
	1972	1.11	4.64	1.35	0.01	0.91	1.04	2.64
	1973	1.25	5.10	5.78	0.03	0.74	1.04	1.05
	1974	0.51	6.00	1.79	0.02	0.43	0.19	1.88
	1975	1.60	8.59	0.65	0.51	0.86	1.71	8.28
Ambrosia dumosa	1971	0.62	0.68	0.50	0.20	0.30	0.20	0.64
	1972	1.56	1.70	13.30	0.05	0.30	0.36	1.04
	1973	0.71	4.50	28.80	0.06	0.24	0.49	6.94
	1974	0.27	0.40	1.71	0.02	0.06	0.04	0.11
	1975	0.42	1.11	0.64	0.69	0.14	0.19	0.82
Lycium andersonii	1971	1.74	1.32	0.15	0.08	0.22	0.35	0.92
	1972	1.11	0.54	0.39	0.01	0.20	0.55	1.12
	1973	0.62	1.95	0.86	0.01	0.19	0.54	1.33
	1974	0.24	0.14	0.02	0.01	0.04	0.03	0.20
	1975	0.69	1.12	1.49	0.42	0.15	0.36	0.94
L. pallidum	1971	1.07	0.37	0.07	0.07	0.10	0.13	0.50
	1972	3.07	0.77	0.20	0	0.20	0.47	0.47
	1973	2.81	3.73	0.38	0.15	0.19	0.88	0.65
	1975	1.04	1.28	0.10	0.21	0.15	0.22	0.65
Grayia spinosa	1971	0.50	0.47	0.07	0	0.17	0.17	0.93
	1972	2.47	1.00	3.53	0	0.32	0.41	1.15
	1973	0.45	0.92	3.52	0.09	0.22	0.51	0.72
	1974	0.14	0.23	0.10	0	0.02	0.02	0.13
	1975	0.28	0.76	0.40	1.18	0.13	0.26	0.53
Krameria erecta	1971	2.64	0.57	0.09	1.09	0.42	0.43	0.95
	1972	1.46	0.23	0.27	0.17	0.21	0.40	0.77
	1973	2.00	4.91	1.43	0.20	0.17	0.54	1.31
	1974	0.58	0.13	0.05	0.09	0.10	0.05	0.25
	1975	1.67	0.52	0.10	1.16	0.11	0.32	1.14

Based on D-vac collections, the five major families were Coleophoridae, Cosmopterigidae, Gelechiidae, Geometridae, and Yponomeutidae. Larvae of other families, such as Phalaenidae, Lasiocampidae, Nymphalidae, and Pyralidae, were common in 1972–3 samples from *A. dumosa*, *K. erecta*, and *L. pallidum*. Greatest abundances in samples were exhibited by gelechiids on *L. pallidum* (1972, 1973), coleophorids on *G. spinosa* (1972), and yponomeutids on *Krameria* (1971, 1973, 1975) and *L. andersonii* (1971). Relative densities of lepidopteran larvae among all defoliators (1972–5) were particularly high in *K. erecta* (79–96%), *L. andersonii* (64–85%), *L. pallidum* (78–97%; based on only two years of data), and *E. nevadensis* (76%, based on only one year of data; Tables 10.19–10.22).

Gelechiid larvae were found to be the most significant defoliators of both species of *Lycium*. These larvae also consumed floral parts and were most abundant during the flowering of host shrubs. Their ecological effects were particularly evident in 1971, when practically no fruits matured on *L. pallidum* and 40–80% of the leaves

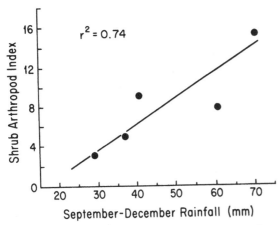

Figure 10.2 Relationship between arthropod abundance and fall rainfall from the previous year in Rock Valley, 1971–1975.

Table 10.18. *Families of Lepidotera sampled in 1971 and 1972 at Rock Valley using pitfall (P) and D-vac (V) techniques of samplings.*

Family	Common name	Sampling technique
Alucitidae	many-plumed moths	V
Arctiidae	tiger moths	P
Blastobasidae	blastobasid moths	P, V
Coleophoridae	casebearer moths	P, V
Cosmopterigidae	cosmopterigid moths	V
Gelechiidae	gelechid moths	P, V
Geometridae	measuring worm moths	P, V
Glyphipterigidae	sedge moths	P, V
Gracillarlidae	leaf blotch miner moths	P, V
Heliozelidae	shield bearer moths	P
Lasiocampidae	tent caterpillar moths	V
Liparidae	liparid moths	P
Lyonetiidae	lyonetiid moths	P, V
Oecophoridae	oecophorid moths	V
Olethreutidae	olethreutidae	V
Phalaenidae	phalaenid moths	P, V
Plutellidae	diamondback moths	V
Psychidae	bagworm moths	P, V
Pterophoridae	plume moths	P, V
Pyralidae	pyralid moths	P, V
Saturniidae	giant silkworm moths	other
Sesiidae	clear-winged moths	V
Springidae	sphinx moths	other
Tineidae	clothes moths	P, V
Tortricidae	tortricid moths	P, V
Yponomeutidae	ermine moths	P, V

were consumed. In 1972 gelechiid larvae constituted 78% of all defoliators on *L. pallidum* and 35% on *L. andersonii* (Table 10.19). These larvae were less abundant but still important on other shrub species; they comprised 20% of the defoliators on *A. dumosa* and *E. nevadensis*, 17% on *K. erecta* and *Larrea tridentata*, and 14% on *G. spinosa*. Large numbers of adult gelechiid moths were attracted to cones of *Ephedra* just before peak of pollen production. Larvae of the family Yponomeutidae (ermine moths) appear to be in competition with gelechiid larvae, given that both heavily used the same shrub species, and their relative numbers seemed to vary inversely. Greatest numbers of yponomeutids on the three primary host shrubs, *L. pallidum*, *L. andersonii*, and *K. erecta*, occurred in 1971, although samples were also relatively rich in 1973 and 1975. Populations on *L. andersonii* reached an estimated maximum density of about 10 000 individuals ha^{-1} during the first week of May, 1971, and about 15 000 individuals ha^{-1} on *K. erecta* at the end of June. Yponomeutid larvae were the most numerous defoliators on *Krameria*, comprising 56–80% of the total. In 1971 they comprised 81% of defoliators that were collected from *L. andersonii*, but between 1972 and 1975 yponomeutid larvae were 21–54% of this group. Community biomass of

larvae in 1972 peaked at 0.92 kg ha^{-1} in May, when individuals were still small but density was high (Table 10.23).

Although casebearer larvae of family Coleophoridae were present on all shrub species, they were most abundant on *G. spinosa*, and in 1972 coleophorids composed 55% of defoliators on this shrub. By comparison, coleophorid larvae composed only 8–10% of defoliators on *Ambrosia*,

Table 10.19. *Relative population size (%) of defoliating insects collected on major shrubs at Rock Valley in 1972.*

	Larrea tridentata	Ambrosia dumosa	Lycium andersonii	Grayia spinosa	Krameria erecta	Lycium pallidum	Ephedra nevadensis	Atriplex confertifolia
Chrysomelidae	4.8	5.0	7.8	5.0	1.7	0.0	0.0	46.7
Curculionidae	6.7	35.8	6.9	2.0	0.0	3.3	25.0	0.0
Total Coleoptera	11.5	40.8	14.7	7.0	1.7	3.3	25.0	46.7
Coleophoridae	0.6	8.6	8.6	55.4	1.7	0.0	10.0	6.7
Cosmopterigidae	0.0	5.4	0.0	17.8	2.0	0.0	0.0	0.0
Gelechiidae	17.4	20.1	34.9	13.9	16.9	78.0	21.0	20.0
Semiothisa larreana	18.4	0.0	0.0	0.0	0.0	0.0	0.0	0.0
Misc. Geometridae	4.2	2.7	7.3	0.0	2.6	0.0	0.0	0.0
Phalaenidae	0.2	1.7	6.5	0.0	4.9	3.3	5.0	0.0
Yponomeutidae	0.8	4.7	16.8	2.0	57.9	5.5	30.0	6.7
Misc. Lepidoptera	2.3	15.7	11.2	2.0	8.0	9.9	10.0	6.7
Total Lepidoptera	43.8		85.3		78.8	96.7	76.0	40.0
Tanaocerus koebelei	0.2	7.4	0.0	2.0	2.6	0.0	0.0	0.0
Bootettix argentatus	29.3	0.0	0.0	0.0	0.0	0.0	0.0	0.0
Misc. Orthoptera	2.1	4.4	0.0	0.0	1.7	0.0	0.0	13.3
Total Orthoptera	31.7	11.8	0.0	2.0	4.3	0.0	0.0	13.3

Table 10.20. *Relative population size (%) of defoliating insects collected on major shrubs at Rock Valley in 1973.*

	Larrea tridentata	Ambrosia dumosa	Lycium andersonii	Grayia spinosa	Krameria erecta
Chrysomelidae	9.7	5.0	7.8	5.0	1.7
Curculionidae	6.2	12.6	6.5	2.6	0.9
Misc. Coleoptera	8.2	6.3	4.3	7.9	1.2
Total Coleoptera	24.1	21.0	17.3	13.1	2.4
Coleophoridae	–	3.5	3.2	15.8	0.3
Cosmopterigidae	–	0.7	–	7.9	1.5
Gelechiidae	0.3	0.1	1.0	1.0	2.7
Semiothisa larreana	19.0	–	–	–	–
Misc. Geometridae	1.6	3.5	7.5	–	2.6
Phalaenidae	1.2	17.7	4.3	7.9	2.0
Yponomeutidae	6.2	10.5	34.4	7.9	56.4
Misc. Lepidoptera	8.2	25.2	9.7	18.4	10.5
Total Lepidoptera	39.3	66.5	64.5	69.4	84.8
Tanaocerus koebelei	0.4	3.5	–	–	0.6
Bootettix argentatus	10.8	–	–	–	–
Misc. Orthoptera	1.6	2.8	1.1	5.3	1.2
Total Orthoptera	12.8	6.3	1.1	5.3	1.8
Misc. gall-formers	22.5	3.5	14.0	10.5	10.8
Misc. Diptera	1.2	2.8	3.2	2.6	–
Misc. Hymenoptera	–	–	–	–	0.1

Table 10.21. *Relative population size of defoliating insects and arachnids collected on major shrubs at Rock Valley in 1974.*

	Larrea tridentata	Ambrosia dumosa	Lycium andersonii	Grayia spinosa	Krameria erecta
Chrysomelidae	16.8	14.9	2.5	9.1	0.0
Curculionidae	4.4	27.7	12.5	18.2	2.0
Misc. Coleoptera	0.9	2.1	2.5	0.0	9.0
Total Coleoptera	22.1	44.7	17.5	27.3	11.0
Coleophoridae	1.8	12.8	7.5	9.1	1.0
Gelechiidae	7.1	12.8	10.0	22.7	2.0
Misc. Geometridae	18.6	0.0	0.0	4.5	2.0
Phalaenidae	0.0	2.1	0.0	0.0	3.0
Yponomeutidae	1.8	6.4	45.0	4.5	72.0
Misc. Lepidoptera	7.9	17.0	7.5	18.2	1.0
Total Lepidoptera	37.2	51.1	77.5	59.1	81.0
Tanaocerus koebelei	0.0	0.0	2.5	0.0	1.0
Misc. Orthoptera	6.2	2.1	2.5	9.1	1.0
Total Orthoptera	11.5	2.1	2.5	9.1	2.0
Misc. gall-formers	21.2	2.1	2.5	4.5	2.0
Misc. Diptera	0.0	0.0	0.0	0.0	0.0
Misc. Hymenoptera	0.0	0.0	0.0	0.0	2.0
Acarina (mites)	8.0	0.0	0.0	0.0	2.0

Table 10.22. *Relative population size (%) of defoliating insects collected on major shrubs at Rock Valley in 1975.*

	Larrea tridentata	Ambrosia dumosa	Lycium andersonii	Grayia spinosa	Krameria erecta	Lycium pallidum
Chrysomelidae	21.0	7.0	8.0	12.5	0.0	2.7
Curculionidae	8.3	19.0	6.0	12.5	0.0	5.3
Total Coleoptera	29.3	26.0	14.0	25.0	0.0	8.0
Lepidoptera						
Coleophoridae	0.0	10.0	4.0	25.0	1.0	2.7
Gelechiidae	1.9	5.0	8.0	9.4	11.4	30.7
Geometridae	31.8	6.0	0.0	0.0	2.5	2.7
Yponomeutidae	1.3	9.0	54.0	12.5	79.7	40.0
Misc. Lepidoptera	0.6	2.0	2.0	0.0	1.0	2.7
Total Lepidoptera	35.6	32.0	68.0	46.9	95.5	78.7
Tanaocerus koebelei	0.0	13.0	2.0	3.1	1.0	0.0
Misc. Orthoptera	35.0	29.0	14.0	21.9	2.5	13.3
Total Orthoptera	35.5	42.0	16.0	25.0	3.5	13.3
Misc. Diptera	0.0	0.0	2.0	3.1	0.0	0.0
Misc. Hymenoptera	0.0	0.0	0.0	0.0	1.0	0.0

Table 10.23. *Estimated monthly biomass (kg dry weight ha^{-1}) of yponomeutid larvae (Lepidoptera) in*

Table 10.23. *Estimated monthly biomass (kg dry weight ha^{-1}) of yponomeutid larvae (Lepidoptera) in 1972 at Rock Valley.*

Month	Estimated density, indi- viduals ha^{-1}	Estimated mean dry mass of an individual, mg	Estimated standing stock, kg dry weight ha^{-1}
March	1942	0.075	0.15
April	1708	0.080	0.14
May	9245	0.100	0.92
June	3126	0.250	0.78
July	567	0.350	0.20
August	386	0.450	0.17
September	304	0.450	0.14

Ephedra, and *L. andersonii* and less than 2% on *Krameria* and *Larrea*. Estimated densities on *Grayia* were about 28 000 individuals ha^{-1} within the sample stand from early to mid-April, 1971, at a time of fruit production.

Semiothisa larreana, a geometrid moth with larvae that are restricted to *L. tridentata*, and related measuring worms composed from 18–32% of all defoliators taken from *Larrea*, and they were therefore the second most important family of moth larvae during late May and early June. Larvae were most abundant in 1975 and least in 1974. Other geometrid larvae were present at very low densities on *K. erecta*, *Lycium andersonii*, and *L. pallidum*. Cosmopterigid larvae were collected only from *Grayia* in 1971–3.

Coleoptera

The second important group of defoliating insects in Rock Valley was Coleoptera, largely weevils (Curculionidae) and leaf beetles (Chrysomelidae), and they commonly exhibited distinctive host preferences. Relative population densities of defoliating Coleoptera ranged from a high of 21–45% on *A. dumosa* to a low of 0–9% on *K. erecta* (Tables 10.19–10.22).

At Rock Valley weevil populations were especially large on *A. dumosa*, comprising 13–36% of all defoliating insects (Tables 10.19–10.22). *Smicronyx imbricata* was the important species on *A. dumosa*, where it developed as larvae in floral tissues, and larval density appeared to be linked

with abundance of floral tissues of *A. dumosa* during spring. Although larvae of *Smicronyx* were relatively host specific, adults were widespread.

Leaf beetles (Chrysomelidae) fed as larvae and adults on dominant shrubs, particularly *A. dumosa*, *Larrea tridentata*, *Lycium andersonii*, and *K. erecta*. Most chrysomelids were host specific, but it was difficult to determine identifications below the genus level. Several genera were extremely abundant on *A. dumosa* and *G. spinosa*. In 1974 relative abundance of chrysomelids reached 15% on *Ambrosia* and 12.5% on *Grayia*, but in other years they were less common, with ranges of 2–7% and 3–9%, respectively (Tables 10.19–10.22). On *Larrea* these beetles formed 5–21% of the defoliators. Larvae of leaf beetles had an estimated density of 1768 individuals ha^{-1} in 1976 (Table 10.14).

Population studies of small broad-nosed weevils were conducted in Rock Valley on species of *Eucyllus*, *Miloderes*, and *Ophryastes* (1971–5). They were best sampled in pitfall traps, and their population sizes and biomasses were reported in our discussion of ground-dwelling arthropods.

Orthoptera

Orthoptera, including grasshoppers and crickets, form an ecologically significant group of phytophagous insects, but they are also important as detritivores. Studies at NTS have identified 58 species in nine families and described patterns of seasonal abundance and habitat distribution (Barnum

1964). Highest species diversity occurs in transitional desert communities of *Coleogyne* (Allred et al. 1963).

In Rock Valley orthopterans comprised the third most important group of insect defoliators, but population densities differed considerably between shrub species. Orthopterans were relatively important as herbivores on *Larrea* but were present on all species (Tables 10.19–10.22) and were important herbivores on *A. dumosa*, *Atriplex confertifolia*, and *G. spinosa*.

Detailed studies were conducted in Rock Valley on *Bootettix argentatus*, which is one of the major defoliators of creosote bush and is restricted to this species. Peak densities of this grasshopper from 1971 to 1973 were estimated as 600–950 individuals ha^{-1}. In 1974 populations dropped sharply to 20% of high 1972 levels (Table 10.19); because in 1971–3 numbers of first instar nymphs exceeded those of later stages, much of the 1974 population consisted of adult survivors from previous years. High densities of 1375 individuals ha^{-1} occurred in 1975. This orthopteran was the most abundant of the chewing defoliators.

Mispagel (1974, 1978) worked intensively with different Rock Valley populations of *B. argentatus* in 1973 and estimated a density of 1440 individuals ha^{-1}. He related population abundance and frequency to soil moisture and soil temperatures required to break diapause and insure adequate survivorship of the egg stage. Larvae of bombyliid flies of *Mythicoymia* were a major predator of egg masses for this and other grasshoppers. Mortality of nymphs was dependent upon air temperature and precipitation, which regulated levels of predation and parasitism on this vulnerable life stage; years with high precipitation were generally poor years for *Bootettix* because of such biotic pressures. Adults were most prominent in late summer, when air temperatures were above 30°C. *Ligurotettix coquilletti*, another defoliating orthopteran that is largely restricted to creosote bush, was much less abundant than *Bootettix*, but is a dominant herbivore on *Larrea* in the Sonoran Desert (Greenfield, Shelley, & Gonzalez-Coloma 1989).

Mean individual biomass was measured for seven orthopterans of the 23 species that were encountered during the IBP studies (Table 10.8). These ranged in size from 33 mg dry weight for *L. coquilletti* and *Insara covilleae* to 838 mg for *Capnobotes fuliginosus*. *Bootettix argentatus* varied in weight between sexes and from year to year; adult males ranged from 15 to 22 mg (1971–3 means) where adult females were 27–64 (1971–4 means), with the largest means for both sexes occurring in 1973, when food resources were excellent. Estimated community biomass of adult and immature *B. argentatus* for 1971–4 ranged from 29.6 g ha^{-1} in 1972 to 11.0 g ha^{-1} in 1971.

Sap-feeding insects

Sap-feeding insects of Rock Valley, including members of Homoptera, Hemiptera, and Thysanoptera, were much more abundant than defoliating insects, but they were often individually very small. These insects feed especially on phloem sap of host plants, although sap-feeding in thrips may be more generalized. Because phloem and xylem sap have relatively low nitrogen content, sap-feeders must process large quantities of fluid for protein requirements and often secrete a carbohydrate-rich honeydew. For this reason many populations of sap-feeders are tended by ants.

In Rock Valley thrips were most abundant, followed by Homoptera, which included representatives of 21 families. Ecologically the most significant homopterans were mealybugs (Pseudococcidae) and treehoppers (Membracidae), but leafhoppers (Cicadellidae) and psyllids (Psyllidae) were also important. Among Hemiptera Miridae and Lygaeidae were important sap-feeders at Rock Valley plots, and Coreidae, Pentatomidae, and Tingidae were also present.

Thrips

Thrips (Thysanoptera) were largely found feeding on flowers of Rock Valley plants, although a few predatory species also occurred. Small sap-feeders scrape and suck liquids from plant surface tissues with a conical beak at the base of the head. Thrips reached exceedingly high densities on many shrubs (Tables 10.24–10.27), especially during

Table 10.24. *Relative population size (%) of sap-feeding insects collected on major shrubs at Rock Valley in 1972.*

	Larrea tridentata	*Ambrosia dumosa*	*Lycium andersonii*	*Grayia spinosa*	*Krameria erecta*	*Lycium pallidum*	*Ephedra nevadensis*	*Atriplex confertifolia*
Hemiptera								
Miridae	3.4	0.2	3.3	4.6	1.6	6.7	0.2	6.1
Misc. Hemiptera	0.0	7.0	0.0	0.0	0.0	0.0	0.0	0.0
Subtotal	3.4	7.2	3.3	4.6	1.6	6.7	0.2	6.1
Homoptera								
Cicadellidae	3.6	0.7	11.4	0.0	17.1	3.3	0.0	1.2
Pseudococcidae	20.1	2.8	25.8	5.8	8.2	33.3	1.8	57.6
Misc. Homoptera	30.9	0.5	10.1	11.0	11.4	33.3	1.3	17.5
Subtotal	74.6	4.0	47.3	16.8	36.7	70.0	3.0	76.3
Thysanoptera	21.7	88.8	49.4	78.6	61.6	23.3	96.8	17.6

Table 10.25. *Relative population size (%) of sap-feeding insects collected on major shrubs at Rock Valley in 1973.*

	Larrea tridentata	*Ambrosia dumosa*	*Lycium andersonii*	*Grayia spinosa*	*Krameria erecta*
Hemiptera					
Lygaeidae	16.5	11.4	53.5	11.9	74.5
Miridae	3.4	0.7	0.9	1.12	0.4
Misc. Hemiptera	0.7	0.0	1.7	0.5	0.2
Subtotal	20.6	12.1	56.1	13.5	75.1
Homoptera					
Cicadellidae	1.7	0.4	8.1	2.4	1.7
Pseudococcidae	1.1	0.3	2.4	1.6	–
Membracidae	21.9	–	–	–	–
Misc. Homoptera	0.7	0.5	2.4	3.2	0.5
Subtotal	25.4	1.2	12.9	7.2	2.2
Thysanoptera	53.9	86.5	31.0	79.4	22.7

flowering periods, when population growth was very rapid. Overall, thrips provided the most numerous sap-feeders, comprising as much as 97% of all individuals on *E. nevadensis* in 1972 (the only year with studies of this host plant), and 50% or more at times on *L. tridentata* (1973), *A. dumosa* (greater than 80% in 1972–4), *Lycium andersonii* (1972, 1974), *G. spinosa* (1972–5), and *K.*

erecta (1972, 1974, 1975). Thrip populations were highest in 1972 and 1973, when they comprised over half of all sap-feeders, declining to 35% in 1974 and 21% in 1975. Causes for these large changes were not investigated, but they hypothetically resulted from changes in the nutritional quality of saps or nectars that were associated with changes in plant water balance.

Table 10.26. *Relative population size (%) of sap-feeding insects collected on major shrubs at Rock Valley in 1974.*

	Larrea tridentata	Ambrosia dumosa	Lycium andersonii	Grayia spinosa	Krameria erecta
Hemiptera					
Lygaeidae	0.4	0.3	0.0	1.9	0.0
Miridae	1.0	0.5	0.0	0.0	2.9
Misc. Hemiptera	0.1	0.0	0.0	0.0	8.6
Subtotal	1.5	0.8	0.0	1.9	11.4
Homoptera					
Cicadellidae	0.7	1.9	7.4	0.0	5.7
Pseudococcidae	54.3	15.8	51.9	63.5	34.3
Misc. Homoptera	0.6	0.5	29.6	3.8	14.3
Membracidae	19.6	0.0	0.0	0.0	0.0
Subtotal	75.2	18.2	88.9	67.3	54.3
Thysanoptera	23.2	80.9	11.1	30.8	34.3

Table 10.27. *Relative population size (%) of sap-feeding insects collected on major shrubs at Rock Valley in 1975.*

	Larrea tridentata	Ambrosia dumosa	Lycium andersonii	Grayia spinosa	Krameria erecta	Lycium pallidum
Hemiptera						
Lygaeidae	0.1	0.0	0.5	0.0	1.3	0.0
Miridae	6.8	3.4	3.7	5.3	2.7	3.0
Misc. Hemiptera	0.1	0.0	0.0	0.0	0.0	0.0
Subtotal	7.1	3.4	4.3	5.3	4.0	3.0
Homoptera						
Cicadellidae	5.1	36.9	14.9	16.0	34.7	16.2
Pseudococcidae	43.9	14.2	12.8	20.6	34.7	28.3
Misc. Homoptera	1.8	8.9	10.1	24.4	10.7	43.4
Membracidae	35.1	0.0	1.1	0.0	0.0	2.0
Subtotal	85.9	60.0	38.9	64.9	80.0	89.9
Thysanoptera	7.1	36.6	56.9	33.8	16.0	7.1

Mealybugs

Mealybugs, most notably *Pseudococcus helianthi*, were generally abundant on all shrubs in Rock Valley, and activity ranged from as many as 20 weeks per year on *Larrea* to up to three weeks on *Ephedra nevadensis* (Table 10.28). Dispersions of newly hatched crawlers were highly contagious. A few sampled shrubs had large numbers of crawlers while most plants supported a few or none. Because this distribution was extremely patchy, sampling data often exhibited pronounced fluctu-

Table 10.28. *Seasonal occurrence of pseudococcids (mealybugs) on shrubs in Rock Valley during 1972.*

Shrub species	Total number of weeks of pseudococcid activity	First/last calendar week of primary activity	Relative population of all pseudococcids collected, %
Larrea tridentata	20	10/36	58.9
Ambrosia dumosa	14	11/35	13.1
Lycium andersonii	10	12/24	6.1
Grayia spinosa	6	10/37	1.1
Krameria erecta	6	20/36	1.3
Lycium pallidum	7	10/19	1.3
Ephedra nevadensis	3	19/25	2.5
Atriplex confertifolia	6	10/38	12.0
Ceratoides lanata	5	14/23	3.7

ations that were unrelated to true changes in numbers (Sleeper & Mispagel 1975). Even allowing for sampling problems, numbers of mealybugs from mid-March to September, 1973, a wet year, were low relative to those taken in other years. Mean number of individuals per shrub of *L. tridentata* was 0.11 in 1973, 4.2 in 1974, and 4.1 in 1975. A similar pattern was expressed in samples from other shrubs. High rainfall and therefore large fluid fluxes in phloem and xylem undoubtedly dilute saps and thus make shoots less nutritious for young sap-feeders. Because 1973 was the wettest of the IBP years, this phenomenon may have depressed numbers of mealybugs in samples of summer, 1973.

Treehoppers

Three species of treehoppers (Membracidae), *Centrodontus atlas*, *Multareis cornutus*, and *Multareoides bifurcatus*, were obligate feeders on *Larrea* (Mispagel 1974; Stave & Shiff 1981). These species comprised from 25.5 (1974) to 64.5% (1971) of sap-feeders in *Larrea* samples. Mean numbers of adults per sample ranged from 2.2 (1972, 1973) to 3.5 (1975; Table 10.27), and although they co-occur on *Larrea*, timing of peak abundances affected temporal separation. Each of the three species also overwintered in a different stage:'*C. atlas* as eggs, *M. cornutus* as nymphs, and *M. bifurcatus* as adults (Stave & Shiff 1981). Rates of nymphal development varied as well;

those of *M. cornutus* first appeared in September and grew slowly through the winter until March when adults appeared, whereas nymphs of the other two species appeared during spring or early summer and developed rapidly to adult stage (Stave & Shiff 1981). Field studies in the Mojave Desert near Barstow, California, have shown that in 1980 adults of *M. cornutus* were most abundant in April, those of *C. atlas* in May, and those of *M. bifurcatus* in late June and early July. Stave and Shiff (1981) argued that the temporal separation of the three species was associated with phenophases of the host plant. This hypothesis is supported by the fact that peaks of adult treehopper adundance in Rock Valley occurred about two weeks to a month later than in Barstow and at different times for different years.

Of the Membracidae in Rock Valley, *C. atlas* was almost always the most abundant, composing 44–61% of all adults taken in 1971–3 and 1975, but only 29% in 1974. Maximum estimated adult densities were ordinarily about 3000 individuals ha^{-1}, peaking during June, but in 1975 they exceeded 4600 individuals ha^{-1}. *Multareis cornutus* comprised 13–28% of sampled adults during 1971–4 and 39% in 1975; its population size peaked early in the season, just before *Centrodontus* increased (cf. Stave & Shiff 1981). However, in 1975 peak numbers of both species coincided during June, when densities of *M. cornutus* were 2700 individuals ha^{-1}, at least twice typical June levels. *Multareoides bifurcatus* was the only membracid that overwintered in Rock Valley as adults, these being mostly females, and its estimated densities peaked in July and early August and rarely exceeded 1000 individuals ha^{-1}.

Leafhoppers

Cicadellidae were notably important in 1972 and 1975 on *K. erecta* and in 1975 on *A. dumosa* (Tables 10.24–10.27). From a low of only 1% of total sap-feeders in 1974, leafhoppers rose to 16% in 1975. One species, *Spathanus excavatus*, was restricted to *Larrea*, and secreted copious quantities of honeydew. Population samples of *S. excavatus* were highly variable both spatially and temporally, due in part to its active movements, and it was frequently encountered in pitfall traps as well as in D-vac samples in foliage.

Hemipterans

Mirid bugs (Miridae) were collected most frequently from *Larrea*, *Ambrosia*, and *Grayia* and much less commonly from *Lycium andersonii* and *Krameria*. Apparent abundance peaks were on *Grayia* in 1972, *Ambrosia* in 1973, and *Larrea* in 1975 (Tables 10.24–10.27). *Larrea* was the sole host for a species of *Phytocoris*, which experienced greatest abundance early or late in the year when mean minimum air temperatures dropped below 10°C. No *Phytocoris* specimens were collected in 1972 during the warm summer months, when mean minimum temperatures exceeded 17°C.

Lygaeid bugs (Lygaeidae) were relatively uncommon in Rock Valley except in 1973, when large numbers of *Nysius ericae* were collected on all shrubs (Table 10.25). Estimated densities of *Nysius* reached 150 000 individuals ha^{-1} in late July and early August, 1973. Large numbers of these migrating bugs were also taken in pitfall traps.

Pollinating insects

Community-level studies of pollination biology are largely lacking for desert regions. The Convergent Ecosystem program of the IBP conducted such studies to compare the Sonoran Desert with the Monte in Argentina (Orians, Cates, Mares, Moldenke, Neff, Rhoades, Rosenzweig, Simpson, Schultz, & Tomoff 1977) and these studies identified relative abundance both of insect and vertebrate groups that utilized flower resources of pollen, nectar, and oils. Among the insects were included solitary bees, solitary wasps, Diptera, Lepidoptera, Coleoptera, and less commonly Hemiptera and Thysanoptera. For *Larrea* plants in those two desert regions, investigators estimated that there were 1.0–2.9 kg ha^{-1} of sugar available in flower nectars of perennials during spring and 0.9–6.9 kg ha^{-1} of pollen. They also discussed the community-level implications of differing methods of pollen transfer.

No intensive studies of pollination biology have been done at NTS, and thus little is known about the nature and specificity of individual groups of pollinators. Lepidoptera, as already discussed as defoliating agents in their juvenile (larval) stages, are certainly important pollinators as adult moths and butterflies. The 26 families of Lepidoptera from Rock Valley (Table 10.18) have numerous species that are likely pollinating agents of annual and some perennial species.

Extensive faunal studies on bees at NTS reported 71 species in 356 genera (Allred 1969a). These bees were collected from more than 40 species of vascular plants while assessing the range of plants that are pollinated by each bee species. However, bees were not studied in any detail at Rock Valley during the IBP program.

Insect predators and parasitoids

During the IBP study in Rock Valley organisms that were collected in D-vac samples of foliage included seven orders of predatory arthropods and two orders of parasitoids, and especially parasitoids were more prevalent. The most common predators were spiders (especially Salticidae and Thomisidae), beetles (Melyridae, Cleridae, and Coccinellidae), neuropterans (Hemerobiidae and Myrmeleontidae), hemipterans (Anthocoridae, Nebidae, Pentatomidae, Phymatidae, and Reduviidae), and predatory thrips (Phloeothripidae).

Most predatory species feed on all available soft-bodied insects, but some are more specific, for example, coccinellids specialize on aphids and scale insects.

Predators were most common on *Larrea*, particularly during June and July of very dry years. *Larrea* also supported more kinds of predators, and relative abundance and diversity of predatory types on *Larrea* probably reflected the greater number of other arthropods serving as potential prey. On some shrubs, e.g., *A. dumosa*, *G. spinosa*, and *L. andersonii*, numbers of predators were conspicuously reduced in 1974 samples, probably reflecting the decreased abundance of prey insects, but even so in 1974 numbers of predators taken from *Larrea* and *K. erecta* were not obviously reduced.

Beeflies

Diptera comprise a large and ecologically diverse order of insects with many important parasitoids and predators of arthropods, but others are pollinators, blood-feeders, or, less commonly, herbivores. Rock Valley studies concentrated on beeflies, which showed high species diversity and abundance as parasites and predators in *Larrea-Ambrosia* communities (Allred et al. 1963).

Mythicomyia comprised greater than 80% of the beeflies in IBP samples; these organisms attacked the egg masses of acridid grasshoppers, principally those of *Bootettix argentatus*. Flies of *Mythicomyia* laid eggs in late spring, and first instar planidia became attached to female grasshoppers during feeding and then were transferred to their eggs during oviposition. Later, larval instars of the fly fed on eggs, entered a brief period of pupation, and emerged as adults in early summer. Estimated densities varied greatly between study years, with maximum monthly values of 1600 and 2400 individuals ha^{-1} in 1971 and 1972, respectively, but surged to more than 23 000 in 1973, tracking a significant population increase of *B. argentatus* in 1972. The population of *Mythicomyia* crashed to zero in 1974 and returned in 1975 to a maximum monthly density of 1100 individuals ha^{-1}.

Wasps

Hymenoptera were important parasitoids or predators on other arthropod species, and at NTS those of Mutillidae and Tiphiidae have been surveyed (Ferguson 1967; Allred 1973; Wasbrauer 1973). Rock Valley studies were conducted on densities of parasitic wasps of superfamilies Chalcidoidea and Ichneumonoidea occurring on shrub foliage. Larvae of these groups are true parasitoids, and larval survival strongly influences size of the adult population, which instead are often nectar-feeders. For all shrub species that were sampled in 1971-2, the relative proportion of chalcidoid wasps was 31.9% Pteromalidae, 23.2% Eulophidae, 7.7% Eupelmidae, 6.1% Encyrtidae, and 4.5% Chalcidae. Eupelmid and pteromalid wasps were notably abundant on *L. tridentata*, mymarid wasps were restricted to *K. erecta*, and mutillid wasps were restricted to *K. erecta* and *Lycium andersonii*. The two common families of ichneumon wasps were Ichneumonidae and Braconidae, and in 1972 braconids formed more than 87% of that wasp subtotal, having particularly large populations on *Lycium*.

Host–parasite relationships of wasps at Rock Valley are not well known, but marked changes in population sizes were common. Large declines from the wet year 1973 were observed for eulophids (−41%), encyrtids and braconids (−32%), and myrmerids (−26%), these comparable to levels in 1971 and 1972. From a relative abundance of 4.8% of all arthropods collected in D-vac samples for 1973, total insect predators and parasitoids decreased to 3.3% in 1974 following declines in wasp populations. In 1975 populations of wasps increased overall, most notably for braconids and ichneumonids, while densities of tachinids, encyrtids, eulophids, and mymarids declined.

Beetles

Predatory Coleoptera were also present among foliage insects in Rock Valley. Melyridae, a family of aggressive predatory beetles, feed on soft-bodied arthropods, and seven species of these were studied. All had similar behavior and feeding habits, both as immatures and adults. However, population densities were difficult to evaluate, because these beetles were strong fliers and could readily migrate between sites. In 1972 high relative densities of suitable arthropod prey led to higher densities of melyrid beetles. Peak densities in 1971 occurred four weeks earlier than in 1972, the reverse of a pattern observed in many other arthropod species.

Ticks and mites

The Acarina, consisting of ticks and mites, have received relatively little ecological attention as compared with other taxa of soil and plant desert arthropods. Ticks have received some biological study because they are important vectors for pathogens and parasites in vertebrates.

O'Farrell and Emery (1976) provided a list of ticks and mites from NTS. Parasitic mites on vertebrates at NTS have been documented for mammals (Allred 1962a, b, 1963; Goates 1963; Allred & Goates 1964a, b; Paran 1966) and reptiles (Allred & Beck 1962, 1964), and ticks and chiggers have also been collected (Beck, Allred, & Brinton 1963; Brennan 1965; Herrin & Beck 1965; Brinton, Beck, & Allred 1965). More recently, ticks collected on desert tortoise (*Gopherus agassizii*) from Rock Valley were identified as *Ornithodoros parkeri* and *O. tunicata* (Medica, pers. comm.). Approximately 400 species of free-living mites were collected on 111 plant species (Jorgensen 1970), and coworkers categorized these as predaceous, phytophagous, and detritivorous (scavenging) species and identified 17 phenotypically distinct groups.

No identifications below the family level were attempted with mites that were collected during the IBP studies at Rock Valley, where seven families of mites were collected in soil and vegetation samples (Table 10.1). Some of these mites are discussed in Chapter 11.

ANTS

Ants and termites are important faunal elements in most deserts. Their importance can be seen immediately by noting large numbers of species and biomass, but they also play critical roles in many ecosystem processes (Mackay 1991). Many ants are important seed predators and compete with granivorous birds and especially small rodents (Brown & Davidson 1977; Brown, Reichman, & Davidson 1979). Different species of ants and the termites are involved in processing organic matter for decomposition processes. At NTS, however, termites while present are of minor ecological significance as compared with those in other arid regions such as the Chihuahuan Desert (Mackay, Zak, & Whitford 1989).

Ants as competitors of vertebrate granivores

Granivorous ants act as competitors with heteromyid rodents, seed-eating birds, and other generalist vertebrates for soil reserves of seeds (see Chapter 11). Interactions of ants and vertebrates, two vastly different life forms, has been studied in fine detail in the Sonoran and Chihuahuan deserts, less so in the Mojave Desert (Brown & Davidson 1977; Brown et al. 1979; Mackay 1991). Although most granivorous ants are generalist feeders, resource partitioning among ant species appears to occur involving microhabitat differences, preferences in food sizes and quality, and temporal patterns of foraging (Whitford, Johnson, & Ramirez 1976; Davidson 1977a,b; Lopez Moreno, & Diaz Betancourt 1986; Rissing 1988). Investigators have also perceived indications that granivorous ants are food limited (Bernstein 1974; Chew 1977), and ant density has been shown to be a function of net primary production (Davidson 1977a). Studies on ants from an Australian desert suggest that many factors probably affect ant diversity (Briese 1982).

Coexistence of competing granivorous ants may be aided by differing periods of foraging activity, both seasonally and diurnally. Whereas seed resources are available at all hours, offset periods of foraging minimizes aggressive encounters between species (Whitford et al. 1976; Chew 1977; Hansen 1978; Mackay 1981; Mehlhop & Scott 1983). Seasonal variations in foraging activity of granivorous desert ants have been described in numerous studies, and omnivorous ants may also temporarily partition resource use (Whitford et al. 1976; Whitford 1978; Hansen 1978; Anderson 1984).

Desert ants of North America appear to partition resources in a variety of ways. Harvester ants of the genus *Pogonomyrmex*, often the most abundant granivores, partition seeds on bases of size and nutritional quality (Chew 1977; Davidson 1977a, b; Hansen 1978; Whitford 1978; Chew & De Vita 1980). Worker body size probably sets the limits for preferential seed choice, with small ants (2 mm) foraging for small seeds and larger ants (7–10 mm) taking the larger ones. Nutritional quality also seems to be important in preferences for seeds in a given size class (Kelrick, MacMahon, Parmenter, & Sisson 1986).

Evidence for competition when seed resources are presumably limited comes from studies of spatial patterning of ant colonies. Territoriality is common in desert ants (Creighton 1966; De Vita 1979), and spacing of colonies increases with increasing aridity of the habitat (Davidson 1977b). Colonies of harvester ants have been shown to be overdispersed and interspecifically aggressive in both field and experimental studies in the Sonoran Desert (Ryti & Case 1984, 1986, 1988).

Ants of the Nevada Test Site and Rock Valley

Early ecological studies conducted at NTS identified 57 species of ants (Cole 1966). Greatest diversity occurred in pinyon (*Pinus*)–juniper (*Juniperus*) woodland communities of NTS, where 32 species were reported (Allred et al. 1963). The most widely distributed ants overall were *Myrmecocystus mexicanus*, *Pheidole bicarinata*, and *Pogonomyrmex californicus*. *Myrmecocystus* was active throughout the year in most plant communities. Important granivorous species of *Pogonomyrmex*,

Table 10.29. *Species of ants (Formicidae) collected in Rock Valley during 1971 and 1972 or from documented specimens in existing reference collections at NTS. Determinations were by Roy Snelling, Los Angeles County Museum of Natural History.*

Taxon	Feeding guild
Subfamily Dolichoderinae	
Conomyrma insana[a]	Predatory
Iridomyrmex	
pruinosus	Honeydew feeder
Subfamily Formicinae	
Camponotus hyattii	Honeydew feeder
Myrmecocystus koso[b]	Honeydew feeder
M. lugubris	Honeydew feeder
M. mexicanus	Honeydew feeder
M. placodos	Honeydew feeder
Subfamily Myrmicinae	
Aphenogaster	
megommatus	Honeydew feeder
Crematogaster depilis	Honeydew feeder
C. mutans	Honeydew feeder
C. nocturna	Honeydew feeder
Leptothorax nitens	Honeydew feeder
Pheidole bicarinata	
paiute	Granivorous
P. desertorum	Granivorous
Pogonomyrmex	
californicus	Granivorous
P. rugosus	Granivorous
Solenopsis molesta	Granivorous
Veromessor pergandei	Granivorous
V. smithi	Granivorous

[a]Synonymous with *Dorymyrmex pyramicus*
[b]Previous identified as *M. flaviceps* and *M. merdax*.

Pheidole, and *Veromessor* were generally present from early spring until fall.

Studies at Rock Valley in 1971 and 1972 identified 19 species of ants in 11 genera and 3 subfamilies of Formicidae (Table 10.29). These included the granivorous ants of *Pheidole*, *Pogonomyrmex*, and *Veromessor*, ants of *Crematogaster*, *Iridomyrmex*, and *Myrmecocystus* that feed on honeydew, and the predatory genus *Conomyrma*.

Relative seasonal abundance of ants was studied

Table 10.30. *Seasonal change in relative abundance (RA) and relative frequency (RF) of ant species at Rock Valley on four dates in 1972. All values are in percent.*

Species	March 31		June 9		July 24		October	
	RA	RF	RA	RF	RA	RF	RA	RF
Pheidole desertorum	–	–	52.8	41.7	56.8	50.7	96.5	68.8
Iridomyrmex pruinosus	47.8	13.7	38.5	38.5	35.3	16.5	0.5	
Conomyrma insana	26.0	4.5	4.0	2.3	3.9	2.0	1.7	
Myrmecocystus koso	4.4	0.2	2.3	1.3	3.9	2.4	1.3	0.6
Pheidole bicarinata	–	–	1.0	0.3	0.1		–	–
Leptothorax nitens	4.4	0.2	0.8	0.2	–	–	–	–
Pogonomyrmex californicus	–	–	0.3		–	–	–	–
Myrmecocystus lugubris	17.3	1.6	0.2		–	–	–	–
Veromessor smithi	–	–	0.1		–	–	–	–
Crematogaster depilis	–	–	–		–		–	
Individuals board^{-1}	1.10		65.1		109.0		13.2	
Number of boards	23		19		23		48	
Number of species	5		9		5		4	

in the southern half of Rock Valley plot 3. Sampling was done on four dates in 1972 using bait boards that were distributed in a grid (Turner 1973). The limited data from the sample suggested that greatest diversity of ants occurred in June, whereas greatest diversity of ants was observed six weeks later in July (Table 10.30). Only five species were present in March or October, but nine were present in June. Mean number of ants per board was only 1 in March, 65 in June, and 109 in July. The two most abundant ant species showed peak relative abundance in different months. *Pheidole desertorum* increased in relative importance from March to its highest abundance in October, whereas *Conomyrma insana* showed the opposite pattern, decreasing in relative abundance each month.

Four important granivorous ant species at Rock Valley showed a gradient in body size and hence appeared to minimize competitive interactions by gathering mean seed sizes in proportion to their body sizes. From largest to smallest, these species were *Pogonomyrmex rugosus*, *P. californicus*, *Pheidole desertorum*, and *P. bicarinata*. Honeydew feeders were also graded in size, in decreasing order *Myrmecocystus mexicanus*, *M. koso*, *M. lugubris*, *Crematogaster depilis*, and *Iridomyrmex*

pruinosus. No data were obtained to calculate ant biomass at the IBP validation site.

CREOSOTE BUSH AS AN ARTHROPOD HOST

Creosote bush (*Larrea tridentata*) has been studied intensively because this is a model shrub host for phytophagous insects across the warm desert regions on North America. Detailed studies from the IBP site in Silverbell, Arizona, in the Sonoran Desert, characterized the varied resources used for feeding, oviposition, and hiding by 22 species of phytophages on *L. tridentata* (Schultz, Otte, & Endes 1977). Included were 6 species of Orthoptera, 6 of Coleoptera, 4 of Heteroptera, 3 of Lepidoptera (larvae), and 3 Membracidae. All but four of these were obligatory feeders on *Larrea*, and of the 15 species that fed on leaves alone, 4 were specialists on young leaves, 6 were specialists on old leaves, and the remainder fed on both (Fig. 10.3). *Bootettix argentatus*, described earlier in this chapter as the dominant orthopteran on the Rock Valley plots, was also an important herbivore at Silverbell, and it restricted its feeding to young leaves. Seven species fed only on flowers, and only three consumed all types of aboveground organs.

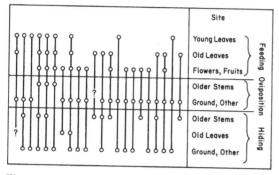

Figure 10.3 Feeding, oviposition, and concealment sites used by 22 species of phytophagous insects on *Larrea tridentata* at the IBP validation site, Silverbell, in the Sonoran Desert of Arizona. From Schultz et al. (1977).

Only three of the phytophagous insects on *L. tridentata* at Silverbell oviposited on old stems, whereas most of the others oviposited in other sites, particularly in soil (Schultz et al. 1977). Although young leaves were a very important food source, they were only rarely used as hiding substrates to avoid predators. Older stems were the single most common substrate for hiding from predators.

Predatory arthropods were also common on *L. tridentata* at Silverbell. Schultz et al. (1977) reported three predatory insects, two mantids and one reduviid bug, but 27 species of spiders in samples. Asilid flies, pentatomid bugs, and mirid bugs were occasionally present on the site.

Studies of Orthoptera

Ecological studies of two important orthopteran herbivores of *Larrea* have been published. Mispagel (1978) studied the ecology and bioenergetics of *B. argentatus* at NTS, with most data from Rock Valley. Using data on energy budgets for this species, he estimated that at Rock Valley from 1971 to 1973 *B. argentatus* consumed 0.42, 0.86, and 0.31 kg ha^{-1} yr^{-1}, respectively. These amounts were approximately 1.1, 2.0, and 0.8% of available leaf biomass for those three years.

The behavioral ecology and population biology of the common creosote bush herbivore *Ligurotettix coquillettii* was studied in Sonoran Desert localities (Greenfield, Shelley, & Downun 1987;

Greenfield, Shelley, & Gonzalez-Coloma 1989). Population densities of *L. coquillettii* varied greatly between shrubs, with quality of the leaf resource influenced positively by nitrogen content and negatively by resin content.

Other Rock Valley observations

During the IBP studies at Rock Valley, *Larrea* was used for sampling seasonal patterns of arthropod abundance, and data were collected from March through August, 1971–5. The seasonal course of population numbers of defoliating and sap-feeding

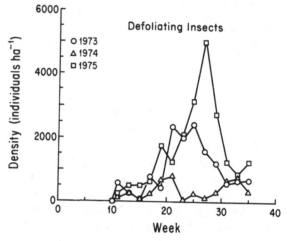

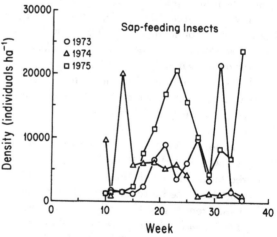

Figure 10.4 Seasonal course of population numbers for defoliating and sap-feeding insects on *Larrea tridentata* in Rock Valley from 1973 to 1975. Early and late season peaks in the latter peaks resulted from large influxes of Pseudococcidae.

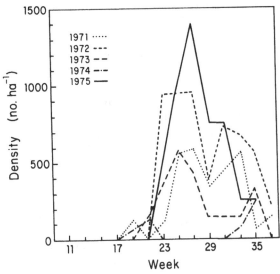

Figure 10.5 Estimated seasonal densities (no. ha^{-1}) of *Bootettix argentatus* on *Larrea tridentata* in Rock Valley from 1971 to 1975..

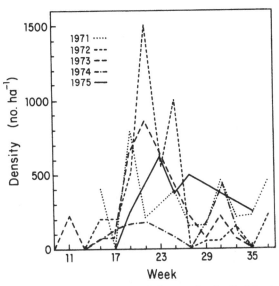

Figure 10.6 Estimated seasonal densities (no. ha^{-1}) of *Semiothisa larreana* on *Larrea tridentata* in Rock Valley from 1971 to 1975.

insects for 1973–5 are shown in Figure 10.4. Peak numbers of defoliating insects occurred in May and June (weeks 21–27) in these three years. Maximum population of defoliating insects, however, changed greatly between years. Peak numbers were only 804 individuals ha^{-1} in 1974 as compared with 5000 individuals ha^{-1} in 1975.

The two most significant defoliators on *Larrea* were the orthopteran *B. argentatus* and the lepidopteran larvae of *Semiothisa larreana*. From 1971 to 1973, *B. argentatus* showed a bimodal peak of abundance, having high population sizes from weeks 25 to 28 and again from weeks 31 to 35 (Fig. 10.5). This pattern was produced by the appearance first of nymphs followed in 6–8 weeks by the emergence of adults. During 1974 maximal densities of this species never exceeded 1700 individuals ha^{-1}, but in 1975 densities rose sharply to 10 000 individuals ha^{-1}. The beefly *Mythicomyia*, a major predator of the eggs of *Bootettix*, was an important regulator of the grasshopper population size.

Semiothisa larreana, a measuring worm, reached maximum population densities on *Larrea* a few weeks earlier than *Bootettix* (weeks 19–23; Fig. 10.6). Peak population densities of this herbi-

vore were greater than 1500 individuals ha^{-1} in 1971, but declined to much smaller populations in subsequent years. Parsitic eulophid and pteromelid wasps may have been responsible in regulating those population densities.

Sap-feeding arthropods on *Larrea* included treehoppers, leafhoppers, mealybugs, plant bugs, thrips, and mites. Treehoppers and mealybugs appeared to be ecologically most important at Rock Valley. Seasonal studies of treehoppers typically showed sharp increases in populations of immature stages from weeks 15 to 20, depending on the year. Population densities of adult treehoppers peaked from weeks 23 to 28, with typical maximum densities from 3000 to 4000 individuals ha^{-1} in 1971–3, but in 1974 densities were 10% of those densities and in 1975 there were nearly 7000 individuals ha^{-1} (Fig. 10.7). Three species, *Multareis cornutus*, *Centrodontus atlas*, and *Multareoides bifurcatus*, combined to form the great majority of treehoppers.

Mealybugs showed highly variable patterns of seasonal abundance on *L. tridentata*, with three or more individual peaks in population density per year. These sharp fluctuations may have

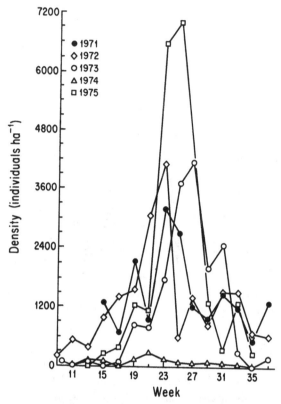

Figure 10.7 Estimated seasonal densities of adult treehoppers (Membracidae) on *Larrea tridentata* in Rock Valley from 1971 to 1975.

related to changes in soil moisture availability; unusually high densities of about 20 000 individuals ha^{-1} were observed in spring, 1974, at the time that soil moisture content had peaked and then started to decrease. Highest population density of mealybugs was recorded in late summer, 1975, when about 23 000 individuals ha^{-1} were present.

Although a detailed model, like that for Silverbell, was not published for creosote bush at Rock Valley, the data clearly showed that *Larrea* is a key host plant for arthropods in this Mojave desert scrub community.

SUMMARY

Arthropods specimens of Rock Valley included two major groups, those collected with soil samples and buried pitfall traps for soil phases of life cycles versus those sampled using vacuum procedures for aboveground phases. Ants, which daily can travel through both environments, were not effectively sampled by any of those methods and were censused in a general way with bait boards.

Of ground-dwelling arthropods that were sampled in Rock Valley, detritivores comprised 42% of total biomass, predators were second, and phytophages, with 8–18% were third. Tenebrionid beetles were the most abundant detritivores of soil and litter and produced 75% of total estimated biomass of that category, with about half of spring totals being contributed by *Eusattus dubius*. Other important detritivores were desert cockroaches (*Arenivaga erratica*), camel cricket (*Ceuthophilus fossor*), and the winter-active desert millipedes. Among phytophagous soil arthropods small weevils were especially abundant, and in 1976 densities of *Eucyllus vagans* reached 85 000 individuals ha^{-1}. Predatory soil arthropods, including scorpions, solpugids, spiders, some mites, and centipedes, were collected in pitfall traps, but were also using aboveground food resources. Of the soil predators scorpions comprised only about 4% of individuals, but greater than 40% of total biomass, with peaks in August and September, particularly due to fairly high numbers of small scorpions. Solpugids were most abundant in late spring and summer, when they comprised about half of total soil predatory arthropods. Population densities of the spider *Psilochorus utahensis* were estimated as 40 000 individuals ha^{-1}, but these small spiders contributed little to community biomass.

Aboveground collections at Rock Valley showed that the community was dominated by relatively few but large defoliating insects, and at the same time very abundant but very small sap-feeders. Lepidopteran larvae were very important defoliators, and in given years were especially destructive on selected shrub species. For example, in 1971 larvae of ermine moths reached peak populations of 15 000 individuals ha^{-1} on the host shrub *Krameria erecta* and high density on *Lycium andersonii*, while Coleophoridae had estimated densities on *Grayia spinosa* of 28 000 individuals ha^{-1}.

Leaf beetles and especially weevils were important defoliating insects. Chrysomelids were mostly host specific, with heaviest infestations on *Ambrosia dumosa* and *G. spinosa*. The small weevil *Smicronyx imbricata* was also abundant on *A. dumosa*, specifically floral tissues. Among orthopterans the leaf-eating grasshopper *Bootettix argentatus* achieved densities of 600–950 individuals ha^{-1} on its host *Larrea tridentata*. In contrast, thrips, which largely feed on flower nectar, reached exceedingly high densities on many shrubs during flowering periods, and they comprised one-fifth to more than one half of all sapfeeders in a given year. Mealybugs, for example, *Pseudococcus helianthi*, were common on shrubs, on some for more than four months per year. Leafhoppers and treehoppers were important sapfeeders on particular host shrubs. The most common predators in shrubs were spiders, beetles, hemipterans, neuropterans, and predatory thrips, but parasitoid Diptera and Hymenoptera were apparently very important predators for the community in how they regulated population sizes of several major defoliators.

Relatively little data were collected at Rock Valley on pollinators, ticks and mites, ants, and termites, which all play important roles in the structure of desert ecosystems. For none was abundance or biomass determined. Ants in particular have been shown to be important granivores, which compete at other desert sites with vertebrate granivores for soil seed reserves. Relative abundance of ants determined that ant samples peaked in July, but species abundance was highest in June. *Pheidole desertorum*, the most abundant ant granivore, had highest relative abundance in October samples, whereas *Conomyrma insana*, the common predatory ant there, was most abundant in spring.

Larrea tridentata was a host plant for important defoliators in Rock Valley, including the grasshopper *B. argentatus* and the geometrid moth larvae of *Semiothisa larreana*. Among sap-feeders, treehoppers and mealybugs were ecologically most important; in 1975 treehoppers reached peaked populations of 7000 individuals ha^{-1} and mealybugs reached 23 000 individuals ha^{-1}. Clearly, creosote bush is a key shrub species in Mojave desert scrub, supporting a number of host-specific arthropods as well as their predators.

11 Soil organisms and seed reserves

The soil environment contains a variety of organisms not treated in previous chapters that yet play major ecological roles in many processes of carbon and nutrient fluxes. These organisms include cryptogamic plants living on the soil surface, such as soil algae and cyanobacteria, lichens, and bryophytes. Although at Rock Valley they do not form a large cover or biomass, such cryptogams provide important nitrogen inputs to many other desert ecosystems. Within the soil matrix occur many types of organisms, including a microflora of bacteria, actinomycetes, and a variety of fungi, and a microfauna that is dominated by such microarthropods as collembolans and mites, nematodes, and protozoa, which have a significant influence on decomposition processes in soil. In addition, soil reserves of vascular plant seeds provide a critically important food resource for granivorous animals (Chapters 7 and 9) as well as a critical pool for future establishment of herbs (Chapter 5) and shrub species (Chapter 4) when favorable rains occur.

CRYPTOGAMIC PLANTS

Cryptogamic plants are generally much less abundant in dry desert ecosystems than in more mesic habitats. Nevertheless, even small biomasses of some of these groups may be ecologically important. This is particularly true for soil crusts and lichens with cyanobacteria (blue-green algae) symbionts, which are able to fix atmospheric nitrogen. General structure and function of soil crusts in the stability of soils from arid and semiarid lands has been reviewed in detail by West (1980).

Shields and Drouet (1962) described the general distribution of terrestrial algae and cyanobacteria of NTS from soil samples taken from 1957 to 1959 in Yucca Flat, Frenchman Flat, and Jackass Flats. In the samples these authors identified 12 species, 10 cyanobacteria and 2 green algae. They concluded that soil algae were best represented on fine-textured clay-loam soils adjacent to playas, and also described algal development as poorest on gently sloping terrain at intermediate elevations.

Rock Valley soil algae, lichens, and bryophytes

Soils of Rock Valley, which contain very little clay (Chapter 2), do not provide a suitable habitat for soil crusts to form, and as such these soils have low diversity and abundance of soil microbes. In Rock Valley overall algal biomass of soils has been estimated by Skujins (1977) using chlorophyll extracts from soil samples. This procedure led to an estimate of 33.4 kg ha^{-1} of algal biomass in soils of Rock Valley.

Twenty-one species of lichens were identified in transect studies conducted in Rock valley (Nash, White, & Marsh 1977; Nash & Moser 1982). Ten of these species were present on the bajada of the IBP validation site, and 16 species occurred on north and south rocky slopes above the site (Table 11.1). Dominant species on bajada transects was *Acarospora strigata*, comprising 78% of total lichen coverage of 1.23 m^2 ha^{-1}. Highest diversity and coverage of lichens was found on north-facing slopes, having 16 species and 13 m^2 ha^{-1} coverage,

Table 11.1. *Estimated cover (m² ha⁻¹) of lichens along three transects in Rock Valley. Data from Nash, White, and Marsh (1977).*

Species	Bajada	North-facing slope	South-facing slope
		Transects	
Acarospora fuscata	0.108	0.450	
A. strigata	0.965	4.900	0.058
Caloplaca amabilis		0.025	
Candelariella aurella	<0.001	0.050	
Collema coccophorum	0.015		
C. polycarpon	0.001		
Dermatocarpon lachneum		6.117	2.475
D. miniatum		0.025	
D. plumbeum		0.033	
Fulgensia desertorum		0.025	
Heppia lutosa		0.092	0.658
Lecanora calcarea	0.014	0.067	
L. cinerea	0.013	0.083	
Lecidea decipiens		0.392	
Peltula obscurans var. *deserticola*	0.001		
P. obscurans var. *hassei*	0.013		
P. polyspora		0.108	
Physica orbicularis		0.008	
Thyrea nigritella		0.342	
Toninia caeruleonigricans		0.158	
Unknown black crust	0.104		
Total cover	1.23	12.98	3.19

of which *Dermatocarpon lachneum* and *A. strigata* combined to form 85% of that cover. Only 3 species of lichens were encountered on south-facing slopes, and *D. lachneum* formed 78% of that cover. Estimated lichen biomass was 0.88 kg ha⁻¹ on the bajada, 3.7 kg ha⁻¹ on the south-facing slope, and a high of 12 kg ha⁻¹ on the north-facing slope.

As is common in most warm desert ecosystems, crustose and squamulose growth-forms of lichens predominated at Rock Valley (Table 11.2). Ninety percent of the lichen species found along bajada

transects occurred on rock substrates, and the remainder was present on soil. One-third of the species occurring on north-facing slopes were present on soil.

Bryophytes were neither diverse nor abundant in Rock valley. Nash et al. (1977) observed no mosses along two 1000-m transects laid out across the IBP validation site. Three species of bryophytes, *Crossidium aberrans*, *Tortula bistratosa*, and an unidentified species of *Grimmia* were sampled on north-slope transects above the site. Highest coverage, an estimated 5.48 m² ha⁻¹, was formed by *T. bistratosa*.

MICROORGANISMS AND INVERTEBRATES OF SOIL MATRIX

General features

The dynamics of soil microorganisms in desert ecosystems have not been widely investigated. In many respects, however, desert soils are interesting model systems for such research in that they contain all of the major interacting groups of microflora and microfauna, but lack the complexity and species diversity of soil taxa found in more mesic ecosystems.

Decomposition processing terrestrial ecosystems are strongly influenced by actions of invertebrates (Chapter 10) and microflora. Microfloral components of decomposer organisms include bacteria, actinomycetes, and fungi. These species secrete extracellular enzymes to promote decomposition of plant tissues. In most ecosystems soil fungi constitute the primary decomposer population for plant litter (Swift, Heal, & Anderson 1979), and fungal mycelia are well adapted for colonizing and permeating dead structures of recently abscised plant tissues. Bacterial decomposers are also important elements, but they are generally best adapted to acting on particulate detritus having a high surface/volume ratio. Freelacing actinomycetes form a characteristic element of soil microflora, but are generally lower in competitive ability than either soil fungi or bacteria (Goodfellow & Cross 1974).

Soil invertebrates generally fall into five func-

Table 11.2. *Relative growth-form distribution of warm desert lichen floras.*

	Number of species	Fruticose	Foliose	Crustose and squamulose
Algerian Sahara Desert (Faurel et al. 1953)	114	0	2	98
Negev Desert, Israel (Galun 1970)	44	5	2	93
Chihuahuan Desert – Jornada Range (Nash et al 1977)	48	0	40	60
Sonoran Desert – Silverbell (Nash, et al. 1977)	21	0	19	81
Sonoran Desert – Maricopa County (Nash et al. 1977)	78	0	28	72
Mojave Desert – Rock Valley (Nash et al. 1977)	21	0	19	81

tional groups: annelids, particularly earthworms in families Lumbricidae and Enchytraeidae; macroarthropods, for example, isopods, millipedes, and insects; microarthropods, especially mites and collembolans; nematodes; and protozoa. These soil animals have three important effects on the decomposition process. First, they have a physical effect in redistributing organic matter. During this process large litter components are fragmented, thereby exposing greater surface area for colonization and attack by microorganisms. This redistribution may also involve movement of organic matter into deeper soil horizons, most notably by earthworms and microarthropods. A second significant influence results from the chemical effect of concentration of certain elements within animal bodies, thereby accelerating nutrient cycling processes. For nitrogen, a critical element in many ecosystems, action of soil fauna tends to promote net mineralization of nitrogen rather than immobilization of nitrogen (Anderson, Gleman, & Cole 1981). One more role of soil fauna in decomposition lies in its biological effect on regulating levels of microbial activity. Protozoa, nematodes, and collembolans feed on microorganisms and thus regulate those population sizes, composition, and activities (Anderson et al. 1981).

Desert soil biotas

Whitford and coworkers have produced a series of important studies of soil decomposers and decomposition in a creosote bush (*Larrea tridentata*)–grassland ecosystem at the Jornada Range Experiment Station in southern New Mexico (Santos, DePree, & Whitford 1978; Santos & Whitford 1981; Santos, Phillips, & Whitford, 1981; Whitford, Freckman, Elkins, Parker, Permalee, Phillips, & Tucker 1981; Whitford, Repass, Parker, & Elkins 1982; Santos, Elkins, Steinberger, & Whitford 1984). In a long-term study of decomposition of buried plant litter and roots, they found that bacteria and yeasts were the initial decomposers and that fungi became more important only at an intermediate stage during the decomposition process. Psocopterans and collembolans also entered the system, feeding either directly on plant litter or on decomposer fungi. Soil microflora entered the trophic system at all stages. Protozoa and nematodes grazed on bacteria and yeasts and were in turn fed upon by tydeid mites. Fungi were consumed by tarsonemid mites and fungiphagous nematodes, which themselves were preyed on by gamasina mites. These trophic relationships are shown in Figure 11.1.

At Jornada researchers found a similar pattern of trophic relationships for soil fauna in surface litter. The most important difference found was the significant role that was played by oribatid mites, which fragmented organic debris at the soil surface. Subterranean termites had an influential role on both soil nitrogen and water availability (Parker, Fowler, Ettershank, & Whitford 1982; Elkins, Sabol, Ward, & Whitford 1984; Gutier-

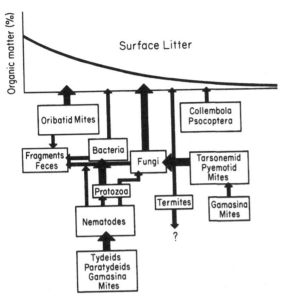

Figure 11.1 Hypothetical trophic relationships of soil fauna and microorganisms in the decomposition of surface litter in the Chihuahuan Desert. Width of the arrows represents relative quantity of organic matter processed by the taxa in each box. From Whitford (1986).

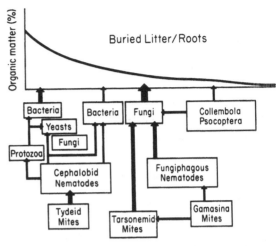

Figure 11.2 Hypothetical trophic relationships of soil fauna in buried litter of the Chihuahuan Desert. Width of the arrows represents the relative quantity of organic matter or biomass consumed by the taxa in each box. From Whitford (1986).

rez & Whitford 1989). Termites also played a role in fragmenting and moving surface litter, but their influence was seasonal and site-specific. Much action of fungi and bacteria was on fragmented organic matter and feces rather than on surface litter (Fig. 11.2). Patterns of predation of soil fauna were similar to those characterizing buried litter.

Field methods in Rock Valley

The soil biotas of Rock Valley and other portions of NTS have not been studied in great detail. During the IBP studies, however, diversity of soil flora and fauna was surveyed to provide comparative data on the biological groups present.

Microflora sampling

In Rock Valley soil microfloras were studied using soil samples that were collected over seasonal cycles between summer, 1971, and December, 1975. These samples were taken beneath and near

shrubs, and extracts from those samples were used for plate counts of microorganisms (Turner 1972; Skujins 1976; Vollmer & Bamberg 1977). In June and September, 1971, duplicate soil samples were collected from around roots of three shrub species, *Ceratoides lanata*, *Krameria erecta*, and *Lycium andersonii*) as well as from intershrub spaces (Turner 1972). Extracts were cultured on peptone glucose and Rose Bengal agars treated with aureomycin for fungi and on soil extract and sodium albuminate agars for actinomycetes and bacteria.

In spring, 1973, samples were taken from soil adjoining *L. andersonii* from each of nine locations. The 9 sampling sites were from 3 positions, at the shrub base, beneath the outer margin of the canopy, and at a distance of two shrub radii beyond the canopy margin in intershrub space, each at depths of 0–0.10, 0.11–0.20, and 0.21–0.30 m. Plate counts were made for each of three taxa, actinomycetes, bacteria, and fungi. Additionally, soil samples were collected from 9 locations adjoining 4 shrub species from May 7 to June 24 and October 1 to November 25, 1974. These were analyzed using sodium caseinate agar for culturing actinomycetes, soil extract agar with PCNB for bacteria, and Rose Bengal agar with streptomycin for fungi.

In March, July, and December, 1975, soil samples were collected from intershrub spaces in Rock Valley. Most of these were from the upper 0.05 m of the soil profile, but some cores were taken at greater depths. Dilution extracts were plated onto a variety of culture media, and counts of microorganisms were made as in earlier studies.

Nematode sampling

Soil nematodes were extracted from 200-cm³ soil samples that were collected three to four times per month from March through December, 1973. These samples were taken from depths of 0–0.10, 0.11–0.20, and 0.21–0.30 m from beneath the center of shrub canopies for *Ambrosia dumosa, K. erecta, L. tridentata,* and *L. andersonii* and at the same depths at the canopy margin and at three shrub radii from the canopy margin in intershrub space. In all, 2880 samples were analyzed. Nematodes were recovered from soil by a sugar flotation sieving method adapted for use in desert soils (Freckman, Mankau, & Ferris 1975). Nematodes were counted, identified, and categorized into four trophic groups: fungal feeders, plant feeders, bacterial feeders, and omnivore-predators (Freckman et al. 1975). Nematode production (*P*) was estimated using methods of McNeill and Lawton (1970).

Microarthropod sampling

The abundance and diversity of soil microarthropods at Rock Valley were studied in 1974 using extractions of 500-cm³ soil samples (Edney, McBrayer, Franco, & Phillips 1974). Samples were collected weekly at 9 positions (3 canopy positions at 3 depths each), as with microflora and nematode studies, for 4 shrub species, *A. dumosa, K. erecta, L. tridentata,* and *L. andersonii.* Soil samples were extracted for three days in a refrigerated room to increase the temperature gradient. Organisms were sorted and then stored in an aqueous solution of 70% ethanol and 5% glycerine.

Soil microflora of Rock Valley

For 1971, mean numbers of colonies of fungi, actinomycetes, and bacteria that were obtained from cultured soils of Rock Valley are given in Table 11.3. Numbers of cultured colonies per gram of dry soil weight were 15–36 times higher in soils adjacent to plant roots as compared with those adjoining shrubs. Mean number of bacteria-actinomycete colonies was 140–1200 times the number of fungal colonies. Counts on different culture media were generally similar. Plate counts of cultures from soils adjoining shrubs were similar regardless of collection date or shrub species, whereas counts of cultures from soils around plant roots were more variable. Some of the duplicate samples differed greatly, and there were also apparent differences correlated with season and shrub species. Most isolated fungi were identified as species of *Aspergillus, Mucor,* and *Penicillium.* Other fungal genera were *Cephalosporium* and *Tichoderma,* and actinomycete species belonged predominantly to the genus *Streptomyces.*

Soil microflora cultured from the 1973 soils adjoining *L. andersonii* were generally similar to those found earlier. As in 1971, species of *Penicillium* and *Aspergillus* were the predominant molds, and more than 95% of isolated actinomycetes belonged to the genus *Streptomyces.* More detailed taxonomic descriptions of the microflora were given by Vollmer & Bamberg (1977). Abundance of microorganisms, as inferred from plate counts from 1971, 1973, and 1974, was reasonably consistent with samples taken in different seasons or years as well as from beneath any of the four common shrub species.

More detailed information on relative abundances of microorganisms at different depths and distances from shrub base is given in Table 11.4. Here the values for 0.10 m depths are mean values based on counts of colonies that were cultured from soil samples from the bases of shrubs, at the edges of shrub canopies, and in intershrub spaces. Values for sample positions are means based on samples at the three soil depths. The majority of organisms in these soil samples were either acti-

Table 11.3. *Mean numbers (colonies g^{-1} dry weight of soil) of fungi, bacteria, and actinomycetes cultured from soils close to plant roots and soils adjoining Rock Valley shrubs (1971). Ranges are given in parentheses.*

Source of sample	Type of microorganism	Shrub species		
		Ceratoides lanata	*Krameria erecta*	*Lycium andersonii*
Close to roots	Fungi ($\times 10^3$)	328 (96–659)	1232 (545–2194)	1233 (263–2721)
	Bacteria and actinomycetes ($\times 10^6$)	398 (212–545)	243 (149–366)	170 (110–225)
Soils adjoining	Fungi ($\times 10^3$)	22 (8–39)	43 (25–53)	37 (20–63)
	Bacteria and actinomycetes ($\times 10^3$)	11 (5–16)	10 (9–11)	9 (6–11)

Table 11.4. *Mean numbers of microorganisms (colonies g^{-1} dry weight of soil) cultured from nonrhizosphere soils taken from near shrubs in Rock Valley. Numbers of fungal colonies are given in thousands, of bacterial colonies in millions.*

Group	Shrub species	Soil date	Sample depth, m			Sample position		
			0.0–0.10	0.11–0.20	0.21–0.30	Base	Canopy edge	Intershrub space
Fungi	*Lycium andersonii*	Spring, 1973	0.21	0.20	0.13	0.31	0.16	0.07
	Larrea tridentata	Summer, 1974	0.21	0.15	0.12	0.21	0.18	0.10
		Fall, 1974	0.23	0.19	0.18	0.24	0.22	0.14
Actinomycetes	*L. andersonii*	Spring, 1973	0.041	0.052	0.047	0.058	0.046	0.035
	L. tridentata	Summer, 1974	0.027	0.027	0.027	0.030	0.031	0.020
		Fall, 1974	0.040	0.045	0.038	0.051	0.036	0.035
Bacteria	*L. andersonii*	Spring, 1973	0.024	0.029	0.017	0.036	0.020	0.014
	L. tridentata	Summer, 1974	0.039	0.019	0.021	0.033	0.029	0.016
		Fall, 1974	0.047	0.042	0.040	0.052	0.041	0.036

nomycetes or other bacteria, whereas fungi composed only a fraction of a percent of all colonies counted. Actinomycetes comprised 68% of all bacterial colonies that were cultured from extractions of soil around *L. andersonii* and about 50% of the colonies from soil samples around *Larrea*. All cultured organisms were most abundant immediately adjacent to shrubs in the top 0.20 m of soil; numbers of microorganism colonies decreased in samples taken at greater distances from the shrub bases, and numbers of fungi and bacteria were conspicuously lower in samples from intershrub spaces. Fungi and bacteria were generally more abundant in samples taken from the upper soil layers than those from the 0.21–0.30 m depth, but actinomycetes were equally well represented at all depths. Vollmer and Bamberg (1977) attributed differences in composition of the microbial populations to substrate availability and utilization and to interspecific interactions. As soils became drier and warmer in spring, 1973, total number of colony counts decreased (Vollmer & Bamberg 1977). Most of this decrease was observed in counts of actinomycetes and bacteria, whereas fungal counts remained relatively unchanged.

Mean numbers of colonies of fungi and bacteria that were cultured from soils of Rock Valley for three seasons during 1975 were reported by Skujins (1977), who used a range of substrates to separate specific groups of bacteria. No consistent pattern of change with season or soil depth was evident, and numbers of fungal and total bacterial colonies found were lower than those found for intershrub spaces in earlier studies (Table 11.4). For example, Durrell and Shields (1960) reported diversity of soil fungi from Frenchman Flat and Yucca Flat at NTS of 41 taxa, these from a range of 550 soil samples collected at depths of 0.075–0.15 m and cultured on Rose Bengal agar containing streptomycin.

In all studies done at Rock Valley, actinomycetes and other bacteria far outnumbered fungi in terms of colony numbers. These results are consistent with those reported for other desert soils (Fuller 1974); however, plate count techniques alone do not necessarily reflect the true abundances of all microorganisms. Naturally, only those organisms that grow on the media used were expressed, and methods used tended to select for fungi that sporulated readily and against those existing in the fragile mycelial stage. Even though Rock Valley data showed few fungi, other workers have reported elsewhere that fungal biomass often exceeds that of other microbes in other desert ecosystems (Went & Stark 1968; Fuller 1974).

Nematodes of Rock Valley

Abundance and distribution of soil nematodes in Rock Valley have been described by Freckman (1978; Freckman & Mankau 1977, 1979). Nematode numbers decreased with soil depth and were highest near or beneath shrubs, but densities were not significantly influenced by any particular species of shrub. Annual mean nematode density (to

0.30 m), corrected for extraction efficiency, was 1.24×10^6 individuals m^{-2} (range $0.6–1.7 \times 10^6$ individuals m^{-2}). Estimated dry biomass of nematodes was 70 mg m^{-2}.

Microbial feeders, mainly Cephalobidae (*Acrobeles complexus*, *Elaphonema* sp., and *Leptonchus* sp.), were the most numerous nematodes, having a mean of 2648 individuals per 500 cm^3 soil (37%) in upper soil surfaces beneath a shrub canopy. Population densities of these microbial feeders decreased sharply both with soil depth and distance outward into intershrub space (Fig. 11.3). The same general pattern of abundance was present in omnivore-predator nematodes, including *Eudorylaimus monohystera*, *Eudorylaimus* sp., and *Pungentus* sp., and fungal feeders such as *Aphelenchus avenae*, *Aphelenchoides* sp., and *Ditylenchus* sp. Plant parasites *Tylenchorhynchus* spp. and *Tylenchida*, however, retained similar densities to 0.20 m soil depths (Fig. 11.3).

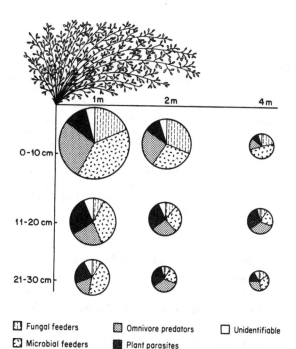

Figure 11.3 Spatial distribution of nematode trophic groups associated with desert shrubs in Rock Valley. Values are number of nematodes per 500 cm^2 soil averaged for four shrub species. From Freckman and Mankau (1977).

In 1974 seasonal population densities of soil nematodes declined beginning in April as soil temperatures increased and soil moisture availability decreased, and they fluctuated at lower densities until more favorable conditions returned in October (Fig. 11.4). Generally, nematodes remained metabolically inactive in Rock Valley soils wherever soil moisture content was below 2.7%. Moderately heavy rainfall in mid-July, 1974, increased soil moisture and led to a sharp increase in population densities of all four trophic groups of nematodes, thereby suggesting that moisture availability is much more critical than soil temperature in controlling nematode populations.

Table 11.5. *Representative summer population densities of nematodes beneath the canopies of four shrub species in Rock Valley. Data from Freckman and Mankau (1977).*

	Depth, m		
Shrub species	0–0.0	0.11–0.20	0.21–0.30
Ambrosia dumosa	5371	1720	775
Krameria erecta	4497	1985	865
Larrea tridentata	6317	2197	883
Lycium andersonii	5810	2535	1422

Total densities of nematode trophic groups varied relatively little between the shrub species that were studied (Table 11.5). Whereas densities of omnivore-predators were not significantly different between species, densities of microbial feeders and fungivores were slightly higher beneath *Ambrosia dumosa* and *Krameria erecta* than other shrubs. Plant parasites were more numerous beneath *Larrea* and *Lycium andersonii*.

Although nematodes are extremely small – requiring two to four million individuals to yield a gram of dry weight – their abundance is great enough to make them a significant component of soil biomass. Calculations for summer samples of Rock Valley suggested a range of biomass from about 4 mg l^{-1} soil in surface soils beneath plant canopies to about 0.2 mg l^{-1} for soil at 0.20–0.37 m depth in intershrub spaces. Overall, nematode biomass was estimated to be about 250 g ha^{-1}, a level 2.5–5 times the biomass of large microarthropods and almost equalling the biomass of all soil arthropods.

Monthly respiration estimates for nematodes were made for soils in which moisture level was high enough to permit metabolic activity. Annual cumulative respiration was estimated as 51 kJ m^{-2}. Bacterial feeders and omnivore-predators contributed 72% of annual energy expenditure for respiration. Estimated production was 26.8 kJ m^{-2} yr^{-1}, and, assuming an assimilation efficiency of 50% (Kitazawa 1967), total consumption by all nematodes may be estimated as 155.2 kJ m^{-2} yr^{-1}.

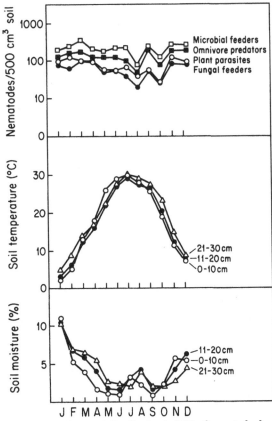

Figure 11.4 Seasonal distribution in 1974 of nematode density for four trophic groups in relation to mean monthly soil temperature and moisture. From Freckman and Mankau (1977).

Freckman (1978) originally reported annual respiration of nematodes in Rock Valley to be 4.84 kJ m^{-2} and estimated annual production as 0.96 kJ m^{-2} using a model developed by Engelmann (1966). In reviewing this work, Turner and Chew (1981) recalculated production using the reported estimate of respiration and the McNeill and Lawton (1970) equation relating production to respiration. The revised annual production estimate was 3.81 kJ m^{-2}. Older estimates differ from present values partly because of recent revisions of the model for estimating nematode densities and partly because the respiration value reported by Freckman (1978) was computed incorrectly.

Soil microarthropods of Rock Valley

Using extracts of soil samples from Tullgren funnels, a wide diversity of both macro- and microarthropods were collected in Rock Valley. Macroarthropods encountered in those samples included many of the groups of soil insects that were already discussed and listed in Chapter 10 and other macroarthropods, such as centipedes, isopods, and a variety of macroarachnids. A taxonomic listing of microarthropod samples is presented in Table 11.6, which includes Collembola, Psocoptera, Prostigmata (9 families), Astigmata (1 family), and Cryptostigmata (7 families).

As with nematodes, microarthropods were more abundant in upper than in lower soil horizons, and beneath a shrub canopy as compared with intershrub spaces (Table 11.7). However, no significant difference in total numbers of soil microarthropods was found between the four shrub species in the study. During 1974 microarthropod abundance increased from early February until early April, at a time when soil moisture was relatively high and mean soil temperatures were increasing. As soil moisture decreased in late spring and soil temperatures increased, populations of microarthropods declined. Estimated microarthropods in the upper 0.30 m of open soil ranged from about 2000 individuals m^{-2} in February to a peak of about 3450 individuals m^{-2} in March and April. Local densities at the base of shrubs, where population size was generally lar-

Table 11.6. *Taxonomic list of Rock Valley micro-arthropods extracted from soil samples with Tullgren funnels.*

Insecta	
Collembola	
Entomobryidae	
Sminthuridae	
Onychiuridae	
Pocturidae	
Psocaptera	
Arachnida	
Acarina	
Prostigmata	
Bdellidae	– *Spinobdella*
Cunaxidae	– *Cunaxa* sp.
	– *Cunaxoides* sp.
Caligonellidae	– *Molothiognathus* sp.
	– *Neothrognathus* sp.
Erythraeidae	– *Hauptmannia* sp.
Linotetranidae	– *Linotetrans* sp.
Nanorchestridae	– *Spelorchestes* sp.
Neophyllobiidae	– *Neophyllobius* sp.
Caeculidae	
Teneriffidae	– *Tarsolarkus* sp.
Trombidiidae	– *Allothrombium* sp.
Tuckerllidae	
Tydeidae	
Astigmata	
Acaridae	
Cryptostigmata	
Belbidae	– *Belba* sp.
Cosmocthoniidae	– *Trichthonius* sp.
Ctenacaridae	– *Aphelacarus acarinus*
Eremaeidae	– *Eremaeus* sp.
Gymnodamaeidae	– *Joshuella striata*
Oribatullidae	– *Multoribates* sp.
Passalozetidae	– *Passalozetes*

gest, ranged from 6680 individuals m^{-2} in January to nearly 18 000 individuals m^{-2} in March (Table 11.7). Nonpredatory prostigmate mites were generally the most abundant group of soil microarthropods, in spring months outnumbering all other mites and collembolans combined (Table 11.8).

SOIL SEED RESERVES

The characteristic abundance and distribution of both perennial and annual plants is a fundamental

Table 11.7. *Mean absolute densities of soil microarthropods collected at three radial positions around shrubs in Rock Valley. Densities (individuals m^{-2}) are averaged to represent three depths of soil sample collections at each position during 1974. Standard deviations are shown in parentheses. Intershrub space samples were collected at three canopy radii from the shrub center. Data from Edney et al. (1975).*

Month	Beneath canopy	Edge of canopy	Intershrub space
January	6680 (1560)	3590 (1030)	1670 (390)
February	7070 (2270)	2660 (1210)	1990 (860)
March	17 950 (8710)	6510 (2940)	3450 (1680)
April	7250 (2390)	4790 (3050)	3440 (1840)
May	6080 (1440)	3460 (680)	2400 (510)
June	4600 (560)	2780 (990)	1930 (950)

Table 11.8. *Mean monthly densities for major soil microarthropods in Rock Valley. Densities have been averaged for samples from all depths and canopy positions for 1974 samples and expressed as individuals m^{-2} of soil surface; standard deviations are shown in parentheses. Data from Edney et al. (1975).*

Month	Density, individuals m^{-2}		Nonpredatory	Predatory
	Collembola	Cryptostigmata	Prostigmata	Prostigmata
January	1043 (775)	470 (118)	573 (275)	108 (39)
February	260 (99)	675 (240)	692 (242)	214 (74)
March	166 (73)	941 (435)	2508 (1132)	411 (197)
April	28 (14)	578 (410)	2378 (1746)	353 (231)
May	10 (3)	400 (92)	1163 (248)	394 (86)

trait of Mojave Desert communities at NTS, as discussed in Chapters 4 and 5. The dynamics of these plant communities are a function of reserves of viable seeds in the soil, where they survive for long periods before germinating under favorable conditions for growth and establishment. As already discussed, rainfall is a key factor influencing plant productivity and reproductive allocation and the replenishing of these soil seed reserves. Much less is known, however, about either the spatial distribution and density of such seed reserves or the biological factors that influence seed populations.

Uses of the seed reserve

Vascular plant seed pools provide the primary resource for many animal populations, especially granivorous heteromyid rodents, seed-eating birds, and harvester ants (Brown, Reichman, & Davidson 1979). The relative importance of individual groups of granivorous animals has been shown to be a function of both the availability of the seed resource and recent history of resource availability (Brown & Lieberman 1973; Reichman 1975, 1979; Ludwig & Whitford 1981). Whitford (1976) showed that in Chihuahuan desert heteromyid rodents, which are capable of maintaining water balance living on a seed diet alone, were much less subject to year-to-year population variations than were cricetid rodents, which require succulent vegetation in their diets. Cricetid rodents may be important granivores in years with high productivity of plant growth and abundant seed resources. Granivorous ants (Chapter 10) are also important resident granivores in North America deserts; these harvester ants include members of the genera *Pheidole*, *Pogonomyrmex*, *Novomessor*, *Solenopsis*, and *Veromessor*. Resident granivorous birds are present in many desert areas, and these species behave much like heteromyid rodent populations by maintaining relatively stable numbers of adults from year-to-year despite drought conditions (Ludwig & Whitford 1981). When rainfall in the desert produces high levels of plant productivity and abundant soil seed reserves in excess of amounts that can be readily consumed by resident granivores, migratory seed-eating birds may become important components of the system and heavily exploit seed reserves during the nongrowing season (Raitt & Pimm 1976). Cody (1971, 1974) has developed optimization models to explain flocking behavior of granivorous finches in the Mojave Desert and resource use by these birds.

Sampling seed reserves in Rock Valley soils

Two groups of IBP studies were conducted in Rock Valley to provide a better understanding of spatial distribution of seed reserves in the soil and dynamics of these reserves. Because these studies required excavation of soil profiles, samples were taken in plots A and C, not on the validation site. In the first group of studies (1971), spatial patterns of seed distribution were analyzed by examining gradients from shrub bases into intershrub spaces and with increasing depth in the soil profile (Childs & Goodall 1973). A second set of investigations, based on samples collected from 1972 to 1974, focused on interrelationships between soil seed reserves at Rock Valley and population dynamics of heteromyid rodents (Nelson & Chew 1977). In 1971 soil and litter samples were collected for seed surveys from a variety of micro-

sites. Distribution of seeds was studied at soil depths of 0–0.01, 0.01–0.02, 0.02–0.05, and 0.05–0.10 m. For each soil sample, 100 g of soil was collected, and surface litter was sampled for 100-cm³ quantities. Soil was collected from beneath each of the six most dominant shrub species (Chapter 4) and from intershrub spaces. Laboratory techniques for separating seeds from soil samples involved first soaking samples in a calgon solution to disperse aggregates. Seeds were then harvested in two successive flotations using a potassium carbonate solution (Goodall, Childs, & Wiebe 1972). Selected additional flotations in zinc chloride solution (specific gravity 1.9) did not increase seed yield. Tests of efficiency for seed recovery demonstrated yields above 90%.

Position of seed reserves beneath the shrub canopy was investigated by sampling four radial distances at 0–0.1, 0.2–0.3, 0.4–0.5, and 0.6–0.7 m from the center of the shrub canopy. For each sequence of samples at these distances estimated seed density was ranked, and probability of each possible ranking was calculated on the assumption of random distribution. These probabilities (P_i), derived from each sampling sequence, were then combined using chi-squared (χ^2) estimate with n degrees of freedom on the null hypothesis of no trend:

$$\chi^2 = -2^n \ln P_i$$

From among the samples collected, 31 species or species groups were identified and an additional four species were found but not identified. Included were 4 shrub species and 31 herbaceous taxa overall. Mean dry weight of individual seeds was measured for 31 taxa, although sample sizes on some of these were small.

For studies on the dynamics of seed reserves, soil samples were collected in plots A and C in February, June, and October, 1972, these dates representing sample times before, during, and after periods of activity by *Perognathus formosus*, the most important heteromyid granivore (Nelson & Chew 1977). Samples were obtained in February from intershrub spaces and beneath canopies of five shrub species, *Ambrosia dumosa*, *Larrea tridentata*, *Lycium andersonii*, *Ephedra*

Table 11.9. *Proportion of seeds (%) at different depths in soils beneath shrub canopies in Rock Valley. Data from Childs and Goodall (1973).*

Species	Percentage of total seeds at:				
	Surface	0–0.01 m	0.01–0.02 m	0.02–0.05 m	0.05–0.10 m
Astragalus lentiginosus	0	22	28	33	17
Chaenactis carphoclinia	24	56	16	0	4
Cryptantha circumscissa	40	43	12	5	0
C. recurvata	56	19	13	12	0
Descurainia pinnata	12	55	23	9	1
Larrea tridentata	68	22	4	3	3
Lycium andersonii	49	44	2	5	0
Mentzelia obscura	0	13	69	6	12
Oryzopsis hymenoides	3	97	0	0	0
Pectocarya spp.	48	48	0	4	0
Phacelia vallis-mortae	18	53	29	0	0
Streptanthella longirostris	43	40	9	8	0
Vulpia octoflora	11	56	20	9	4
Unknown	10	58	20	15	0
All species	29	46	14	8	3

nevadensis, and *Krameria erecta*. Only soil collections beneath the first 3 species were taken on the last 2 dates. Individual soil samples were collected for a 0.20 × 0.25 m area (500 cm^3) and included the surface litter and upper 0.02 m of mineral soil from one position beneath the canopy of each shrub. Because it was difficult to position the metal sampler entirely between the canopies of smaller shrubs, the procedure was changed for June and October to include a pooled collection of four cylindrical samples totalling 497 cm^3 beneath each shrub. Additional samples were taken in October, 1973, and from intershrub spaces only in October, 1974. Methods for treating soil samples and extracting seeds were very similar to those described above and are presented in detail by Nelson and Chew (1977). Their procedures were determined to have an efficiency of recovery of 86%, with equal efficiencies for both large and small seeds.

Depth distribution of seeds

Greatest overall seed densities at the plots in Rock Valley occurred in the uppermost soil, with 46% of total seeds (Table 11.9). Soil surface contained 29% of total soil seed pools, whereas seed densities decreased sharply at soil depths greater than 0.01 m. Among 13 common taxa, only two shrub species, *Larrea tridentata* and *Lycium andersonii*, and the annual *Cryptantha recurvata* had greater seed densities on the surface of soil (Childs & Goodall 1973). Two common herbaceous species were unusual for having relatively deep seeds; *Astragalus lentiginosus* had 50% of its seeds below 0.02 m soil depth and 17% below 0.05 m, and *Mentzelia obscura* had 12% of its seeds below 0.05 m (Table 11.9). Where litter was present, most species had more seeds in the surface soil than in litter. However, *Larrea* was a major exception with a large proportion of its large seeds

Table 11.10. *Estimates of seed densities (seeds m^{-2}) beneath different shrub species at Rock Valley. Data from Childs and Goodall (1973).*

	Between shrubs	*Ephedra nevadensis*	*Lycium andersonii*	*Lycium pallidum*	*Ambrosia dumosa*	*Krameria erecta*	*Larrea tridentata*
Canopy species:							
Number of canopy							
individuals:	5	1	3	4	4	2	5
Number of 0.01 m^2							
samples:	5	4	9	12	12	5	19
Seed species							
Chaenaetis carphoclinia	0	0	0	70	30	20	250
Cryptantha circumscissa	0	40	0	0	0	30	270
Descurainia pinnata	30	0	520	390	110	30	430
Larrea tridentata	0	0	0	0	150	0	650
Lycium andersonii	0	0	670	250	30	0	20
Oryzopsis hymenoides	0	0	0	0	110	260	120
Streptanthella							
longitrostris	0	0	0	20	10	100	250
Vulpia octoflora	30	180	360	160	300	1590	690
Other species	40	50	320	310	170	1280	210
Total	100	270	1870	1200	1210	3310	2890

(actually segments of a schizogenous fruit) remaining in the litter.

Seed densities relative to shrub canopies

Seed densities beneath shrub canopies varied considerably among individual shrub species (Childs & Goodall 1973). Highest densities were found beneath *K. erecta* (fruits) and *L. tridentata*, with 3310 and 2890 units m^{-2}, respectively (Table 11.10). Seed densities beneath *L. andersonii*, *L. pallidum*, and *A. dumosa* were intermediate, with 1200–1870 seeds m^{-2}. Limited sampling beneath the canopy of *E. nevadensis* produced 270 seeds m^{-2}. Densities for intershrub spaces was only about 100 seeds m^{-2}. By comparison, studies on seed distribution in the Sonoran Desert of Arizona have reported 7000–15 000 seeds m^{-2} beneath shrubs and 4000–5625 seeds m^{-2} from open areas (Reichman 1984; Price & Reichman 1987).

Total seed populations had patchy distributions beneath shrub canopies. With *Larrea tridentata* and *Lycium pallidum* there was a statistically sig-

nificant tendency for seeds to be concentrated adjacent to shrub bases, but that pattern was not found for other species (Childs & Goodall 1973; Table 11.11). Seed densities near the center of *Larrea* canopies were twice the levels found at 0.6–0.7 m from that center. Beneath *Krameria*, however, highest total seed densities occurred at the canopy periphery.

Based on calculations by Nelson and Chew (1977), mean seed densities were notably variable between sample plots at Rock Valley and between years. Highest total seed density was 6222 seeds m^{-2} in June, 1972, with significantly lower densities of 3042 and 2814 seeds m^{-2}, respectively, for samples in February and October beneath canopies of three shrub species (Table 11.12). Seed densities beneath shrub canopies were ten times greater than those found in intershrub spaces, but the same significant seasonal change in abundance was present (Table 11.12).

Soil seed densities also varied beneath individual shrub canopies (Nelson & Chew 1977). Total seed density was significantly lower beneath *A.*

Table 11.11. *Mean seed densities (seeds m^{-2}) at radial distances from the center of the canopies of six shrub species at Rock Valley in November, 1971. Data from Childs and Goodall (1973).*

Shrub species	0–0.10 m		0.20–0.30 m		0.40–0.50 m		0.60–0.70 m	
	n	Density, seeds m^{-2}	n	Density. seeds m^{-2}	n	Density, seeds m^{-2}	n	Density, seeds m^{-2}
Larrea tridentata	6	3867	5	4164	4	3163	4	1945
Ambrosia dumosa	4	2293	4	318	3	1243	1	940
Lycium andersonii	3	1987	3	2357	2	1063	1	790
Krameria erecta	2	1215	1	1580	1	5310	1	7620
Ephedra nevadensis	1	460	1	0	1	0	1	630
Lycium pallidum	4	2495	3	787	4	515	1	0

Table 11.12. *Mean densities and ranges of densities (95% confidence limit) encountered of seeds collected beneath shrub canopies and in intershrub space on three dates in 1972 at Rock Valley. Data from Nelson and Chew (1977).*

Month	Number of samples	Number of shrub species	Mean seed density, number m^{-2}	95% confidence limit for seed density, number m^{-2}
Beneath shrub canopies				
February	30	5	2435	1740–3246
February	18	3	3042	1980–4331
June	30	3	6222	5019–7555
October	30	3	2814	2248–3444
Intershrub spaces				
February	34	–	360	266–469
June	53	–	570	444–713
October	35	–	231	171–300

dumosa than either *L. tridentata* or *Lycium andersonii* (Table 11.13). Among the five dominant annuals, which composed 71–73% of herb seed pool (Chapter 5), only *Vulpia octoflora* showed no significant difference in density for *A. dumosa* versus other shrub canopies. Densities of seeds from each shrub species was highest beneath its parent canopy. As with data for total seed densities, all of the dominant annuals showed significant declines in populations for June to October, 1972 (Table 11.13). Shrub canopy volume, and its related effect on populations of annual plants beneath it, was significantly correlated with seed densities.

Annual variation in seed density

Year-to-year variations in seed density seem to be related both to seed production in the previous growing season and to population sizes of granivorous rodents. Within plot C, where densities of granivorous rodents were high, seed densities were ten times greater in October, 1973, following a highly productive year for annual plant growth, than in October, 1972, following a poor growth year for plants (Table 11.14). Seed densities within plot A in October, 1973, when few granivorous rodents were present, were 16 times 1972 densities found in plot C. Seed densities for 1973 in

Table 11.13. *Densities of seeds in the upper 0.02 m of soil in June and October, 1972, for three shrub species in Rock Valley. Ten samples are averaged for each shrub species. Data from Nelson and Chew (1977).*

Month	Seed source	Ambrosia dumosa	Larrea tridentata	Lycium andersonii
June	*Bromus rubens*	118	498	368
	Pectocarya heterocarpa	888	1223	604
	P. platycarpa	132	452	146
	Thelypodium lasiophyllum	104	232	1464
	Vulpia octoflora	2782	4101	2572
	Other annuals	844	1239	946
	Total annuals	4268	7745	6100
	Total shrubs	318	283	1297
	Total all seeds	4586	8028	7397
October	*Bromus rubens*	65	242	140
	Pectocarya heterocarpa	172	228	234
	P. platycarpa	44	113	88
	Thelypodium lasiophyllum	67	318	455
	Vulpia octoflora	925	2545	851
	Other annuals	550	379	892
	Total annuals	1823	3825	2660
	Total shrubs	125	164	506
	Total all seeds	1948	3989	3166

Table 11.14. *Mean seed densities (seeds m^{-2}) beneath shrub canopies (Larrea tridentata and Lycium andersonii) and in intershrub spaces for sample dates in October of 1972–1974 in Rock Valley. Plot C is an area with relatively high densities of Perognathus formosus and is compared with Plot A, which had very low populations of this granivore. Plant productivity was high in 1973, as compared with relatively poor growth years in 1972 and 1974. Adapted from Nelson and Chew (1977).*

| Year | Plot | Seed density, individuals m^{-2} | | |
		Total herbs	Total shrubs	Total seeds
Beneath shrub canopies				
1972	C	3 243	335	3 578
1973	C	36 395	864	37 259
1973	A	55 859	1 318	57 177
Intershrub spaces				
1972	C	243	26	269
1973	C	6 046	105	6 151
1973	A	9 474	84	9 558
1974	C	3 636	28	3 664
1974	A	7 200	60	7 260

intershrub spaces were 23 and 37 times greater for plots C and A, respectively, than for 1972 in plot C. In 1974 seed number decreased by about 40% in the plots (Table 11.14).

Total seed population and biomass

Using data on coverage of shrub canopies at Rock Valley (19.3%) and mean population densities of soil seed pools for plant species beneath and between canopies, overall estimates of seed populations were made by Childs and Goodall (1973). Adding measurements of mean seed dry weight for important species, biomass of soil seed pools for the plots can be calculated.

Eight of 31 species that were studied by Childs and Goodall (1973) had seed weights of 1 mg or more (Table 11.15). These included four shrub (*Larrea tridentata*, *Lycium andersonii*, *L. pallidum*, and *K. erecta*; all of these seeds are actually fruits) and four herbaceous species (*Amsinckia tessellata*, *Astragalus lentiginosus*, *Oryzopsis hymenoides*, and *Erioneuron pulchellum*). Five species (*Cryptantha micrantha*, *Eriogonum trichopes*, and three unidentified taxa) had small seeds of 0.10 mg or less in weight. Overall, 70% of the 26 herbaceous species in the study had seed weights less than 0.40 mg.

Estimates by Childs and Goodall of total soil seed reserve in the study plots suggested that total population density (all surface area) averaged about 427 seeds m^{-2} (Table 11.16). The eight most abundant species accounted for 63% of the seed population. Eliminating three shrub species from that list showed that five common herbaceous species comprised 59% of the total seed pool. Assuming a mean seed weight of 0.35 mg for other species comprising the seed pool (Table 11.15), estimated total seed biomass at plots A and C in Rock Valley was 3.69 kg ha^{-1} (Childs & Goodall 1973). The most abundant eight species produced 74% of that total, whereas the five dominant herbaceous species made up only 24%.

In contrast to that analysis, Nelson and Chew (1977) estimated higher community levels of seed densities and biomass. Their February, 1971, values of mean overall densities of 943 seeds m^{-2} and biomass of 4.67 kg ha^{-1} were not consistent

Table 11.15. *Mean weight (mg) of individual seeds of annual and perennial plants at Rock Valley. Data from Childs and Goodall (1973).*

Species	Seed or fruit weight, mg
Amsinckia tessellata	1.99
Astragalus lentiginosus	3.17
Chaenactis carphoclinia	0.32
Chorizanthe rigida	0.57
Cryptantha circumscissa	0.14
C. micrantha	0.02
C. nevadensis	0.24
C. recurvata	0.37
Descurainia pinnata	0.15
Eriogonum maculatum	0.39
E. trichopes	0.10
Erioneuron pulchellum	1.00
Gilia spp.	0.98
Grayia spinosa	0.25
Ipomopsis polycladon	0.75
Krameria erecta	9.71
Larrea tridentata	6.63
Lycium andersonii	2.21
L. pallidum	14.00
Mentzelia obscura	0.30
Oryzopsis hymenoides	1.72
Pectocarya spp.	0.41
Phacelia fremontii	0.22
P. vallis-mortae	0.64
Streptanthella longirostris	0.24
Stylocline micropoides	0.21
Vulpia octoflora	0.37

Table 11.16. *Population density and biomass (dry weight) of soil seed reserves in study plots at Rock Valley. Data from Childs and Goodall (1973).*

Species	Seed density, m^{-2} number	Biomass, mg m^{-2}
Chaenactis carphoclinia	12.0	0.38
Cryptantha circumscissa	11.0	1.5
Descurainia pinnata	71.9	10.8
Larrea tridentata	25.7	170.4
Lycium andersonii	31.3	69.2
Oryzopsis hymenoides	15.8	21.2
Streptanthella longirostris	12.2	2.9
Vulpia octoflora	129.0	48.0
Other species	117.2	41.0
Total	426.9	369[a]

[a]Equals 3.69 kg ha^{-1}

Table 11.17. *Overall habitat density and biomass of soil seed reserves between seasons and between years at Rock Valley. These calculations are based on a shrub cover of 19.3% and intershrub space (open areas) of 80.7%. Data from Nelson and Chew (1977).*

Sample date	Plot	Seed density, seeds m^{-2}	Seed dry weight, kg ha^{-1}	Ratio annuals to total seed biomass
February, 1972	C	943	4.67	0.81
June, 1972	C	1 805	12.82	0.57
October, 1972	C	802	5.30	0.56
October, 1973	C	12 150	52.3	0.83
October, 1973	A	18 750	84.3	0.85

with estimates by Childs and Goodall (1973) for November, 1971, because there was no significant recruitment of new seeds over this period. Calculations of Childs and Goodall required a mean overall seed size of 1.2 mg as compared with 0.43–0.71 mg for the dates studied by Nelson and Chew. Both differences in sampling methods and site differences may explain discrepancies between the two studies. Price and Reichman (1987) estimated a mean standing biomass of soil seeds of 1.15 g m^{-2} in the Sonoran Desert of Arizona. Similar values have been reported in other studies (Reichman & Oberstein 1977; Reichman 1979, 1984).

Observed variation in overall density and biomass of seed reserves between seasons and years was large, as might be expected for a dynamic desert ecosystem (Nelson & Chew 1977). Overall seed density and biomass increased two and three times, respectively, from February to June, 1972, and then fell to lower levels again in October following peak foraging activity by granivorous rodents (Table 11.17). For plot C, with high predation by *Perognathus formosus*, overall seed density increased 15 times and biomass 10 times from October, 1972 to October, 1973, a period with very high plant production. In nearby plot A, with much lower predation, both seed densities and biomass were 50% higher, with an estimated 18 750 seeds m^{-2} and 84.3 seeds ha^{-1} (Table 11.17).

Seed demography

Seed demography for a theoretical Mojave Desert ecosystem has been modelled by Wilcott (1973) using climatic data for Las Vegas, Nevada, and herb data from Beatley (1969b). This model suggested that a minimum seed dormancy of one year and a germination threshold of 15–25 mm of rain are important components of desert annual survival strategies. Given these traits, seed predators could consume up to 90% of current seed crops and 50% of older seed cohorts (or a 70% loss of all cohorts) without change in population structure of adult plants. That model predicted a high volume of 13 300 seeds ha^{-1}, not much different from those observed in 1973, and his model, generally supported therefore by empirical sampling data, suggested that seed reserves can persist well for at least 80 years despite heavy pressure by granivores.

SUMMARY

Desert soil crusts and lichens that contain cyanobacteria are able to fix atmospheric nitrogen and contribute to overall biomass of soil organisms. In Rock Valley, however, estimated biomass in soil crusts was only 33.4 kg ha^{-1}, and that for lichens, which were not abundant, was 0.88 kg ha^{-1} on the bajada.

The soil microflora consists of numerous heterotrophic decomposers, including fungi, actinomycetes, bacteria, nematodes, and microarthropods. Based on an extensive program of culturing soil samples, in Rock Valley bacteria and actinomycetes appeared to be much more abundant than fungi. Greatest biomass in the soil matrix was contributed by nematodes, having an estimated dry biomass of 70 mg m^{-2}, including a high number of individuals that fed on microbes. Nematodes produced a large biomass in soils beneath shrub canopies and next to roots and had a small contribution in intershrub spaces. Nematode population size changed drastically as soil temperatures increased and soil moisture availability decreased, and these organisms were limited by moisture and generally were inactive at very low soil water content. Microarthropods of Rock Valley were most abundant in upper soil horizons, and local densities at bases of shrubs reached 18 000 m^{-2} in March, of which prostigmate mites were very common.

The desert soil environment provides long-term storage for seeds and fruits of annual and perennial plants, for when enough water is available for growth, and serves as a huge cache of food resources for granivorous animals. Almost half of the seeds in Rock Valley occurred in the litter and uppermost two centimeters of soil. Highest densities were observed beneath shrub canopies, being ten times greater than in intershrub spaces, and in Rock Valley seed densities were greatest beneath *Larrea tridentata*. Densities of seeds from each shrub tended to be highest beneath its own canopy. Among winter annuals, eight species accounted for nearly two-thirds of all seeds found in soil. Total seed biomass in Rock Valley was estimated by various investigators as 3.69–4.67 kg ha^{-1}. Lower seed biomass and abundance was correlated with high density of the heteromyid *Perognathus formosus*, one of the important granivores in the Rock Valley community. However, soil seed reserves were so large that they can persist even with heavy pressure by granivores year after year. Evidence from other desert habitats suggests that different groups of granivores are selective on the types of seeds that are taken and may influence the relative densities of annual species. Seed reserves, however, are a highly variable resource in both time and space.

12 Nitrogen cycling

Because of the relatively low biomass in most desert scrub communities, the proportion of total system nitrogen held by the biomass is relatively small compared with that of other ecosystems (Skujins 1981). Typically 70–98% of total system nitrogen in deserts is contained within the soil compartment, but the absolute levels of soil nitrogen are still very low. Environmental factors sharply limit nitrogen inputs to desert soils. Precipitation is low and restricts atmospheric inputs of nitrogen as wetfall, and significant amounts of nitrogen fixation only occur in areas where cyanobacterial or lichen crusts are well developed or where nodulated woody legumes comprise a major portion of the vegetation cover. The extensive size of most desert regions acts to limit dryfall of particulate or gaseous nitrogen, as well as animal inputs of nitrogen. When system inputs of nitrogen through the atmosphere (wetfall and dryfall), fixation, and animal activities are low, then it is very difficult to build up large pools of nitrogen. Hence, whereas drought limitations on primary production are very pervasive in desert ecosystems, nitrogen may also be an important limiting factor in years when rainfall is plentiful (Ettershank, Ettershank, Bryant, & Whitford, 1978; Lauenroth, Dodd, & Sims 1978; Romney, Wallace, & Hunter 1978; Sharifi, Meinzer, Nilsen, Rundel, Virginia, Jarrell, & Herman 1988; Fisher, Zak, Cunningham, & Whitford 1988).

The dispersed pattern of perennials produces a mosaic of nitrogen distribution with relatively high concentrations of nitrogen occurring beneath shrubs, and low concentrations in open areas between shrubs (Chapter 2). This characteristic patterning of nitrogen distribution has been termed 'islands of fertility' (Muller 1953; Garcia-Moya & McKell 1970). After nitrogen begins to accumulate in one site, shrub growth, litterfall, and re-establishment tend to maintain the fertility. In many more mesic ecosystems one finds that runoff and leaching would remove accumulated nitrates, but in deserts low levels of precipitation make soil nitrate a relatively stable storage pool.

INTERSYSTEM NITROGEN FLUXES

To produce a nitrogen model for the Mojave desert scrub community at the Rock Valley site, fluxes of nitrogen into and out of the system are important elements to quantify. Inputs to the system can occur as both wet and dry atmospheric inputs and fixation of atmospheric nitrogen. System outputs would include erosional losses, denitrification, and ammonia volatilization.

Atmospheric inputs

Precipitation inputs of ammonium and nitrate nitrogen were measured in 1976 in Mercury Valley at NTS. These values extrapolated to an input of 2.2 kg N ha^{-1} yr^{-1}, with 0.4 kg ha^{-1} as ammonium and 1.8 kg ha^{-1} as nitrate. Rainfall in 1976 was higher than normal, but these values of nitrogen nevertheless appear to be relatively high in comparison with careful studies of nitrogen input to chaparral ecosystems in California, in which measured input was estimated to be about 1 kg N ha^{-1} yr^{-1} (Schlesinger & Hasey 1980).

Adding a reasonable estimate for dry depositional inputs of nitrogen in the form of ammonium nitrate or other particulates, a reasonable estimate for a northern Mojave desert scrub community would be 1–2 kg N ha^{-1} yr^{-1} arriving from a combination of wet and dry deposition.

Nitrogen fixation in soil crusts

Nitrogen fixation by free-living organisms in soil crusts, particularly in cyanobacteria and lichens, appears to be an important nitrogen input to many desert ecosystems (Shields 1957; Rychert, Skujins, Sorenson, & Porcella 1978; West 1990). Among cyanobacteria the most important nitrogen-fixing genera are *Nostoc*, *Scyotonema*, and *Anabaena*, although other fixers may also be present. The lichen genus *Collema*, having *Nostoc* as its photosynthetic symbiont, is extremely important in many desert regions, and one common species, *C. tenax*, was found to have a 32.4% coverage in Curlew Valley, Utah (Pearson 1972).

Warm desert ecosystems characteristically have much lower covers of lichens and cyanobacteria than cold deserts. MacGregor and Johnson (1971) reported a 4.25% 'algal' crust in the Sonoran Desert of southern Arizona. Nash, White, and Marsh (1977) analyzed transects at IBP study sites in the Sonoran, Mojave, and Chihuahuan deserts and found that coverage by lichens with photosynthetic symbionts was extremely low, ranging from 0 to 6 m^2 ha^{-1}. However, grazing and other activities may have a major impact on soil crust stability, so that on sites where human activity occurred there may have been more extensive algal and lichen crusts in the past.

Although there have been numerous studies of fixation in soil crusts, many of these have been superficial and have tended to overestimate the magnitude of fixation present. There is no question, however, that many desert cyanobacteria and lichens with cyanobacterial symbionts are able to fix significant amounts of atmospheric nitrogen (Mayland & McIntosh 1966; Mayland, McIntosh, & Fuller 1966; Rogers, Lange, & Nicholas 1966; MacGregor & Johnson 1971; Skujins & Klubek 1978; Eskew & Ting 1978). The magnitude

of this fixation in terms of an ecosystem input of nitrogen is much less certain, because there are obvious uncertainties in calculating annual inputs of nitrogen through free-living fixation. Current estimates of such fixation are extrapolated from laboratory incubations of soil crusts under relatively optimal conditions. Most studies have found that water, more so than temperature, is the most important limiting factor for this type of fixation (Henriksson & Simu 1971; Hitch & Stewart 1973). Without careful field and experimental measurements and controls, it is very difficult to extrapolate from fixation rates obtained over a few minutes or an hour period to seasonal *in situ* values.

Rychert et al. (1978) provided a table of estimated annual nitrogen inputs by fixation in soil crusts from a variety of desert areas, but for a variety of reasons it is likely that most of those estimates were much too high. Using data from Mayland et al. (1966) they calculated 7.11 kg N ha^{-1} yr^{-1} of fixation for a Sonoran Desert site in southern Arizona, but the estimates assumed 100% coverage. Data from MacGregor and Johnson (1971) from the same region were extrapolated to 13–18 kg N ha^{-1} yr^{-1} fixation. However, those same two source papers estimated fixation of up to 4 g ha^{-1} hr^{-1} after saturating rain and using a measured value for crust coverage of 4%. Assuming 200 hours each year of saturated surface soil having optimal conditions for fixation (see Rychert & Skujins 1974), less than 1 kg N ha^{-1} yr^{-1} of fixation would likely be achieved. Estimates of up to 100 kg N ha^{-1} yr^{-1} by soil crusts at Curlew Valley, Utah (Porcella, Fletcher, Sorenson, Pidge, & Dugan 1973; Rychert & Skujins 1974), also seem to be unjustifiably high.

Nitrogen fixation in desert soils by species of free-living bacteria has not been widely studied. Although total amounts are undoubtedly small, they could still be quite significant in the long-term nitrogen balance of nitrogen-poor desert soils. Both *Azotobacter* and *Clostridium*, important aerobic and anaerobic nitrogen-fixing bacteria, respectively, are commonly found in desert soils, particularly in association with plant roots

(Mahmoud, Abou El-Fadl, & Elmofty 1964; Heth-ener 1967; Abd-El-Malck 1971; Farnsworth, Romney, & Wallace 1978). Studies using ^{15}N and acetylene reduction techniques have produced estimates that in soils of semiarid grasslands in California fixation by free-living bacteria could account for up to $2\,kg\,N\,ha^{-1}\,yr^{-1}$ (Steyn & Delwiche 1970). However, in most desert soils, which have low organic matter and carbon amend-ments, there generally is negligible fixation by free-living microorganisms. Sandy soils generally have less heterotrophic fixation than clay soils. Often, however, there is a potential for significant fixation. When surface soils were moistened with a 10% glucose solution in Curlew Valley they showed high rates of nitrogen fixation (Rychert et al. 1978). It is only under specialized micro-environmental conditions in desert soils, such as in the rhizosphere zone or in algal–lichen crusts, that organic carbon is naturally high enough to promote significant bacterial fixation.

Soil crusts and lichens in Rock Valley

For Rock Valley the very low cover of lichens with cyanobacterial symbionts (Nash, White, & Marsh 1977) and the relative absence of algal soil crusts suggests that free-living fixation is negligible. In Rock Valley there is probably less than 1% cover by algal crusts (Hunter, pers. comm.). A small amount of bacterial fixation is certainly present, but it seems unlikely that total annual nitrogen fixation by free-living organisms could exceed $1\,kg\,ha^{-1}$, and amount fixed is probably signifi-cantly lower.

Legumes with rhizobium nodules

Symbiotic fixation may provide a significant input of nitrogen in many desert ecosystems where leg-umes with rhizobium nodules or certain other woody vascular plants with actinomycete nodules are major components of the plant cover. Woody legumes are particularly significant in this respect, and many of the warm desert, arid regions of the world have large areas that are dominated by spe-cies of *Prosopis* (mesquite) or *Acacia*, two notable

legume genera with nitrogen-fixing potential (Rundel, Nilsen, Sharifi, Virginia, Jarrell, Kohl, & Shearer 1982; Rundel 1989).

For Rock Valley symbiotic fixation of nitrogen is unimportant, because the flora of the IBP vali-dation site contained only 6 species of legumes, 5 species of annuals (species of *Astragalus* and *Lupinus*) and the shrubby *Psorothamnus arbores-cens* (Mojave indigo bush). None of these occurred with a significant amount of biomass, even during years with unusual amounts of precipitation. No vascular plant species with the generally accepted presence of actinomycete nodules was present in Rock Valley.

Estimates of nitrogen fixation in Rock Valley

Our overall estimates of nitrogen fixation in Rock Valley total less than $1\,kg\,ha^{-1}\,yr^{-1}$. This value compares well with an estimate of $0.5\,kg\,N\,ha^{-1}\,yr^{-1}$ for the northern Mojave Desert (Wallace et al. 1978). Hunter, Wallace, Romney, and Wieland (1975) estimated a higher rate of fixation, $3.5\,kg\,N\,ha^{-1}\,yr^{-1}$, for the same area by 'semisymbiotic' bacteria and less than $1\,kg\,N\,ha^{-1}\,yr^{-1}$ fixed by free-living and symbiotic bacteria. These values may well be too high due to the techniques used (Hunter, pers. comm.).

Erosional and dust outputs for desert regions

Erosional fluxes of nitrogen out of desert ecosys-tems through action by wind and water are poten-tially large, but there has been very little research to quantify such losses in natural arid, disturbed arid, or semiarid environments (see Brandson, Gifford, & Owen 1972; Sturges 1975). The effects of erosional transfers are relative, of course, in the sense that transfers may be intersystem or intra-system fluxes. For example, wind may move soil and litter particles kilometers to the nearest topo-graphic barrier or only a few centimeters. Water runoff may carry dissolved sediment hundreds of kilometers to a major river or may be a local transfer within the community (Fletcher, Sorenson, & Porcella 1978).

Wind is probably the most important form of erosion in regions where precipitation is low. The potential for wind erosion can be calculated on the basis of equations developed by the U.S. Department of Agriculture that include considerations of soil erodibility, local wind erosion climatic factor, soil surface roughness, vegetation cover, and length of wind fetch (Woodruff & Siddoway 1965; Skidmore & Woodruff 1968). In general, wind erosion is highest where the following climatic conditions occur: (1) extremely variable and frequently high wind velocities, (2) low and variable precipitation, (3) high frequency of drought, (4) rapid extreme changes in temperature, and (5) high evaporation rates (Fletcher et al. 1978). All of these conditions are typical of desert ecosystems (Chapter 2).

On the basis of nitrogen composition data for soil hummocks (coppice mounds) and interhummock depositions at sites in Arizona and Utah, Fletcher et al. (1978) used the USDA wind soil loss equation to calculate wind losses of nitrogen. They estimated a particulate nitrogen loss in Arizona of $0.28–5.1$ kg ha^{-1} yr^{-1}, with a mean of 1.7 kg N ha^{-1} yr^{-1}. For Curlew Valley, Utah they estimated a slightly higher range of $1.0–5.6$ kg N ha^{-1} yr^{-1}, with a mean of 3.4 kg N ha^{-1} yr^{-1}.

Erosional losses by water can be estimated from a universal soil loss equation developed by the U.S. Department of Agriculture (Wischmeier 1959; Wischmeier & Smith 1965). Average annual loss of soil (on a probability level) is calculated from a rainfall erosion factor, a soil erodibility factor, vegetation cover, slope angle, slope length, and a management history factor. Biotic factors affect this equation in determining soil erodibility and vegetation cover.

Whereas the above equation was developed for use east of the Rocky Mountains, Fletcher et al. (1978) adapted it to use for arid and semiarid regions of the western United States. Using runoff nitrogen contents of samples that were collected in small study plots in their Arizona study area, they estimated runoff losses of nitrogen to be $0.24–24.7$ kg ha^{-1} yr^{-1}, with a mean value of 7.2 kg ha^{-1} yr^{-1}. Local areas of disturbed soil surface produced the high end of the loss range. For Curlew Valley, they estimated runoff losses of nitrogen as $0.24–3.1$ kg ha^{-1} yr^{-1}, with a mean of 1.62 kg ha^{-1} yr^{-1}, which was close to measurements of runoff nitrogen in a series of small study plots at that site.

Projected erosional losses in Rock Valley

Direct measurements or quantified calculations of erosional losses of nitrogen by wind or water out of the Rock Valley IBP validation site have not been made. Because the site has a rocky surface of desert pavement, we feel that wind erosion is probably relatively small at undisturbed sites in comparison with the estimated losses of particulate nitrogen from other desert sites described above. A reasonable estimate would be less than 1 kg N ha^{-1} yr^{-1}. Because this site is situated on the lower bajada, it is difficult to estimate accurately net erosional losses of nitrogen by water. Most of such outputs would be intrasystem fluxes within the study site itself. Losses from the downslope edges of the site would largely be balanced by inputs from above across the upslope margin of the slope. Only in times of unusually intense rainfall would there be sufficient surface flow to cause significant outputs from the system. A time-averaged guess at annual erosional fluxes of nitrogen through surface flow would be less than 1 kg ha^{-1} yr^{-1}. Under conditions of extreme precipitation intensity, at intervals of decades or more, major sheet flow of water likely occurs at Rock Valley, with major impacts on erosional losses of nitrogen from soil and litter pools.

Denitrification

Gaseous losses of nitrogen through denitrification in the soil are potentially a significant output for this element in desert ecosystems. In agricultural systems denitrification may be the predominant factor in nitrogen loss (Delwiche 1956; Broadbent & Clark 1965; Payne 1973; Focht 1978; Knowles 1981). Experimental studies have shown that soils in the Sonoran Desert have a high potential for denitrification, particularly in upper soil horizons with the highest carbon content (Bowman & Focht 1974; Westerman & Tucker, 1978a, b). Given appropriate conditions, denitrification rates may be increased by higher soil temperature, higher levels of organic carbon, and greater soil moisture content. Complete soil saturation is not necessary for denitrification to occur because anaerobic microsites that promote this flux may occur at low soil water contents than field capacity (-0.03 MPa).

Direct measurements of denitrification fluxes were made in a mesquite woodland with high soil nitrate concentration (Virginia, Jarrell, & Franco-Vizcaino, 1982), and peak rates were similar to those reported for irrigated vegetable crops in southern California (Ryden & Lund 1980). An estimated 0.5 kg N ha^{-1} might be lost from those mesquite stands following a major rainfall event, but such events are so infrequent that mean annual denitrification flux is still quite small. It should be expected that denitrification may be a significant flux of nitrogen loss in washes and other seasonally flooded desert soils, where organic carbon and nitrate nitrogen levels are favorable for this process to occur.

For Rock Valley, where soil nitrate levels are very small and rainfall is low and irregular, denitrification certainly must be negligible, although it has never been quantified.

Output from ammonia volatilization

Volatile losses of nitrogen from soil in the form of ammonia may be a significant flux in soils with a high pH. Many desert soils formed from calcareous parent materials, as occur over much of NTS (Chapter 2), may be subjected to such losses to some degree; however, almost nothing is known about whether significant ecosystem losses occur in arid regions. The only pertinent data were from studies in cold desert areas of Utah, from which Dutt and Marion (1974) reported that significant ammonia volatilization occurred from soils treated with high levels of nitrogen fertilizer. Experimental incubation studies of nitrogen losses from soils

in the same area were also conducted by Klubek, Eberhardt, and Skujins (1978), who found that about 7% of the ^{15}N added to decomposing litter of *Artemisia tridentata* was lost to ammonia volatilization after a 10-week incubation and a moisture content near field capacity. However, this was less than 20% of the measured loss due to incubation. Almost no measurable volatilization occurred under the same conditions from cyanobacterial crusts. These results suggest that ammonia volatilization is not a significant output of nitrogen under field conditions in these soils. New studies are needed to access potential for ammonia volatilization in calcareous soils with high organic matter contents, where significant output fluxes may be present.

For Rock Valley, the low rainfall and very low organic matter content of the soil are conditions indicating that ammonia volatilization is probably negligible.

NITROGEN POOL SIZES

Soil nitrogen

Soil pools of nitrogen in desert ecosystems are not only low, but also strikingly inhomogeneous because there is low biomass of vegetation. Stable vegetation cover produces clear patterns of both vertical and horizontal distribution of nitrogen in desert soils. In Yucca Flat studies of soil nitrogen distribution around shrubs have shown the pattern very clearly. Total nitrogen levels beneath *Larrea tridentata* were about 0.13% at 0.10 m depth, but decreased to less than 0.04% below 0.20 m (Nishita & Haug 1973). In the open areas between shrubs total soil nitrogen was a relatively constant 0.02–0.03%, and slightly lower levels in the surface soil in these interspaces may represent nitrogen mining by shallow roots of *Larrea* as well as the lack of litter accumulation.

In cold desert communities, where root/shoot ratios are high, vertical gradients of nitrogen concentration beneath shrubs are more gradual due to the large proportion of biomass concentrated in the root system. Shadscale (*Atriplex confertifolia*) communities in the Great Basin showed relatively

Table 12.1. *Nitrogen concentration in soil beneath shrubs from four profiles at the corners of the IBP validation site in Rock Valley. Estimates of total storages were made assuming a shrub cover of 22.6%.*

Soil pit	Profile	Thickness, m	N, %,	Concentration of N, kg ha^{-1}
59	A11	0.06	0.091	114.8
	A12	0.06	0.047	59.3
	C1	0.11	0.041	94.8
	C2	0.11	0.034	78.6
	C3c	0.23	0.032	154.7
Total				502
60	A1	0.05	0.12	126.1
	A2	0.13	0.03	82.0
	A3	0.20	0.032	134.5
	C1	0.25	0.03	157.7
Total				500
61	A1	0.09	0.193	365.1
	A2	0.10	0.030	63.0
	B	0.18	0.023	87.0
	C1	0.10	0.036	75.6
Total				591
62	A1	0.09	0.081	153.2
	A2	0.12	0.048	121.1
	C1	0.11	0.027	62.4
	C2	0.18	0.048	181.6
Total				518

low enrichment of nitrogen in surface soils over deeper soils (Bjerregaard 1971; Charley & West 1975). A steeper gradient occurred in sagebrush (*Artemisia tridentata*) communities in the Great Basin, which have somewhat lower root/shoot ratios (Wallace & Romney 1972; Charley & West 1975).

Soil nitrogen pools in Rock Valley

Soil nitrogen pools in Rock Valley were studied using excavations of soil profiles at sites both beneath and between shrubs. Near the corners of the IBP validation site (Fig. 2.15), 4 profiles were dug beneath shrub canopies, and 9 more were sampled in intershrub openings within the validation site (Romney et al. 1973). The C horizons of the 4 shrub profiles extended to depths of between 0.47 and 0.63 m, and thus the biologi-

cally active layer of soil was assumed to be a mean depth of 0.55 m. Throughout Rock Valley a strong calcrete layer was found at 0.30–0.70 m depth. Concentrations and pool sizes of nitrogen in the Rock Valley soil profiles were calculated on the basis that 62% of the soil weight occurred in the less-than-2 mm fraction and that the bulk density of that fraction was 1.5 mg mm^3 (Mehuys et al. 1973).

Nitrogen concentrations in surface A1 horizons (0.05–0.09 m) at Rock Valley ranged from 0.08 to 0.19% N by weight (Table 12.1). Using a mean shrub cover of 22.6%, for the validation site, this surface horizon below shrub canopies contained 115–365 kg N ha^{-1}, approximately 30% of total soil nitrogen pools. For the full profiles, nitrogen pools were relatively similar, ranging from 480 to 591 kg N ha^{-1} beneath shrubs (Table 12.1). For the nine profiles in open areas between shrubs, only 3 exhibited the A1 horizon and 4 had A2 horizons (Romney et al. 1973). Overall, nitrogen concentrations in the open areas averaged 0.025% above the caliche layer, a comparable value to that found at Yucca Flat by Nishita and Haug (1973). Using the mean open area of 77.4% at the Rock Valley validation site, total soil nitrogen in the intershrub spaces is here estimated to be about 990 kg N ha^{-1}. Combining this value with mean nitrogen pool beneath shrub canopies, 52.3 g N m^2, yields an estimated total soil nitrogen pool at the Rock Valley site to be 1533 kg ha^{-1}. These nitrogen values do not include nitrate nitrogen, which was measured separately.

Nitrate nitrogen and ammonium in soils of Rock Valley

Nitrate nitrogen is the most important form of soil nitrogen for plant uptake, and soil concentrations of inorganic nitrate were measured from numerous soil profiles throughout NTS (Romney et al. 1973).

Open areas between shrubs at Mercury Valley and Rock Valley were found to have a low concentration of about 2 µg NO$_{3\,g}^{-1}$ soil, with little variation seasonally or with depth in the soil profile (Fig. 12.1; Hunter, Romney, & Wallace 1982). Nitrate measurements beneath shrubs are more variable. At Yucca Flat Nishita and Haug (1973)

Soil Nitrogen

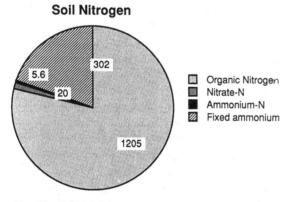

Figure 12.1 Distribution of soil nitrogen components in Rock Valley. Total soil nitrogen pools were estimated to be 1533 kg ha^{-1}.

reported enrichment of nitrate concentrations to a depth of 0.15 m beneath *Larrea tridentata* and to 0.08 m beneath *Krameria erecta*. Measurements given by Wallace, Romney, and Hunter (1978) suggested possible nitrate concentrations up to 17.8 µg g^{-1} of soil in spring samples taken to depths of 0.225 m in Mercury Valley. Observations by Hunter et al. (1982) in Rock Valley showed nitrate concentrations ranging from 4 to 31 µg g^{-1} to depths of 0.03. m beneath various species of shrubs. The highest measurement beneath *Lycium andersonii* was exceptional, in that the majority of nitrate values were 4–10 µg g^{-1}.

If we accept 2 µg g^{-1} as the nitrate concentration at all depths in bare areas, we can then estimate about 7.9 kg ha^{-1} of nitrate to a depth of 0.55 m in these intrashrub areas, which is similar to the estimate of 5–10 kg ha^{-1} suggested by Wallace et al. (1978a). Estimating nitrate nitrogen pools beneath shrubs requires additional assumptions. If mean concentrations of nitrate nitrogen are adopted from Hunter et al. (1982) for the upper soil horizons beneath shrubs (to a depth of 0.15 m for *L. tridentata* and 0.08 m for the other species) and the bare zone concentration of 2 µg g^{-1} is used for deeper portions of the beneath-shrub soil profile to 0.55 m, then the shrub soil nitrate pool can be estimated to be about 12 kg ha^{-1}. This results in an estimate of the total pool nitrate nitrogen of 20 kg ha^{-1}, only 1.3% of total soil nitrogen.

Fixed inorganic ammonium has been reported to form a large component of soil nitrogen at NTS; Wallace et al. (1978) estimated a soil storage of about 500 kg N ha^{-1} in the form of fixed ammonium, but no measurement of this form of nitrogen was made at Rock Valley. Other analyses have shown that fixed ammonium concentrations are about the same beneath shrubs, in the open, and at all soil depths. Because of this uniformity, Hunter et al. (1982) have suggested that distribution of soil ammonium reflects nonbiological controls.

Nishita and Haug (1973) reported mean fixed ammonium concentrations of about 49 µg g^{-1} of soil beneath *K. erecta* and about 69 µg g^{-1}

Table 12.2. *Estimated storages (kg ha^{-1}) of nitrogen in Rock Valley soils contrasted with general estimates for the northern Mojave Desert.*

Type of nitrogen	Rock Valley 0–0.55 m	Northern Mojave Desert	
		Hunter et al. (1975) 0–0.90 m	Wallace et al. (1978a)
Fixed ammonium	302	450–650	500?
Available ammonium	5.6	5–15	–
Nitrate	20	10–300	20–60
Organic nitrogen	1205	500–1500	500–1500
Total	1533	965–2465	1020–2060

beneath *L. tridentata* in Jackass Flats. We have used a median concentration of 59 µg g^{-1} to estimate fixed ammonium storages at Rock Valley. If all of the material underlying an area of one hectare to a depth of 0.55 m was composed of soil with a bulk density of 1.5 mg mm^3 and an ammonium content of 59 µg g^{-1}, storage of ammonium would be about 487 kg ha^{-1}. But if we take into account that about one-third of this material is composed of rocks, then the adjusted storage is 302 kg ha^{-1}. This bound ammonium forms 20% of the total soil nitrogen pool, but fixed ammonium is chemically fixed and unavailable for plant uptake. This fraction would normally be included as part of the organic nitrogen pool.

Available ammonium in Mojave Desert soils at NTS is low. From Jackass Flats mean values were relatively consistent at 1–2 µg g^{-1} (Nishita and Haug 1973). Hunter et al. (1982) measured available ammonium concentrations of about 0.5 µg g^{-1} at Mercury Valley, but these values are probably not directly comparable to those measured earlier at Jackass Flats. If available ammonium at Rock Valley is estimated to have a mean concentration of 1 µg g^{-1} of soil throughout the profile, ignoring obvious patterns of heterogeneity, the IBP validation site would have 5.6 kg NH$_4$–N ha^{-1}. This pool is highly labile because plant uptake and microbial immobilization deplete the pool at the same time that decomposition and mineralization processes produce more ammonium.

Only a very small part of soil nitrogen at Rock Valley is in the form of nitrate nitrogen or ammonium available for plant uptake. Together these compose only 1.7% of soil nitrogen. Organic soil nitrogen forms the largest pool with nearly 80%, four times the size of the fixed ammonium pool (Table 12.2). Total soil nitrogen pool for Rock Valley, estimated to be 1533 kg ha^{-1}, is in the middle range of values estimated for other soil nitrogen studies from the northern Mojave Desert.

Nitrogen concentrations in perennial plants

Standing pools of nitrogen in plant tissues, both living and dead, can be estimated from data on nitrogen concentrations in individual tissue types and from estimates of biomass for each of these tissues, presented in Chapters 4 and 5. Concentrations of nitrogen in shrub root tissues were taken from Wallace and Romney (1972) and Wallace, Romney, Kleinkopf, and Soufi (1978). Mean value of those analyses for six perennials studied was, 1.47%, a significantly higher value than the value of 0.8% reported for shrubs from the western Mojave Desert by Garcia-Moya and McKell (1970). More studies are needed to verify these high Rock Valley concentrations of root nitrogen. Concentrations of nitrogen in living and dead stems at Rock Valley were reported (Hunter, Wallace, Romney, & Wieland 1975); mean values were very similar, with 0.94% N in dead stems and 0.85% in live stems. Nitrogen concentrations of new shrub growth in our calculations was also based on data from analyses on dominant shrubs by Hunter et al. (1975) and Wallace et al. (1978a).

Table 12.3. *Nitrogen concentrations (% dry weight) in tisssue pools of dominant shrubs in the northern Mojave Desert at NTS. Data are from Hunter, Romney, Childress, and Kinnear (1975) and Wallace et al. (1978a); ND = no data.*

Species	Roots	Dead stems	Old stems	New stems	Leaves	Flowers	Fruits
Ambrosia dumosa	1.03	0.80	0.80	0.78	2.47	2.10	2.29
Atriplex confertifolia	ND	1.20	0.48	1.31	1.75	1.79	ND
Ceratoides lanata	1.54	0.51	0.65	1.11	2.45	2.16	ND
Ephedra nevadensis	ND	0.77	0.84	2.18	none	2.00	2.47
Grayia spinosa	ND	0.53	0.57	1.56	1.65	ND	1.72
Krameia erecta	0.78	1.06	0.81	1.66	2.19	ND	2.07
Larrea tridentata	2.04	1.25	1.28	1.66	2.34	2.06	2.67
Lycium andersonii	1.65	1.04	1.07	1.98	2.36	2.50	2.29
L. pallidum	1.83	1.34	1.11	1.19	3.01	3.40	ND
Mean value	1.47	0.94	0.85	1.49	2.29	2.29	2.25

Table 12.4. *Nitrogen pool size (kg ha^{-1}) for shrub biomass compartment at Rock Valley. Data for current year growth is based on biomass for 1973, an unusually favorable year for production.*

Shrub species	Dead stems	Old stems	New stems	Leaves	Flowers and fruits
Ambrosia dumosa	1.63	0.82	0.21	0.94	1.35
Atriplex confertifolia	0.52	0.17	0.06	0.33	0.30
Ceratoides lanata	0.11	0.07	0.05	0.10	0.04
Ephedra nevadensis	0.30	0.75	0.99	–	0.34
Grayia spinosa	0.37	0.43	0.16	0.45	0.21
Krameia erecta	0.60	0.94	0.24	1.24	0.76
Larrea tridentata	2.74	1.28	0.15	0.56	0.22
Lycium andersonii	1.50	3.73	0.68	1.75	0.68
L. pallidum	0.70	1.63	0.20	1.77	0.78
Other species	0.14	0.12	n/a	0.10	n/a
Total	8.6-0;	9.90	2.79	7.24	4.76

Mean shrub concentrations for leaves (2.29%), new stems (1.49%), flowers (2.29%), and fruits (2.25%) were used for uncommon species, which were not analyzed (Table 12.3).

Aboveground perennial tissues contained about 33 kg N ha^{-1} in spring, 1973, with 18.5 kg N ha^{-1} in old and dead stems and 14.8 kg N ha^{-1} in new tissues (Table 12.4). The four most important spe-

cies for total aboveground nitrogen, paralleling their relative order in biomass, were *Lycium andersonii* (8.3 kg ha^{-1}), *L. pallidum* (5.1 kg ha^{-1}), *Larrea tridentata* (5.0 kg ha^{-1}), and *Ambrosia dumosa* (5.0 kg ha^{-1}). However, among these species *A. dumosa* and *L. tridentata* had a large fraction of their nitrogen tied up in standing dead stems. Considering live tissues only, both *Kram-*

eria erecta and *Ephedra nevadensis* had equal or greater importance than *L. tridentata*. Overall, new leaves were the greatest sink for nitrogen among new tissues, having nearly 50% of the total (Fig. 12.2).

The pool of aboveground nitrogen in perennial plants varied from year to year with the relative primary production of those shrubs. Because nitrogen content of new tissues was not measured in all years of the IBP studies, we do not know whether tissue concentration was another important variable. Such variation was small between 1971 and 1972, and it is likely that year-to-year variation was relatively unimportant. Nitrogen content of new growth varied from a low of 3.9 kg N ha^{-1} in 1971 to the high of 14.8 kg N ha^{-1}, cited above, for 1973, an exceptional year (Table 12.5). A typical year of relatively poor growing conditions (i.e., 1971, 1972, 1974, and 1975) would be expected to have 4–5 kg N ha^{-1} in new tissues. Values measured for nitrogen content of new growth amounted to 2.1% of the net production of this growth. The data in Table 12.5 for 1971 and 1972 differ slightly from values reported by Turner (1973) and Hunter et al. (1975) because ours incorporated later corrections to the IBP data base. A pool size for aboveground biomass of shrubs during the six IBP years is therefore estimated to be about 25 kg N ha^{-1}.

Table 12.5. *Estimated mobilization of nitrogen in aboveground spring growth of shrubs in Rock Valley, 1971–1976.*

Year	Aboveground net production, kg ha^{-1}	Mobilization of nitrogen in new growth, kg ha^{-1}
1971	183	3.9
1972	206	4.4
1973	682	14.8
1974	220	4.7
1975	210	4.4
1976	380	9.7

Nitrogen concentrations in annual plants

Nitrogen pools in annual plants of Rock Valley varied greatly with their growth stage of development and between years. These nitrogen pools were sampled in the spring, when annual plant biomass had reached its peak of development and seeds had not been dispersed. In calculating total nitrogen pools, a mean tissue concentration of 0.9% N was multiplied by values of biomass production (Table 5.6). The mean nitrogen concentration was derived from mean values for shoot nitrogen from 12 species of annuals, reported by Wallace et al. (1978b), and mean values from five species for leaves (1.4% N), root (0.05% N), and reproductive tissues (1.8% N; Wallace & Romney 1972; Turner 1972). The five representative species, *Camissonia munzii*, *Chaenactis carphoclinia*, *Mentzelia obscura*, *Phacelia fremontii*, and *P. vallis-mortae*, were relatively abundant in every IBP spring sample.

Biomass-weighted mean nitrogen concentrations of annual plants on the Rock Valley site were relatively constant during the six years of IBP studies, with values from 1.33 to 1.44% N (Table 12.6). Because concentration of nitrogen for each tissue type was only determined for a single year, variation present resulted from changes in relative allocation between tissue types. Greatest amount of nitrogen by annual plants occurred under the unusually favorable growing conditions of 1973, when the annual pool was 9.15 kg N ha^{-1}.

Nitrogen Allocation to New Shrub Growth kg ha^{-1}

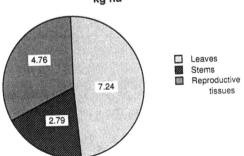

Leaves
Stems
Reproductive tissues

4.76
7.24
2.79

Figure 12.2 Relative allocation of nitrogen to new growth of shrub species in Rock Valley during 1973. Total allocation was estimated to be 14.8 kg ha^{-1}.

Table 12.6. *Estimated amounts of N mobilized by winter annual floras in Rock Valley, 1971–1976.*

Year	Net production, g m^{-2}	Weighted mean N concentration, %	Nitrogen, kg ha^{-1}
1971	0.46	1.37	0.063
1972	0.30	1.42	0.043
1973	68.8	1.33	9.15
1974	1.75	1.44	0.25
1975	5.3	1.41	0.75
1976	14.7	1.40	2.06

At the low extreme, 1971 and 1972 each had less than 0.07 kg N ha^{-1}, less than 1% of 1973 levels (Table 12.6). Total value for all components of nitrogen pools in herbs and shrubs at Rock Valley is shown in Tables 12.4–12.6 for 1971–6.

Litter nitrogen in Rock Valley

As previously described, litter biomass beneath shrubs was measured in 1973 (Strojan, Turner, & Castetter 1979), and additional crude estimates were made of litter biomass in open areas between shrubs as well as litter buried in the soil. Totally dead but standing shrubs can logically also be included as surface litter.

Nitrogen content of plant litter components was not measured during the IBP studies. A reasonable estimate would be that following translocation of nitrogen from leaves and other tissues before they abscise, the mean litter nitrogen content would be about 0.5%. Using this value and estimates of both biomass pools, previously discussed, surface litter would comprise about 3 kg N ha^{-1}, buried litter no more than 0.1–0.6 kg N ha^{-1}, and standing dead shrubs about 0.4 kg N ha^{-1}. Combining these values plus a small amount for other surface litter in intershrub spaces leads us to an estimated total nitrogen pool in litter of about 4 kg ha^{-1}.

Wallace et al. (1978) suggested a litter compartment of 20–80 kg N ha^{-1} for the northern Mojave Desert. This range of values was not based on any direct measurements, but rather on the assumption of a long-term equilibrium with postulated high levels of annual nitrogen uptake by vascular plants. Limited direct measurements of litter pools at Rock Valley indicate that such estimates were far too high.

Nitrogen in animal biomass

Amount of nitrogen in live animal tissues fluctuates widely both seasonally and between years as animal population rates of growth or decline change. Nitrogen contents of individual orders of animals from our calculations are based on a variety of literature reports: birds (Evans 1973), mammals (Robbins 1983), reptiles (Boyd & Goodyear 1971), grasshoppers (Karasov 1982), and soil arthropods and non-grasshopper phytophagous insects (Prints & Williams, pers. comm.).

Because biomass pools of individual animals orders were not measured during each year of the IBP study, it is not possible for us to develop a full understanding of the temporal dynamics of animals as a sink for nitrogen, but relative importance of some animal pools of nitrogen can be described. Relative importance of individual nitrogen pools in animal populations closely parallels biomass data, primarily because tissue nitrogen contents do not vary greatly among those animals. Tortoises formed the largest single vertebrate group, with about 23.8 g N ha^{-1} (Table 12.7). Other groups that exceeded 10 g N ha^{-1} under favorable growth conditions were jackrabbits (5.5–20.2 g N ha^{-1}), lizards (19.0 g N ha^{-1}), snakes (11.9 g N ha^{-1}), heteromyid rodents (8.3–11.8 g N ha^{-1}), and pocket gophers (10.5 g N ha^{-1}). Invertebrate groups, with their surprisingly high total biomass, far exceeded vertebrates in size of their standing pools of nitrogen. Surface-dwelling arthropods were the largest single pool with 115 g N ha^{-1} followed in importance by soil arthropods (34.5 g N ha^{-1}) and soil nematodes (27.5 g N ha^{-1}).

Estimates of total animal nitrogen pools ranged from 156 to 190 g N ha^{-1} (Table 12.7). These data are based in many cases on relatively optimal animal population size in a given year, and thus

Table 12.7. *Estimated dry biomass, nitrogen content, and standing pool of nitrogen in heterotrophic animal populations at Rock Valley. An asterisk indicates that estimated biomass was not based on quantitative data.*

Group	Biomass, g ha^{-1}	Nitrogen content, %	Nitrogen pool, g ha^{-1}
Vertebrates	707–876		81.9–100.5
Breeding birds (1971 and 1973 data)	5.5–9.2	10.7	0.6–1.0
Tortoises	200	11.9	23.8
Snakes	100*	11.9	11.9
Lizards	160	11.9	19.0
Jackrabbits (Fall, 1971 and 4 other censuses)	50–184	11	5.5–20.2
Cricetid rodents (excl. *Onychomys*)	5*	11	0.6
Heteromyid rodents and *Onychomys* (1971–1973, summer 1974 and 1975)	75–107	11	8.3–11.8
Pocket gophers	96	11	10.5
Ground squirrels	15*	11	1.7
Invertebrates	1558–1715		74.3–89.2
Grasshoppers on shrubs (1974 and 1975)	5–135	9.8	0.5–13.2
Phytophagous insects (excl. grasshoppers)	3–30	8.4	0.25–2.5
Surface-dwelling arthropods (1976)	1000	11.5	11.5
Soil arthropods	300	11.5	34.5
Soil nematodes	250	11	27.5
Animal total	2265–2591		156–190

animal pools of nitrogen would be expected to be considerably smaller for much of the year.

INTRASYSTEM NITROGEN FLUXES

Having considered pool sizes of nitrogen at Rock Valley, it is appropriate to return to an analysis of the magnitude of nitrogen transfers between compartments within the Rock Valley system. These intrasystem fluxes include plant uptake, litterfall, decomposition, and mineralization. Herbivory, a biologically important process but one of limited importance in the nitrogen cycle, will not be discussed here.

Plant uptake

Direct measurements made at Rock Valley of primary production by vascular plants provide a means to estimate root uptake of nitrogen from the soil. Our means for making these calculations involves data on mean pool size of nitrogen in new plant tissues multiplied by a coefficient reflecting translocation efficiency for nitrogen in perennial shrubs. The assumptions being made are that mean pool size for the six years of IBP studies represents 60% new nitrogen taken up from soil and 40% retranslocated nitrogen from stem tissues that have matured and leaf tissues that have been shed. No translocation is assumed from fruits before they are dispersed.

Spring growth of new tissues of shrubs averaged 4.5 kg N ha^{-1} over the six-year period (Table 12.8). Multiplying this value by a retranslocation coefficient of 0.60 gives an estimate of about 4 kg N ha^{-1} yr^{-1} taken up from soil. Annual plants have no translocation of nitrogen, and thus their mean annual uptake of soil nitrogen was 2.5 kg N ha^{-1} yr^{-1} from 1971 to 1976. Together shrubs and annuals accounted for about 6.5 kg N ha^{-1} yr^{-1}. Poor years for plant growth, e.g., 1971 and 1972, would have provided only

Table 12.8. *Rock Valley pool size and flux rates for nitrogen cycle of Mojave desert scrub.*

		Mean size of N pool, kg ha^{-1}, or flux, kg ha^{-1} yr^{-1}
Pool		
Soil	total	1533
	nitrate nitrogen	20
	ammonium	6
	fixed ammonium	302
	organic nitrogen	1205
Litter		4
Standing dead wood		8.5
Perennial shrubs	total	40
	belowground	15
	aboveground	25 (22.4–33.3)
Annuals		2.5 (0.04–9.15)
Heterotrophic animals	total	173 (156–190)
	vertebrates	91.2 (81.9–100.5)
	invertebrates	81.7 (74.3–89.2)
		Mean size of N pool, kg ha^{-1}
Intersystem flux		
Atmospheric inputs		2
Nitrogen fixation	(total)	<1
	free-living	<1
	symbiotic	+
Erosional losses		2
Denitrification		+
Ammonia volitilization		+
Intrasystem flux		
Litterfall	total	5
	perennial shrubs	2.5
	annuals	2.5
Plant uptake	total	6.5
	perennial shrubs	4.5
	annuals	2.5
Decomposition		5.0
Mineralization		6.5

about 2 kg N ha^{-1} yr^{-1} uptake from soil. In a year such as 1973, which had much precipitation, nitrogen uptake would have exceeded 20 kg N ha^{-1} yr^{-1}, as storage pools would have

provided only a small part of necessary nitrogen for new tissue growth of shrubs. Our mean value ranges for plant uptake are significantly lower than the estimate of 10–40 kg N ha^{-1} yr^{-1} made by Wallace et al. (1978), who made no allowance for retranslocation of nitrogen from storage pools.

Litterfall

Litterfall measurements beneath shrubs were carried out in Rock Valley during 1975 and 1976 (Strojan et al. 1979). With a biomass flux of 199 and 541 kg ha^{-1} yr^{-1} in these two years, respectively, and a mean litter nitrogen content of 0.5%, annual transfer of nitrogen in litterfall from shrubs was 1.0 and 2.7 kg N ha^{-1} yr^{-1}, respectively. In years of unusually favorable growing conditions, as in 1973, litterfall would certainly be much larger. Indirect calculations of year-to-year variations in litterfall fluxes can be made from calculations of the minimum litterfall loss of nitrogen, which would have occurred with loss of all leaf tissues and reproductive structures, assuming a 50% efficiency of translocation of this element from leaf tissues before they are shed. Using all of these direct and indirect calculations, we estimate that litterfall fluxes of nitrogen from perennial shrubs averages about 2.5 kg N ha^{-1} yr^{-1}.

By definition, all productivity by annuals becomes litter during the same year. This pool of nitrogen ranged from 0.06 to 9.15 kg N ha^{-1} yr^{-1} during the six-year study in Rock Valley, with a mean of 2.5 kg N ha^{-1} yr^{-1}. It is interesting to observe that this flux is identical to that estimated for uptake by perennial plants.

Decomposition

The process of decomposition in relation to nutrient cycling processes in arid regions has been studied intensively in the Chihuahuan Desert of New Mexico (Whitford 1986; Whitman, Stinnett, & Anderson 1988). Limited studies in Rock Valley have provided some data on rates of litter decomposition for three shrubs species (Strojan et al. 1987). Comanor and Staffeldt (1978) and Santos, Steinberger, and Whitford (1984) have provided some limited data on decomposition rates for other parts of the Mojave Desert.

For the Chihuahuan Desert, soil microflora, microfauna, and termites all play major roles in the biological processing of plant litter and operate somewhat independently from direct climatic controls (Whitford 1986). Early stages of decomposition of buried soil litter and dead roots involve activity of bacteria and yeasts, which are grazed on by protozoans and nematodes, which in turn are preyed on by tydeid mites (see Chapter 11). Soil fungi become increasingly important as decomposition proceeds, and tarsonemid mites and fungiphagous nematodes enter the system feeding on these fungi. Gamasina mites act as predatory species whereas collembolans and psocopterans enter at later stages, feeding directly on the decomposing litter and on soil fungi.

For surface litter, the biological decomposition process for the Chihuahuan Desert is somewhat different (Whitford 1986). Fragments and feces of oribatid mites and other microarthropods provide a major substrate for decomposition by bacteria, yeasts, and other fungi (see Chapter 11). In this model grazing and predation by protozoans, nematodes, and several groups of mites help to release mineral nutrients into soil. Termites, relatively unimportant in the Rock Valley ecosystem, collembolans, and psocopterans all feed on surface litter in the final stages of decomposition. The microarthropod fauna of Rock Valley, however, is quite different than that of the Chihuahuan Desert, with both prostigmatid and oribatid mites being abundant (Chapter 10).

Santos, Elkins, Steinberger, and Wilkins (1984) measured rates of biomass loss in decomposing litter of *Larrea tridentata* near Boulder City, Nevada, as well as other sites in the Sonoran and Chihuahuan deserts, and suggested that litter disappearance was better correlated with long-term average precipitation than with actual precipitation over the period of study. They concluded that long-term climatic patterns, rather than short-term climatic fluctuations, shape the structure of desert soil communities. Rock Valley studies by Strojan et al. (1987), however, contradict those hypotheses; these authors pointed out statistical problems in treating data sets on decomposition pooled from different regions where the biotic environment of soil microarthropods differs greatly.

Wallace et al. (1978) suggested that the northern Mojave desert scrub community has a decomposition flux for nitrogen of 5–10 kg ha^{-1} yr^{-1}, but they had no experimental data base to support this estimate. For our model, we use a mean value of 5 kg ha^{-1} yr^{-1} for decomposition based on an assumption of a steady-state relationship between litterfall and decomposition. This assumption needs to be tested, but absence of large pools of nonwoody litter and low C/N ratios of litter suggest that our assumptions are reasonable.

Mineralization

Detailed studies of mineralization rates of organic nitrogen to ammonium and rates of nitrification of ammonium to nitrate have not been carried out at Rock Valley or other parts of NTS. Wallace et al. (1978) made the assumption that total nitrate pools turn over once annually, with most of this change occurring rapidly in March and April. This suggestion came from a crude guestimate of 2% of soil organic nitrogen pool mineralizing annually. On this basis, they derived a value of 10–30 kg ha^{-1} yr^{-1} of nitrogen mineralization. Such a rate of mineralization is considerably higher than our calculation of mean nitrogen uptake by vascular plants (6.5 kg N ha^{-1} yr^{-1}) and would lead to nitrate accumulation in soil, if there were not other balancing losses. If soil nitrates are relatively stable through time, then a lower mean mineralization rate than 6.5 kg N ha^{-1} yr^{-1} is to be expected.

NITROGEN CYCLE

Whereas much remains to be learned about the dynamics of nitrogen cycling processes at the Rock Valley study site, available data and estimates of nitrogen pool sizes and fluxes, described in this chapter, can be integrated into a preliminary model of how this cycle operates in a mean year (Table 12.8; Fig. 12.3).

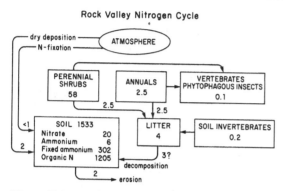

Figure 12.3 Hypothetical nitrogen cycle for the IBP validation site in Rock Valley. See text for a discussion of the assumptions made in this model.

The largest system pool is soil nitrogen with 1533 kg ha^{-1}. More than 98% of this nitrogen occurs as organic nitrogen or fixed ammonium unavailable for plant growth. Almost nothing is known about the dynamics of mineralization within this pool. Small pools of nitrate and available ammonium that are present at a single point in time may well be in a dynamic change with regular inputs from mineralization, plant uptake, and microbial immobilization during the spring season. Wallace, Romney, and Hunter (1978) provided data on seasonal dynamics of soil nitrate pools at NTS, suggesting that mineralization is a dynamic process throughout the year. Nevertheless, the standing pool of nitrates alone is three times the estimated annual uptake rate for vascular plants. This situation suggests that nitrogen should not be limiting for plant growth under normal conditions. However, in exceptional, favorable years for growth, such as 1973, when plant nitrogen uptake can exceed 30 kg ha^{-1}, nitrogen availability would be expected to be potentially limiting. Fertilization experiments at Mercury Valley support this suggestion, with nitrogen additions increasing net primary production only under conditions of irrigation (Hunter, Romney, Childress, & Kinnear 1975; Hunter, Romney, Hill, Wallace, Ackerman, & Kinnear 1976).

Overall our model suggests that the soil nitrogen pool may not be in short-term equilibrium.

Inputs to this pool from dry and wet atmospheric deposition, nitrogen fixation, and litter decomposition total less than 7 kg N ha^{-1} yr^{-1}, but mean output levels from plant uptake and erosional losses are about 8.5 kg N ha^{-1} yr^{-1}. This small difference may relate to the imprecision associated with some of our estimates. It is possible, however, that net losses are occurring and that the large soil nitrogen pool had its origin in the past under more mesic conditions and greater nitrogen inputs through fixation of atmospheric nitrogen. Rare events of major sheet flow erosion likely occur at irregular intervals of decades or longer, and such events would have sharp impacts on erosion and litter pools.

Our model also indicates that nitrogen pools in perennial plants may be increasing with time. Shrub productivity over the mean year studied takes up 6 kg N ha^{-1} yr^{-1} from soil, but litterfall accounts for only 5 kg N ha^{-1} yr^{-1}. Additional nitrogen is retained in new stem growth of shrubs. This accumulation is equivalent to a 3% annual increase in the aboveground pool of nitrogen in shrubs. We anticipate that this accumulation would reach steady-state level when shrub canopy growth on a stand basis reaches its limits of water available for growth maintenance. Alternating cycles of wet and multiple dry years undoubtedly provide a basis for considerable long-term variation in the dynamics of shrub nitrogen balance between uptake and loss.

SUMMARY

A model for nitrogen cycling at the Rock Valley site can be proposed using estimated inputs and outputs based on the IBP studies and extrapolated data from similar desert ecosystems. Atmospheric inputs as ammonium and nitrate were estimated to be 2.2 kg N ha^{-1} yr^{-1} in Mercury Valley. Because algal crusts, which harbor nitrogen-fixing cyanobacteria, and lichens with nitrogen-fixing symbionts were poorly represented in Rock Valley, they contributed only about 1 kg N ha^{-1} yr^{-1}, and a similar value was generated for nitrogen fixation of free-living bacteria, *Azotobacter* and *Clostridium*, in those desert soils. Erosional transfers by wind and water would occur in Rock Valley, a very gently sloping bajada, but neither intersystem nor intrasystem fluxes of nitrogen by erosional forces were quantified. Projected erosional losses in Rock Valley, given average wind speeds, amount of precipitation, and site slope, were about 2 kg ha^{-1} yr^{-1}. Denitrification, which utilizes soil nitrate, was estimated to be negligible because soils of Rock Valley had exceedingly low concentrations of soil nitrate. Similarly, ammonia volatilization was judged to be negligible, because soils had very low organic matter and received low rainfall.

The nitrogen pool was low because at Rock Valley there is low biomass of vegetation. Soil nitrogen was about 990 kg N ha^{-1} in intershrub spaces (77.4% of surface area) and 480–591 kg ha^{-1} beneath shrubs (22.6%), yielding an estimate of total soil nitrogen pool of 1513 kg ha^{-1}. Most nitrogen occurred in the surface horizon. Low concentrations of soil nitrate were measured in Rock Valley and Mercury Valley at 2 µ g^{-1} soil, or about 20 kg ha^{-1} and therefore about 1.3% of the total nitrate nitrogen pool. Fixed ammonium, which composes rocks and also is unusable by plants, represented 302 kg ha^{-1}, and free ammonium in soil was estimated at 6 kg ha^{-1}.

Biomass measurements from IBP studies provided data to estimate that component of the nitrogen pool in Rock Valley. For plants, typical aboveground biomass of shrubs consisted of about 25 kg N ha^{-1}, belowground biomass of about 15 kg N ha^{-1}, and herb biomass of about 2.5 kg N ha^{-1}. Litter on soil surface was scanty and comprised only 4 kg N ha^{-1}. The total nitrogen pool of heterotrophic animals was 156–190 kg N ha^{-1}, with slightly more invested in vertebrates than invertebrates.

Uptake of soil nitrogen by roots was estimated as 4.5 kg N ha^{-1} yr^{-1} after assuming that 40% of nitrogen used was retranslocated nitrogen from tissues before they were shed. An estimate of mean nitrogen uptake by annuals was 2.5 kg N ha^{-1} yr^{-1}. Litterfall flux therefore was 2.5 kg N ha^{-1} yr^{-1}, and it averaged the same for perennial shrubs. Decomposition produced an estimated 5 kg N ha^{-1} yr^{-1}. No reasonable approach could be taken to estimate intrasystem flux of mineralization with any degree of reliability, and existing published values for Rock Valley were judged to be too high, given known values for biomass.

Two interpretations for the model of nitrogen cycle in Rock Valley are possible to explain why outputs appear to exceed inputs. Current estimates of the nitrogen components are imprecise, so that a refinement may show that the system is in equilibrium. However, if the components are relatively accurate, then the soil nitrogen pool is not in equilibrium and is experiencing at least a short-term net loss of nitrogen.

13 Human impacts on Mojave Desert ecosystems

Desert ecosystems have in recent times been subjected increasingly to untested contacts with humans and their characteristic activities. Within the past 40 years there has been heavy use of desert scrub communities for recreational purposes, military testing, and construction projects, all of which have modified physical characteristics of soil and altered the composition of desert scrub vegetation and impacted animal populations. Because deserts, by their very definition, receive little annual rainfall, effects from human disturbance can persist for long periods of time, and research has shown that impacts to the ecosystem are extremely long term. Most applied ecological studies in desert regions have been conducted for environmental impact reports, which can access the biological resources at a site relative to undisturbed parcels of desert vegetation, such as at Rock Valley. Such unpublished analyses make recommendations for minimizing impacts of human activities and mitigating what damage is done. There are relatively few articles that describe carefully designed experiments for individual types of human impacts, analyzing short- and long-term effects on desert ecosystems using treatments of differing durations under wet and dry conditions and on various soil types. One goal is to learn which activities forever alter characteristics of natural desert communities versus those that, especially by unaided processes, return a habitat to predisturbance conditions.

SECONDARY PLANT SUCCESSION

Early analyses by desert plant ecologists provided no convincing evidence that plant succession occurred in desert ecosystems (Shreve & Hinckley 1937; Muller 1940; Shreve 1942). Many studies in the Mojave Desert have focused on the topic of secondary plant succession, i.e., recruitment of a new resident plant community following complete removal of shrubs leading to the return of a mature, relatively stable assemblage of long-lived shrubs (Vasek 1979/80, 1983; Rowlands 1980). None the less, the relatively few studies on secondary plant succession still have not yielded a clear synthesis on the process. When a desert habitat is stripped of its aboveground plant biomass, revegetation is slow, and perennials that recolonize may not be the same ones that were dominant before disturbance. Investigators have attempted to determine how human activities modify desert soils and so influence plant succession and length of time it may take to restore the original plant community, if, in fact, the original community is ever fully restored.

Revegetation of a desert ghost town

Wells (1961) reported on secondary succession in the rediscovered townsite of Wahmonie, Nevada, located on the bajada just east of Jackass Flats at 1320 m elevation on NTS. This settlement was started in 1928 and then hastily abandoned after clearing shrubs to construct a street system.

Using a distance method for sampling perennial vegetation, Wells identified pronounced changes in species composition and relative abundance on disturbance plots in the ghost town as compared with adjacent vegetation, which was *Larrea–Grayia–Lycium* desert scrub (Chapter 4). Only 2 of the 5 dominant desert scrub species, *Lycium*

Table 13.1. *Perennial plants in desert scrub and on revegetated street system in the ghost town of Wahmonie, Nevada, in 1961, 33 years after the townsite was abandoned. Data from Wells (1961).*

Species	Desert scrub		Revegetated street system	
	Density, individuals ha⁻¹	Frequency in quadrats, %	Density, individuals ha⁻¹	Frequency in quadrats, %
Grayia spinosa	2674	100	111	20
Lycium andersonii	1396	96	867	74
Larrea tridentata	850	80	44	6
Ephedra nevadensis	652	70	934	86
Coleogyne ramosissima	497	42	35	4
Stipa speciosa	341	44	2595	96
Thamnosma montana	52	8	378	48
Acamptopappus shockleyi	12	2	79	8
Ceratoides lanata	12	2	10	2
Krameria erecta	12	2	10	2
Succulents	25	4	0	0
Hymenoclea salsola	0	0	267	32
Salazaria mexicana	0	0	200	20
Other species	0	0	30	6
Total	6523		5560	

andersonii (desert thorn) and *Ephedra nevadensis* (Mormon tea), were also very abundant on disturbed sites (Table 13.1), probably because those species resprout from belowground parts after disturbance. Three other regional dominants were present on disturbed sites at only 4.2–7.0% of typical desert scrub densities. The most abundant perennial in the ghost town was the bunchgrass *Stipa speciosa*, comprising 47% of the total, and also common were *Thamnosma montana* (turpentine bush), *Hymenoclea salsola* (cheesebush). and *Salazaria mexicana* (bladder sage). These three species are characteristically found along desert washes and naturally disturbed habitats and are apparently pioneers in plant succession at that elevation.

Wahmonie was revisited in 1978–9 by Webb and Wilshire (1980), who studied soil compaction by quantifying bulk densities, soil strengths, and infiltration rates of the revegetated street system and adjacent desert scrub plots. After 51 years of revegetation, the southwestern streets showed almost no residual compaction of soil, but the northeastern townsite streets, which had a soil bulk density of 1.66 mg mm³, still showed significant compaction to a soil depth of 0.20 m. Infiltration rate of townsite streets was 128 mm hr⁻¹ as compared with 199 mm hr⁻¹ for southwestern streets and avenues and 166–227 mm hr⁻¹ for controls. Given enough time, therefore, recovery of compacted desert soils is possible (Webb, Steiger, & Wilshire 1986). Following more than half a century of recovery, coverage of *Stipa* and *Hymenoclea* at Wahmonie was still much greater than in control areas, while coverage of *Larrea* and *Grayia* was much lower relatively (Webb & Wilshire 1980).

Both studies at Wahmonie used control sites consisting of *Larrea–Grayia–Lycium* desert scrub; however, Webb and Wilshire (1980) obtained significantly higher density (8700 individuals ha⁻¹) and vastly different counts for most species, for example, as shown in *Stipa speciosa* (Tables 13.1–13.2). Although such differences in species density and relative cover may have involved real changes and plant succession, they also could be attributed to other factors, such as differences in measurement techniques (point quarter method versus

Table 13.2. *Perennial plants in the ghost town of Wahmonie, Nevada, in 1979, 51 years after the townsite was abandoned. Comparisons are made between control plots on the southwestern corner of the ghost town and streets and avenues nearby, showing density (individuals ha^{-1}), percent composition, and percent cover. Data from Webb and Wilshire (1980).*

Species	Southwestern control			Revegetated street and avenues		
	Density, individuals ha^{-1}	Cover, %	Composition, %	Density, individuals ha^{-1}	Cover, %	Composition, %
Larrea tridentata	350	9.1	4	25	1.4	0
Grayia spinosa	1350	8.4	15	40	0.4	1
Ephedra nevadensis	1140	3.6	13	510	3.5	8
Stipa speciosa	3800	2.0	43	4160	4.3	68
Lycium andersonii	1060	1.7	12	210	1.4	3
Thamnosma montana	400	0.4	5	540	1.8	9
Acamptopappus shockleyi	60	0.4	1	40	0.2	1
Hymenoclea salsola	50	0.2	1	360	1.5	6
Coleogyne ramosissima	250	0.1	3	6	0.1	0
Salazaria mexicana	240	0.0	3	190	0.9	3
Total	8700	25.9	100	6140	15.9	99

belt transects) or because the same specific study areas were not used.

Between 1961 and 1979 densities and relative covers of species on revegetated streets had also changed, but disturbed sites were, as before, populated primarily by *Stipa, Ephedra, Hymenoclea,* and *Thamnosma,* suggesting that succession had yielded little change in the 18 years between studies. Moreover, species densities, relative cover, and composition on revegetated streets, avenues, and the main road differed significantly from those of the controls. Webb and Wilshire suggested that soil compaction was a major limiting factor in recolonization of long-lived species, such as *L. tridentata;* in other disturbance systems soil compaction can greatly restrict growth rate at bulk densities greater than 1.60 mg mm^3. On the other hand, perennial cover on the townsite was actually greater than on the northeastern control, and the townsite had therefore achieved predisturbance densities but with vastly altered species composition. Slow reestablishment of *L. tridentata* may therefore be due in part to poor germination traits (Barbour 1968).

Using least-squares linear regression analysis produced a crude approximation for soil recovery time of 70–100 years, but a four-parameter model, exponential decay curves produced a range from 70–680 years, based on only four time-dependent data points (Webb & Wilshire 1980). Hence, there is still considerable uncertainty on time needed to restore soil to its former state, but probably at least a century. None the less, even if a soil is restored to its predisturbance condition within a century after clearing, revegetation may not result in the restoration of the original plant community. Certainly at Wahmonie the original plant community had not been recovered by natural processes after over 50 years.

Another recent study in the Panamint Mountains of California described the recovery of a higher elevation *Grayia–Lycium* desert scrub 75 years after the townsite of Harrisburg was abandoned (Webb, Steiger, & Turner 1987). The investigators determined that soil compaction on the disturbed site had been completely ameliorated. Yet even though the disturbed site had the same shrub density as nearby *Grayia-Lycium* desert

Table 13.3. *Densities of shrubs (individuals ha⁻¹) in controls and former tent areas on three abandoned military camps in the eastern Mojave Desert. Data from Prose, Metzger, and Wilshire (1987).*

Shrub species	Ibis		Clipper		Iron Mountain	
	Control	Tent area	Control	Tent area	Control	Tent area
Ambrosia dumosa	2959	3043	1274	1501	233	2150
Encelia frutescens	331	452	0	0	0	0
Hymenoclea salsola	0	249	0	0	0	100
Larrea tridentata	584	127	726	299	200	167
Porophyllum gracile	0	167	0	0	32	0
Stephanomeria pauciflora	207	41	0	0	49	132
Other perennials	1080	0	75	0	17	50
Total	5161	4084	2075	1800	467	2599

scrub (21 000 individuals ha⁻¹), the disturbed site had a markedly different type of vegetation; 85% of its perennials and cover consisted of *Chrysothamnus viscidiflorus*, *Ephedra nevadensis*, and *Hymenoclea salsola*. Bordering the abandoned townsite was a natural debris-flow from 1976 that had a totally different, third set of dominant perennials. Consequently, no simple model for succession was diagnosed for the upper bajada desert vegetation belt.

Natural revegetation of military encampments

Revegetation was investigated at three abandoned World War II military encampments in the eastern Mojave Desert (Prose, Metzger, & Wilshire 1987). These three camps were used for military training exercises from 1942 to 1944 and were placed in areas of *Larrea-Ambrosia* desert scrub. Camp Clipper (575 m elevation) was active for the shortest time, and its fine-textured soils had the lowest bulk densities in control and disturbed samples; in fact, the tent and parking lot at Camp Clipper showed no significant compaction and were less dense than controls near camps Ibis (525 m) and Iron Mountain (150 m). Tent areas at Ibis and Iron Mountain had very high bulk densities of 1.69 and 1.74 mg mm³, respectively, and on parking lots and roads bulk density exceeded 1.80 mg mm³.

At all three camps *Ambrosia dumosa* was the most abundant perennial on control sites and former tent areas, comprising at least half of the total shrubs (Table 13.3). This species is an opportunistic, long-lived perennial with good colonizing properties. *Larrea tridentata*, the other dominant long-lived shrub, occurred in all samples but was greatly reduced in abundance and percent cover on tent areas; this species has poor colonizing properties and does not tolerate highly compacted soils. *Hymenoclea salsola*, a short-lived perennial and a good colonizer, occurred on the tent areas with highly compacted soil, and this species developed a fairly dense population on the parking lot at Camp Iron Mountain, which was a very arid habitat. *Encelia frutescens*, another short-lived perennial with good colonizing properties, at Camp Ibis was relatively abundant on the tent area, but also was more successful on the parking lot (1501 individuals ha⁻¹), where it flourished with *A. dumosa*, together comprising 94% of the perennial flora. These data show that certainly 40 years was not enough time under arid conditions for highly compacted soils to reestablish characteristic *Larrea* stands on disturbed sites. However, Prose et al. (1987) showed that *L. tridentata* and *A. dumosa* could become established on scraped barren desert within 40 years where compaction was not a problem. This is good evidence that succession can occur in the arid regions of the Mojave Desert.

Succession on plowed fields

Between 1913 and 1920 homesteaders in the eastern Mojave Desert were required to plow a portion of each ranch for crops, whether or not produce was obtained. Of course, without water and fertilizer those crops failed, and fields were abandoned to return to desert vegetation. Fields (last plowed from 1913 to 1930) were sampled in Lanfair Valley during winter, 1982, from four vegetational belts to determine species diversity and density of shrubs on old fields and in adjacent unplowed areas (Carpenter, Barbour, & Bahre 1986).

In the zone of creosote bush desert scrub (1100 m elevation), off-field plots were dominated by relatively tall specimens of *L. tridentata* with subdominant shrubs of *Lycium andersonii*, *Gutierrezia microcephala*, and *Acamptopappus sphaerocephalus* and abundant clumps of the grass genus *Hilaria*. On old fields density, cover, and canopy size of *Larrea* was typically about the same, and the most abundant understory perennials were *Hilaria*, *Sphaeralcea*, and *Hymenoclea salsola*, but *Lycium* and *Acamptopappus* had similar important values as in off-field stands. There was only 20% difference in species richness between paired plots. Thus, in 65 years lowland desert scrub community had essentially been recovered.

At 1280 m and 1430 m undisturbed sites supported Joshua tree woodland having *Yucca brevifolia* as the most conspicuous plant along with *Y. schidigera* (Mojave yucca) and *Opuntia acanthocarpa* (buckhorn cholla), and understory consisted of *Hilaria*, *Haplopappus cooperi*, and *Leucelene ericoides*. Old fields essentially lacked the large succulents, whereas *H. salsola* and *S. ambigua* doubled in importance. At higher elevations *L. ericoides* and *G. sphaerocephalus* shared dominance with several other perennials. These analyses showed that *G. microcephala*, *H. salsola*, *L. ericoides*, and *S. ambigua* responded positively to past disturbance, *Hilaria* and *Ephedra nevadensis* responded negatively, and *A. sphaerocephalus*, *Lycium andersonii*, and *Haplopappus cooperi* had no strong response. At least in lowland desert scrub, revegetation of plowed fields naturally developed into a reconstitution of creosote bush

desert scrub within 65 years, but Joshua tree woodland did not form naturally during the same period.

Construction of utility corridors

Clearing a barren right-of-way for a utility, for example, pipelines and power transmission lines, obviously creates a short-time elimination of vegetation and disruption of soil horizons. Several scientific studies have attempted to quantify long-term biological effects of utility corridors across desert landscapes.

In 1960 a large natural gas pipeline was laid in a trench from Newberry to Lucerne Valley, San Bernardino County, California, and in 1972 belt transects were used to sample species at 10 intervals along the 33.8-km pipeline right-of-way (Vasek, Johnson, & Eslinger 1975). Sampling was done directly over the trench, on the scraped berm, and from two controls 50 m east and west of the pipeline in undisturbed desert scrub. Species were categorized as long-lived perennials, short-lived pioneer shrubs, pioneer perennial herbs, and an unassigned group of herbaceous perennials and suffrutescent species that can occur in either disturbed or undisturbed habitats. Jaccard's coefficient of community similarity was used to compare degree of identity of samples, and community ecological quality was also quantified as an index.

Control sites on the two sides of the pipeline were essentially identical, but differed markedly from disturbed transects of trench and berm. Coefficients of similarity for controls versus trench and controls versus berm were extremely low in 9 of 10 sample areas. Berm and trench often had a similar flora, typically dominated by short-lived pioneer shrubs and pioneer herbs. Community quality of berm and trench samples was poor, often rated as zero and with means of 4.06 and 5.61, respectively, whereas that of controls was 25.06. Regarding succession, Vasek et al. observed that after 12 years disturbed transects in four of sample areas, those with high production, had significant revegetation by long-lived shrubs that were typical of adjacent desert scrub community; this indicated that secondary succession was already progressing toward recovery of the original community. Sample areas with low production, caused by low precipitation and poor, rocky soil, had low community quality on disturbed soils and very little revegetation by long-lived shrubs. Under optimal natural conditions, the investigators estimated that succession could be completed at points along the pipeline by 30–40 years if revegetation and early growth rates were extrapolated, but a slowing of the process would be expected later in the successional process.

A parallel study, using similar methods, was conducted on two power transmission lines passing through Lucerne Valley in San Bernardino County (Vasek, Johnson, & Brum 1975). These power lines were close to each other and parallel along much of the distance, at one point they merged, and at another point they diverged sharply. Construction was completed in 1937 for one and 1970–1 for the other, so that Vasek and coworkers had an opportunity to compare patterns of revegetation and succession in comparable habitats between recent disturbance and that performed 33 years earlier. Six study areas were used, four with paired transects where powerlines were roughly parallel and passed through essentially the same type of terrain and vegetation, one where they merged, and one with paired transects at the farthest point of divergence. Samples were taken in controls 50 m away from the wires, midway between the pylons and directly beneath wires, along the edge of the access road, and directly beneath pylons.

Plant abundance was greater in disturbed area than controls for the 1937 powerline and less for the 1970 powerline, and short-lived perennials dominated in highly disturbed sites (beneath pylons) for the newer line. Community quality was actually judged to be higher beneath wires and along the access road than on the control transect, and in most cases enhancement of plant growth and diversity probably resulted because those microhabitats received extra moisture via drip or runoff, respectively. Vegetation of the two powerlines was judged on statistical evidence to be the same, and greatest variability was observed beneath pylons, where revegetation responses lacked predictability. Drastic disturbance of soil when pylons were constructed was undoubtably

the explanation for decreased vegetation beneath pylons, but after 33 years vegetation beneath pylons approached that of controls, although signs of disturbance were still evident.

Corridors of five natural gas pipelines (constructed 1956–73) and seven power transmission lines (constructed 1924–77) across the Mojave Desert in southern California were studied by Lathrop and Archbold (1980b) using the transect methods of earlier investigators (Vasek, Johnson, & Eslinger 1975; Vasek, Johnson, & Brum 1975). Having additional corridors through different regions of the Mojave Desert introduced more variability in the analysis, but also permitted Lathrop and Archbold to observe that corridor age affected type of vegetation found along the utility corridor. When different-age corridors of utilities are compared, for example, 1956 and 1963 pipelines (Fig. 13.1) and 1924 and 1968 powerlines

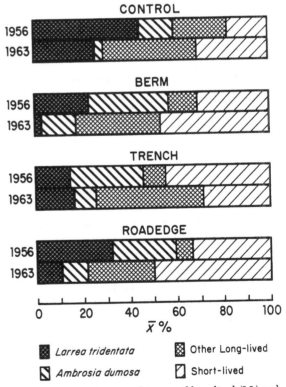

Figure 13.1 Mean percent biomass of long-lived (LL) and short-lived (SL) shrubs along pipeline corridor in 1956 and 1963 across the Mojave Desert of southern California and examined in 1974. After Lathrop and Archbold (1980a).

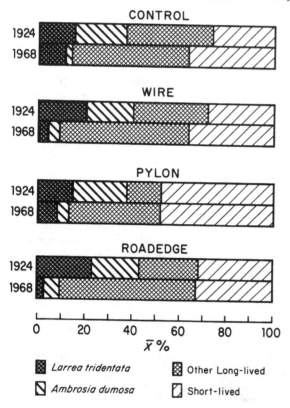

Figure 13.2 Mean percent biomass of long-lived (LL) and short-lived (SL) shrubs along powerline corridors erected in San Bernardino County, California, in 1924 and 1964 and examined in 1974. After Lathrop and Archbold (1980b).

(Fig. 13.2) passing through creosote bush desert scrub, one observes that percent biomass of long-lived dominants, *Larrea tridentata* and *Ambrosia dumosa*, were being restored through succession along trenches and berms of pipelines and pylons of powerlines. Consequently, it seems likely that long-lived perennial shrubs are capable of reestablishment as the dominant vegetation along utility corridors given many decades of time. These authors projected recovery times of about 100 years for the most disturbed transects of pipelines and trenches and 20 years in the relatively undisturbed strips under the powerline wires.

Lathrop and Archbold (1980a) also compared revegetation along two aqueducts (constructed 1913 and 1970) across Mojave desert scrub in Kern County, California. Aqueducts, when compared with pipeline data for stands of similar age (Lathrop & Archbold 1980b), had mean population densities 2–3 times higher, and biomass of the communities was also relatively high (Table 13.4). Perhaps because sample areas of the aqueducts had good production, diversity and percent ground cover were also greater than on sites having pipelines. Species diversity on disturbed aqueduct transects were lower as compared with controls (Table 13.4), and measures of species diversity and vegetation production were higher for the 1913 aqueduct. Along disturbed transects percent of long-lived shrubs was significantly higher on the 1913 aqueduct. The 1913 aqueduct had high biomasses on undisturbed and disturbed transects and nearly equal percent of long-lived shrubs, but the relative biomass of species differed because *Chrysothamnus* spp. were dominant over *Larrea* and *Ambrosia* on disturbed sites (Fig. 13.3). The 1970 aqueduct had significantly lower biomass of long-lived species on disturbed transects, with 51.9–56.6% short-lived species, notably *Hymenoclea salsola*, and 19.5–21.7% of the long-lived shrubs were *Chrysothamnus* spp. These data indicated that succession over these wide, mostly untrenched corridors followed stages of dominance of short-lived to long-lived shrubs, but did not involve significant recruitment of *Larrea* and *Ambrosia* from the surrounding desert scrub community even over a 65-year period. Hence, a stable long-lived community is established, but one not having predisturbance composition. Lathrop and Archbold suggested that competitive interactions for water may allow *Chrysothamnus* and other long-lived colonizers to maintain a long-term, stable community composition that is different from that of control areas.

Recolonization of a borrow pit

Vasek (1979/80) reported on recolonization of a borrow pit, where highway crews had removed soil and rock to a depth of 1–2 m from a 70 × 30 m plot. Only a few resident large shrubs were missed by the excavation when the operation closed in

Table 13.4. *Perennial vegetation at five natural gas pipelines in San Bernardino County and two aqueducts in Kern County. For each type of transect is recorded mean number of species per site, total ground cover (%), relative ground cover by long-lived shrubs (LL; %), shrub density (individuals ha^{-1}), and shrub biomass (kg ha^{-1}). Data from Lathrop and Archbold (1980a, b).*

Transect type	Mean number of species site^{-1}	Total ground cover, %	Ground cover by LL, %	Shrub density, individuals ha^{-1}	Shrub biomass, kg ha^{-1}
Pipelines (35 sites)					
Control	5.4	3.5	78	3 990	656
Berm	4.7	1.9	60	3 680	230
Trench	4.3	1.8	57	3 370	166
Roadedge	3.9	2.3	57	4 220	244
1913 aqueduct (11 sites)					
Control	9.7	7.0	53	10 150	744
Right-of-way	6.5	6.9	65	8 100	788
1970 aqueduct (15 sites)					
Control	7.9	5.8	62	7 600	672
Right-of-way	5.1	2.8	35	3 800	298
Roadedge	3.9	2.8	43	4 150	302

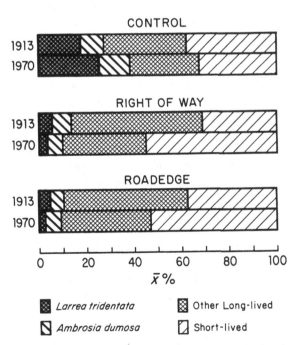

Figure 13.3 Mean percent biomass of long-lived (LL) and short-lived (SL) shrubs along aqueduct corridors across the eastern Mojave Desert, California, in 1913 and 1970 and examined in 1974. After Lathrop and Archbold (1980b).

summer, 1971. The pit bottom consisted of rock, gravel, and some bedrock, none of the typical upper horizon of a desert soil.

Vasek followed revegetation of the borrow pit bottom and its sides beginning in November, 1973, when density on the borrow was 5377 individuals ha^{-1} and slightly higher on the partially disturbed pit sides (Table 13.5). About 85% of the plants in disturbed sites were *Encelia frutescens*, *Ambrosia dumosa*, and *Stephanomeria pauciflora*. Little change was observed on the pit sides from 1973 to 1979, except small increases in the long-lived *Larrea tridentata* and *Eriogonum fasciculatum* and a small but real decrease in *S. pauciflora*, which is a pioneer species with a short life span. On the borrow pit bottom similar changes were observed where the succulent *Opuntia bigelovii* doubled in frequency, exceeding its density on pit sides; its establishment was facilitated by dispersal of vegetative shoots (joints), mostly by packrats (*Neotoma lepida*), and subsequent asexual reproduction via rooted joints. In 1979 an undisturbed control area had *L. tridentata* and *O. bigelovii* as codominants and at relatively high densities. How-

Table 13.5. *Shrub densities (individuals ha⁻¹) on a lower bajada desert site (793 m elevation) located 32 km west of Needles, California, where a rectangular borrow pit was excavated and subsequently studied during revegetation. Observations were made on an undisturbed control of adjacent desert scrub, the borrow pit bottom, and the partially disturbed pit sides. Data from Vasek (1979/80).*

	Density, individuals ha⁻¹						
	Control	Borrow pit bottom			Pit sides		
Perennial species	1979	1973	1975	1979	1973	1975	1979
Encelia frutescens	162	24	2800	2837	1464	1514	1500
Ambrosia dumosa	47	725	857	1002	2732	2757	2750
Stephanomeria pauciflora	31	1412	1371	1116	678	649	375
Porophyllum gracile	31	343	334	586	54	0	89
Optuntia bigelovii	1703	290	543	580	518	595	518
Larrea tridentata	1125	13	114	164	54	54	143
Eriogonum fasciculatum	16	40	0	76	54	0	71
Krameria grayi	375	0	0	0	18	0	018
Other species	377	113	58	145	176	216	286
Total	3721	5377	6085	6506	5750	5785	5750

ever, plant density on the control was significantly lower than on the pit borrow.

Vasek (1979/80) also documented the first seedling of *Larrea* on the pit borrow in late 1975, presumably stimulated by heavy rains in September. After favorable summer rains in 1976 and 1977, additional creosote bush seedlings were observed, often beneath the canopy of established plants, i.e., the initial pioneer species. This pattern helps to qualify this as an example of succession, with replacement of colonizing species by long-lived shrubs of the dominant vegetation type.

Vegetation along road edges

Roads traversing the desert, paved and unpaved, are water harvesting systems that redistribute rainwater from the road surface to its edges, and this runoff pattern produces an 'edge effect', wherein a very lush strip of vegetation forms along the road edge (Frenkel 1970; Johnson, Vasek, & Yonkers 1975). Johnson et al. (1975) compared roadedge vegetation with that of adjacent control desert shrub and quantified that, indeed, all measures of productivity were much higher for roadside perennials for an unpaved road; edges had higher

density, greater percent ground cover, and greater aboveground biomass due especially to development of larger, fuller shrub canopies (Table 13.6). Even more pronounced was enhancement of shrub growth along a paved desert road (11 m wide), for which ground cover was 37.6%, shrub volume was exceedingly high, and aboveground biomass was 25 times that of controls. *Larrea tridentata* was the most important roadedge shrub, comprising 86% of total aboveground biomass along the unpaved road and 95% along the paved road, indicating that road edge can support a long-lived and very stable type of vegetation.

Road edges also supported relatively productive crops of winter annuals. For the unpaved road, roadedge and control plots each had 13 species, whereas the road edge of the paved highway had 23 species versus 17 for the control (Table 13.7). The single largest contributor to annual biomass was the introduced *Erodium cicutarium* (Fig. 13.4), comprising 55–75% of biomasses. Two introduced grasses, *Bromus rubens* and *Schismus barbatus* (Fig. 13.5), were also large contributors to biomass. Most native ephemeral species occurred at very low densities.

Roadedge vegetation that was studied by John-

Table 13.6. *Perennial vegetation within 2-m-wide strips of disturbed roads (D) and nearby relatively undisturbed control strips (U) in Lucerne Valley, San Bernardino County, California (Johnson, Vasek, and Yonkers 1975).*

Species	Shrub density, individuals 100-m^{-2}		Ground cover, %		Shrub volume, m^3 100-m^{-2}		Aboveground biomass, g m^{-2}	
	D	U	D	U	D	U	D	U
Larrea tridentata	8.5	6.0	26.3	3.8	13.55	1.67	490.55	60.56
Ambrosia dumosa	18.5	1.5	2.6	0.3	0.31	0.03	22.59	2.39
Ephedra nevadensis	2.5	1.0	1.6	0.3	0.25	0.03	33.28	3.31
Acamptopappus sphaerocephalus	0.5	10.0	0.2	0.2	0.03	0.02	2.77	1.88
Hymenoclea salsola	2.5	1.5	0.6	0.9	0.12	0.23	5.76	11.23
Eriogonum fasciculatum	3.0	2.0	1.1	0.6	0.16	0.10	8.62	5.48
D/U transect	2.93		5.43		6.95		6.69	

Table 13.7. *Species composition and density of winter annuals along the edge of an unpaved road, a paved road, and in adjacent undisturbed control sites for each in the Lucerne Valley of the western Mojave Desert. Exotic species are indicated by an asterisk. From Johnson, Vasek, and Yonkers (1975)*

Species	Unpaved road		Paved road	
	Control	Road edge	Control	Road edge
Amsinckia tessellata	14.8	30.4	0.2	1.0
*Bromus rubens**	2.4	12.4	–	8.3
Camissonia campestris	–	0.8	–	6.2
Camissonia boothii	0.4	–	3.0	1.0
Camissonia sp.	–	–	1.0	2.3
Chaenactis fremontii	–	–	–	1.9
Chaenactis stevioides	–	0.8	0.6	–
Chamaesyce sp.	–	–	–	0.2
Coreopsis bigelovii	–	–	0.8	0.4
Cryptantha spp.	6.4	2.8	–	1.0
Descurainia pinnata	24.8	30.8	1.6	3.8
Eremalche exilis	0.4	–	0.8	–
Eriastrum eremicum	–	–	1.6	0.4
Eriogonum spp.	–	0.4	0.8	2.3
Eriophyllum wallacei	5.6	0.4	2.0	6.4
*Erodium cicutarium**	158.8	144.8	117.6	156.0
Gilia sp.	2.4	2.8	13.6	2.1
Langloisia sp.	–	–	0.4	5.2
Lepidium sp.	0.8	0.4	–	1.2
Linanthus parryae	–	–	0.2	0.2
Malacothrix californica	–	–	–	0.2
Mentzelia affinis	0.4	–	4.2	11.0
Pectocarya sp.	1.6	0.4	0.8	1.2
*Schismus barbatus**	21.6	35.2	9.6	52.3
*Sisymbrium altissimum**	–	–	–	0.4
Total	240.4	262.4	158.2	264.6
No. of plots	25	25	30	32
No. of species	14	13	17	23

Figure 13.4 The introduced winter annual *Erodium cicutarium.*

Figure 13.5 The introduced grass *Schismus barbatus,* which grows successfully in soils compacted by vehicle tires.

son et al. (1975) did not resemble invasions by pioneer species that characterize other disturbed habitats described in this chapter. Instead, roadside microsites appeared to have increased establishment of *L. tridentata,* which under typical desert conditions has poor germination (Barbour 1968) and tends to recolonize very slowly.

Soil disturbance, erosion, and hydrologic balance

Disturbance of the desert pavement on soil surface (Chapter 2) may have profound impacts on sterility of desert scrub ecosystems that may not be immediately visible. Such disturbance can impact rates of soil erosion as well as hydrologic balance between moisture infiltration and runoff. Studies of erosional impacts on aridland ecosystems, using rainfall simulators in the field, have been carried

out at a number of sites, including NTS (Lane 1986). Field simulations of rainfall were conducted at NTS in 1983 and 1984 using a Swanson rotating boom simulator, which could deliver 60–130 mm hr^{-1} with drop sizes similar to natural rainfall (Simanton, Johnson, Nyhan, & Romney 1986). Two test plots at NTS were used in these studies: the site in Area 11 had coarse, loamy soil and was dominated by *Lycium andersonii* and *Atriplex confertifolia;* the other site near Mercury had a loamy soil dominated by *Menodora spinescens* and *A. confertifolia* (Romney, Hunter, & Wallace 1986). Soil surface structure was measured as the coverage of rock and gravel particles greater than 5 mm diameter. Analyses showed an exponential relationship of erosion rate and gravel cover, with highest rates of erosion occurring where rocky pavement cover was least (Fig. 13.6). Erosion rate is presented as metric tons of sediment per hec-

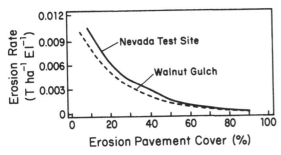

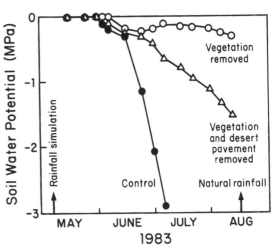

Figure 13.6 Relationship of erosion rate to gravel cover in field simulations of rainfall conducted at the Nevada Test Site using a Swanson rotating boom. After Simanton, Johnson, Nyhan, and Romney (1986).

Figure 13.7 Relationship of soil water potential at 0.15 m and 0.30 m during drying following removal of vegetation or vegetation and desert pavement. After Romney, Hunter, and Wallace (1986).

tare per unit of erosivity (EI). The EI factor measures physical force of the rainfall event in MJ mm^{-1} ha^{-1} hr^{-1}. The exponential pattern of relationship between erosion rate and pavement cover was very similar to that observed at Walnut Gulch in the Sonoran Desert in Arizona (Simanton et al. 1986). Compared with the Arizona site, however, vegetation was more effective in reducing erosion rates at NTS.

Relative importance of vegetation and desert pavement cover in affecting soil hydrologic patterns was investigated using rainfall simulator experiments. Soil psychrometers and fiber glass soil moisture cells were used to sequentially follow conditions of soil water potential at 0.15 m and 0.30 m depths following simulated rainfall in plots at Mercury and Area 11 (Romney et al. 1986). In May, 1983, a simulated rainfall event of about 100 mm was applied to paired plots with one of three types of treatments: (a) all vegetation removed, (b) all vegetation and desert pavement (2 mm) removed, and (c) control with natural cover. The strikingly different patterns of soil moisture depletion that resulted are shown for a soil depth of 0.30 m in the Mercury plots (Fig. 13.7). In control plots soil water potential began to drop sharply six weeks after the treatment and reached 3.0 MPa by early July, when perennials continued to utilize soil moisture. In contrast, plots without vegetation showed very little drop in soil water potential even in August. In plots without vegetation and desert pavement, soil

moisture experienced a steady decline in soil water potential to about −1.6 MPa by mid-August.

Erosional studies at NTS demonstrated quantitatively that removal of vegetation and desert pavement greatly increased the rates of moisture runoff from NTS soils, and this had profound impacts on hydrologic balance (Table 13.8). Erosion rates on removal plots in Area 11 were up to seven times greater than those for control plots. Overall, bare plots averaged runoff rates 2.7 times higher than control plots and 1.3 times higher for plots with vegetation removed than controls (Table 13.8). For control plots, mean runoff for all seasons and sites was 41.3% of rainfall applied in the simulations (Romney et al. 1986). In experimental plots, because runoff was increased less water was available for soil moisture recharge. Such soil moisture recharge is of critical importance for maintaining primary production by perennial shrubs (Lane, Romney, & Hakonson 1984).

Whereas soil runoff was fairly predictable for experimental manipulations used in rainfall simulation studies at NTS, actual runoff may vary considerably with amount and intensity of precipitation. Heavy desert rains can produce flash floods

Table 13.8. *Relative runoff rates of simulated rainfall from experimental plots at Mercury, Nevada and Area 11 at NTS during spring (SP) and fall (F) treatments at three levels of soil moisture content. All values are presented as % of control value at the same season and soil moisture content for each site. Adapted from Simanton et al. (1986).*

Site	Soil moisture	Treatment			
		Vegetation removal		Vegetation and pavement removal	
		Spring	Fall	Spring	Fall
Mercury					
	Dry	178	135	230	162
	Wet	159	153	183	155
	Very Wet	147	133	171	157
Area 11					
	Dry	33	144	700	317
	Wet	100	162	391	331
	Very Wet	118	115	245	188

in relatively short periods of time. Wallace and Romney (1972) documented effects of flood damage at Rock Valley and other nearby areas in southern Nevada following heavy rains in January and February of 1969.

Soil disturbance may result in increased wind erosion as well as erosional losses of sediment via runoff. Windblown or aeolian dust has major environmental and economic consequences (Goudie 1978, 1983; Prospers 1981; Pye 1987) and thus is a subject of considerable concern in arid lands, where high winds are frequent and there is little vegetational cover. Studies using a portable wind tunnel have demonstrated erodibility of various undisturbed and disturbed soils within the Mojave Desert (Gillette, Adams, Endo, & Smith 1980; Gillette, Adams, Muhs, & Kihl 1982). Gillette and coworkers found that for undisturbed soils even a weak surface crust, having a modulus of rupture less than 0.07 MPa, will protect the soil from wind erosion. Disturbed soils, however, were readily eroded unless the undisturbed crust had a modulus of rupture greater than 0.1 MPa. Threshold velocity of winds for erosion of disturbed soils ranged from only 0.47–0.62 m sec^{-1} as compared with 1.43 to greater than 2.0 m sec^{-1} for the same soils without disturbance (Table 13.9). For sand dunes the threshold velocities for

Table 13.9. *Mean threshold velocities (m sec^{-1}) of wind erosion for sediment types in the Mojave Desert. Data were determined by Gillette et al. (1980, 1982) using a portable wind tunnel.*

Sediment type	n	Threshold velocity, m sec^{-1}	
		Undisturbed	Disturbed
Desert pavement	4	216	58
Playa crust	14	>200	62
Alluvial fan	5	143	47
Sand dune	2	50	31

sediment transport were 0.31 and 0.50 m sec^{-1} for disturbed and undisturbed sediments, respectively.

IMPACTS OF OFF-ROAD VEHICLES ON THE MOJAVE DESERT ECOSYSTEM

Any first-time visitor to desert ecosystems can observe that driving a vehicle through desert scrub causes great damage to shrubs and produces tracks on the soil surface (Carter 1974). Moreover, early studies showed low plant production at disturbed sites. For example, in Kern County, Cali-

Table 13.10. *Estimated densities of winter annuals in Rock Valley in control and test areas for vehicle impacts in 1973–1974. Data from Vollmer et al. (1976).*

Sample area	Number of quadrats	Size of quadrats, m²	Estimated density of annuals, individuals m⁻²
Control			
Zone 22	52	0.10	112
Zone 23	36	0.10	80
Zone 25	136	0.10	46
Test area			
Ruts of regular track	50	0.25	8
Hump of regular track	50	0.25	24
Randomly driven areas	400	0.25	39

fornia, Davidson and Fox (1974) observed that at a heavily used site for motorcycle races total number of annuals was three to four times greater in adjacent, undisturbed desert scrub than in pit areas or along the race trail of April, 1973, where abundance of native species was extremely low and about 80% of the annuals were the introduced species *Erodium* and *Schismus* spp. (Figs. 13.4–13.5). Scientists have attempted to quantify long-term effects of vehicular traffic across these communities (Wilshire & Nakata 1976; Webb & Wilshire 1983).

An IBP study in Rock Valley

To evaluate direct effects of vehicular compaction on a Mojave desert scrub community, a study was conducted in Rock Valley during the IBP study (Vollmer, Maza, Medica, Turner, & Bamberg 1976). A 300×300 m test area was established outside the northeast corner of the validation site in an area of comparable vegetation on the validation site. Before starting the experiment, both areas were trapped to determine densities of rodents and lizards. Two 4-wheel drive trucks, weighing 2544 and 2274 kg and having different tire width and wheelbases, were used on the test area. In November, 1973, the heavier truck was driven along a fixed course to establish a track, and between late November, 1973, and early May, 1974, trucks drove that track on 18 occasions, each time circling the track twice. From early

December, 1973, to early May, 1974, there were also 17 random excursions across the test area, not using the track. During spring and summer, 1974, Vollmer and coworkers counted rodents and lizards on the experimental and control plots, sampled winter annual populations, and assessed damage to shrubs within small sampling quadrats in randomly driven areas, in the middle of the track, and in tire track ruts to compare with quadrats having undisturbed desert scrub on the validation site.

Mean density of annuals on the validation site was 79 individuals m⁻² (46–112 m⁻² in three zones, 224 quadrats total) as compared with 39 m⁻² in areas exposed to random driving (400 quadrats), 24 m⁻² in the middle of the track (50 quadrats), and 8 m⁻² in ruts (50 quadrats; Table 13.10). Many quadrats beneath tire tracks were devoid of annuals, but even on the control plot over two-thirds of the quadrats had 0–4 plants. Analyses showed that only densities of annuals in ruts were significantly different from all other areas. About 54% of the shrubs in the track sustained 90% or more damage, and 56% of the shrubs in randomly driven areas received damage, which was estimated to be no more than 10%. There was no difference in the vulnerability of individual shrub species to damage by random driving. Densities of seven rodent species at Rock Valley showed no apparent differences between the test area and control plots, at least none that could be attributed to vehicular disturbance. The

Table 13.11. *Densities of lizard species on IBP control driven plots in Rock Valley, 1974. From Vollmer, Maza, Medica, Turner, and Bamberg (1976).*

Sample date	Cnemidophorus tigris individuals ha^{-1}		Callisaurus draconoides individuals ha^{-1}	
	IBP Site	Driver plots	IBP Site	Driver plots
13–14 May	2.3	3.6	1.6	0.3
15–16 May	2.9	1.7	1.8	0.6
23–24 May	1.8	1.1	1.2	0.3
May mean	2.3	2.1	1.5	0.4

most striking change was a temporary reduction of *Perognathus longimembris* (little pocket mouse) on the validation site from April to July, 1974, during a very dry year and when food availability and quality was low. For *Uta stansburiana* (side-blotched lizard), populations in both sampled areas declined between October, 1973, and October, 1974; age distributions were essentially identical during March, but by October, 1974, there were fewer juveniles on the validation plot than in the test area. No significant differences were observed between sampled areas for *Cnemidophorus tigris* (whiptail lizard), but for *Callisaurus draconoides* (zebra-tailed lizard) density was extremely low on the test area (0.4 individuals m^{-2}) and significantly lower than on the IBP site (1.5 individuals m^{-2}; Table 13.11).

Soil compaction

Soil compaction causes a reduction in the waterholding capacity of a soil and a decrease in soil gas volume, and when soil strength is very high, i.e., soil is compact, it becomes difficult or impossible for roots to penetrate and animals to burrow. Establishment and regrowth of most plants is slow on compacted soils.

Studies on soil compaction were conducted on Mojave Desert soils at five sites in California using a 4-wheel drive vehicle (2190 kg) and a motorcycle (188 kg; Adams, Endo, Stolzy, Rowlands, & Johnson 1982). Soil strength was measured with a penetrometer at 0.05-m intervals in intershrub spaces when soil was wet and later when it was dry.

The heavy 4-wheel vehicle produced greater soil strength following trials on wet soil, especially those with much loam, and certain soils showed significant compaction at a soil depth of 0.25 m after only three passes. High compaction (150 g mm^{-2}) was observed at 0.05–0.10 m soil depth within 20 passes on wet soil and after 5 passes on dry soil. This level of compaction is comparable to that observed in abandoned townsites and developments, as described earlier in this chapter.

With the lighter motorcycle, soil strength values in the uppermost soil horizon of a wet soil did not differ from controls following a single pass, but after 5–10 passes soil strength at 0.05 and 0.10 m depth was double that of untreated controls at the same soil depths. Compaction of wet soil by the motorcycle was not observed below 0.20 m, even after 100 passes, but with 20 or fewer passes significant compaction had occurred at 0.10 m soil depth. After 10 motorcycle passes over a dry soil, soil strength at 0.10 m had also doubled and exceeded 250 g mm^{-2}, producing a soil that most roots cannot penetrate.

Adams et al. (1982) suggested that off-road vehicle driving would cause a significant reduction in populations of desert annuals in subsequent years. Although data were not presented in the paper, they noted significant reductions in cover of annuals in vehicular tracks, especially *Chaenactis fremontii* and the introduced *E. cicutarium*, which have prominent taproots, and conversely higher cover by the introduced grass *Schismus*, which has many thin, fibrous roots near the surface. Experiments were not performed to determine why annual cover was reduced relative to that on control plots.

Working in a desert scrub dominated by *Larrea* and *Atriplex torreyi* in western San Bernardino, California, Webb (1982) studied effects of four motorcycle trails created with 1, 10, 100, and 200 passes, driven at a constant rate to avoid acceleration. Measurements were made to quantify soil bulk density, saturated conductivity, soil compaction, penetration resistance, response to rainfall, and draining properties. Soils at that site had

desert pavement, which was noticeably disrupted by the initial impact of the motorcycle, and most annuals on the track were destroyed by pass 10. Soil of the upper 60 mm showed an increase in bulk density as a logarithmic function of the number of passes (Fig. 13.8), and compaction, as estimated from several parameters, was significant. Compaction strongly affected infiltration rates, which decreased as a logarithmic function of the number of passes (Fig. 13.9). Hence, soil properties varied inversely with number of motorcycle passes and had lower capacity to support plant growth. One year after the experiment, when the 1- and 10-pass trails had disappeared, multiple-pass samples showed that they had significantly higher densities than the original undisturbed soils, but above 30 mm had a lower density from that immediately following the motorcycle treatments. Webb (1982; Webb & Stielstra 1980) found that *S. barbatus* (Fig. 13.5) grew exceedingly well on the compacted soils of tire tracks, because its roots penetrated and loosened surface soil.

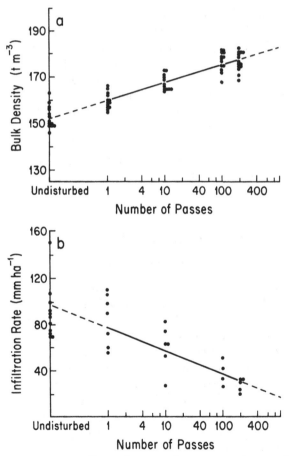

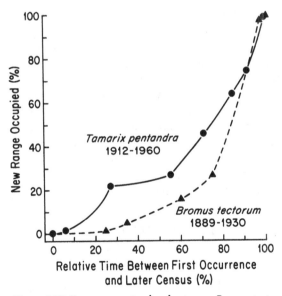

Figure 13.8 Effects of motorcycle passes on desert soil characteristics: (a) Change in soil bulk density as a function of number of passes by a motorcycle. From Webb (1982). (b) Soil infiltration rate as a function of number of passes by a motorcycle. From Webb (1982).

Figure 13.9 Range expansion by cheatgrass, *Bromus tectorum*, in the western United States and saltcedar, *Tamarix pentandra*, in the southwestern United States. From Mack (1985) and Robinson (1965).

Vehicles on sand dunes

Outside the boundary of the Mojave Desert but applicable to its biological studies are observations on effects of off-road vehicles at the Algodones Dunes in extreme southeastern California. These dunes are structurally comparable to dune systems in the Mojave Desert. Because of their shifting nature and granular composition, sand dunes experience few of the long-term effects of soil compaction. The problem here is that dune plants (psammophytes) often incur lethal damage to the root systems, by exposure to air, sudden high soil temperature, or mechanical breakage. A survey on paired plots of undisturbed and ORV-used areas showed dramatic reduction in numbers and diversity of herbaceous and woody plants as well as the animals living in those dune communities, and herbaceous cover was virtually absent where vehicle use had been heavy (Luckenbach & Bury 1983).

Loss of plants at the Algodones Dunes apparently resulted in lowered production of arthropods, especially noticeable for the phytophagous Coleoptera. Denuding effects were observed even where vehicular activity was relatively low. Lizards and rodents were significantly less abundant and had extremely low body weight on the impacted plots, and rodent diversity was also lower. Luckenbach and Bury (1983) found a relatively good linear relationship between number of terrestrial vertebrate individuals at a site and density of perennial vegetation. Evidence of physical damage to animals was demonstrated by a higher incidence of tail loss in lizards on impacted dunes. In short, on desert dunes all parameters characterizing biotic diversity and abundance decreased significantly under stresses of vehicular traffic.

RADIATION BIOLOGY AND ENERGY DEVELOPMENT PROJECTS

Following detonations of the first atomic bombs over Japan to end World War II, the world was horrified by not only the power of those nuclear devises, but also the grotesque changes to humans and the environment caused by radioactive fallout. Immediate answers were needed on lethal and toxic levels of radioactive materials for humans, animals, plants, and microbes. Scientists launched an initial phase of nuclear research that was focused on physiology and had the task to determine levels of toxicity. Such experiments were best performed under relatively controlled laboratory conditions. However, a new field, called radiation ecology or radioecology, evolved to understand what happened to radioisotopes after their release into the environment, how they become diluted or concentrated and move through ecosystems, affecting food chains (Odum, 1959).

Beginning in the 1950s ecological research was conducted at the Nevada Test Site and other national facilities, where radioactive materials could be released into the environment for studying impacts on natural systems (see Historical Prologue). At NTS the challenge to researchers was to discriminate the separate effects of physical damage versus exposure to radioactivity, and it was equally complex to evaluate effects of the long list of beta and gamma radionuclides that were encountered by organisms (Nishita 1965; Wallace & Romney 1972; Cannon, Strobell, Bush, & Bowles 1981).

Close-in effects of nuclear testing

Aboveground testing

After eight years (1951–8) of aboveground testing at NTS, there was surprisingly little destruction of desert vegetation (Shields & Wells 1962, 1963; Shields, Wells, & Rickard 1963). For example, when a 40-kiloton nuclear fission device was detonated at 90 m above ground, desert vegetation was totally destroyed within 0.8 km from the tremendous shock and heat that was generated, but much of the potential and imagined destruction to vegetation was minimized because weapons were detonated near the centers of barren playas and therefore at a considerable distance from surrounding vegetation. Radioactive materials for the most part were drawn upward with the ascending fireball and subsequently dispersed at high altitude from the fallout cloud (Wallace & Romney 1972).

From aboveground explosions the most severely damaged shrub was *Grayia spinosa*, which was selectively killed beyond 1.1 km by blast damage because its wood is extremely brittle. Shields and Wells (1962, 1963) observed that death of shrubs from a typical aboveground test in Yucca Flat was not symmetrical around ground zero, and shrubs were killed and damaged at greater distances when growing on loose, sandy alluvium without desert pavement than on compact, unsorted alluvium having a well-developed pavement.

Disturbance of the landscape around ground zero opened up the site for colonization by pioneer species (Shields & Wells 1962, 1963; Shields et al. 1963; Wallace & Romney 1972). Chief among initial invaders were short-lived perennials, for example, bunchgrasses *Stipa speciosa* and *Oryzopsis hymenoides*, and such weedy native species as *Atriplex canescens*, *Hymenoclea salsola*, *Mirabilis pudica*, and *Sphaeralcea ambigua*. Very few plants initially were able to invade within 0.5 km from ground zero, the most notable exceptions being *Salsola kali* (Russian thistle), a very aggressive C_4 annual weed, and a native C_3 annual, *Mentzelia albicaulis* (stickleaf). *Bromus rubens* and species of *Chaenactis* were pioneers that progressively invaded denotation sites in the direction of ground zero.

Belowground testing

Damage to desert scrub from a belowground detonation in soft sediments at NTS was extensive with the formation of a huge crater and deposition of expelled throwout (ejected soil and parent material). The Project Sedan explosion in July, 1962, for example, provided a detonation that yielded 100 + 15% kilotons. An average 6 m of debris was deposited at 0.3 km from ground zero, up to 1 m at 0.6 km, and up to 0.3 m at 0.77 km (Turner 1963; Turner & Gist 1965). Within 0.6 km from ground zero all vegetation was totally destroyed; from 0.6 to 1.5 km shrubs were damaged but not totally destroyed, whereas beyond 1.5 km vegetation was not destroyed but instead covered with radioactive dust.

With Project Sedan a thriving adult lizard pretest population living within 1.2 km of ground zero was exterminated, and for six months after the blast only two adults were observed within 1.8 km (Turner & Gist 1965). At 1.4 km there were few adult lizards of *Cnemidophorus tigris* one year after the blast, but densities of *Uta stansburiana* were similar to pretest values by the summer of 1963. At 2.6 km the lizard community had a surviving but reduced posttest population, and one year later there was no evidence of a persistent change that was attributable to the test. Virtually all juvenile lizards were killed within 0.5 km of ground zero. None the less, one month after the blast young lizards hatched and emerged at 0.80 km, suggesting that buried eggs were not completely destroyed by the explosion; however, mortality of those hatchlings was very high. There was no evidence of juveniles as close as 2.6 km. Based on evidence that prey items were available in posttest plots and irradiated lizards in the laboratory showed no lethal or visible effects after treatment with 635–1450 rads, Turner and Gist (1965) interpreted very high posttest mortality of juveniles to be a function of lack of shrub cover from summer heat.

Effects of radioactive and nonradioactive dust

During the decade of underground detonations, venting of materials resulted in exposure of desert plants and animals to high radiation levels, and shrub foliage retained fallout particles for a significant period (Romney, Lindberg, Hawthorne, Bystrom, & Larson 1963). Damage due to fallout on leaves typically was visually expressed 3–6 months later (Wallace & Romney 1972). Much of that damage was attributed to beta rather than gamma radiation, and investigators showed that shrubs covered during and for two weeks after the explosion with sheets of 6-mil polyethylene plastic escaped fallout dust damage. Martin (1965) determined that most radioactive particles on leaves were smaller than 5 µm, and less than 2% were larger than 44 µm.

Beatley (1965) reported on a study done in conjunction with the July, 1962 detonations of Project Sedan at Yucca Flat and Small Boy on the playa in southeastern Frenchman Flat. She had been monitoring plots of desert scrub communities near those testing events and affixed dosimeters to shrubs prior to the detonations. At one plot adjacent to Yucca Flat shrubs and soil were heavily blanketed with radioactive dust. Cumulative dosage at that plot averaged 4737 R + 15% of gamma radiation. Many of the leaves of the evergreen leaves of *L. tridentata* were greenish-gray from the dust and had abscised by mid-winter so that by spring all branchlets were defoliated. No flowers were observed that spring in those plots. Undusted plots around Yucca Flat showed no defoliation of *Larrea* and normal spring flowering. Heavy rains in September, 1963, stimulated basal sprouting from buds on old main stems.

At plots located 8 km east of the Small Boy ground zero, the site received neither blast damage nor radioactive dust and had relatively low (600 R) exposure to radiation. However, during construction of the detonation device vehicle traffic on the playa produced much nonradioactive (clay) dust, which coated the resinous leaves of *Larrea*. Beatley discovered that *Larrea* with nonradioactive dust experienced the same type of defoliation as those around Yucca Flat with radioactive dust. Moreover, she observed that defoliation was proportional to dust cover along a gradient from the unpaved access road of the detonation site and was greatest on the side of the shrub facing the road. She concluded, therefore, that defoliation was not related to irradiation, but was strictly associated with heavy spring and summer dust cover of the leaves, i.e., a mechanical and not a radiation problem for cause of leaf and bud death.

Larrea tridentata typically does not occur in heavy desert soil with high clay content, and Beatley suggested that this pattern means that its evergreen leaves were not exposed to heavy dust consisting of small soil particles. This species commonly grows near margins of playas, but probably can do so only if dust from the playa is not excessive. Dust only becomes a problem when the smooth playa surface is broken and disturbed.

Effects of ionizing radiation on plants

Rock Valley field studies and associated laboratory tests were conducted to evaluate the direct effects of ionizing radiation on the common desert shrubs, and methods and results have been published (Kaaz, Wallace, & Romney 1971; Harvey 1972; Wallace & Romney 1972; Vollmer & Bamberg 1975). In plot B of Rock Valley, a team of researchers followed the effects of ionizing radiation on numerous growth parameters for six species of dominant shrubs. Chosen for study were examples of several different types of shrubs (Chapter 3), including *L. tridentata*, *Ephedra nevadensis*, with apyllous photosynthetic stems, and *Ambrosia dumosa*, *Grayia spinosa*, *Krameria erecta*, and *Lycium andersonii* with drought-deciduous leaves.

Generally speaking, radiation effects on shrubs were weak and often difficult to document. Two species were identified as showing some radiation damage, a class III type damage in which there is reduction of lateral shoot production, leaf production, and reproductive structures. *Ephedra* was the most sensitive species (Kaaz et al. 1971; Harvey 1972). Shrubs in plot B had a cumulative radiation exposure of 3.9–9.8 kR, and in 1969, when rainfall was higher than average and *E. nevadensis* grew well in Rock Valley, plants in plot B showed marked reduction in vegetative and reproductive growth (Table 13.12). Plants showing very good vegetative growth comprised 25.8% in the control plot D as compared with 0% in plot B, and at the other those showing only poor growth were 3.0% in control versus 29% in plot B. Total percent of plants with reproductive structures was 35.9 on the control plot versus 4.7 on the irradiated plot. Seed production was 26.9% of total plants versus 1.3%, respectively, and control plants had more seeds per plants. Because *E. nevadensis* spreads predominantly by means of underground shoots rather than by seeds, a reduction in seed production probably had less impact on the community than did the sharp limitation by

Table 13.12. *Quantitative differences in vegetative and reproductive growth of* Ephedra nevadensis *growing in control and gamma-irradiated plots at Rock Valley, 1965–1969 (After Kaaz, Wallace, and Romney, 1971). All results showed statistically significant radiation effect growth at P = 0.01 level.*

	Control		Irradiated plot	
	n	% total plants	n	% total plants
Shoot growth				
luxuriant	495	25.8	0	0.0
heavy with many shoots 0.15–0.30 m long	1066	55.7	505	31.0
sparse, shoots less than 0.07 m long	296	15.5	650	40.0
little or none	58	3.0	472	29.0
Staminate cones per male plant				
over 100	208	10.9	3	0.2
26–100	242	12.6	11	0.7
25 or fewer	238	12.4	63	3.8
Fertile cones per female plant				
over 100	109	5.7	0	0.0
26–100	169	8.8	1	0.1
25 or fewer	238	12.4	19	1.2

radiation on shoot growth and, hence, photosynthetic capacity.

Vollmer and Bamberg (1975) reported on irradiance of *K. erecta* for 3331 exposure days, and during the 10-yr experiment plants received from 3.7 to 29.6 kR of gamma radiation. Plants that received greater than 6 R day^{-1} had low leaf production and almost complete inhibition of fruit production. At plot B, 16% of the shrubs died as compared with only 1.2% in the nonirradiated plot, and plants that received greater than 25 kR cumulative exposure were dead or tended to have less than 10% live shoots, indicative that death would have occurred by prolonging exposure to radiation.

In intensive laboratory tests *E. nevadensis* exhibited mild radiation damage above 0.2 kR, whereas *L. tridentata* and *Lycium andersonii* showed similar damage starting at 4.5 kR and 5.0 kR, respectively (Harvey 1972). Those same experiments demonstrated that stem chlorosis occurred in *Ephedra* at 0.6 kR chronic exposure, and this was accompanied by cessation of terminal growth, shriveling of buds, and failure of sexual reproduction. *Ephedra* was judged to be most susceptible of the shrubs to radiation damage because its cells had the largest chromosomal volume, similar to results of earlier studies in forest plants of eastern North America (Woodwell & Sparrow 1962; Sparrow & Woodwell 1963). For Mojave Desert plants that were tested, germination was mostly unaffected by relatively intense exposure to gamma radiation (Wallace & Romney 1972).

Effects of radiation on animals

When the field of comparative radiosensitivity was in its infancy, estimates for acute doses of radiation were based almost exclusively on responses of laboratory-reared organisms (Odum 1959; Hines 1962; Carlson & Gassner 1964). Early studies suggested that acute doses for mammals were much smaller than those required for lower vertebrates, and that serious symptoms may be manifested in mammals at doses from 0.1 to 1.0 kR (Odum 1959). Certainly the initial study on *Perognathus formosus*, which used 1.35 kR in treatments, provided data showing that wild-collected animals from NTS had fairly high levels of resistance to damage from ionizing radiation (Gambino & Lindberg 1964; French 1965).

Studies on *Perognathus*

To obtain some data on the possible effects of radiation on vertebrates, a study was performed at NTS to document visible chromosomal aberrations of *P. formosus* (Towner 1965). From 1963 to 1964 Towner live-trapped 123 individuals from Yucca Flat on a site where amount of radioactive fallout had been very high during the aboveground testing program from 1951 to 1958, and he similarly trapped 84 individuals from Jackass Flats, where no nuclear testing had been done. To his surprise, an examination of 60 individuals from each site yielded negative results, excluding one possible exception, a female from near Yucca Flat that had consistently aberrant karyotypes in all leukocytes.

Studies in Rock Valley with the ^{137}Cs source (see Historical Prologue) involved chronic gamma irradiation on a large, tagged population of *P. formosus* in plot B for comparisons with nonirradiated pocket mice in other plots (French et al. 1974; Turner 1975). Chronic exposure, measured with tiny implanted dosimeters, was 211–360 R yr^{-1}, and most of the dose was received from April to September. The irradiated rodent population had a reduced survival of juveniles less than six months of age, an instantaneous death rate of 0.219, more than twice that of controls (0.104 and 0.075), and, consequently, a short life expectancy. For the irradiated population its computed intrinsic rate of increase was 0.314 as compared with 0.493 and 0.498 for controls. Nevertheless, even though there were detectable differences in the life expectancy curves of irradiated versus control rodents, the experiment provided no evidence that reproduction was impaired or that reproductive organs were rendered sterile.

Lizards of Rock Valley

Radioecologists were very surprised to discover that lizards of Rock Valley were more sensitive to ionizing radiation than were rodents (Turner 1975). Initially herpetologists quantified significant differences in growth rates of irradiated and nonradiated populations (Turner, Hoddenbach, & Lannom 1965; Turner, Lannom, Kania, & Kowalewsky 1967), and after four years of irradiation there appeared to be no lethal effect on these populations when individuals received 2 R day^{-1} of radiation (Turner & Lannom 1968; Turner, Medica, Lannom, & Hoddenbach 1969a, b). But, as sampling continued, a high incidence of female sterility was diagnosed among the irradiated populations of several different species, initially in *Crotaphytus wislizenii* and *Cnemidophorus tigris*, then *Phyrnosoma platyrhinos* and *Uta stansburiana* (Turner, Licht, Thrasher, Medica, & Lannom 1973; Medica, Turner, & Smith 1973a, b; Turner 1975; Turner & Medica 1977). Female sterility was fully documented from striking demographic differences of irradiated and control plots when in plot B summer hatchlings and spring yearlings were virtually absent, suggesting reproductive failure of irradiated populations. Laparotomy of lizards demonstrated that in 1968 female leopard lizards (*C. wislizenii*) had enormous hypertrophy of pleuroperitoneal fat bodies and apparent loss of ovarian tissue (Turner et al. 1973). Female sterility in that species occurred after a cumulative dose of 1.5 + 15% kR. At a daily dose of 0.4–0.5 R day^{-1} and with periods of inactivity within burrows, this means that female sterility takes several years before being manifested. *Crotaphytus*, unlike most smaller lizards, has delayed reproduction and does not breed until it is 20–21 months of age. Thus, if female reproduction maturity does not start until the third year and sterility can occur within the next two years, then a serious drop in population size is expected, and, in fact, the irradiated population of leopard lizards declined precipitously for those reasons. Ironically, this same population of irradiated animals has been observed to have estimated ages of nearly 12 years (Medica & Turner 1984), so that a substantial portion of the breeding female population became sterile.

Female sterility was investigated in great detail for *U. stansburiana* (Turner & Medica 1977). Females that were born on plot B possessed normal reproductive organs as juveniles and participated in normal reproduction for at least one season, but experienced abnormal increases in ovary weight and mean fat-body weight around month 20. For irradiated females, during the second reproductive year an increase in mean ovary weight began in March and diverged sharply from control animals in April until sterility was noted in July. Weight of fat bodies in irradiated females increased in the fall months to exceedingly high levels in sterile or half-sterile (one functional ovary) individuals. By month 32 almost all females of *U. stansburiana* became sterile. Cumulative dose to produce sterility was from 0.5 kR in certain individuals to 1.2 kR in others, but empirical data showed that 83% of sterile females were 20 months old.

Follow-up studies on irradiated versus non-irradiated lizards documented several other attributes of chronic gamma radiation on *U. stansburiana* in Rock Valley. Trying to pinpoint sites of radiation damage, detailed analyses conducted on individuals having doses 1.5–10 R day^{-1} showed that low levels of radiation affected the spermatogenic process directly and produced no detectable interference with pituitary function (Pearson, Licht, Nagy, & Medica 1978). Sterile females also had significantly lower rates of energy intake by feeding, lower rates of energy expenditure via respiration, and exceedingly large energy storage as fat deposits (Nagy & Medica 1985).

Solar thermal power systems

An energy development project with potential for considerable alteration of Mojave Desert ecosystems is the solar thermal power system. The primary environmental impact of this type of system comes from required clearance of large areas of desert land. Of course, desert landscapes are model candidates for such stations because they receive high levels of annual solar irradiance and land is inexpensive and easy to clear and level.

In 1980–1 a pilot 10-MWe solar thermal power system was constructed near Barstow in the Mojave Desert by a consortium of Southern California Edison, the U.S. Department of Energy, the Los Angeles Department of Water and Power, and the California Energy Resources Conservation and Development Commission. Baseline ecological studies were conducted by the Laboratory of Biomedical and Environmental Sciences at UCLA on this pilot project (Turner 1979, 1981, 1982). UCLA investigators found little or no evidence of significant changes in biological communities outside the area directly impacted by construction of the plant. Downwind micrometeorological measurements revealed that only small effects on air temperature (<0.05°C), wind speed (<0.4 m s^{-1}), and evaporation rates (<1.5 ml hr^{-1}) were present. These alterations are within the range of normal spatial heterogeneity in desert microclimates, and are less than those to be expected downwind of irrigated agricultural fields (de Vries 1959; Rider, Philip & Bradley 1963).

Effects of coal ash on desert plants

Coal-burning power plants produce large quantities of coal ash as a biproduct of generating electricity. Power plants in desert regions have used landfill disposal techniques to bury this ash material, totalling tens of thousands of tons, beneath desert soils around these plants. Experimental studies have investigated ecological effects of aerial deposition of coal ash on vegetation in the Mojave Desert of California (Vollmer, Turner, Straughan, & Lyons 1982) and Nevada (Turner & Vollmer 1982).

At California sites, Vollmer et al. (1982) applied precipitator ash to field plots at concentrations of 5, 10, 30, and 100 mt ha^{-1}. The lowest dose was about 28 times the worst-case field situations. Observations from five consecutive growing seasons found no effect on annuals from the 5 mt ha^{-1} treatment, but significant negative impacts from 10 mt ha^{-1}. Increased surface soil pH, increased soil salinity, and trace-metal contamination may have been responsible for ecological problems resulting from heavy applications of coal ash to desert soils.

At Jackass Flats on NTS effects on annuals were not observed at 5 mt ha^{-1} dose, but reduction in numbers and abundance of species were observed at 10 mt ha^{-1} and especially the two highest doses (Turner & Vollmer 1982). At ash deposition of 100 mt ha^{-1} annual density was reduced to 2–4% of control values. Analyses of rank abundance also suggested that precipitator ash may affect species composition of the annual flora. *Bromus rubens*, an introduced grass, was exceedingly common in treated plots with 30 or 100 mt ha^{-1} dose in 1979 and all plots in 1980 and 1981. Under control conditions *Amsinckia tessellata* and *Caulanthus cooperi* were uncommon in desert scrub, but relatively abundant in plots with high ash content. No perceptive differences were observed between test and control plots on success of seed germination. The majority of measurements on shrubs, *Larrea tridentata*, *Ambrosia dumosa*, and *Lycium andersonii*, during four growing seasons showed no statistically significant effects of ash application as compared with pretest values.

INTRODUCED SPECIES AND REVEGETATION

The creosote bush desert scrub that characterizes most areas within the Mojave Desert is actually a subtly changing assemblage of plants and animals. At each site species density, relative and total plant cover, species composition, and biomass are controlled not only by factors of microclimate (Chapters 4 and 5), but also through biotic interactions, such as predation and competition. Natural selection tends to maintain producers and users in reasonable proportions, and, if one exceeds its normal limits, other elements of the community are augmented or decreased accordingly. For example, when population sizes of a prey increase in a given year, predators commonly increase their own populations to make use of excess prey items. Conversely, if in a bad year certain prey items become scarce, dependent predators also become

scarce and often shift to lower quality resources until more productive times occur. This dynamic equilibrium holds true for species that have belonged to the communities for thousands of years, but some serious problems can arise when nonnative species, new introductions that evolved under different selection regimes, enter the habitat.

Biological invasions of desert regions

Exotic plants and animals, i.e., those not native, are now widely distributed in arid regions of western North America, and present a potential for significant changes in the structure and function of these ecosystems. The most successful species are Old World forms that evolved in regions with long histories of human disturbance. Hence, when potential niches are opened, introduced species are already well adapted as pioneer species. Exotics can outcompete native species and pose a major threat if they simultaneously cause shifts in the interactions within the native community.

Invaders of western deserts

Alien plant invasions into the Great Basin Desert have been documented in detail by Mack (1986). He pointed out how sudden introduction of agriculture and livestock grazing in the last half of the nineteenth century swiftly changed desert steppe environment throughout the Intermontane West. Large native grazers produced comparatively little landscape disturbance before that time, and promoted dominance of perennial bunchgrasses and shrubs. With livestock and farming came weedy annuals from Eurasian steppe regions; weeds became permanent occupants and even dominants in the new environments. Once established, exotic annual grasses, especially *Bromus tectorum* (cheatgrass) provided a continuity of biomass that has radically altered the natural fire cycle. Areas with little or no natural fire now burn regularly, and such fire cycles further promote growth of annual grasses at the expense of native perennials and forbs.

The invasion of *B. tectorum*, beginning in the

late 1800s, provides the classic example of a logistic pattern of population growth in a successful exotic (Mack 1981, 1985). Following an initial lag phase when small colonies of this species were founded in isolated localities, a log phase of growth began soon after 1910, accompanied by rapid range expansion. By 1930, *B. tectorum* had colonized much of its potential range.

Detailed accounts of range expansions into arid zones are relatively limited. Robinson (1965) has provided some data for *Tamarix ramosissima* (saltcedar), a phreatophytic tree that has colonized water courses throughout the southwestern United States. This invader showed a slow growth of areal expansion from 1925 to 1939 followed by a logarithmic increase in its range.

Feral domestic animals are also found in certain desert regions of western United States. These include wild burro (*Equus asinus*) and horse (*E. caballus*) in the Mojave Desert Region (McKnight 1961), although they are generally restricted to moister upland sites. Concern has been expressed that wild burros may compete with desert bighorn sheep (*Ovis canadensis*) for limited food resources (Woodward 1976). Approximately 65 wild horses, divided into about 15 bands of 2–25 individuals each, are present on NTS (Greger 1990).

Alien plants at NTS

Approximately 125 alien species, nearly 12% of the total flora, are present at NTS (Beatley 1976a), and composition of the MDR flora are similar (Chapter 1; Rowlands et al. 1982). The great majority of these are annual species of Eurasian origin, largely confined to moist, disturbed sites or areas around human habitation. Without irrigation or artificial water supplies few of these species would survive.

A small list of weedy annuals have been able to invade relatively undisturbed communities of Mojave desert scrub. Included here are *Bromus rubens*, *B. trinii*, *Schismus* spp. (Fig. 13.5), and *Erodium cicutarium* (Fig. 13.4). Beatley (1976a) suggested that the mentioned grasses are so well integrated into native plant communities that it would be inferred that they are natives if their

history of introduction had not been documented. *Erodium cicutarium* is also widespread, but appears to prefer disturbed microsites when it invades native communities. None the less, this species is now widely utilized by native animals as a food resource (Reichman 1977; Inouye, Byers, & Brown, 1980). *Bromus tectorum* is essentially ubiquitous on disturbed sites in northern parts of NTS, but occurs sparingly in undisturbed areas (Beatley 1976a).

In the last decade the relative dominance and diversity of introduced weedy annuals have increased sharply at NTS, with significant increases in densities of *B. rubens* in Rock Valley and other Mojavean areas over the past three decades (Hunter & Medica 1989a, b; Hunter 1990, 1991). Population densities in the 1960s were never higher than 14 individuals m^{-2} before peaking at 91 individuals m^{-2} in 1976, the final year of the IBP study. In 1988 densities of *B. rubens* reached a remarkable 2034 individuals m^{-2}, 25 times the density of all native annuals! That year biomass of *B. rubens* was 34.2 g m^{-2}, 97% of annual biomass total, to be compared with a high of 3.2 g m^{-2} during the high of 1976 during the IBP studies. *Bromus tectorum*, previously restricted largely to Great Basin habitats, has expanded greatly at NTS in recent years and is now well established at Yucca Flat, where it invaded detonation sites (Hunter 1990, 1991).

Whereas introduced annuals do particularly well following habitat disturbance, their increasing ecological success in the Mojave Desert appears to be a widespread phenomenon that is independent of human actions. Hunter (1991) suggested that increased abundance of *Bromus* at NTS has resulted from establishment during the relatively wet cycle that occurred from 1973 to 1988. Once high densities of *Bromus* are present, their dead culms and litter persist for several years and serve to carry fire across intershrub spaces that would otherwise have served as barriers. Introduction of fire as an environmental factor into these desert habitats may profoundly affect community structure for both ephemerals and perennial shrubs in future years. Moreover, zoologists are concerned because some species, particularly lizards, have decreased abundances where brome grasses thrive because normal types of foraging behavior are prohibited (Cowles & Bogert 1944).

Species of Russian thistle (*Salsola*) are very common invaders of disturbed sites of NTS (Wallace & Romney 1972). These C_4 summer annuals rapidly pioneer disturbed roadsides and cleared areas; their rapid growth rates and effective tumbleweed mechanism of seed dispersal greatly improve the invasiveability of these species on revegetated NTS sites, even those having relatively undisturbed soils but where shrubs were killed by radiation (Wallace & Romney 1972). Hunter and Romney (1989) reported that *S. kali* has become increasingly widespread at NTS in recent years. It is interesting that these tumbleweeds, emblematic of our idea of the 'Old West', were only introduced very late in the nineteenth century, well after the West was 'tamed'.

Grazing impacts on Mojave desert scrub

Although the Mojave Desert region, like other arid lands of North America, has been subjected to grazing pressures for 50–100 years or more, surprisingly little quantitative information is available on the environmental impacts of grazing by domestic animals. This lack of data is unexpected; given that information it is critical for resource management, especially, to quantify sustainable yield and carrying capacity of grazed desert ecosystems.

Studies from various desert regions have shown that grazing has the potential to alter both the structural nature of soils and biotic composition of communities (Knoll & Hopkins 1959). In arid regions heavy grazing tends to cause soil compaction and breakup of surface crust, thereby impacting hydrologic inputs, nutrient stability, and erodibility (Klemmedson 1956; Ellison 1960; Sharp, Bond, Neuberger, Kulmon, & Lewis 1964; Arndt 1966; Gifford & Hawkins 1978). Other investigations established that cover of perennial shrubs and grasses increased rapidly in response to an elimination of grazing pressure (Gardner 1950; Blydenstein, Hungerford, Day, & Humphrey 1957).

Quantitative studies on sheep grazing in the Mojave Desert were conducted in four areas between California City and Inyokern, California (Webb & Stielstra 1980). Webb and Stielstra established 'grazed' and 'ungrazed' plots at sites in *Larrea* desert scrub at which comparable measurements were made of perennial vegetation (cover, shrub height, and shrub volume), annual plant biomass, and soil conditions (bulk density and moisture content). Mean shrub cover was significantly reduced by sheep in comparisons between grazed and ungrazed areas (Table 13.13). Likewise, grazed stands had significantly shorter shrubs and smaller shrub volumes. In grazed plots mean area cover of *Ambrosia dumosa* was 16–19% lower than in control plots, whereas volume was 21% and 65% less (Table 13.13).

In addition to grazing on shrubs, sheep studied by Webb and Stielstra fed heavily on the dense annual plants in 1978. Heavy grazing reduced aboveground biomass of annuals by 69% beneath plants of *L. tridentata*. Densities of annuals between shrubs were reduced by 24% and 28%, respectively, in two comparisons of grazed and ungrazed samples on lightly grazed plots. Whereas density of annual species was not affected by grazing, data from these studies revealed that the exotic species of *Erodium* and *Schismus* were more resistant to grazing pressure than were native species of annuals. This observation parallels notes by Davidson and Fox (1974) that *Schismus* increases in density in areas disturbed by offroad vehicles; hence, environmental disturbance actively promotes expansion of exotic plant species.

Grazing activities may also have a variety of impacts on wildlife populations. Busack and Bury (1974) found that grazed areas of the western Mojave Desert supported lower numbers and biomass of lizards than did control areas. Grazing may also negatively impact desert tortoise (*Gopherus agassizii*) populations through destruction of burrows and competition of livestock with tortoises for limited plant food reserves (Berry 1978).

Table 13.13. *Impact of sheep grazing on paired plots at two sites in the western Mojave Desert between Inyokern and California City, California. Significant diffferences in mean (5% level) are indicated by an asterisk after the value for lightly grazed plots. Adapted from Webb and Stielstra (1980).*

	Site 1		Site 4
	Ambrosia dumosa	*Acamptopappus sphaerocephalus*	*Ambrosia dumosa*
Mean shrub height (m)			
Heavily grazed	0.21	0.20	0.32
Lightly grazed	0.37*	0.38*	0.32
Mean shrub cover (m^2 $shrub^{-1}$)			
Heavily grazed	0.085	0.068	0.137
Lightly grazed	0.178*	0.143*	0.189*
Mean shrub volume (m^3 $shrub^{-1}$)			
Heavily grazed	0.0167	0.0125	0.0525
Lightly grazed	0.0481*	0.0387*	0.0664*

Restoration of denuded desert soils

Land disturbance activities that remove native vegetation and disperse fertile mounds, which formed beneath shrubs, produce areas where the soil surface will remain bare for long periods and be subjected to wind and water erosion, unless revegetation steps are taken. Sites at NTS that were denuded for mining activity near the end of the last century show some evidence of natural recovery, but a return to typical desert scrub conditions needs a much longer period of time (Wallace, Romney, & Hunter 1980a). Nuclear explosion sites on Yucca Flat from mid-1950s aboveground tests are now developing as sparse grasslands, having natural shrub recovery only along drainage channels. Probably another century must pass before test localities can return to near original vegetation cover (Romney, Wallace, & Childress 1971; Wallace, Romney, & Hunter 1980b).

Revegetation studies at the Nevada Test Site

NTS, which has had a long history of localized impacts by humans, has served as a model system for investigating approaches to problems for revegetating disturbed ecosystems of winter-rainfall warm deserts. Romney, Hunter, and Wallace (1990) estimated that about 5% of the 3500 km^2 property has been visibly impacted by wildfires and overgrazing by native herbivores, largely pocket gophers (*Thomomys*), during the last four decades. Nuclear testing and radiation experiments accounted for another 3.3% disturbance and road construction another 1%.

Successful restoration of vegetation has occurred on managed denuded desert soils by transplanting native perennials during spring months and using nitrogen fertilizer (Romney, Wallace, Kaaz, Hale, & Childress 1977). These plantings, which must be protected from deer and rabbit grazing by fencing, have been nurtured by periodic irrigation of each transplanted specimen during dry months of the first summer. Once transplanted shrubs have experienced normal winter dormancy and recharge of soil moisture from fall and winter precipitation, there usually is no further need for supplemental irrigation (Wallace et al. 1980). Fencing did not prevent loss of transplanted shrubs from activities of pocket gophers (*Thomomys*) at NTS plots (Hunter et al. 1980).

There are two major limiting factors that restrict

natural regeneration of Mojave Desert shrublands at NTS (Romney et al. 1990). The first of these is lack of sufficient rainfall for seedling establishment (Ackerman 1979), and the second is grazing by jackrabbit (*Lepus californicus*) populations. Jackrabbits are opportunistic feeders that have a heavy impact on new shrub seedlings. Impacts of their feeding on new seedlings and transplants have been described in detail by Hunter (1987) and Hunter, Wallace, and Romney (1980). Those studies at NTS, using 14 species of native shrubs, found that fencing individual seedlings significantly increased survivorship (42–23%) and shrub size over unfenced controls, even though grazing by native burrowing animals continued. Experiments with supplemental water have produced increased shrub production in dry years and led to remarkably high levels of jackrabbit grazing in these local areas (Romney, Hunter, & Wallace 1986). If protected from jackrabbits, an 80% success rate in shrub transplants was consistently obtained (Romney et al. 1990).

Studies on revegetation at NTS (for example, Wallace, Romney, & Hunter 1980a, 1990; Romney, Wallace, & Hunter 1989a, b; Romney, Hunter, & Wallace 1990) have been of great significance in establishing that the challenge of revegetating disturbed desert lands is not an impossible one, and that human management can effectively speed this normally slow process.

SUMMARY

A variety of studies on disturbed Mojave desert scrub have provided convincing evidence that secondary succession is a very long process, requiring many decades. Soil compaction from vehicular traffic can produce very high soil bulk densities in the upper soil profile and thereby modify the ability of soil to absorb and retain water from rainfall. Pioneer perennials tend to be short-lived forms that are replaced eventually by long-lived shrubs, but the resultant collection of long-lived shrubs may not be identical to the composition of the original community. *Ambrosia dumosa* is a long-lived shrub that was a pioneer species in many of the disturbed sites, but *Larrea tridentata*

is established very slowly via rare germination and success of its seedlings. Studies of revegetation along utility corridors in California have estimated recovery times of about 100 years for the most disturbed sites and within 30 years for others.

Vehicular damage to lowland desert scrub is dependent upon many variables, including weight of vehicles, frequency of driving, and moisture content of the soil at the time of vehicular traffic. Obvious negative impacts are major damage to shrub canopies and destruction of animal burrows, but studies of driven versus control plots have revealed that soil compaction from vehicular traffic can drastically reduce the production and composition of annual plants and probably certain reptile species. Alien annuals, such as species of *Schismus* and *Erodium cicutarium*, tend to be common on disturbed soils.

Studies at NTS attempted to distinguish between close-in destruction of shrubs by blast damage and radiation effects. Colonization around ground zero by pioneer species included short-lived perennials, exotic weeds, and native herbs. Eggs of some animals may survive certain types of blasts in zones where vegetation is killed, but then mortality of hatched individuals is very high, perhaps owing to a lack of shrub cover from summer heat. In one study it was shown that either radioactive or nonradioactive dust that coats leaves causes defoliation of even evergreen species.

At Rock Valley, long-term chronic exposure of plants to ionizing gamma radiation resulted in few documented effects on shrubs, being most noticeable for *Ephedra nevadensis* and *Krameria erecta*, but those species persisted in the irradiated populations. Likewise an irradiated population of the rodent *Perognathus formosus* showed differences in parameters of life history tables, but was sustained in the fenced enclosure during the five-year study. In contrast, four species of lizards exhibited an onset of female sterility due to continuous exposure to gamma radiation, and this sterility resulted in sharp declines in these populations following several years of exposure.

Introduced plant species are not only colonizers of disturbed sites, but also may be found outcompeting natives on relatively undisturbed desert scrub habitats. One of the most serious invaders are brome grasses (*Bromus*), which can achieve high densities and thereby change the composition of a typical annual flora. Grazing in the desert, mostly by sheep, disturbs the soil profile and can vastly lower production by annuals, and in a number of ways interferes with lives of native animals. Great damage to plant communities can be caused by local abundance of herbivores, such as pocket gophers (*Thomomys*) and jackrabbits (*Lepus*). Protecting seedlings and young shrubs from herbivores by using fences results in large increases in survivorship and shrub size, which means that management practices can be used to accelerate the recovery of desert vegetation after it has been damaged or removed.

References

ABD-EL-MALEK, Y. (1971). Free-living nitrogen-fixing bacteria in Egyptian soils and their possible contributions to soil fertility, pp. 377–91. *In* T. A. Lie & E. G. Mulder (eds.), *Biological Nitrogen Fixation in Natural and Agricultural Habitats.* Martinus Nijhoff, The Hague.

ACKERMAN, T. L. (1979). Germination and survival of perennial plant species in the Mojave Desert. *The Southwestern Naturalist* 24:399–408.

ACKERMAN, T. L. & BAMBERG, S. A. (1974). Phenological studies in the Mojave Desert at Rock Valley (Nevada Test Site), pp. 215–26. *In,* H. Lieth (ed.), *Phenology and Seasonality Modeling.* Springer-Verlag, New York.

ACKERMAN, T. L., ROMNEY, E. M., WALLACE, A., & KINNEAR, J. E. (1980). Phenology of desert shrubs in southern Nye County, Nevada. *Great Basin Naturalist Memoirs* (4):4–23.

ADAMS, J. A., ENDO, A. S., STOLZY, L. H., ROWLANDS, P. G., & JOHNSON, H. B. (1982). Controlled experiments on soil compaction produced by off-road vehicles in the Mojave Desert, California. *Journal of Applied Ecology* 19:167–75.

ADAMS, M. S. & STRAIN, B. R. (1968). Photosynthesis in stems and leaves of *Cercidium floridum.* Spring and summer diurnal field response and relation to temperature. *Oecologia Plantarum* 3:285–97.

ADAMS, M. S. & STRAIN, B. R. (1969). Seasonal photosynthetic rates in stems of *Cercidium floridum* Benth. *Photosynthetica* 3:55–62.

ADAMS, M. S., STRAIN, B. R., & TING, I. P. (1967). Photosynthesis in chlorophyllous stem tissue and leaves of *Cercidium floridum*: accumulation and distribution of ^{14}C from $^{14}CO_2$. *Plant Physiology* 42:1797–9.

ALBEE, B. J., SHULTZ, L. M., & GOODRICH, S. (1988). *Atlas of the Vascular Plants of Utah. Utah Museum of Natural History Occasional Publications* 7:1–670.

ALEXANDER, C. E. & WHITFORD, W. G. (1968). Energy requirements of *Uta stansburiana. Copeia* 1968:678–83.

ALLRED, D. M. (1962a). Mites on grasshopper mice at the Nevada atomic test site. *Great Basin Naturalist* 22:101–4.

ALLRED, D. M. (1962b). Mites on squirrels at the Nevada atomic test site. *Journal of Parasitology* 48:817.

ALLRED, D. M. (1963). Mites from pocket mice at the Nevada Test Site (Acarina). *Proceedings of the Entomological Society of Washington* 65:231–3.

ALLRED, D. M. (1965). Note of phalangids at the Nevada Test Site. *Great Basin Naturalist* 25:37–8.

ALLRED, D. M. (1969a). Bees of the Nevada Test Site. *Great Basin Naturalist* 29:20–4.

ALLRED, D. M. (1969b). Lepidoptera of the Nevada Test Site. *Great Basin Naturalist* 29:42.

ALLRED, D. M. (1973). Additional records of mutillid wasps from the Nevada Test Site. *Great Basin Naturalist* 33:156–62.

ALLRED, D. M. & BECK, D. E. (1962). Ecological distribution of mites on lizards at the Nevada atomic test site. *Herpetologia* 18:47–51.

ALLRED, D. M. & BECK, D. E. (1963). Ecological distribution of some rodents at the Nevada Atomic Test Site. *Ecology* 44:211–14.

ALLRED, D. M. & BECK, D. E. (1964). Mites on reptiles at the Nevada atomic test site. *Transactions of the American Microscopical Society* 83:266–8.

ALLRED, D. M. & BECK, D. E. (1965). A list of Scarabaeidae beetles of the Nevada Test Site. *Great Basin Naturalist* 25:77–9.

ALLRED, D. M. & BECK, D. E. (1967). Spiders of the Nevada Test Site. *Great Basin Naturalist* 27:11–25.

ALLRED, D. M., BECK, D. E., & JORGENSEN, C. D. (1963). Biotic communities of the Nevada Test Site. *Brigham Young University Science Bulletin, Biology Series* 2(2):1–52.

ALLRED, D. M. & GOATES, M. A. (1964a). Mites from mammals at the Nevada Test Site. *Great Basin Naturalist* 24:71–3.

ALLRED, D. M. & GOATES, M. A. (1964b). Mites from wood rats at the Nevada nuclear test site. *Journal of Parasitology* 50:171.

ALLRED, D. M & MULAIK, S. (1965). Two isopods of the Nevada Test Site. *Great Basin Naturalist* 25:43–7.

AMUNDSON, R. G., CHADWICK, O. A., & SOWENS, J. M. (1989). A comparison of soil climate and biological activity along an elevation gradient in the eastern Mojave Desert. *Oecologia* 80:395–400.

ANDERSON, A. N. (1984). Community organization of ants in the Victoria mallee. *The Victorian Naturalist* 101:248–51.

ANDERSON, R. Y., GLEMAN, D. C., & COLE, C. V. (1981). Effects of saprophytic grazing on net mineralization, pp. 201–6. *In* F. C. Clark & T. Rosswall (eds.), *Terrestrial Nitrogen Cycles. Ecological Bulletin* 33.

ARMOND, P. A., SCHREIBER, U., & BJÖRKMAN, O. (1977). Photosynthetic acclimation to temperature in *Larrea divaricata*: light harvesting efficiency and capacity of photosynthetic electron transport reactions. *Carnegie Institution of Washington Yearbook* 76:335–41.

ARNDT, W. (1966). IV. The effect of traffic compaction on a number of soil properties. *Journal of Agricultural Engineering Research* 11:182–7.

ASPLUND, K. K. (1974). Body size and habitat utilization in whiptail lizards (*Cnemidophorus*). *Copeia* 1974: 695–703.

AUSTIN, G. T. & BRADLEY, W. G. (1971). The avifauna of Clark County, Nevada. *Journal of the Arizona Academy of Science* 6:283–303.

BALPH, D. F., ed. (1971). *Sampling Small Mammal Populations in Curlew Valley. US/IBP Desert Biome Research Memorandum* 72–1. Utah State University, Logan.

BALPH, D. F., ed. (1973). *Curlew Valley Validation Site Report. US/IBP Desert Biome Research Memorandum* 73–1. Utah State University, Logan.

BAMBERG, S. A., KAAZ, H. W., MAZA, B. G., & TURNER, F. B. (1974). Abiotic measurements, pp. 11–24. *In* F. B. Turner & J. F. McBrayer (eds.), *Rock Valley Validation Site Report. US/IBP Desert Biome Research Memorandum* 74–2. Utah State University, Logan.

BAMBERG, S. A., KLEINKOPF, G. E., WALLACE, A., & VOLLMER, A. (1975). Comparative photosynthetic production of Mojave Desert shrubs. *Ecology* 56:732–6.

BAMBERG, S. A., VOLLMER, A. T., KLEINKOPF, G. E., & ACKERMAN, T. L. (1976). A comparison of seasonal primary production of Mojave Desert shrubs during wet and dry years. *American Midland Naturalist* 95:398–405.

BAMBERG, S. A., WALLACE, A., KLEINKOPF, G. E., & VOLLMER, A. (1974). *Plant Productivity and Nutrient Interrelationships of Perennials in the Mojave Desert. US/IBP Desert Biome Research Memorandum* 73–10. Utah State University, Logan.

BAMBERG, S. A., WALLACE, A., ROMNEY, E. M., & HUNTER, R. E. (1980). Further attributes of the perennial vegetation in the Rock Valley area of the northern Mojave Desert. *Great Basin Naturalist Memoirs* (4):39–41.

BANTA, B. H. (1960). Notes on the feeding of the western collared lizard, *Crotaphytus collaris baileyi* Stejneger. *Wasmann Journal of Biology* 18:309–11.

BANTA, B. H. (1961). Herbivorous feeding of *Phrynosoma platyrhinos* in southern Nevada. *Herpetologica* 17:136–7.

BARBOUR, M. G. (1968). Germination requirements of the desert shrub *Larrea divaricata*. *Ecology* 49:915–23.

BARBOUR, M. G. (1969). Age and space distribution of the desert shrub *Larrea divaricata*. *Ecology* 50:679–85.

BARBOUR, M. G. (1973). Desert dogma reexamined: root/shoot productivity and plant spacing. *American Midland Naturalist* 89:41–57.

BARNUM, A. H. (1964). Orthoptera of the Nevada Test Site. *Brigham Young University Science Bulletin, Biology Series* 4(3):1–134.

BARRETT, S. L. (1990). Home range and habitat of the desert tortoise (*Xerobates agassizi*) in the Picacho Mountains of Arizona. *Herpetologica* 46:202–6.

BARTHOLOMEW, G. A. (1964). The roles of physiology and behavior in the maintenance of homeostasis in the desert environment. *Symposium by the Society of Experimental Biology* 18:7–29.

BARTHOLOMEW, G. A. (1972). The water economy of seed-eating birds that survive without drinking, pp. 237–54. *In* K. H. Voous (ed.), *Proceedings of the XVth International Ornithological Congress*. E. J. Brill, Leiden.

BARTHOLOMEW, G. A. (1982a). Energy metabolism, pp. 46–93. *In* M. S. Gordon, G. A. Bartholomew, A. D. Grinnell, C. B. Jorgensen, & N. N. White, *Animal Physiology: Principles and Adaptations*, 4th edn. Macmillan Publishing Company, New York.

BARTHOLOMEW, G. A. (1982b). Body temperature and energy metabolism, pp. 333–406. *In* M. S. Gordon, G. A. Bartholomew, A. D. Grinnell, C. B. Jorgensen & N. N. White, *Animal Physiology: Principles and Adaptations*, 4th edn. Macmillan Publishing Company, New York.

BARTHOLOMEW, G. A. (1982c). Physiological control of body temperature, pp. 167–211. *In* C. Gans & F. H. Pough (eds.), *Biology of the Reptilia*, vol. 12. Academic Press, London.

BARTHOLOMEW, G. A. & CADE, T. J. (1956). Water consumption of house finches. *The Condor* 58:406–12.

BARTHOLOMEW, G. A. & CADE, T. J. (1957). Temperature regulation, hibernation, and aestivation in the little pocket mouse, *Perognathus longimembris*. *Journal of Mammalogy* 38:60–72.

BARTHOLOMEW, G. A. & DAWSON, W. R. (1954). Body temperature and water requirements in the mourning dove, *Zenaidura macroura marginella*. *Ecology* 35:181–7.

BARTHOLOMEW, G. A. & HUDSON, J. W. (1961). Desert ground squirrels. *Scientific American* 205(5):107–16.

BARTHOLOMEW, G. A., LAWIEWSKI, R. C. & CRAWFORD, E. C., JR. (1968). Patterns of panting and gular flutter in cormorants, pelicans, owls and doves. *The Condor* 70:31–4.

BARTHOLOMEW, G. A. & MacMILLEN, R. E. (1960). The water requirements of mourning doves and their use of sea water and NaCl solutions. *Physiological Zoology* 33:171–8.

BEATLEY, J. C. (1965). Effects of radioactive and non-radioactive dust upon *Larrea divaricata* Cav., Nevada Test Site. *Health Physics* 11:1621–5.

BEATLEY, J. C. (1966). Ecological status of introduced brome grass (*Bromus* spp.) in desert vegetation of southern Nevada. *Ecology* 47:548–54.

BEATLEY, J. C. (1967). Survival of winter annuals in the northern Mojave Desert. *Ecology* 48:745–50.

BEATLEY, J. C. (1969a). Dependence of desert rodents on winter annuals and precipitation. *Ecology* 50:721–4.

BEATLEY, J. C. (1969b). Biomass of desert winter annual plant populations in southern Nevada. *Oikos* 20:261–73.

BEATLEY, J. C. (1970). Perennation in *Astragalus lentiginosus* and *Tridens pulchellus* in relation to rainfall. *Madroño* 20:326–32.

BEATLEY, J. C. (1974a). Effects of rainfall and temperature on the distribution and behavior of *Larrea tridentata* (creosote-bush) in the Mojave Desert of Nevada. *Ecology* 55:245–61.

BEATLEY, J. C. (1974b). Phenological events and their environmental triggers in Mohave Desert ecosystems. *Ecology* 55:856–63.

BEATLEY, J. C. (1975). Climates and vegetation pattern across the Mojave/Great Basin Desert transition of southern Nevada. *American Midland Naturalist* 93:53–70.

BEATLEY, J. C. (1976a). *Vascular Plants of the Nevada Test Site and Central-Southern Nevada: Ecologic and Geographic Distributions*. Energy Research and Development Administration TID-26881. Technical Information Center, Office of Technical Information, Springfield, Virginia.

BEATLEY, J. C. (1976b). Rainfall and fluctuating plant populations in relation to distributions and numbers of desert rodents in southern Nevada. *Oecologia* 24:21–42.

BEATLEY, J. C. (1976c). Environments of kangaroo rats (*Dipodomys*) and effects of environmental change on populations in southern Nevada. *Journal of Mammalogy* 57:67–93.

BEATLEY, J. C. (1979). *Shrub and Tree Data for Plant Associations Across the Mojave/Great Basin Desert Transition of the Nevada Test Site, 1963–1975*. U.S. Department of Energy Contract EY-76-S-02-2307,

National Technical Information Service, U.S. Department of Commerce, Springfield, Virginia.

BECAMP (1990). *Mammal List for the Nevada Test Site*. Nevada Test Site, Mercury.

BECAMP (1991a). *Reptile List for the Nevada Test Site*. Nevada Test Site, Mercury.

BECAMP (1991b). *Bird List for the Nevada Test Site*. Nevada Test Site, Mercury.

BECK, D. E., ALLRED, D. M., & BRINTON, E. P. (1963). Ticks of the Nevada Test Site. *Brigham Young University Science Bulletin, Biology Series* 4(1):1–11.

BELKIN, D. A. (1961). The running speeds of the lizards *Dipsosaurus dorsalis* and *Callisaurus draconoides*. *Copeia* 1961:223–4.

BELL, K. L., HIATT, H. D., & NILES, W. E. (1979). Seasonal changes in biomass allocation in eight winter annuals of the Mojave Desert. *Journal of Ecology* 67:781–7.

BENDER, G. L., ed. (1982). *Reference Handbook on the Deserts of North America*. Greenwood Press, Westport, Connecticut.

BENNERT, H. W. & MOONEY, H. A. (1969). The water relations of some desert plants in Death Valley, California. *Flora* 168:405–27.

BENNETT, A. F. (1972). The effect of activity on oxygen consumption, oxygen debt, and heart rate in the lizards *Varanus gouldii* and *Sauromalus hispidus*. *Journal of Comparative Physiology* 79:259–80.

BENNETT, A. F. & DAWSON, W. R. (1976). Metabolism, pp. 127–223. *In* C. Gans & W. R. Dawson (eds.), *Biology of the Reptilia*, vol. 5. Academic Press, London.

BENSON, L. & DARROW, R. A. (1981). *Trees and Shrubs of the Southwestern Deserts*. University of Arizona Press, Tucson.

BERNSTEIN, R. A. (1974). Seasonal food abundance and foraging activity in some desert ants. *American Naturalist* 108:490–8.

BERRY, K. H. (1974). The ecology and social behavior of the chuckwalla, *Sauromalus obesus*. *University of California Publications in Zoology* 101:1–60.

BERRY, K. H. (1978). *Livestock Grazing and the Desert Tortoise*. Proceedings of the 43rd North American Wildlife and Natural Resources Conference, Phoenix, Arizona.

BERRY, K. H. (1984). *The Status of the Desert Tortoise (Gopherus agassizii) in the United States*. Desert Tortoise Council Report to U.S. Fish and Wildlife Service, Sacramento, California.

BERRY, K. H. (1986a). Desert tortoise (*Gopherus agassizii*) research in California, 1976–1985. *Herpetologica* 42:62–7.

BERRY, K. H. (1986b). Desert tortoise (*Gopherus agassizii*) relocation: implications of social behavior and movements. *Herpetologica* 42:113–25.

BERRY, K. H. & NICHOLSON, L. L. (1984). Attributes

of populations at twenty-seven sites in California, pp. 5.1–5.85. *In* K. H. Berry (ed.), *The Status of the Desert Tortoise (Gopherus agassizii) in the United States*. Desert Tortoise Council Report to U.S. Fish and Wildlife Service, Sacramento, California.

BERRY, K. H. & TURNER, F. B. (1986). Spring activities and habits of juvenile desert tortoise, *Gopherus agassizii*, in California. *Copeia* 1986:1010–2.

BILLINGS, D. W. (1949). The shadscale vegetation zone of Nevada and eastern California in relation to climate and soils. *American Midland Naturalist* 42:87–109.

BILLINGS, D. W. & MORRIS, R. J. (1951). Reflection of visible and infrared radiation from leaves of different ecological groups. *American Journal of Botany* 38:327–31.

BJERREGAARD, R. S. (1971). *The Nitrogen Budget of Two Salt Desert Shrub Plant Communities of Western Utah*. Ph.D. dissertation, Utah State University, Logan.

BJÖRKMAN, O. (1973). Comparative studies on photosynthesis in higher plants, pp. 1–63. *In* A. Giese (ed.), *Current Topics in Photobiology, Photochemistry, and Photophysiology*, vol. 8. Academic Press, London.

BJÖRKMAN, O. & BADGER, M. R. (1977). Thermal stability of photosynthetic enzymes in heat- and cool-adapted C_4 species. *Carnegie Institution of Washington Yearbook* 76:346–54.

BJÖRKMAN, O., BADGER, M. R., & ARMOND, P. A. (1980). Response and adaptation of photosynthesis to high temperature, pp. 233–49. *In* N. C. Turner & P. J. Kramer (eds.), *Adaptations of Plants to Water and High Temperature Stress*. Wiley–Interscience, New York.

BJÖRKMAN, O., BOYNTON, J., & BERRY, J. (1976). Comparison of the heat stability of photosynthesis, chloroplast membrane reactions, photosynthetic enzymes, and soluble protein in leaves of heat-adapted and cold-adapted C_4 species. *Carnegie Institution of Washington Yearbook* 75:400–7.

BJÖRKMAN, O., MOONEY, H. A., & EHLERINGER, J. (1975). Photosynthetic responses of plants from habitats with contrasting thermal environments. *Carnegie Institution Washington Yearbook* 74:743–59.

BJÖRKMAN, O., PEARCY, R. W., HARRISON, A. T., & MOONEY, H. A. (1972). Photosynthetic adaptation to high temperatures: a field study in Death Valley, California. *Science* 175:786–9.

BLANKENNAGEL, R. K. & WEIR, J. E. JR. (1973). Geohydrology of Pahute Mesa, Nevada Test Site, Nye County, Nevada. *U.S. Geological Survey Professional Paper* 712–B:1–35.

BLYDENSTEIN, J., HUNGERFORD, C. R., DAY, G. I., & HUMPHREY, R. R. (1957). Effect of domestic livestock exclusion on vegetation in the Sonoran Desert. *Ecology* 38:522–6.

BOWERS, M. A. (1982). Foraging behavior of heteromyid rodents: field evidence of resource partitioning. *Journal of Mammalogy* 63:361–7.

BOWERS, M. A. (1986). Geographic comparison of microhabitats used by three heteromyids in response to rarefaction. *Journal of Mammalogy* 67:46–52.

BOWERS, M. A. (1987). Precipitation and the relative abundance of desert winter annuals: a 6-year study in the northern Mojave Desert. *Journal of Arid Environments* 12:141–9.

BOWERS, M. A. (1988). Seed removal experiments on desert rodents: the microhabitat by moonlight effect. *Journal of Mammalogy* 69:201–4.

BOWERS, M. A. & BROWN, J. H. (1982). Body size and coexistence in desert rodents: chance or community structure? *Ecology* 63:391–400.

BOWMAN, R. A. & FOCHT, D. D. (1974). The influence of glucose and nitrate concentrations upon denitrification rates in sandy soils. *Soil Biology and Biochemistry* 6:297–321.

BOWNS, J. E. & WEST, N. E. (1976). *Blackbrush (Coleogyne ramosissima Torr.) on Southwestern Utah Rangelands*. Utah Agricultural Experiment Station Research Report 22.

BOYD, C. E. & GOODYEAR, C. P. (1971). The protein content of some common reptiles and amphibians. *Herpetologica* 27:317–20.

BRADLEY, W. G. (1968). Food habits of the antelope ground squirrel in southern Nevada. *Journal of Mammalogy* 49:14–21.

BRADLEY, W. G. & DEACON, J. E. (1967). The biotic communities of southern Nevada. *Anthropology Papers of the Nevada State Museum* 14:201–95.

BRADLEY, W. G. & MAUER, R. A. (1971). Reproduction and food habits of Merriam's kangaroo rat, *Dipodomys merriami*. *Journal of Mammalogy* 52:497–507.

BRADLEY, W. G., MILLER, J. S., & YOUSEF, M. K. (1974). Thermoregulatory patterns in pocket gophers: desert and mountain. *Physiological Zoology* 47:172–9.

BRADSHAW, S. D. (1988a). *Ecophysiology of Desert Rodents*. Academic Press, Sydney.

BRADSHAW, S. D. (1988b). Desert reptiles: a case of adaptation or pre-adaptation? *Journal of Arid Environments* 14:155–74.

BRANSON, F. A., GIFFORD, G. F., & OWEN, J. R. (1972). *Rangeland Hydrology*. Range Science Series 1. Society of Range Management, Denver, Colorado.

BRAUN, E. J. & DANTZLER, W. H. (1972). Function of mammalian-type and reptilian–type nephrons in kidney of desert quail. *American Journal of Physiology* 222:617–29.

BRAUN, E. J. & DANTZLER, W. H. (1974). Effects of ADH on single-nephron glomerular filtration rates in the avian kidney. *American Journal of Physiology* 226:1–8.

BRENNAN, J. M. (1965). Five new chiggers from

southwestern United States (Acarina: Trombiculidae). *Journal of Parasitology* 51:108–13.

BRIESE, D. T. (1982). Partitioning of resources amongst seed-harvesters in an ant community in semi-arid Australia. *Australian Journal of Ecology* 7:299–307.

BRIGGS, L. J. & SHANTZ, H. L. (1914). Relative water requirement of plants. *Journal of Agricultural Research* 3:1–63

BRINTON, E. P., BECK, D. E., & ALLRED, D. M. (1965). Identification of the adults, nymphs and larvae of ticks of the genus *Dermacentor* Koch (Ixodidae) in the western United States. *Brigham Young University Science Bulletin, Biology Series* 5(4):1–44.

BROADBENT, F. E. & CLARK, F. G. (1965). Denitrification, pp. 344–59. *In* W. V. Bartholomew & F. G. Clark (eds.), *Soil Nitrogen.* American Society of Agronomy, Madison, Wisconsin.

BROWER, J. H. (1964). *Changes in the Insect Populations of the Low Shrub Synusia in a Natural Forest Community Exposed to Chronic Gamma Radiation Stress.* M.S. thesis, University of Massachusetts, Amherst.

BROWN, D. E., ed. (1982). Biotic communities of the American Southwest-United States and Mexico. *Desert Plants* 4:1–432.

BROWN, D. E., CARMONY, N. B, LOWE, C. H., & TURNER, R. M. (1976). A second locality for native California fan palms (*Washingtonia filifera*) in Arizona. *Journal of the Arizona Academy of Science* 11:37–41.

BROWN, D. E., LOWE, C. H., & PASE, C. P. (1979). A digitized classification for the biotic communities of North America, with community (series) and association examples for the Southwest. *Journal of the Arizona–Nevada Academy of Science* 14 (Supplement 1):1–16.

BROWN, J. H. (1971). The desert pupfish. *Scientific American* 225(5):104–10.

BROWN, J. H. (1973). Species diversity of seed-eating desert rodents in sand dune habitats. *Ecology* 54:775–87.

BROWN, J. H. (1975). Geographical ecology of desert rodents, pp. 315–41. *In* M. L. Cody & J. M. Diamond (eds.), *Ecology and Evolution of Communities.* Harvard University Press, Cambridge.

BROWN, J. H. & DAVIDSON, D. W. (1977). Competition between seed–eating rodents and ants in desert ecosystems. *Science* 196:880–2.

BROWN, J. H. & GIBSON, A. C. (1983). *Biogeography.* C. V. Mosby, St. Louis, Missouri.

BROWN, J. H. & LIEBERMAN, G. A. (1973). Resource utilization and coexistence of seed-eating rodents in sand dune habitats. *Ecology* 54:788–97.

BROWN, J. H., REICHMAN, O. J., & DAVIDSON, D. W. (1979). Granivory in desert ecosystems. *Annual Review of Ecology and Systematics* 10:201–27.

BURGE, B. L. (1978). Physical characteristics and patterns of utilization of cover sites by *Gopherus agassizii* in southern Nevada. *Proceedings, Symposium of the Desert Tortoise Council* 1978:80–111.

BUSACK, S. D. & BURY, R. B. (1974). Some effects of off-road vehicles and sheep grazing on lizard populations in the Mojave Desert. *Biological Conservation* 6:179–83.

CALDER, W. A. (1981). Diuresis on the desert? Effects of fruit- and nectar-feeding on the house finch and other species. *The Condor* 83:267–8.

CALDER, W. A. (1984). *Size, Function, and Life History.* Harvard University Press, Cambridge.

CALDER, W. A. & KING, J. R. (1974). Thermal and caloric relations of birds, pp. 259–413. *In* D. S. Farner & J. R. King (eds.), *Avian Biology,* vol. 4. Academic Press, New York.

CALDER, W. A. & SCHMIDT-NIELSEN, K. (1967). Temperature regulation and evaporation in the pigeon and the roadrunner. *American Journal of Physiology* 213:883–9.

CALDWELL, M. M. (1985). Cold desert, pp. 198–212. *In* B. F. Chabot & H. A. Mooney (eds.), *Physiological Ecology of North American Plant Communities.* Chapman and Hall, New York.

CALDWELL, M. M. & RICHARDS, J. M. (1986). Competing root systems: morphology and models of absorption, pp. 251–73. *In* T. Givnish (ed.), *On the Economy of Plant Form and Function.* Cambridge University Press.

CALDWELL, M. M., WHITE, R. S., MOORE R. T., & CAMP, L. B. (1977). Carbon balance, productivity and water use of cold-winter desert shrub communities dominated by C_3 and C_4 species. *Oecologia* 29:275–300.

CALKIN, H. W. & PEARCY, R. W. (1984). Seasonal progressions of tissue and cell water relation parameters in evergreen and deciduous perennials. *Plant, Cell and Environment* 7:347–52.

CANNON, H. L., STROBELL, M. E., BUSH, C. A., & BOWLES, J. M. (1981). *Effects of Nuclear and Conventional Chemical Explosions on Vegetation.* U.S. Geological Survey Open File Report 81–1300.

CANNON, W. A. (1911). The root habits of desert plants. *Carnegie Institution of Washington Publication* 131.

CANNON, W. A. (1949). A tentative classification of root systems. *Ecology* 30:542–8.

CARLSON, W. D. & GASSNER, F. X., eds. (1964). *Effects of Ionizing Radiation on the Reproductive System.* Pergamon Press, New York.

CARPENTER, D. E., BARBOUR, M. G., & BAHRE, C. J. (1986). Old field succession in Mojave Desert scrub. *Madroño* 33:111–22.

CARTER, L. J. (1974). Off-road vehicles: a compromise plan for the California desert. *Science* 183:396–9.

CASE, C., ed. (1984). Site characterization in connection with the low level defense waste management site in Area 5 of the Nevada Test Site, Nye County, Nevada –

final report. Department of Energy DOE/NV/10162–13.

CASTETTER, R. C. & HILL, H. O. (1979). Additions to the birds of the Nevada Test Site. *Western Birds* 10:221–3.

CATES, R. G. & RHOADES, D. F. (1977). Patterns in the prediction of antiherbivore chemical defenses in plant communities. *Biochemical Systematics and Ecology* 5:185–93.

CHAMBERLIN, R. V. (1962). Millipedes from the Nevada test areas. *Proceedings of the Biology Society of Washington* 75:53–5.

CHAMBERLIN, R. V. (1963). A new genus in the chilopod family Tampiyidae. *Proceedings of the Biological Society of Washington* 76:33–5.

CHAMBERLIN, R. V. (1965). A new genus and species in the chilopod family Tampiyidae. *Great Basin Naturalist* 25:39–42.

CHAPPELL, M. A. & BARTHOLOMEW, G. A. (1981). Standard operative temperatures and thermal energetics of the antelope ground squirrel *Ammospermophilus leucurus. Physiological Zoology* 54:81–93.

CHARLEY, J. L. & WEST, W. E. (1975). Plant-induced soil chemical patterns in some shrub-dominated semi-desert ecosystems of Utah. *Journal of Ecology* 63:945–64.

CHEW, R. M. (1961). Ecology of spiders in a desert community. *Journal of the New York Entomological Society* 69:5–41.

CHEW, R. M. (1975). *Effect of Density on the Population Dynamics of* Perognathus formosus *and its Relationships within a Desert Ecosystem. US/IBP Desert Biome Research Memorandum* 75–18. Utah State University, Logan.

CHEW, R. M. (1977). Some ecological characteristics of the ants of a desert-shrub community in southeastern Arizona. *American Midland Naturalist* 98:33–49.

CHEW, R. M. & BUTTERWORTH, B. B. (1964). Ecology of rodents in Indian Cove (Mojave Desert), Joshua Tree National Monument, California. *Journal of Mammalogy* 5:203–25.

CHEW, R. M. & CHEW, A. E. (1965). The primary productivity of a desert shrub (*Larrea tridentata*) community. *Ecological Monographs* 33:355–75.

CHEW, R. M. & CHEW, A. E. (1970). Energy relationships of the mammals of a desert shrub (*Larrea tridentata*) community. *Ecological Monographs* 40:1–21.

CHEW, R. M. & DE VITA, J. (1980). Foraging characteristics of a desert ant assemblage: functional morphology and species separation. *Journal of Arid Environments* 3:75–83.

CHEW, R. M. & TURNER, F. B. (1974). *Effect of Density on the Population Dynamics of* Perognathus formosus *and its Relationships within a Desert Ecosystem. US/IBP Desert Biome Research Memorandum* 74–20. Utah State University, Logan.

CHILDS, S. & GOODALL, D. W. (1973). *Seed Reserves of Desert Soils. US/IBP Desert Biome Research Memorandum* 73–5. Utah State University, Logan.

CHRISTMAN, S. P. (1984). Plot mapping: estimating densities of breeding bird territories by combining spot mapping and transect techniques. *The Condor* 86:237–41.

CLARK, S. B., LETEY, J., Jr., LUNT, O. R., WALLACE, A., KLEINKOPF, G. E., & ROMNEY, E. M. (1980). Transpiration and CO_2 fixation of selected desert shrubs as related to soil water potential. *Great Basin Naturalist Memoirs* (4):110–6.

CLOKEY, I. W. (1951). Flora of the Charleston Mountains, Clark County, Nevada. *University of California Publications in Botany* 24:1–274.

CLOUDSLEY-THOMPSON, J. L. (1977). The black beetle paradox. *Entomological Monographs Magazine* 113:19–22.

CLOUDSLEY-THOMPSON, J. L. (1979). Adaptive functions of the colours of desert animals. *Journal of Arid Environments* 2:95–104.

CLOUDSLEY-THOMPSON, J. L. (1988). *Evolution and Adaptation of Terrestrial Arthropods.* Springer-Verlag, Berlin.

CLOUDSLEY-THOMPSON, J. L. & CHADWICK, M. J. (1964). *Life in Deserts.* Dufour Editing, Philadelphia.

CODY, M. L. (1971). Finch flocks in the Mojave Desert. *Theoretical Population Biology* 2:142–58.

CODY, M. L. (1974). Optimization in ecology. *Science* 183:1156–64.

CODY, M. L. (1986a). Structural niches in plant communities, pp. 381–405. *In* J. Diamond & T. J. Case (eds.), *Community Ecology.* Harper and Row, New York.

CODY, M. L. (1986b). Spacing patterns in Mojave Desert plant communities: near neighbor analyses. *Journal of Arid Environments* 11:199–217.

COLE, A. C., JR. (1966). Ants of the Nevada Test Site. *Brigham Young University Science Bulletin, Biology Series* 7(3):1–27.

COLLINS, A. (1988). Natural history of the desert horned lizard *Phrynosoma platyrhinos* in the central Mojave Desert, pp. 29–37. *In* H. F. De Lisle, P. R. Brown, B. Kaufman, & B. M. McGurty (eds.), *Proceedings of the Conference on California Herpetology.* Southwestern Herpetologists Society, Van Nuys, California.

COMANOR, P. L. & STAFFELDT, E. G. (1978). Decomposition of plant litter in the western North American deserts. Pp. 31–49 *in* N. E. West & J. Skujins (eds.), *Nitrogen in Desert Ecosystems.* Dowden, Hutchinson and Ross, Inc., Stroudsbury, Pennsylvania.

EHLERINGER, J. R. (1988). Seasonal patterns of canopy development and carbon gain in nineteen warm desert shrub species. *Oecologia* 75:327–35.

COMSTOCK, J. P. & EHLERINGER, J. R. (1988). Contrasting photosynthetic behavior in leaves and twigs of *Hymenoclea salsola*, a green-twigged warm desert shrub. *American Journal of Botany* 75:1360–70.

CONGDON, J. D., DUNHAM, A. E., & TINKLE, D. W. (1982). Energy budgets and life histories of reptiles, pp. 155–99. *In* C. Gans & F. H. Pough (eds.), *Biology of the Reptilia*, vol. 13. Academic Press, London.

CONNOR, E. F. & SIMBERLOFF, D. (1979). The assembly of species communities: chance or competition? *Ecology* 60:1132–40.

COOKE, R. U. & WARREN, A. (1973). *Geomorphology in Deserts.* University of California Press, Berkeley.

CORNWALL H. R. (1972). Geology and mineral deposits of southern Nye County, Nevada. *Nevada Bureau of Mines Bulletin* 77:1–49.

COSTA, W. R., NAGY, K. A., & SHOEMAKER, V. H. (1976). Observations of the behavior of jackrabbits (*Lepus californicus*) in the Mojave Desert. *Journal of Mammalogy* 57:399–402.

COWLES, R. B. (1946). Note on the arboreal feeding habits of the desert iguana. *Copeia* 1946:172–3.

COWLES, R. B. (1979). *Desert Journal.* University of California Press, Berkeley.

COWLES, R. B. & BOGERT, C. M. (1944). A preliminary study of the thermal requirements of desert reptiles. *Bulletin of the American Museum of Natural History* 83:265–96.

CRAWFORD, C. S. (1972). Water relations in a desert millipede, *Orthoporus ornatus* (Girard) (Spirostreptidae). *Comparative Biochemistry and Physiology* 42A:521–35.

CRAWFORD, C. S. (1978). Seasonal water balance in *Orthoporus ornatus*, a desert millipede. *Ecology* 59:996–1006.

CRAWFORD, C. S. (1979). Desert detritivores: a review of life history patterns and trophic roles. *Journal of Arid Environments* 2:31–42.

CRAWFORD, C. S. (1981). *Biology of Desert Invertebrates.* Springer-Verlag, Berlin.

CRAWFORD, E. C., JR. (1962). Mechanical aspects of panting in dogs. *Journal of Applied Physiology* 17:249–51.

CREIGHTON, W. S. (1966). The habits of *Pheidole ridicula* Wheeler with remarks on habit patterns in the genus *Pheidole* (Hymenoptera: Formicidae). *Psyche* 73:1–7.

CRONQUIST, A., HOLMGREN, A. H., HOLMGREN, N. H., & REVEAL, J. L. (1972). *Intermountain Flora*, vol. 1. Hafner Publishing Company, New York.

CUNNINGHAM, G. L. & STRAIN, B. R. (1969). Ecological significance of seasonal leaf variability in a desert shrub. *Ecology* 50:400–8.

DANTZLER, W. H. & SCHMIDT-NIELSEN, K.

(1966). Excretion in fresh-water turtle (*Pseudemys scripta*) and desert tortoise (*Gopherus agassizii*). *American Journal of Physiology* 210:198–210.

DAVIDSON, D. W. (1977a). Species diversity and community organization in desert seed-eating ants. *Ecology* 58:711–24.

DAVIDSON, D. W. (1977b). Foraging ecology and community organization in desert seed-eating ants. *Ecology* 58:725–37.

DAVIDSON, R. & FOX, M. (1974). Effects of off-road motorcycle activity on Mojave Desert vegetation and soil. *Madroño* 22:381–90.

DAWSON, W. R. (1967). Interspecific variation in physiological responses of lizards to temperature, pp. 230–57. *In* W. W. Milstead (ed.), *Lizard Ecology. A Symposium.* University of Missouri Press, Columbia.

DAWSON, W. R. (1984). Physiological studies of desert birds: present and future considerations. *Journal of Arid Environments* 7:133–55.

DAWSON, W. R. & BARTHOLOMEW, G. A. (1968). Temperature regulation and water economy of desert birds, pp. 357–94. *In* G. W. Brown, jr. (ed.), *Desert Biology*, vol. 1. Academic Press, New York.

DAWSON, W. R. & TEMPLETON, J. R. (1963). Physiological responses to temperature in the lizard *Crotaphytus collaris*. *Physiological Zoology* 36:219–36.

DeDECKER, M. (1984). *Flora of the Northern Mojave Desert, California.* California Native Plant Society Special Publication 7, Berkeley, California.

DEGEN, A. A., PINSHOW, B., & ALKON, P. U. (1982). Water flux in chukar partridges (*Alectoris chukar*) and a comparison with other birds. *Physiological Zoology* 55:64–71.

DEGEN, A. A., PINSHOW, B., & SHAW, P. J. (1983). Must desert chukars (*Alectris chukar sinaica*) drink water? Water influx and body mass changes in response to dietary water content. *The Auk* 101:47–52.

DELWICHE, C. C. (1956). Denitrification, pp. 233–56. *In* W. D. McElroy & B. Glass (eds.), *Inorganic Nitrogen Metabolism.* Johns Hopkins Press, Baltimore, Maryland.

DEPARTMENT OF WATER RESOURCES (1975). *Vegetative Water in California, 1974.* State of California, The Resources Agency Bulletin 113–3 (April, 1975).

DePUIT, E. J. & CALDWELL, M. M. (1975). Gas exchange of three cool semi-desert species in relation to temperature and water stress. *Journal of Ecology* 63:835–58.

DE VITA, J. (1979). Mechanisms of interference and foraging among colonies of the harvester ant *Pogonomyrmex californicus* in the Mojave Desert. *Ecology* 60:729–37.

DE VRIES, D. A. (1959). The influence of irrigation on the energy balance and climate near the ground. *Journal of Meteorology* 16:256–70.

DINGMAN, R. E. & BANDOLI, J. (1973). *Density and Dietary Habits of Pocket Gophers* (Thomomys bottae centralis) *in Rock Valley. US/IPB Biome Research Memorandum 73–21.* Utah State University, Logan.

DINGMAN, R. E. & BYERS, L. (1974). *Interaction between a Fossorial Rodent (the Pocket Gopher), Thomomys bottae, and a Desert Plant Community. US/IBP Desert Biome Research Memorandum 74–22.* Utah State University, Logan.

DIXON, K. L. (1959). Ecological and distributional relations of desert scrub birds of western Texas. *The Condor* 61:397–409.

DOBROWOLSKI, J. P., CALDWELL, M. M., & RICHARDS, J. H. (1990). Basin hydrology and plant root systems, pp. 243–92. *In* C. B. Osmond, L. F. Pitelka, & G. M. Hidy (eds.), *Plant Biology of the Basin and Range.* Springer-Verlag, New York.

DORN, R. I. & OBERLANDER, T. M. (1981). Rock varnish origin, characteristics, and usage. *Zeitschrift für Geomorphologie* 25:420–36.

DOWNTON, W. J. S., BERRY, J. A., & SEEMANN, J. R. (1984). Tolerance of photosynthesis to high temperature in desert plants. *Plant Physiology* 74:786–90.

DUFFY, C. J. & AL-HASSAN, S. (1988). Groundwater circulation in a closed desert basin: topographic scaling and climatic forcing. *Water Resources Research* 24:1675–78.

DURRELL, L. W. & SHIELDS, L. M. (1960). Fungi isolated in culture from soils of the Nevada Test Site. *Mycologia* 52:636–41.

DUTT, G. R. & MARION, G. M. (1974). *Predicting Nitrogen Transformations and Osmotic Potentials in Warm Desert Soils. US/IBP Desert Biome Research Memorandum 74–47.* Utah State University, Logan.

EDNEY, E. B. (1966). Absorption of water vapor from unsaturated air by *Arenivaga* sp. (Polyphagidae, Dictyoptera). *Comparative Biochemistry and Physiology* 19:387–408.

EDNEY, E. B. (1974). Desert arthropods, pp. 311–85. *In* G. W. Brown, jr. (ed.), *Desert Biology,* vol. 2. Academic Press, New York.

EDNEY, E. B. (1977). *Water Balance in Land Arthropods. Zoophysiology and Ecology,* vol. 9. Springer-Verlag, Berlin.

EDNEY, E. B., FRANCO, P. J., & McBRAYER, J. F. (1976). *Abundance and Distribution of Soil Microarthropods in Rock Valley, Nevada. US/IBP Desert Biome Research Memorandum 76–24:11–27.*

EDNEY, E. B., HAYNES, S., & GIBO, D. (1974). Distribution and activity of the desert cockroach *Arenivaga investigata* (Polyphagidae) in relation to microclimates. *Ecology* 55:420–7.

EDNEY, E. B., McBRAYER, J. F., FRANCO, P. J., & PHILLIPS, A. W. (1974). Distribution of soil arthropods in Rock Valley, Nevada, pp. 44–5. *In* F. B. Turner & J. F. McBrayer (eds.), *Rock Valley Validation Site Report. US/IBP Desert Biome Research Memorandum 74–32.* Utah State University, Logan.

EDNEY, E. B., McBRAYER, J. F., FRANCO, P. J., & PHILLIPS, A. W. (1975). Abundance and distribution of soil microarthropods in Rock Valley, Nevada, pp. 39–45. *In,* F. B. Turner (ed.), *Rock Valley Validation Site Report. US/IBP Desert Biome Research Memorandum 75–29.* Utah State University, Logan.

EDWARDS, W. R. & EBENHARDT, L. (1967). Estimating cottontail abundance from live-training data. *Journal of Wildlife Management* 3:87–96.

EHLERINGER, J. R. (1980). Leaf morphology and reflectance in relation to water and temperature stress, pp. 295–308. *In* N. C. Turner & P. J. Kramer (eds.), *Adaptations of Plants to Water and Temperature Stress.* Wiley-Interscience, New York.

EHLERINGER, J. R. (1981). Leaf absorptances and Mohave and Sonoran Desert plants. *Oecologia* 49:366–70.

EHLERINGER, J. R. (1982). The influence of water stress and temperature on leaf pubescence development in *Encelia farinosa. American Journal of Botany* 69:670–5.

EHLERINGER, J. R. (1983a). Ecology and ecophysiology of leaf pubescence in North American desert plants, pp. 113–32. *In* E. Rodriguez, P. Healey & I. Mehta (eds.), *Biology and Chemistry of Plant Trichomes.* Plenum Press, New York.

EHLERINGER, J. R. (1983b). Ecophysiology of *Amaranthus palmeri,* a Sonoran Desert summer ephemeral. *Oecologia* 57:107–12.

EHLERINGER, J. R. (1984). Intraspecific effects on water relations, growth, and reproduction in *Encelia farinosa. Oecologia* 63:153–8.

EHLERINGER, J. R. (1985). Annuals and perennials of warm deserts, pp. 162–80. *In* B. F. Chabot & H. A. Mooney (eds.), *Physiological Ecology of North American Plant Communities.* Chapman and Hall, New York.

EHLERINGER, J. R. & BJÖRKMAN, O. (1978a). Pubescence and leaf spectral characteristics in a desert shrub, *Encelia farinosa. Oecologia* 36:151–62.

EHLERINGER, J. R. & BJÖRKMAN, O. (1978b). A comparison of photosynthetic characteristics of *Encelia* species possessing glabrous and pubescent leaves. *Plant Physiology* 62:185–90.

EHLERINGER, J. R., BJÖRKMAN, O., & MOONEY, H. A. (1976). Leaf pubescence: effects on absorptance and photosynthesis in a desert shrub. *Science* 192:376–7.

EHLERINGER, J. R. & COOPER, T. A. (1988). Correlations between carbon isotope and microhabitat in desert plants. *Oecologia* 76:562–6.

EHLERINGER, J. R. & FORSETH, I. (1980). Solar tracking by plants. *Science* 210:1094–8.

EHLERINGER, J. R. & MOONEY, H. A. (1978). Leaf

hairs: effects on physiological activity and adaptive value to a desert shrub. *Oecologia* 37:183–200.

EHLERINGER, J. R., MOONEY, H. A., & BERRY, J. A. (1977). Photosynthesis and microclimate of a desert winter annual. *Ecology* 58:280–6.

EHLERINGER, J. R. & WERK, K. S. (1986). Modification of solar radiation absorption patterns and implications for carbon gain at the leaf level, pp. 57–82. *In* T. J. Givnish (ed.), *On the Economy of Plant Form and Function*. Columbia University Press, New York.

EKREN, E. B. (1968). Geologic setting of Nevada Test Site and Nellis Air Force Range. *Memoirs of the Geological Society of America* 110:11–9.

EL-GHONEMY, A. A., WALLACE, A., & ROMNEY, E. M. (1980a). Frequency distribution of numbers of perennial shrubs in the northern Mojave Desert. *Great Basin Naturalist Memoirs* (4):34–8.

EL-GHONEMY, A. A., WALLACE, A., & ROMNEY, E. M. (1980b). Multivariate analysis of the vegetation in a two-desert interface. *Great Basin Naturalist Memoirs* (4):42–58.

EL-GHONEMY, A. A., WALLACE, A., & ROMNEY, E. M. (1980c). Socioecological and soil–plant studies of the natural vegetation in the northern Mojave Desert – Great Basin Desert interface. *Great Basin Naturalist Memoirs* (4):73–88.

EL-GHONEMY, A. A., WALLACE, A., ROMNEY, E. M., & VALENTINE, W. (1980). A phytosociological study of a small desert area in Rock Valley, Nevada. *Great Basin Naturalist Memoirs* (4):59–72.

ELKINS, N. Z., SABOL, G. V., WARD, T. J., & WHITFORD, W. G. (1986). The influence of subterranean termites on the hydrological characteristics of a Chihuahuan Desert ecosystem. *Oecologia* 68:521–8.

ELLISON, L. (1980). Influence of grazing on plant succession on rangelands. *Botanical Review* 26:1–78.

ELSE, P. L. & HULBERT, A. J. (1981). Comparison of the "mammal machine" and the "reptile machine": energy production. *American Journal of Physiology* 240:R3–9.

ENGELMANN, M. D. (1966). Energetics, terrestrial field studies, and animal productivity. *Advances in Ecological Research* 3:73–115.

ESKEW, D. L. & TING, I. P. (1978). Nitrogen fixation by legumes and blue-green algae-lichen crusts in a Colorado Desert environment. *American Journal of Botany* 65:850–6.

ESSINGTON, E. H. & GILBERT, R. O. (1977). Plutonium, americium and uranium in blow-sand mounds of safety-shot sites at the Nevada Test Site and Tonopah Test Range, pp. 81–146. *In, Transuranium in Desert Ecosystems. U.S. Department of Energy Report* NVO–181.

ETTERSHANK, G., ETTERSHANK, J. A., BRYANT, M., & WHITFORD, W. G. (1978). Effects of nitrogen fertilization on primary production in a Chihuahuan Desert ecosystem. *Journal of Arid Environments* 1:135–9.

EVANS, D. M. (1973). Seasonal variations in the body composition and nutrition of the vole *Microtus agrestis*. *Journal of Animal Ecology* 42:1–18.

EVENARI, M., SHANAN, L., & TADMOR, N. (1971). *The Negev: The Challenge of a Desert*. Harvard University Press, Cambridge.

FAHN, A. (1964). Some anatomical adaptations of desert plants. *Phytomorphology* 14:93–102.

FARNSWORTH, R. B., ROMNEY, E. M., & WALLACE, A. (1978). Nitrogen fixation by microfloral-higher plant associations in arid to semiarid environments, pp. 17–19. *In* N. E. West & J. Skujins (eds.), *Nitrogen in Desert Ecosystems*. Dowden, Hutchinson, and Ross, Inc., Stroudsburg, Pennsylvania.

FERGUSON, W. E. (1967). Male sphaeropthalmine mutillid wasps of the Nevada Test Site. *Brigham Young University Science Bulletin, Biology Series* 8(4):1–26.

FISHER, F. M., ZAK, J. C., CUNNINGHAM, G. L., & WHITFORD, W. G. (1988). Water and nitrogen effects on growth and allocation patterns of creosotebush in the northern Chihuahuan Desert. *Journal of Range Management* 41:387–91.

FLETCHER, J. E., SORENSON, D. L., & PORCELLA, D. B. (1978). Erosional transfers of nitrogen in desert ecosystems, pp. 171–181. *In* N. E. West & J. Skujins (eds.), *Nitrogen in Desert Ecosystems*. Dowden, Hutchinson, and Ross, Inc., Stroudsburg, Pennsylvania.

FOCHT, D. D. (1978). Methods for analysis of denitrification in soils, pp. 433–90. *In*, D. R. Nielsen & J. G. MacDonald (eds.), *Nitrogen in the Environment*, vol. 2, *Soil–Plant Nitrogen Relationships*. Academic Press, New York.

FONTEYN, P. J. & MAHALL, B. E. (1978). Competition among desert perennials. *Nature* 275:544–5.

FONTEYN, P. J. & MAHALL, B. E. (1981). An experimental analysis of structure in a desert plant community. *Journal of Ecology* 69:883–96.

FORSETH, I. N. & EHLERINGER, J. R. (1980). Solar tracking response to drought in a desert annual. *Oecologia* 44:159–63.

FORSETH, I. N. & EHLERINGER, J. R. (1982). Ecophysiology of two solar tracking desert winter annuals. II. Leaf movements, water relations, and microclimate. *Oecologia* 54:41–9.

FORSETH, I. N. & EHLERINGER, J. R. (1983a). Ecophysiology of two solar tracking desert winter annuals. III. Gas exchange responses to light, CO_2 and VPD in relation to long-term drought. *Oecologia* 57:344–51.

FORSETH, I. N. & EHLERINGER, J. R. (1983b). Ecophysiology of two solar tracking desert winter annuals. IV. Effects of leaf orientation on calculated daily carbon gain and water use efficiency. *Oecologia* 58:10–18.

FORSETH, I. N., EHLERINGER, J. R., WERK, K. S., & COOK, C. S. (1984). Field water relations of Sonoran Desert annuals. *Ecology* 65:1436–44.

FRANCO, P. J., EDNEY, E. B., & McBRAYER, J. F. (1979). The distribution and abundance of soil arthropods in the northern Mojave Desert. *Journal of Arid Environments* 2:137–49.

FRECKMAN, D. W. (1978). Ecology of anhydrobiotic nematodes, pp. 345–57. *In* J. H. Crowe & H. S. Clegg (eds.), *Dry Biological Systems*. Academic Press, New York.

FRECKMAN, D. W. & MANKAU, R. (1977). Distribution and trophic structure of nematodes in desert soils, pp. 511–14. *In* U. Lohm & T. Persson (eds.), *Soil Organisms as Components of Ecosystems*. Ecological Bulletin, Stockholm.

FRECKMAN, D. W. & MANKAU, R. (1979). Nematodes and microflora in the root rhizosphere of four desert shrubs, pp. 423–32. *In* J. L. Harley & R. Russell (eds.), *The Soil–Root Interface*. Academic Press, London.

FRECKMAN, D. W., MANKAU, R., & FERRIS, H. (1975). Nematode community structure in desert soils: nematode recovery. *Journal of Nematology* 7:343–6.

FRENCH, N. R. (1964). *Description of a Study of Ecological Effects on a Desert Area from Chronic Exposure to Low Level Ionizing Radiation*. U.S. Atomic Energy Commission Report UCLA 12–532, Los Angeles, California.

FRENCH, N. R. (1965). Radiation and animal populations: problems, progress and projections. *Health Physics* 11:1557–68.

FRENCH, N. R. (1976). Selection of high temperatures for hibernation by the pocket mouse, *Perognathus longimembris*: ecological advantages and energetic consequences. *Ecology* 57:185–91.

FRENCH, N. R. (1977). Circannual rhythmicity and entrainment of surface activity in the hibernator, *Perognathus longimembris*. *Journal of Mammalogy* 58:37–43.

FRENCH, N. R., MAZA, B. G., HILL, H. O., ASCHWANDEN, A. P., & KAAZ, H. W. (1974). A population study of irradiated desert rodents. *Ecological Monographs* 44:45–72.

ASCHWANDEN, A. P. (1967). Life spans of *Dipodomys* and *Perognathus* in the Mojave Desert. *Journal of Mammalogy* 48:537–48.

FRENCH, N. R., MAZA, B. G., HILL, H. O., ASCHWANDEN, A. P., & KAAZ, H. W. (1974). A population study of irradiated desert rodents. *Ecological Monographs* 44:45–72.

FRENCH, R. H. (1983). *A Preliminary Analysis of Precipitation in Southern Nevada*. U.S. Department of Energy Contract Report DOE/NV/10162–10. Water Resources Center, Desert Research Institute, Reno, Nevada.

FRENKEL, R. E. (1970). Ruderal vegetation along some California roadsides. *University of California Publications in Geography* 20:1–163.

FULLER, W. H. (1974). Desert soils, pp. 31–101. *In* G. W. Brown (ed.), *Desert Biology*, vol. 2. Academic Press, New York.

GALUN, M. (1970). *The Lichens of Israel*. The Israel Academy of Sciences and Humanities, Jerusalem.

GAMBINO, J. J. & LINDBERG, R. G. (1964). Response of the pocket mouse to ionizing radiation. *Radiation Botany* 22:586–97.

GARCIA-MOYA, E. & McKELL, C. M. (1970). Contributions of shrubs to the nitrogen economy of a desert-wash plant community. *Ecology* 51:81–8.

GARDNER, J. L. (1950). Effects of thirty years of protection from grazing in desert grassland. *Ecology* 51:81–8.

GARLAND, T. (1983). The relation between maximal running speed and body mass in terrestrial mammals. *Journal of Zoology, London* 199:157–70.

GATES, D. M. (1968). Transpiration and leaf temperature. *Annual Review of Plant Physiology* 19:211–38.

GATES, D. M., ALDERFER, R., & TAYLOR, E. (1968). Leaf temperatures of desert plants. *Science* 159:994–5.

GATES, D. M. & PAPIEN, L. E. (1971). *Atlas of Energy Budgets of Plant Leaves*. Academic Press, London.

GERMANO, D. J. (1988). Age and growth histories of desert tortoises using scute annuli. *Copeia* 1988:914–20.

GERTSCH, W. J. (1949). *American Spiders*. D. van Nostrand, New York.

GERTSCH, W. J. & ALLRED, D. M. (1965). Scorpions of the Nevada Test Site. *Brigham Young University Science Bulletin, Biology Series* 6(4):1–15.

GETTINGER, R. D. (1984). Energy and water metabolism of free-ranging pocket gophers, *Thomomys bottae*. *Ecology* 65:740–51.

GIBSON, A. C. (1982). The anatomy of succulence, pp. 1–16. *In* I. P. Ting & Gibbs, M. (eds.), *Crassulacean Acid Metabolism*. American Society of Plant Physiologists, Rockville, Maryland.

GIBSON, A. C. (1983). Anatomy of photosynthetic old stems of nonsucculent dicotyledons from North American deserts. *Botanical Gazette* 144:347–62.

GIBSON, A. C. (1996). *Structure–Function Relationships of Warm Desert Plants*. Springer-Verlag, Berlin.

GIBSON, A. C. & NOBEL, P. S. (1986). *The Cactus Primer*. Harvard University Press, Cambridge.

GIFFORD, G. F. & HAWKINS, R. H. (1978). Hydrologic impact of grazing on infiltration: a critical review. *Water Resources Research* 14:305–13.

GILBERT, R. O. & ESSINGTON, E. H. (1977). Estimating total 239–240 Pu in blow-sand mounds of two safety shot sites, pp. 367–421. *In, Transuranium in Desert Ecosystems*. U.S. Department of Energy Report NVO–181.

GILLETTE, D. A., ADAMS, J., ENDO, L., & SMITH, D. (1980). Threshold velocities for input of soil particles into the air by desert soils. *Journal of Geophysical Research* C85:5621–30.

GILLETTE, D. A., ADAMS, J., MUHNS, D., & KIHL, R. (1982). Threshold friction and rupture moduli for crushed desert soils for input of soil particles into air. *Journal of Geophysical Research* 87:9003–15.

GOATES, M. A. (1963). Mites on kangaroo rats at the Nevada Test Site. *Brigham Young University Science Bulletin, Biology Series* 3(4):1–12.

GOLDSTEIN, D. L. (1984). The thermal environment and its constraint on activity of desert quail in summer. *The Auk* 101:542–50.

GOLDSTEIN, D. L. & NAGY, K. A. (1985). Resource utilization by desert quail: time and energy, food and water. *Ecology* 66:378–87.

GOODALL, D. W., CHILDS, S., & WIEBE, H. (1972). *Methodological and Validation Study of Seed Reserves in Desert Soils. US/IBP Desert Biome Research Memorandum 72–8.* Utah State University, Logan.

GOODFELLOW, M. & CROSS, T. (1974). Actinomycetes, pp. 269–302. *In* C. H. Dickinson & G. J. F. Pugh (eds.), *Biology of Plant Litter Decomposition*. Academic Press, London.

GORDON, M. S. (1982). Water and solute metabolism, pp. 272–332. *In* M. S. Gordon, G. A. Bartholomew, A. D. Grinnell, C. B. Jorgensen, & F. N. White, *Animal Physiology: Principles and Adaptations*, 4th edn. Macmillan Publishing Company, New York.

GOUDIE, A. S. (1978). Dust storms and their geomorphological implications. *Journal of Arid Environments* 1:291–310.

GOUDIE, A. S. (1983). Dust storms in space and time. *Projects in Physical Geography* 7:502–30.

GOUDIE, A. S. (1985). Calcrete, pp. 91–131. *In* A. S. Goudie & K. Pye (eds.), *Chemical Sediments and Geomorphology*. Academic Press, London.

GREENFIELD, M. D., SHELLY, T. E., & DOWNUM, K. R. (1987). Variation in host-plant quality: implications for territoriality in a desert grasshopper. *Ecology* 68:828–38.

GREENFIELD, M. D., SHELLEY, T. E., & GONZALEZ-COLOMA, A. (1989). Territory selection in a desert grasshopper: the maximization of conversion efficiency on a chemically defended shrub. *Journal of Animal Ecology* 58:761–71.

GREGER, P. (1990). Top secret area: managed like a national park. *Humane Society of Southern Nevada* 1:6.

GREGORY, P. T. (1982). Reptilian hibernation, pp. 53–154. *In* C. Gans & F. H. Pough (eds.), *Biology of the Reptilia*, vol. 13. Academic Press, London.

GRUBBS, D. E. (1980). Tritiated water turnover in free-living desert rodents. *Comparative Biochemistry and Physiology* 66A:89–98.

GULMON, S. L. & MOONEY, H. A. (1977). Spatial and temporal relationships between two desert shrubs, *Atriplex hymenelytra* and *Tidestromia oblongifolia* in Death Valley, California. *Journal of Ecology* 65:831–8.

GUTIERREZ, J. R. & WHITFORD, W. G. (1989). Effect of eliminating subterranean termites on the growth of creosotebush, *Larrea tridentata*. *The Southwestern Naturalist* 34:549–51.

HADLEY, N. F. (1972). Desert species and adaptations. *American Scientist* 60:338–47.

HADLEY, N. F. (1974). Adaptational biology of desert scorpions. *Journal of Arachnology* 2:11–23.

HADLEY, N. F. (1979). Wax secretion and colour phases of the desert tenebrionid beetle *Cryptoglossa verrucosa* (Le Conte). *Science* 203:367–9.

HADLEY, N. F. (1982). Cuticle ultrastructure with respect to the lipid waterproofing barrier. *Journal of Experimental Zoology* 222:239–48.

HADLEY, N. F. (1990). Physiological ecology of desert scorpions, pp. 321–40. *In* G. A. Polis (ed.), *Biology of scorpions*. Stanford University Press, Stanford, California.

HADLEY, N. F. & WILLIAMS, S. C. (1968). Surface activities of some North American scorpions in relation to feeding. *Ecology* 49:726–34.

HALVORSON, W. L. & PATTEN, D. T. (1975). Productivity and flowering of winter ephemerals in relation to Sonoran Desert shrubs. *American Midland Naturalist* 93:311–19.

HAMILTON, W. J., III. (1975). Coloration and its thermal consequences for diurnal desert insects, pp. 67–89. *In* N. F. Hadley (ed.), *Environmental Physiology of Desert Organisms*. Dowden, Hutchinson, and Ross, Inc., Stroudsburg, Pennsylvania.

HAMILTON, W. J., III. & SEELY, M. K. (1976). Fog basking by the Namib Desert beetle, *Onymacris unguiculatus*. *Nature* 262:284–5.

HANSEN, S. R. (1978). Resource utilization and coexistence of three species of *Pogonomyrmex* ants in an Upper Sonoran grassland community. *Oecologia* 35:109–17.

HART, J. S. (1971). Calometric determination of average body temperature of small mammals and its variation with environmental conditions. *Canadian Journal of Zoology* 29:224–33.

HARVEY, R. A. (1972). An assessment of the seasonal growth on some desert shrubs subjected to chronic low intensity gamma radiation, pp. 12–48. *In* A. Wallace & E. M. Romney, *Radioecology and Ecophysiology of Desert Plants at the Nevada Test Site*. U.S. Atomic Energy Commission Office of Information Service, TID–25954, Oak Ridge, Tennessee.

HATLEY, C. L. & MacMAHON, J. A. (1980). Spider community organization: seasonal variation and the role of vegetation architecture. *Environmental Entomology* 9:632–9.

HAWBECKER, A. C. (1975). The biology of some desert-dwelling ground squirrels, pp. 277–303. *In* I. Prakash & P. K. Ghosh (eds.), *Rodents in Desert Environments. Monographiae Biologicae*, vol. 28. Dr. W. Junk, The Hague.

HAYWARD, C. L., KILLPACK, M. L., & RICHARDS, G. L. (1963). Birds of the Nevada Test Site. *Brigham Young University Science Bulletin, Biology Series* 3(1):1–27.

HEATH, J. E. (1964). Head–body temperature differences in horned lizards. *Physiological Zoology* 37:273–9.

HEATH, J. E. (1965). Venous shunts in the cephalic sinuses of horned lizards. *Physiological Zoology* 39:129–37.

HEINRICH, B. & BARTHOLOMEW, G. A. (1971). An analysis of pre-flight warmup in the sphinx moth, *Manduca sexta*. *Journal of Experimental Biology* 55:223–39.

HENRIKSSON, E. & SIMU, B. (1971). Nitrogen fixation by lichens. *Oikos* 22:119–22.

HENSLEY, M. M. (1954). Ecological relations of the breeding bird population of the desert biome in Arizona. *Ecological Monographs* 24:185–207.

HERRIN, C. S. & BECK, D. E. (1965). Observations on the biology, anatomy and morphology of *Otobius lagophilus* Cooley and Kohls. *Brigham Young University Science Bulletin, Biology Series* 6(2):1–19.

HETHENER, P. (1967). Activité microbiologique des sols a *Cupressus dupreziana* A. Camus au Tassili N'Ajjer (Sahara central). Bulletin Société Histoire Naturale Afrique du Nord, Algerie 58: 39–100.

HILL, H. O. (1980). Breeding birds in a desert scrub community in southern Nevada. *The Southwestern Naturalist* 25:173–80.

HILL, H. O. & BURR, T. (1974). Birds of the Rock Valley validation site, pp. 51–5. *In* F. B. Turner & J. F. McBrayer (eds.), *Rock Valley Validation Site Report. US/IBP Desert Biome Research Memorandum* 74–2. Utah State University, Logan.

HINDS, D. S. (1973). Acclimatization of thermoregulation in the desert cottontail, *Sylvilagus audubonii*. *Journal of Mammalogy* 54:708–28.

HINES, N. O. (1962). *Proving Ground. An Account of the Radiobiological Studies in the Pacific, 1946–1961.* University of Washington Press, Seattle.

HITCH, C. J. B. & STEWART, D. P. (1973). Nitrogen fixation in lichens in Scotland. *New Phytologist* 72:509–24.

HODDENBACH, G. A. & TURNER, F. B. (1968). Clutch size of the lizard *Uta stansburiana* in southern Nevada. *American Midland Naturalist* 80:262–5.

HOUGHTON, J. G. (1969). *Characteristics of Rainfall in the Great Basin.* Desert Research Institute, Reno, Nevada.

HOUGHTON, J. G., SAKAMOTO, C. M., & GIFFORD, R. O. (1975). *Nevada's Weather and Climate. Nevada Bureau of Mines and Geology Special Publication 2.*

HOWARD, W. E. & CHILDS. H. E. (1959). Ecology of pocket gophers with emphasis on *Thomomys bottae mewa*. *Hilgardia* 29:277–358.

HUBBS, C. L. & MILLER, R. R. (1948). The zoological evidence between fish distribution and hydrographic history in the desert basins of western United States, pp. 17–166. *In, The Great Basin with Emphasis on Glacial and Postglacial Times. University of Utah Bulletin, Biological Series* 107.

HUDSON, J. W. (1962). The role of water in the biology of the antelope ground squirrel, *Citellus leucurus*. *University of California Publications in Zoology* 64:1–56.

HUEY, R. B. (1982). Temperature, physiology, and the ecology of reptiles, pp. 25–91. *In* C. Gans & F. H. Pough (eds.), *Biology of the Reptilia*, vol. 12. Academic Press, London.

HULBERT, A. J. & DAWSON, T. J. (1974). Water metabolism in perameloid marsupials from different environments. *Comparative Biochemistry and Physiology* 47A:617–33.

HULBERT, A. J. & ELSE, P. L. (1981). Comparison of the "mammal machine" and the "reptile machine": energy use and thyroid activity. *American Journal of Physiology* 241:R350–6.

HUMPHREY, R. R. (1975). Phenology of selected Sonoran Desert plants at Punta Cirio, Sonora, Mexico. *Journal of the Arizona Academy of Science* 10:50–67.

HUNT, C. B. (1966). *Plant Ecology of Death Valley, California. U.S. Geological Survey Professional Paper 509.*

HUNT, C. B. (1975). *Death Valley: Geology, Ecology, Archaeology.* University of California Press, Berkeley, California.

HUNTER, R. B. (1987). Jackrabbit–shrub interactions in the Mojave Desert, pp. 88–92. *In* D. Provenza, J. T. Flinders & E. D. McArthur (eds), *Proceedings– Symposium on Plant–Herbivore Interactions. USDA, Forest Service, Intermountain Research Station Report* INT–222. Ogden, Utah.

HUNTER, R. B. (1989). Competition between adult and seedling shrubs of *Ambrosia dumosa* in the Mojave Desert, Nevada. *Great Basin Naturalist* 49:79–84.

HUNTER, R. B. (1990). Recent increases in *Bromus* populations in the Nevada Test Site, pp. 22–5. *In* E. D. McArthur, E. M. Romney, S. D. Smith, & P. T. Tueller (compilers), *Symposium on Cheatgrass Invasion, Shrub Die-off, and Other Aspects of Shrub Biology and Management. General Technical Report* INT–276. U.S. Department of Agriculture, Forest Service, Intermountain Research Station, Ogden, Utah.

HUNTER, R. B. (1991). *Bromus* invasions on the Nevada Test Site: present status of *B. rubens* and *B. tectorum* with notes on their relationship to disturbance and altitude. *Great Basin Naturalist* 51:176–82.

HUNTER, R. B. & MEDICA, P. A. (1989). *Status of the*

Flora and Fauna on the Nevada Test Site. United States Department of Energy Report DOE/NV/102630–2. National Technical Information Service, U.S. Department of Commerce, Springfield, Virginia.

HUNTER, R. B., ROMNEY, E. M., CHILDRESS, J. W., & KINNEAR, J. E. (1975). *Responses and Interactions in Desert Plants as Influenced by Irrigation and Nitrogen Applications.* US/IBP Desert Biome Research Memorandum 75–13. Utah State University, Logan.

HUNTER, R. B., ROMNEY, E. M., & WALLACE, A. (1980). Rodent–denuded areas of the northern Mojave Desert. *Great Basin Naturalist Memoirs* (4):208–11.

HUNTER, R. B., ROMNEY, E. M., & WALLACE, A. (1982). Nitrate distribution in Mojave Desert soils. *Soil Science* 134:22–30.

HUNTER, R. B., ROMNEY, E. M., WALLACE, A., HILL, H. O., ACKERMAN, T. L., & KINNEAR, J. E. (1976). *Responses and Interactions in Desert Plants as Influenced by Irrigation and Nitrogen Applications.* US/IBP Desert Biome Research Memorandum 76–14. Utah State University, Logan.

HUNTER, R. B., WALLACE, A. & ROMNEY, E. M. (1976). *Nitrogen Transformations in Rock Valley and Adjacent Areas of the Mojave Desert.* US/IBP Desert Biome Research Memorandum 76–26. Utah State University, Logan.

HUNTER, R. B., WALLACE, A., & ROMNEY, E. M. (1980). Fencing enhances shrub survival and growth for Mojave Desert revegetation. *Great Basin Naturalist Memoirs* (4):212–15.

HUNTER, R. B., WALLACE, A., ROMNEY, E. M. & WIELAND, P. A. T. (1975). *Nitrogen Transformation in Rock Valley and Adjacent Areas of the Mojave Desert.* US/IBP Desert Biome Research Memorandum 75–35. Utah State University, Logan.

INOUYE, R. S., BYERS, G. S., & BROWN, J. H. (1980). Effects of predation and competition on survivorship, fecundity, and community structure of desert annuals. *Ecology* 61:1344–51.

IVERSON, J. B. (1982). Adaptations to herbivory in iguanine lizards, pp. 60–76. *In* G. M. Burghardt & A. S. Rand (eds.), *Iguanas of the World: Their Behavior, Ecology, and Conservation.* Noyes Publications, Park Ridge, New Jersey.

JACOBSON, E. R. & WHITFORD, W. G. (1971). Physiological responses to temperature in the patch-nosed snake, *Salvadora hexalepis. Herpetologica* 27:289–94.

JOHNSON, H. B. (1975). Gas exchange strategies in desert plants, pp. 105–20. *In* D. M. Gates & R. B. Schmerl (eds.), *Perspectives of Biophysical Ecology. Ecological Studies,* vol. 12. Springer-Verlag, New York.

JOHNSON, H. B. (1976). Vegetation and plant communities of southern California deserts – a functional view, pp. 125–64. *In* J. Latting (ed.), *Plant Communities of Southern California. California Native Plant Society, Special Publication 2,* Berkeley, California.

JOHNSON, H. B., VASEK, F. C., & YONKERS, T. (1975). Productivity, diversity, and stability relationships in Mojave Desert roadside vegetation. *Bulletin of the Torrey Botanical Club* 102:106–15.

JOHNSON, K., JOHNSON, P., & BELLOWS, T. (1975). Invertebrates, pp. 60–82. *In* W. G. Whitford (ed.), *Jornada Validation Site Report. US/IBP Desert Biome Research Memorandum 75–4.* Utah State University, Logan.

JOHNSON, M. S. & HIBBARD, D. E. (1957). Geology of the Atomic Energy Commission Nevada proving grounds area, Nevada. *U.S. Geological Survey Bulletin* 1021-K:1–55.

JOHNSON, S. R. (1965). An ecological study of the chuckwalla, *Sauromalus obesus* Baird, in the western Mojave Desert. *American Midland Naturalist* 73:1–29.

JORGENSEN, C. D. (1970). *Free-living Mites of the Nevada Test Site.* U.S. Atomic Energy Commission Report C00–1731–4. National Technical Information Services, Springfield, Virginia.

JORGENSEN, C. D. & HAYWARD, C. L. (1963). Notes on shrews from southern Nevada. *Journal of Mammalogy* 44:582.

JORGENSEN, C. D. & HAYWARD, C. L. (1965). Mammals of the Nevada Test Site. *Brigham Young University Science Bulletin, Biology Series* 6(3):1–81.

JORGENSEN, C. D. & TANNER, W. W. (1963). The application of the density probability function to determine the home ranges of *Uta stansburiana* and *Cnemidophorus tigris tigris. Herpetologica* 19:105–15.

JUHREN, M., WENT, F. W., & PHILLIPS, E. (1956). Ecology of desert plants. IV. Combined field and laboratory work on germination of annuals in the Joshua Tree National Monument, California. *Ecology* 37:318–30.

KAAZ, H. W., WALLACE, A., & ROMNEY, E. M. (1971). Effect of a chronic exposure to gamma radiation on the shrub *Ephedra nevadensis* in the northern Mojave Desert. *Radiation Biology* 11:33–7.

KAPPEN, L. (1981). Ecological significance of resistance to high temperature, pp. 439–74. *In* O. L. Lange, P. S. Nobel, C. B. Osmond, & H. Ziegler (eds.), *Encyclopedia of Plant Physiology,* new series 12B. Springer-Verlag, Berlin.

KAPPEN, L., OERTLI, J. J., LANGE, O. L., SCHULZE, E.-D., EVENARI, M., & BUSCHBOM, U. (1975). Seasonal and diurnal courses of water relations of the arido-active plant *Hammada scoparia* in the Negev Desert. *Oecologia* 21:175–92.

KARASOV, W. H. (1981). Daily energy expenditure and the cost of activity in a free living mammal. *Oecologia* 51:253–9.

KARASOV, W. H. (1982). Energy assimilation, nitrogen requirement, and diet in free-living antelope ground squirrels, *Ammospermophilus leucurus. Physiological Zoology* 55:378–92.

KARASOV, W. H. (1983). Wintertime energy conservation by huddling in antelope ground squirrels, *Ammospermophilus leucurus*. *Journal of Mammalogy* 64:341–
5.

KARASOV, W. H. & DIAMOND, J. M. (1985). Diges-
tive adaptations for fueling the cost of endothermy. *Science* 228:202–4.

KAY, F. R. (1970). Environmental responses of active lizards at Saratoga Springs, Death Valley, California.
Great Basin Naturalist 30:146–65.

KAY, F. R., MILLER, B. W., & MILLER, C. L.
(1970). Food habits and reproduction of *Callisaurus draconoides* in Death Valley, California. *Herpetologica*
26:431–6.

KEARNEY, T. H. & PEEBLES, R. H. (1960). *Arizona
Flora*, 2nd edn. University of California Press,
Berkeley.

KEAST, A. (1959). Australian birds: their zoogeography
and adaptations to an arid continent, pp. 89–114. *In*
A. Keast, R. L. Crocker, & C. S. Christian (eds.), *Biogeography and Ecology in Australia*. Junk Publishers,
The Hague.

KELRICK, M. I., MacMAHON, J. A.,
PARMENTER, R. R., & SISSON, D. V. (1986).
Native seed preferences of shrub–steppe rodents, birds
and ants: the relationships of seed attributes and seed
use. *Oecologia* 68:327–37.

KENAGY, G. J. (1972). Saltbush leaves: excision of hyper-
saline tissues by a kangaroo rat. *Science* 178:1094–6.

KENAGY, G. J. (1973a). Adaptations for leaf eating in the
Great Basin kangaroo rat, *Dipododys microps*. *Oecologia* 12:383–412.

KENAGY, G. J. (1973b). Daily and seasonal patterns of
activity and energetics in a heteromyid rodent community. *Ecology* 54:1201–19.

KENAGY, G. J. (1976). The periodicity of daily activity
and its seasonal changes in free-ranging and captive
kangaroo rats. *Oecologia* 24:105–40.

KENDEIGH, S. C. (1944). Measurements of bird populations. *Ecological Monographs* 14:67–106.

KISTLER, R. W. (1968). Potassium argon ages of rocks in
Nye and Emeralda Counties, Nevada. *Memoirs of the
Geological Society of America* 110:251–62.

KITAZAWA, Y. (1967). Community metabolism of soil
invertebrates in forest ecosystems of Japan, pp. 649–61.
In K. Petrusewicz (ed.), *Secondary Productivity of Terrestrial Ecosystems*, vol. 2. Panstowe Wydawnictwo
Naukowe, Warsaw.

KLEINKOPF, G. E., HARTSOCK, T. L., WALLACE,
A., & ROMNEY, E. M. (1980). Photosynthetic strategies of two Mojave Desert shrubs. *Great Basin Naturalist Memoirs* (4):100–9.

KLEMNEDSON, J. D. (1956). Interrelations of vegetation, soils, and range conditions induced by grazing.
Journal of Range Management 9:134–8.

KLUBEK, B., EBERJARDT, P. J., & SKUJINS, J.
(1978). *In vitro* ammonia volatilization and denitrification from Great Basin soils, pp. 107–29. *In*
N. E. West & J. Skujins (eds.), *Nitrogen in Desert Ecosystems*. Dowden, Hutchinson, and Ross, Inc., Stroudsburg, Pennsylvania.

KLUGE, M. & TING, I. P. (1978). *Crassulacean Acid
Metabolism: Analysis of an Ecological Adaptation*. *Ecological Studies Series*, vol. 30. Springer-Verlag, Berlin.

KNOLL, G. & HOPKINS, H. M. (1959). The effects of
grazing and trampling upon certain soil properties.
Transactions of the Kansas Academy of Sciences
62:221–31.

KNOWLES, R. (1981). Denitrification, pp. 315–29. *In*
F. E. Clark & R. Rosswell (eds.), *Terrestrial Nitrogen
Cycles. Processes, Ecosystem Strategies and Management Impacts*. Swedish Natural Science Research Council, Stockholm.

KOTLER, B. P. (1984a). Harvesting rates and predatory
risk in desert rodents: a comparison of two communities
on different continents. *Journal of Mammalogy* 65:91–6.

KOTLER, B. P. (1984b). Risk of predation and the structure of desert rodent communities. *Ecology* 65:689–701.

KOTLER, B. P. (1985). Owl predation of desert rodents
which differ in morphology and behavior. *Journal of
Mammalogy* 66:824–8.

KOUR, E. L. & HUTCHINSON, V. H. (1970). Critical
thermal tolerances and heating and cooling rates of lizards in diverse habitats. *Copeia* 1970:219–29.

KRAMER, P. J. (1969). *Plant and Soil Water Relations: A
Modern Synthesis*. McGraw-Hill, New York.

KREKORIAN, C. O. (1976). Home range size and overlap and their relationship to food abundance in the
desert iguana, *Dipsosaurus dorsalis*. *Herpetologica*
32:405–12.

KREKORIAN, C. O. (1983). Population density of the
desert iguana, *Dipsosaurus dorsalis* (Reptilia:
Iguanidae), in southern California. *Copeia* 1983:268–71.

KREKORIAN, C. O. (1984). Life history of the desert
iguana, *Dipsosaurus dorsalis*. *Herpetologica* 40:415–24.

KUCHLER, A. W. (1977). The map of the natural vege-
tation of California, pp. 909–38. *In* M. G. Barbour &
J. Major (eds.), *Terrestrial Vegetation of California*.
John Wiley and Sons, New York.

KUIJT, J. (1969). *The Biology of Parasitic Flowering
Plants*. University of California Press, Berkeley.

LANE, L. J., ed. (1986). *Proceedings of the Rainfall Sem-
inar Workshop*. Society for Range Management,
Denver.

LANE, L. J., ROMNEY, E. M., & HAKONSON T.,
(1984). Water balance calculations and net production of
perennial vegetation in the northern Mojave Desert.
Journal of Range Management 37:12–18.

LANGE, O. L. (1959). Untersuchungen über Warm-
schaushalt und Hitzeresistenz maurestanisher Wüsten-
und Savannenpflanzen. *Flora* 147:595–651.

LANGE, O. L., SCHULZE, E.-D., KAPPEN, L.,
BUSCHBOM, U., & EVENARI, M. (1975). Photosynthesis of desert plants as influenced by internal and

external factors, pp. 121–43. *In* D. M. Gates & R. B. Schmerl (eds.), *Perspectives for biophysical Ecology, Ecological Studies*, vol. 12. Springer-Verlag, New York.

LARSEN, E. (1986). Competitive release in microhabitat use among coexisting desert rodents: a natural experiment. *Oecologia* 69:231–7.

LARSON, K. H., OLAFSON, J. H., MORK H. M., & HOWTON, D. R. (1952). *Field Observations and Preliminary field Data Obtained by the UCLA Survey Group, Operation Jangle, Nov. 1951. U.S. Atomic Energy Commission Report* UCLA-182.

LASIEWSKI, R. C. & BARTHOLOMEW, G. A. (1969). Condensation as a mechanism for water gain in nocturnal desert poikilotherms. *Copeia* 1969:405–7.

LATHROP, E. W. & ARCHBOLD, E. F. (1980a). Plant response to Los Angeles aqueduct construction in the Mojave Desert. *Environmental Management* 4:215–26.

LATHROP, E. W. & ARCHBOLD, E. F. (1980b). Plant response to utility right of way construction in the Mojave Desert. *Environmental Management* 4:215–26.

LAUENROTH, W. K., DODD, J. L., & SIMMS, P. L. (1978). The effects of water- and nitrogen-induced stresses on plant community structure in a semiarid grassland. *Oecologia* 36:211–22.

LEAVITT, V. D. (1970). Soil survey of Area 18, Nevada Test Site. *Southwest Regional Hydrology Laboratory* 74r:1–119. Environmental Monitoring and Support Laboratory, Environmental Protection Agency, Las Vegas.

LEAVITT, V. D. & MASON, B. J. (1971). Soil survey of Area 15, Nevada Test Site. *Southwest Regional Hydrology Laboratory* 106r:1–46. Environmental Monitoring and Support Laboratory, Environmental Protection Agency, Las Vegas.

LE HOUEROU, H. N. (1974). Rain use efficiency: a unifying concept in arid-land ecology. *Journal of Arid Environments* 7:213–47.

LEVITT, J. (1972). *Responses of Plants to Environmental Stresses*, 1st edn.; 2nd edn. (1980). Academic Press, New York.

LILLYWHITE, H. B. & MADERSON, P. F. A. (1982). Skin structure and permeability, pp. 397–442. *In* C. Gans & F. H. Pough (eds.), *Biology of the Reptilia*, vol. 12. Academic Press, London.

LINDBERG, R. G. & HAYDEN, P. (1974). Thermoperiodic entrainment of arousal from torpor in the little pocket mouse, *Perognathus longimembris*. *Chronobiologia* 1:356–61.

LONGLAND, W. S. & JENKINS, S. H. (1987). Sex and age affect vulnerability of desert rodents to owl predation. *Journal of Mammology* 68:746–54.

LONGSTRETH, D. J., HARTSOCK, T. L., & NOBEL, P. S. (1980). Mesophyll cell properties for some C_3 and C_4 species with high photosynthetic rates. *Physiologia Plantarum* 48:494–8.

LOPEZ MORENO, I. & DIAZ BETANCOURT, M.

(1986). Foraging behavior of granivorous ants in the Pincante Desert, pp. 115–18. *In* L. C. Drickamer (ed.), *Behavioral Ecology and Population Biology*. Privel, I. E. C., Toulouse.

LUCKENBACH, R. A. (1982). Ecology and management of the desert tortoise (*Gopherus agassizi*) in California, pp. 1–37. *In* R. B. Bury (ed.), *North American Tortoises: Conservation and Ecology*. U.S. Fish and Wildlife Service Wildlife Research Report 12.

LUCKENBACH, R. A. & BURY, R. B. (1983). Effects of off-road vehicles on the biota of the Algodones Dunes, Imperial County, California. *Journal of Applied Ecology* 20:265–86.

LUDWIG, J. A. & WHITFORD, W. G. (1981). Short-term water and energy flow in arid ecosystems, pp. 271–99. *In*, D. W. Goodall & R. A. Perry (eds.), *Arid Land Ecosystems: Structure, Functioning and Management*, vol. 2. *International Biological Programme Synthesis*, vol. 17. Cambridge University Press, London.

LUNT, O. R., LETEY, J., & CLARK, S. B. (1973). Oxygen requirements for root growth in three species of desert shrubs. *Ecology* 54:1356–62.

MABRY, T., HUNZIKER, J. H., & DIFEO, D. eds. (1977). *Creosote Bush, Biology and Chemistry of Larrea in New World Deserts*. US/IBP Synthesis Series 6. Dowden, Hutchinson and Ross, Inc., Stroudsburg, Pennsylvania.

MacDOUGALL, W. B. (1973). *Seed Plants of Northern Arizona*. Museum of Northern Arizona, Flagstaff, Arizona.

MacGREGOR, A. N. & JOHNSON, D. E. (1971). Capacity of desert algal crusts to fix atmospheric nitrogen. *Proceedings of the Soil Science Society of America* 35:843–4.

MACK, R. N. (1981). Invasion of *Bromus tectorum* L. into western North America: an ecological chronicle. *Agro-Ecosystems* 7:145–65.

MACK, R. N. (1985). Invading plants: their potential contribution to population biology, pp. 127–42. *In* J. White (ed.), *Studies on Plant Demography*. Academic Press, London.

MACK, R. N. (1986). Alpen plant invasion into the Intermontane West, pp. 191–213. *In* H. A. Mooney & J. A. Drake (eds)., *Ecology of Biological Invasions of North America and Hawaii*. Springer-Verlag, New York.

MACKAY, W. P. (1981). A comparison of the nest phenologies of three species of *Pogonomyrmex* harvester ants (Hymenoptera: Formicidae). *Ecological Entomology* 5:353–71.

MACKAY, W. P. (1991). The role of ants and termites in desert communities, pp. 113–50. *In* G. A. Polis (ed.), *The Ecology of Desert Communities*. University of Arizona, Tucson.

MACKAY, W. P., ZAK, J., & WHITFORD, W. G.

(1989). The natural history and role of subterranean termites in the northern Chihuahuan Desert, pp. 50–78. *In* J. Schmidt (ed.), *Special Biotic Relationships in the arid Southwest*. University of New Mexico, Albuquerque.

MacMAHON, J. A. (1979). North American deserts: their floral and faunal components, pp. 21–82. *In* D. W. Goodall & R. A. Perry (eds.), *Arid-Land Ecosystems: Structure, Functioning and Management*, vol. 1. Cambridge University Press.

MacMAHON, J. A. & WAGNER, F. H. (1985). The Mojave, Sonoran and Chihuahuan deserts of North America, pp. 105–202. *In* M. Evenari, I. Noy-Meir & D. W. Goddall (eds.), *Hot Deserts and Arid Shrublands, A. Ecosystems of the world* 12A. Elsevier, Amsterdam.

MacMILLEN, R. E. (1962). The minimum water requirements of mourning doves. *The Condor* 64:165–6.

MacMILLEN, R. E. (1972). Water economy of nocturnal desert rodents pp. 147–74. *In* G. M. O. Maloiy (ed.), *Comparative Physiology of Desert Animals*. Zoological Society of London, London.

MacMILLEN, R. E. & LEE, A. K. (1967). Australian desert mice: independence of exogenous water. *Science* 158:383–5.

MAHMOUD, S. A. Z., ABOU EL-FADL, M., & ELMOFTY, M. KH. (1964). Studies on the rhizosphere microflora of a desert plant.*Folia Mikrobiologica* 9:1–8.

MANNING, S. J. & GROENEVELD, D. P. (1990). Shrub rooting characteristics and water acquisition on xeric sites in the western Great Basin, pp. 2338–44. *In* C. B. Osmond, L. F. Pitelka, & G. M. Hidy (eds.), *Plant Biology of the Basin and Range. Ecological Studies*, vol. 80. Springer-Verlag, New York.

MARDER, J. (1973). Body temperature regulation in the brown-necked raven (*Corus corax ruficollis*). II. Thermal changes in the plumage of ravens exposed to solar radiation. *Comparative Biochemistry and Physiology* 45A:431–40.

MARTIN, W. E. (1965). Interception and retention of fallout by desert shrubs. *Health Physics* 11:1341–54.

MAUTZ, W. J. (1979). The metabolism of reclusive lizards, the Xantusiidae. *Copeia* 1979:577–84.

MAUTZ, W. J. & NAGY, K. A. (1987). Ontogentic changes in diet, field metabolic rate, and water flux in the herbivorous lizard *Dipsosaurus dorsalis*. *Physiological Zoology* 60:640–58.

MAXEY, G. B. (1968). Hydrogeology of desert basins. *Ground Water* 6:10–22.

MAXIMOV N. A. (1929). *The Plant in Relation to Water*. Allen and Unwin, London.

MAYHEW W. W. (1966a). Reproduction in the psammophilous lizard *Uma scoparia*. *Copeia* 1966:114–22.

MAYHEW, W. W. (1966b). Reproduction in the arenicolous lizard, *Uma notata*. *Ecology* 47:9–18.

MAYHEW, W. W. (1968). Biology of desert amphibians and reptiles, pp. 195–356, *In* G. W. Brown (ed.), *Desert Biology*, vol. 1. Academic Press, New York.

MAYLAND, H. F. & McINTOSH, T. H. (1966). Availability of biologically fixed atmospheric nitrogen 15 to higher plants. *Nature* 209:421–2.

MAYLAND, H. F., McINTOSH, T. H., & FULLER, W. H. (1966). Fixation of isotopic nitrogen on a semiarid soil by algal crust organisms. *Proceedings of the Soil Science Society of America* 30:56–60.

MAZA, B. G. (1974). Mammals, pp. 55–9. *In* F. B. Turner & J. F. McBrayer (eds.), *Rock Valley Validation Site Report. US/IBP Desert Biome Research Memorandum 74-2*. Utah State University, Logan.

MAZA, B. G., FRENCH, N. R., & ASCHWANDEN, A. P. (1973). Home range dynamics in a population of heteromyid rodents. *Journal of Mammology* 54:405–25.

McCLEARY, J. A. (1968). The biology of desert plants, pp. 141–94. *In* G. W. Brown, jr. (ed.), *Desert Biology*. Academic Press, New York.

McKNIGHT, T. (1961). A survey of feral livestock in California. *Journal of the Association of Pacific Coast Geographers* 23:28–42.

McNAB, B. K. (1979). Climatic adaptation in the energetics of heteromyid rodents. *Comparative Biochemistry and Physiology* 62A:813–20.

McNEILL, S. & LAWTON, J. H. (1970). Annual production and respiration in animal populations. *Nature* 225:472–4.

MEDICA, P. A. (1967). Food habits, habitat preference, reproduction and diurnal activity in four sympatric species of whiptail lizards (*Cnemidophorus*) in south central New Mexico. *Bulletin of the Southern California Academy of Science* 66:251–76.

MEDICA, P. A. (1973). Reptiles, pp. 168–73. *In* F. B. Turner (ed.), *Rock Valley Validation Site Report. US/IBP Desert Biome Reserach Memorandum 73–2*. Utah State University, Logan.

MEDICA, P. A., BURY, R. B., & LUCKENBACH, R. A. (1980). Drinking and construction of water catchments by the desert tortoise, *Gopherus agassizii*, in the Mojave Desert. *Herpetologica* 36:301–4.

MEDICA, P. A., BURY, R. B., & TURNER, F. B. (1975). Growth of the desert tortoise (*Gopherus agassizi*) in Nevada. *Copeia* 1975:639–43.

MEDICA, P. A., HODDENBACH, G. A., & LANNOM, J. R., Jr. (1971). *Lizard Sampling Techniques. Rock Valley Miscellaneous Publications* (1):1–55. Utah State University, Logan.

MEDICA, P. A., LYONS, C. L., & TURNER, F. B. (1982). A comparison of 1981 populations of desert tortoises (*Gopherus agassizii*) in grazed and ungrazed areas in Ivanpah Valley, California, pp. 99–124. *In* K. A. Hashagen & M. W. Trotter (eds.), *Proceedings of the 1982 Symposium*. Desert Tortoise Council, Long Beach, California.

MEDICA, P. A. & SMITH, D. D. (1974). Reptiles, pp. 49–51. *In* F. B. Turnder & J. F. McBrayer (eds.), *Rock Valley Validation Site Report. US/IBP Desert Biome Research Memorandum 74–2.* Utah State University, Logan.

MEDICA, P. A. & TURNER, F. B. (1976). Reproduction by *Uta stansburiana* (Reptilia, Lacertilia, Iguanidae) in southern Nevada. *Journal of Herpetology* 10:123–8.

MEDICA, P. A. & TURNER, F. B. (1984). Natural longevity of iguanid lizards in southern Nevada. *Herpetological Review* 15:34–5.

MEDICA, P. A., TURNER, F. B., & SMITH, D. D. (1973a). Effects of radiation on a fenced population of horned lizards (*Phrynosoma platyrhinos*) in southern Nevada. *Journal of Herpetology* 7:79–85.

MEDICA, P. A., TURNER, F. B., & SMITH, D. D. (1973b). Hormonal induction of color change in female leopard lizards, *Crotaphytus wislizenii. Copeia* 1973: 658–61.

MEHLHOP, P. & SCOTT, N. J. (1983). Temporal patterns of seed use and availability in a guild of desert ants. *Ecological Entomology* 8:69–85.

MEHUYS, G. B. (1973). *Influence of Stones on Isothermal and Thermally Induced Movement of Water through Relatively Dry Soil.* Ph.D. dissertation, University of California, Riverside.

MEHUYS, B., STOLZY, L. H., LETEY, J., & WEEKS, L. V. (1973). *Evaluation of Critical Soil Properties Needed to Predict Soil Water Flow under Desert Conditions. US/IBP Desert Biome Research Memorandum 73–43.* Utah State University, Logan.

MEINZER, F. C., RUNDEL, P. W., SHARIFI, M. R., & NILSEN, E. T. (1986). Turgor and osmotic relations of the desert shrub *Larrea tridentata. Plant, Cell and Environment* 9:467–75.

MEINZER, F. C., WISDOM, C. S., GONZALEZ-COLOMA, A., RUNDEL, P. W., & SHULTZ, L. M. (1990). Effects of leaf resin on stomatal behaviour and gas exchange of *Larrea tridentata* (DC.) Cov. *Functional Ecology* 4:579–84.

MEYER, S. E. (1986). The ecology of gypsophile endemism in the eastern Mojave Desert. *Ecology* 67:1303–13.

MILLER, A. H. & STEBBINS, R. C. (1964). *The Lives of Desert Animals in Joshua Tree National Monument.* University of California Press, Berkeley.

MILLER, L. (1932). Notes on the desert tortoise (*Testudo agassizii*). *Transactions of the San Diego Natural History Museum* 7:187–208.

MILLER, L. (1955). Further observations on the desert tortoise, *Gopherus agassizii,* of California. *Copeia* 1955: 113–18.

MILLER, M. R. (1951). Some aspects of the life history of the yucca night lizard, *Xantusia vigilis. Copeia* 1951: 114–20.

MINNICH, J. E. (197a). Evaporative water loss from the desert iguana, *Dipsosaurus dorsalis. Copeia* 1970:575–8.

MINNICH, J. E. (1976). Water procurement and conservation by desert reptiles in their natural environment. *Israel Journal of Medical Sciences* 12:740–58.

MINNICH, J. E. (1977). Adaptive responses in the water and electrolyte budgets of native and captive desert tortoise, *Gopherus agassizii,* to chronic drought, pp. 102–29. *In, Proceedings of the Desert Tortoise Council Symposium,* Long Beach, California.

MINNICH, J. E. (1979a). Reptiles, pp. 391–641. *In* G. M. O. Maloiy (ed.), *Comparative Physiology of Osmoregulation in Animals,* vol. 1. Academic Press, London.

MINNICH, J. E. (1979b). Comparison of maintenance electrolyte budgets of free-living desert and gopherv tortoises (*Gopherus agassizii* and *G. polyphemus*), pp. 166–74. In E. St. Amart (ed.), *Proceedings of the 1979 Desert Tortoise Council Symposium,* Long Beach, California.

MINNICH, J. E. (1982). The use of water, pp. 325–95. *In* C. Gans & F. H. Pough (eds.), *Biology of the Reptilia,* vol. 12. Academic Press, London.

MINNICH, J. E. & SHOEMAKER, V. H. (1970). Diet, behavior and water turnover in the desert iguana, *Dipsosaurus dorsalis. American Midland Naturalist* 84:496–509.

MISPAGEL, M. E. (1974). *An Ecological Analysis of Insect Populations of Larrea tridentata in the Mojave Desert.* M. S. thesis, California State University, Long Beach.

MISPAGEL, M. E. (1978). The ecology and bioenergetics of the acridid grasshopper, *Bootettix punctatus* on creosote bush, Larrea tridentata, in the northern Mojave Desert. *Ecology* 59:779–88.

MONSON, R. K. & SMITH, S. D. (1982). Seasonal water potential components of Sonoran Desert plants. *Ecology* 63:113–23.

MOONEY, H. A. (1980). Seasonality and gradients in the study of stress adaptations, pp. 279–94. *In* N. C. Turner & P. J. Kramer (eds.), *Adaptations of Plants to Water and High Temperature Stress.* Wiley-Interscience, New York.

MOONEY, H. A., BJÖRKMAN, O., & COLLATZ, G. J. (1978). Photosynthetic acclimation to temperature in the desert shrub, *Larrea divaricata.* I. Carbon dioxide exchange characteristics of intact leaves. *Plant Physiology* 61:406–10.

MOONEY, H. A., BJÖRKMAN, O., EHLERINGER, J., & BERRY, J. (1976). Photosynthetic capacity of *in situ* Death Valley plants. *Carnegie Institution of Washington Yearbook* 75:410–13.

MOONEY, H. A. & EHLERINGER, J. R. (1978). The carbon gain benefits of solar tracking in a desert annual. *Plant, Cell and Environment* 1:307–11.

MOONEY, H. A., EHLERINGER, J. R., &
BERRY, J. A. (1974). High photosynthetic capacity of a winter desert annual in Death Valley. *Science* 194:322–4.

MOONEY, H. A., EHLERINGER, J. R., &
BJÖRKMAN, O. (1977). The energy balance of leaves of the evergreen desert shrub *Atriplex hymenelytra*. *Oecologia* 29:301–10.

MOONEY, H. A., TROUGHTON, J. H., &
BERRY, J. A. (1974). Arid climates and photosynthetic systems. *Carnegie Institution of Washington Yearbook* 73:793–805.

MOORE, R. G. (1978). Seasonal and daily activity patterns and thermoregulation in the southwestern speckled rattlesnake (*Crotalus mitchelli pyrrhus*) and the Colorado Desert sidewinder (*Crotalus cerastes laterorepens*). *Copeia* 1978:439–42.

MOTT, J. J. & McCOMB, A. J. (1974). Patterns in annual vegetation and soil microrelief in an arid region of Western Australia. *Journal of Ecology* 62:115–26.

MOTT, K. A., GIBSON, A. C., & O'LEARY J. W. (1982). The adaptive significance of amphistomatic leaves. *Plant, Cell and Environment* 5:455–60.

MULLER, C. H. (1940). Plant succession in the *Larrea-Flourensia* climax. *Ecology* 21:206–12.

MULLER, C. H. (1953). The association of desert annuals with shrubs. *American Journal of Botany* 42:53–62.

MULROY, T. W. & RUNDEL, P. W. (1977). Annual plants: adaptations to desert environments. *Bioscience* 27:109–14.

MUMA, M. H. (1963). Solpugida of the Nevada Test Site. *Brigham Young University Science Bulletin, Biology Series* 3(2):1–13.

MUMA, M. H. & MUMA, K. (1988). *The Arachnid Order Solpugida in the United States. Supplement 2, A Biological Review.* Southwest Offset, Silver City, New Mexico.

MUNZ, P. A. (1974). *A Flora of Southern California.* University of California Press, Berkeley.

MUNZ, P. A. & KECK, D. D. (1959). *A California Flora.* University of California Press, Berkeley.

MUSICK, H. B. (1975). Barrenness of desert pavement in Yuma County, Arizona. *Journal of the Arizona Academy of Science* 10:24–8.

MUTH, A. (1977a). Body temperatures and associated postures of the zebra-tailed lizard, *Callisaurus draconoides. Copeia* 1977:122–5.

MUTH, A. (1977b). Thermoregulatory postures and orientation to the sun: a mechanistic evaluation for the zebra-tailed lizard, *Callisaurus draconoides. Copeia* 1977:710–20.

NAGY, K. A. (1972). Water and electrolyte budgets of free-living desert lizard, *Sauromalus obesus. Journal of Comparative Physiology* 79:39–62.

NAGY, K. A. (1973). Behavior, diet and reproduction in a desert lizard, *Sauromalus obesus. Copeia* 1973:93–102.

NAGY, K. A. (1975). Water and energy budgets of free-living animals: measurement using isotopically labeled water, pp. 227–45. *In* N. F. Hadley (ed.), *Environmental Physiology of Desert Organisms.* Dowden, Hutchinson, and Ross, Inc., Stroudsburg, Pennsylvania.

NAGY, K. A. (1980). CO₂ production in animals: analysis of potential errors in the doubly labeled water method. *American Journal of Physiology* 238:R466–73.

NAGY, K. A. (1983). Ecological energetics, pp. 24–54. *In* R. B. Huey, E. R. Pianka, & T. W. Schoener (eds.), *Lizard Ecology.* Harvard University Press, Cambridge.

NAGY, K. A. (1987). Field metabolic rate and food requirement scaling in mammals and birds. *Ecological Monographs* 57:111–28.

NAGY, K. A. (1988a). Seasonal patterns of water and energy balance in desert vertebrates. *Journal of Arid Environments* 14:201–10.

NAGY, K. A. (1988b). Energetics of desert reptiles, pp. 165–86. *In* P. K. Ghosh & I. Prakash (eds.), *Ecophysiology of Desert Vertebrates.* Scientific Publishers, Jodhpur, India.

NAGY, K. A. & MEDICA, P. A. (1977). Seasonal water and energy relations of free-living desert tortoises in Nevada: preliminary report, pp. 52–7. *In, Proceedings of the 1977 Symposium*, Desert Tortoise Council, Long Beach, California.

NAGY, K. A. & MEDICA, P. A. (1985). Altered energy metabolism in an irradiated population of lizards at the Nevada Test Site. *Radiation Research* 103:98–104.

NAGY, K. A. & MEDICA, P. A. (1986). Physiological ecology of desert tortoises in southern Nevada. *Herpetologica* 42:73–92.

NAGY, K. A. & PETERSON, C. C. (1987). Water flux scaling, pp. 131–40. *In* P. Dejours, L. Bolis, C. R. Taylor, & E. R. Weibel (eds.), *Comparative Physiology: Life in Water and on Land. Fidia Research Series* 9. Liviana Press, Padova.

NAGY, K. A. & PETERSON, C. C. (1988). Scaling of water flux rate in animals. *University of California Publications in Zoology* 120:1–172.

NAGY, K. A. & SHOEMAKER, V. H. (1975). Energy and nitrogen budgets of the free-living desert lizard, *Sauromalus obesus. Physiological Zoology* 48:252–62.

NAGY, K. A., SHOEMAKER, V. H., & COSTA, W. R. (1976). Water, electrolyte, and nitrogen budgets of jackrabbits (*Lepus californicus*) in the Mojave Desert. *Physiological Zoology* 49:351–63.

NASH, T. H., III & MOSER, T. J. (1982). Vegetational and physiological patterns of lichens in North American deserts. *Journal of the Hattori Botanical Laboratory* 53:331–6.

NASH, T. H., WHITE, S. L. & MARSH, J. E. (1977). Lichen and moss distribution and biomass in hot desert ecosystems. *The Bryologist* 80:470–5.

NATIONAL ACADEMY OF SCIENCES. (1977). *An Evaluation of the International Biological Program.* National Academy of Sciences, Washington, D.C.

NEAL B. J. (1965). Reproductive habits of round-tailed and Harris antelope ground squirrels. *Journal of Mammalogy* 46:200–6.

NEAL, J. T., ed. (1975). *Playas and Dried Lakes. Occurrence and Development. Benchmark Papers in Geology,* 20. Dowden, Hutchinson, and Ross, Inc., Stroudsburg, Pennsylvania.

NELSON, J. F. & CHEW, R. M. (1977). Factors affecting seed reserves in the soil of a Mojave Desert ecosystem, Rock Valley, Nye County, Nevada. *American Midland Naturalist* 97:300–20.

NEUFELD, H. S., MEINZER, F. C., WISDOM, C. S., SHARIFI, M. R., RUNDEL, P. W., NEUFELD, M. S., GOLDRING, Y., & CUNNINGHAM, G. L. (1988). Canopy architecture of *Larrea tridentata* (DC.) Cov., a desert shrub: foliage orientation and direct beam radiation interception. *Oecologia* 75:54–60.

NEWELL, I. M. (1955). An autosegregator for use in collecting soil-inhabiting arthropods. *Transactions of the American Microscopical Society* 74:389–92.

NILSEN, E. T., MEINZER, F. C., & RUNDEL, P. W. (1989). Stem photosynthesis in *Psorothamnus spinosus* (smoke tree) in the Sonoran Desert of California. *Oecologia* 79:193–7.

NILSEN, E. T., SHARIFI, M. R. & RUNDEL, P. W. (1984). Comparative water relations of phreatophytes in the Sonoran Desert of California. *Ecology* 65:767–78.

NILSEN, E. T., SHARIFI, M. R., RUNDEL, P. W., JARRELL, W. M., & VIRGINIA, R. A. (1983). Diurnal and seasonal water relations of the desert phreatophyte *Prosopis glandulosa* (honey mesquite) in the Sonoran Desert of California. *Ecology* 64:1181–93.

NISHITA, H. (1965). Decay characteristics of neutron induced radioactivity in soils. *Health Physics* 11:1527–41.

NISHITA, H. & HAUG, R. M. (1973). Distribution of different forms of nitrogen in some desert soils. *Soil Science* 116:51–8.

NOBEL, D. C. (1968). Kane Springs Wash volcanic center, Lincoln County, Nevada. *Memoirs of the Geological Society of America* 110:109–16.

NOBEL, P. S. (1976a). Water relations and photosynthesis of a desert CAM plant, *Agave deserti. Plant Physiology* 58:576–82.

NOBEL, P. S. (1976b). Photosynthetic rates of sun *versus* shade leaves of *Hyptis emoryi* Torr. *Plant Physiology* 58:218–23.

NOBEL, P. S. (1978). Microhabitat, water relations and photosynthesis of a desert fern, *Notholaena parryi. Oecologia* 31:293–309.

NOBEL, P. S. (1980). Water vapor conductance and CO_2 uptake for leaves of a C_4 desert grass, *Hilaria rigida. Ecology* 61:252–8.

NOBEL, P. S. (1981a). Wind as an ecological factor, pp. 475–500. *In* O. L. Lange, P. S. Nobel, C. B. Osmund, & H. Ziegler (eds.), *Encyclopedia of Plant Physiology,* New Series 12A. Springer-Verlag, Berlin.

NOBEL, P. S. (1981b). Spacing and transpiration of various sized clumps of a desert grass, *Hilaria rigida. Journal of Ecology* 69:735–42.

NOBEL, P. S. (1982a). Orientation, PAR interception, and nocturnal acidity increases for terminal cladodes of a widely cultivated cactus, *Opuntia ficus-indica. American Journal of Botany* 69:1462–9.

NOBEL, P. S. (1982b). Orientations of terminal cladodes of platyopuntias. *Botanical Gazette* 143:219–24.

NOBEL, P. S. (1985). Desert succulents, pp. 181–97. *In* B. F. Chabot & H. A. Mooney (eds.), *Physiological Ecology of North American Plant Communities.* Chapman and Hall, New York.

NOBEL, P. S. (1988). *Environmental Biology of Agaves and Cacti.* Cambridge University Press.

NOBEL, P. S. (1991a). *Physicochemical and Environmental Plant Physiology.* Academic Press, San Diego.

NOBEL, P. S. (1991b). Achievable productivities of CAM plants; basis for high values compared with C_3 and C_4. *New Phytologist* 119:183–205.

NORRIS, K. S. (1953). The ecology of the desert iguana, *Dipsosaurus dorsalis. Ecology* 34:265–87.

NORRIS, K. S. (1967). Color adaptation and its thermal relationships, pp. 162–229. *In* W. W. Milstead (ed.), *Lizard Ecology. A symposium.* University of Missouri Press, Columbia.

NORRIS, K. S. & DAWSON, W. R. (1964). Observations on the water economy and electrolyte excretion of chuckwallas (Lacertilia, *Sauromalus*). *Copeia* 1964:638–46.

NORRIS, K. S. & KAVANAU, J. L. (1966). The burrowing of the western shovel-nosed snake, *Chionactis occipitalis,* and the undersand environment. *Copeia* 1966:650–64.

NORRIS, K. S. & LOWE, C. H., Jr. (1964). An analysis of background color matching in amphibians and reptiles. *Ecology* 45:565–80.

NORTON, B. E. (1975). IBP studies in the desert biome. *Bulletin of the Ecological Society of America* 55:6–10.

ODENING, W. R., STRAIN, B. R., & OECHEL, W. C. (1974). The effect of decreasing water potential on net CO_2 exchange of intact desert shrubs. *Ecology* 55:1086–95.

ODUM, E. P. (1959). *Fundamentals of Ecology,* 2nd edn. W. B. Saunders Company, Philadelphia.

OECHEL, W. C., STRAIN, B. R., & ODENING, W. R. (1972). Photosynthetic rates of a desert shrub, *Larrea divaricata* Cav., under field conditions. *Photosynthetica* 6:183–8.

O'FARRELL, T. P. & EMERY, L. A. (1976). *Ecology of the Nevada Test Site: A Narrative Summary and Anno-*

desert shrub, *Larrea divaricata* Cav., under field conditions. *Photosynthetica* 6:183–8.

O'FARRELL, T. P. & EMERY, L. A. (1976). *Ecology of the Nevada Test Site: A Narrative Summary and Annotated Bibliography. U.S. Energy Research and Development Administration, Nevada Operations Office Report* NVO–167, Las Vegas, Nevada.

OHMART, R. D. (1972). Physiological and ecological observations concerning the salt-excreting glands of the roadrunner. *Comparative Biochemistry and Physiology* 43A:311–16.

OHMART, R. D. & LASIEWSKI, R. C. (1971). Roadrunners: energy conservation by hypothermia and absorption of sunlight. *Science* 172:67–9.

OLIVIER, H. (1961). *Irrigation and Climate.* Edward Arnold Publishers, Ltd., London.

OPPENHEIMER, H. R. (1960). Adaptations to drought: xerophytism. In, *Plant-Water Relationships in Arid and Semi-Arid conditions. Reviews of Research, UNESCO, Arid Zone Research* 15:105–38.

ORIANS, G. H., CATES, R. G., MARES, M. A., MOLDENKE, M., NEFF, J., RHOADES, D. F., ROZENWEIG, M. I., SIMPSON, B. B., SCHULTZ, J. C., & TOMOFF, C. S. (1977). Resource utilization systems, pp. 164–224. *In* G. H. Orians & O. T.Solbrig (eds.), *Convergent Evolution in Warm Deserts.* Dowden, Hutchinson, and Ross, Inc., Stroudsburg, Pennsylvania.

ORIANS, G. H. & SOLBRIG, O. T. (1977). A cost-income model of leaves and roots with special reference to arid and semiarid areas. *American Naturalist* 111:677–90.

ORSHAN, G. (1954). Surface reduction and its significance as a hydroecological factor. *Journal of Ecology* 42:442–4.

OSBORNE, W. (1975). Invertebrates, pp. 23–47. *In* R. S. Sinn, R. D. Anderson, M. Merritt, W. Osborne, & J. A. MacMahon (eds.), *Curlew Valley Validation Site Report. US/IBP Desert Biome Research Memorandum* 75–4. Utah State University, Logan.

OSMOND, C. B. (1978). Crassulacean acid metabolism, a curiosity in context. *Annual Review of Plant Physiology* 29:379–414.

OSMOND, C. B., WINTER, K., & ZIEGLER, H. (1982). Functional significance of different pathways of CO_2 fixation in photosynthesis, pp. 479–547. *In* O. L. Lange, P. S. Nobel, C. B. Osmond, & H. Ziegler (eds.), *Encyclopedia of Plant Physiology,* New Series 12B. Springer-Verlag, Berlin.

OVERTON, W. S. & DAVIS, D. E. (1969). Estimating the numbers of animals in wildlife populations, pp. 403–55. *In* R. H. Giles (ed.), *Wildlife Management Techniques.* The Wildlife Society, Washington, D.C.

PACKARD, G. C. & PACKARD, M. J. (1970). Eccritic temperatures of zebra-tailed lizards on the Mojave Desert. *Herpetologica* 26:168–72.

PARAN, T. P. (1966). A new fur mite, *Lavoimyobia hughesi* n.g., n. sp. (Acarina: Myobiidae) from a North American rodent. *Journal of Medical Entomology* 3:172–8.

PARKER, L. W., FOWLER, H. G., ETTERSHANK, G., & WHITFORD, W. G. (1982). The effects of subterranean termite removal on desert soil nitrogen and ephemeral flora. *Journal of Arid Environments* 5:53–9.

PARKER, W. S. (1972a). Ecological study of the western whiptail lizard, *Cnemidophorus tigris gracilis* in Arizona. *Herpetologica* 28:360–9.

PARKER, W. S. (1972b). Aspects of the ecology of a Sonoran Desert population of the western banded gecko, *Coleonyx variegatus* (Sauria: Eublepharinae). *American Midland Naturalist* 88:209–24.

PARKER, W. S. & PIANKA, E. R. (1973). Notes on the ecology of the iguanid lizard, *Sceloporus magister.* *Herpetologica* 29:143–52.

PARKER, W. S. & PIANKA, E. R. (1974). Further ecological observations on the western banded gecko, *Coleonyx variegatus. Copeia* 1974:528–31.

PARKER, W. S. & PIANKA, E. R. (1975). Comparative ecology of populations of the lizard *Uta stansburiana. Copeia* 1975:615–32.

PARKER, W. S. & PIANKA, E. R. (1976). Ecological observations on the leopard lizard (*Crotaphytus wislizeni*) in different parts of its range. *Herpetologica* 32:95–114.

PATTEN, D. T. (1978). Productivity and production efficiency of an upper Sonoran Desert ephemeral community. *American Journal of Botany* 65:891–5.

PATTEN D. T. & SMITH, S. D. (1975). Heat flux and the thermal regime of desert plants, pp. 1–19. *In* N. F. Hadley (ed), *Environmental Physiology of Desert Organisms.* Dowden, Hutchinson, and Ross, Inc., Stroudsburg, Pennsylvania.

PAYNE, W. J. (1973). Reduction of nitrogenous oxides by microorganisms. *Biochemical Review* 37:429–52.

PEARCY, R. W. (1977). Acclimation of photosynthetic and respiratory carbon dioxide exchange to growth temperature in *Atriplex hymenelytra* (Torr.) Wats. *Plant Physiology* 59:795–9.

PEARCY, R. W., BJÖRKMAN, O., HARRISON, A. T., & MOONEY, H. A. (1971). Photosynthetic performance of two desert species with C_4 photosynthesis in Death Valley, California. *Carnegie Institution of Washington Yearbook* 70:540–50.

PEARSON, A. K., LICHT, P., NAGY, K. A., & MEDICA, P. A. (1978). Endocrine function and reproductive impairment in an irradiated population of the lizard *Uta stansburiana. Radiation Research* 76:610–23.

PEARSON, L. C. (1972). Survey of lichens, pp. 55–62. *In* D. F. Balph (coord.), *Curlew Valley Validation Site Report, US/IBP Desert Biome Research Memorandum* 72–1. Utah State University, Logan.

PENMAN, H. L. (1956). Estimating evaporation. *Transactions of the American Geophysical Union* 37:43–50.

PETERSEN, K. C. & BEST, C. B. (1985). Nest-site selection by sage sparrows. *The Condor* 87:217–21.

PHILLIPS, A., MARSHALL, J., & MONSON, G. (1964). *The Birds of Arizona*. University of Arizona Press, Tucson.

PHILLIPS, D. L. & MacMAHON, J. A. (1981). Competition and spacing patterns in desert shrubs. *Journal of Ecology* 69:97–115.

PHILLIPS, S. E., MELMES, A. R., & FOSTER, R. C. (1987). Calcified filamenta: an example of biological influences in the formulation of calcrete in South Australia. *Australian Journal of Soil Research* 25:405–28.

PIANKA, E. R. (1971). Comparative ecology of two lizards. *Copeia* 1971:129–38.

PIANKA, E. R. (1986). *Ecology and Natural History of Desert Lizards*. Princeton University Press.

PIANKA, E. R. & PARKER, W. S. (1972). Ecology of the iguanid lizard *Callisaurus draconoides. Copeia* 1972:493–508.

PIANKA, E. R. & PARKER, W. S. (1975). Ecology of horned lizards: a review, with special reference to *Phrynosoma platyrhinos. Copeia* 1975:141–62.

PINSHOW, B., DEGEN, A. A., & ALKON, P. U. (1983). Water intake, existence energy and response to water deprivation in the sand partridge *Ammoperdix heyi* and the chukar *Alectoris chukar*: two phasianids of the Negev Desert. *Physiological Zoology* 56:281–9.

POLIS, G. A. (1988). Foraging and evolutionary responses of desert scorpions to harsh environmental periods of food stress. *Journal of Arid Environments* 14:123–34.

POLIS, G. A., ed. (1990). *Biology of Scorpions*. Stanford University Press, Stanford, California.

POLIS, G. A. & FARLEY, R. D. (1980). Population biology of a desert scorpion: survivorship, microhabitat, and the evolution of life history strategy. *Ecology* 61:620–9.

POLIS, G. A. & McCORMICK, S. J. (1986). Scorpions, spiders and solpugids: predation and competition among distantly related taxa. *Oecologia* 71:111–16.

POLIS, G. A. & YAMASHITA, T. (1991). The ecology and importance of predaceous arthropods in desert communities, pp. 180–222. *In* G. A. Polis (ed.), *The Ecology of Desert Communities*. University of Arizona Press, Tucson.

PORCELLA, D. B., FLETCHER, J. E., SORENSON, D. L., PIDGE, G. C., & DUGAN, A. (1973). *Nitrogen and Carbon Flux in a Soil-Vegetation Complex in the Desert Biome. US/IBP Desert Biome Research Memorandum 73–36*. Utah State University, Logan.

PORTER, W. P. & TRACY, C. R. (1983). Biophysical analyses of energetics, time-space utilization, and distributional limits, pp. 55–83. *In* R. B. Huey, E. R. Pianka, & T. W. Schoener (eds.),

Lizard Ecology. Harvard University Press, Cambridge.

PRICE, M. V. & REICHMAN, O. J. (1987). Distribution of seeds in Sonoran Desert soils: implications for heteromyid rodent foraging. *Ecology* 68:1797–811.

PRICE, M. V., WASER, N. M., & BASS, T. A. (1984). Effects of moonlight on microhabitat use by desert rodents. *Journal of Mammalogy* 65:353–6.

PROSE, D. V., METZGER, S. K., & WILSHIRE, H. G. (1987). Effects of substrate disturbance on secondary plant succession: Mojave Desert, California. *Journal of Applied Ecology* 24:305–13.

PROSPERS, J. M. (1981). Arid regions as sources of mineral aerosols in the marine atmosphere. *Geological Society of America Special Papers* 186:71–86.

PROSSER, C. L., ed. (1973). *Comparative Animal Physiology*, 3rd edn. Saunders, Philadelphia.

PYE, K. (1982). *Aeolian Dust and Dust Deposits*. Academic Press, London.

PYE, K. (1987). *Aeolian Dust and Dust Deposits*. Academic Press, London.

QUIRING, R. F. (1968). *Climatological data. Nevada Test Site and Nuclear Rocket Development Station*. U.S. Department of Commerce Technical Memorandum ERLTM-ARL 7. ESSA Research Laboratories, Air Resources Laboratory, Las Vegas.

RAITT, R. J. (1972). Bajada birds, pp. 124–6. *In* W. G. Whitford (ed.), *Jornada Validation Site Report. US/IBP Desert Biome Research Memorandum 72–4*. Utah State University, Logan.

RAITT, R. J. & MAZE, R. L. (1968). Densities and species composition of breeding birds of a creosotebush community in southern New Mexico. *The Condor* 70:193–205.

RAITT, R. J. & PIMM, S. L. (1976). Dynamics of bird communities of the Chihuahuan Desert, New Mexico. *The Condor* 78:427–42.

RAVEN, P. H. & AXELROD, D. I. (1978). Origin and relationships of the California flora. *University of California Publications in Botany* 72:1–134.

REEVES, C. C. (1976). *Caliche: Origin, Classification, Morphology and Uses*. Estacado Books, Lubbock, Texas.

REGAL, P. J. (1966). Thermophilic response following feeding in certain reptiles. *Copeia* 1966:588–90.

REICHMAN, O. J. (1975). Relationships of desert rodent diets to available resources. *Journal of Mammalogy* 56:731–51.

REICHMAN, O. J. (1976). Relationships between dimensions, weights, volumes, and calories of some Sonoran Desert seeds. *The Southwestern Naturalist* 20:573–86.

REICHMAN, O. J. (1977). Optimization of diets through food preferences by heteromyid rodents. *Ecology* 58:454–7.

REICHMAN, O. J. (1979). Desert granivore foraging and its impact on seed densities and distributions. *Ecology* 60:1085–92.

REICHMAN, O. J. (1984). Spatial and temporal variation

of seed distributions in Sonoran Desert soils. *Journal of Biogeography* 11:1–11.

REICHMAN, O. J. (1991). Desert mammal communities, pp. 311–47. *In* G. A. Polis (ed.), *The Ecology of Desert Communities*. University of Arizona Press, Tucson.

REICHMAN, O. J. & OBERSTEIN, D. (1977). Selection of seed distribution types by *Dipodomys merriami* and *Perognathus amplus*. *Ecology* 58:636–43.

REICHMAN, O. J. & Van de GRAAF K. M. (1975). Association between ingestion of green vegetation and desert rodent reproduction. *Journal of Mammalogy* 56:503–6.

REYNOLDS, H. G. (1958). The ecology of Merriam kangaroo rat (*Dipodomys merriami* Mearns) on the grazing lands of southern Arizona. *Ecological Monographs* 28:111–27.

RHOADES, D. F. & CATES, R. G. (1976). Toward a general theory of plant anti-herbivore chemistry, pp. 168–213. *In* J. W. Wallace & R. L. Mansell (eds.), *Biochemical Interactions between Plants and Insects*. Plenum Press, New York.

RICHARDS, G. L. (1962). Wintering habits of some birds at the Nevada Atomic Test Site. *Great Basin Naturalist* 22:30–1.

RIDER, N. E., PHILIP, J. R., & BRADLEY, E. F. (1963). The horizontal transport of heat and moisture. A micrometeorological study. *Quarterly Journal of the Royal Meteorological Society* 89:527–31.

RISSING, S. W. (1981). Prey preferences in the desert horned lizard: influence of prey foraging method and aggressive behavior. *Ecology* 62:1031–42.

RISSING, S. W. (1988). Seed-harvesting ant association with shrubs: competition for water in the Mojave Desert? *Ecology* 69:809–13.

ROBBERECHT, R., MAHALL, B. E. & NOBEL, P. S. (1983). Experimental removal of intraspecific competitors – effects on water relations and productivity of a desert bunchgrass, *Hilaria rigida*. *Oecologia* 60:21–4.

ROBBINS, C. T. (1983). *Wildlife Feeding and Nutrition*. Academic Press, New York.

ROBINSON, T. W. (1965). *Introduction, Spread, and Areal Extent of Saltcedar* (Tamarix) *in the Western States*. U.S. Geological Survey Professional Paper 491–A.

ROGERS, R. W., LANGE, R. T., & NICHOLAS, D. J. D. (1966). Nitrogen fixation by lichens of arid soil crusts. *Nature* 209:96–7.

ROMNEY, E. M., HALE, V. Q., WALLACE, A., LUNT, O. R., CHILDRESS, J. D., KAAZ, H., ALEXANDER, G. V., KINNEAR J. E., & ACKERMAN, T. L. (1973). *Some Characteristics of Soil and Perennial Vegetation in Northern Mojave Desert Areas of the Nevada Test Site*. U.S. Atomic Energy Commission Report UCLA 12–916. National Technical Information Services, Springfield, Virginia.

ROMNEY, E. M., HUNTER, R. B., & WALLACE, A. (1986). Shrub use of water from simulated rainfall in the Mojave Desert, pp. 25–9. *In* L. J. Lane (ed.), *Rainfall Simulator Workshop*. Society for Range Management, Denver.

ROMNEY, E. M., HUNTER, R. B., & WALLACE, A. (1990). Vegetation on disturbed areas at the Nevada Test Site, pp. 344–9. *In, Proceedings – Symposium on Cheatgrass Invasion, Shrub Die-off and Other Aspects of Shrub Biology and Management*. General Technical Report INT–276. U.S. Department of Agriculture, Forest Service, Intermountain Research Station, Ogden, Utah.

ROMNEY, E. M., LINDBERG, R. G., HAWTHORNE, H. A., BYSTROM, B. G., & LARSON, K. H. (1963). Contamination of plant foliage with radioactive fallout. *Ecology* 44:343–9.

ROMNEY, E. M. & WALLACE, A. (1980). Ecotonal distribution of salt-tolerant shrubs in the northern Mojave Desert. *Great Basin Naturalist Memoirs* (4):134–9.

ROMNEY, E. M., WALLACE, A., & CHILDRESS, J. D. (1971). Revegetation problems following nuclear testing activities at the Nevada Test Site, pp. 1015–22. *In, Proceedings of the Third National Symposium on Radioecology*. Oak Ridge, Tennessee.

ROMNEY, E. M., WALLACE, A., & HUNTER, R. B. (1978). Plant response to nitrogen fertilization in the northern Mojave Desert and its relationship to water manipulations, pp. 232–43. *In* N. E. West & J. Skujins (eds.), *Nitrogen in Desert Ecosystems*. Dowden, Hutchinson, and Ross, Inc., Stroudsburg, Pennsylvania.

ROMNEY, E. M., WALLACE, A., & HUNTER, R. B. (1989a). Transplanting of native shrubs on disturbed land in the Mojave Desert, pp. 50–3. *In* A. Wallace, E. D. McArthur, & M. R. Haferkamp (compilers), *Proceedings – Symposium on Shrub Ecophysiology and Biotechnology; 1987 June 30–July 2; Logan, UT. General Technical Report INT–256*. U.S. Department of Agriculture, Forest Service, Intermountain Research Station, Ogden, Utah.

ROMNEY, E. M., WALLACE, A., & HUNTER, R. B. (1989b). Pulse establishment of woody shrubs on denuded Mojave Desert land, pp. 54–5. *In* E. D. McArthur, E. M. Romney, S. D. Smith & P. T. Tueller (compilers), *Proceedings – Symposium on Shrub Ecophysiology and Biotechnology; 1987 June 30–July 2; Logan, UT. General Technical Report INT–256*. U.S. Department of Agriculture, Forest Service, Intermountain Research Station, Ogden, Utah.

ROMNEY, E. M., WALLACE, A., KAAZ H., & HALE, V. Q. (1980). The role of shrubs on redistribution of mineral nutrients in soil in the Mojave Desert. *Great Basin Naturalist Memoirs* (4):124–33.

ROMNEY, E. M., WALLACE, A., KAAZ, H., HALE, V. Q., & CHILDRESS, J. D. (1977). Effects of shrubs on redistribution of mineral nutrients in zones near roots in the Mojave Desert, pp. 303–10. *In* J. K. Marshall (ed.), *The Below-ground Ecosystem: A Synthesis of Plant Associated Processes. Range Science Series 26*, Colorado State University, Fort Collins.

ROTENBERRY J. T. & WIENS, J. A. (1989). Reproductive biology of shrubsteppe passerine birds: geographical and temporal variation in clutch size, brood size, and fledgling success. *The Condor* 91:1–14.

ROWLANDS, P. G. (1980). Recovery, succession, and revegetation in the Mojave Desert, pp. 75–118. *In* P. G. Rowlands (ed.), *The Effects of Disturbance on Desert Soils, Vegetation and Community Processes with Emphasis on Off-road Vehicles: A Critical Review. Bureau of Land Management Desert Plant Staff, Special Publication U.S. Department of the Interior*, Washington, D.C.

ROWLANDS, P. G., JOHNSON, H., RITTER, E., & ENDO, A. (1982). The Mojave Desert, pp. 103–62. *In* G. L. Bender (ed.), *Reference Handbook on the Deserts of North America*. Greenwood Press, Westport, Connecticut.

RUNDEL, P. W. (1989). Ecological success in relation to plant form and function in the woody legumes, pp. 377–98. *In* C. H. Stirton & J. L. Zarucchi (eds.), *Advances in Legume Biology*. Missouri Botanical Garden, St. Louis.

RUNDEL, P. W., EHLERINGER, J. R., & NAGY, K. A., eds. (1988). *Stable Isotopes in Ecological Research*. Springer-Verlag, New York.

RUNDEL, P. W. & FRANKLIN, T. (1991). Vines in arid and semi-arid ecosystems, pp. 337–56. *In* F. E. Putz & H. A. Mooney (eds.), *The Biology of Vines*. Cambridge University Press.

RUNDEL, P. W. & JARRELL, W. M. (1989). Water in the environment, pp. 29–56. *In* R. W. Pearcy, J. R. Ehleringer, H. A. Mooney, & P. W. Rundel (eds.), *Plant Physiological Ecology: Field Methods and Instrumentation*. Chapman and Hall, London.

RUNDEL, P. W., NILSEN, E. T., SHARIFI, M. R., VIRGINIA, R. A., JARRELL, W. M., KOHL, D. H., & SHEARER, G. B. (1982). Seasonal dynamics of nitrogen cycling for a *Prosopis* woodland in the Sonoran Desert. *Plant and Soil* 67:343–53.

RUNDEL, P. W. & NOBEL, P. S. (1991). Structure and function in desert root systems, pp. 349–78. *In* D. Atkinson (ed.), *Plant Root Growth. An Ecological Perspective*. Blackwell Scientific Publications, Oxford.

RUSH, F. E. (1971). Regional ground-water systems in the Nevada Test Site area, Nye, Lincoln, and Clark Counties, Nevada. *Water Resources-Reconnaissance Series Report* 54:1–25.

RUSSELL, S. M., GOULD, P. J., & SMITH, E. L. (1973). *Population Structure, Foraging Behavior and Daily Movements of Certain Sonoran Desert Birds. US/*

IBP Desert Biome Research Memorandum 73–27. Utah State University, Logan.

RYCHERT, R. C. & SKUJINS, J. (1974). Nitrogen fixation by blue-green algae-lichen crusts in the Great Basin Desert. *Proceedings of the Soil Science Society of America* 38:768–71.

RYCHERT, R. C., SKUJINS, J., SORENSON, D., & PORCELLA, D. (1978). Nitrogen fixation by lichens and free-living microorganisms in deserts, pp. 20–33. *In* N. E. West & J. Skujins (eds.), *Nitrogen in Desert Ecosystems*. Dowden, Hutchinson, and Ross, Inc., Stroudsburg, Pennsylvania.

RYDEN, J. C. & LUND, L. J. (1980). Nature and extent of directly measured denitrification losses from some irrigated vegetable crop production units. *Proceedings of the Soil Science Society of America* 44:505–11.

RYTI, R. T. & CASE, T. J. (1984). Spatial arrangement and diet overlap between colonies of desert ants. *Oecologia* 62:401–4.

RYTI, R. T. & CASE, T. J. (1986). Overdispersion of ant colonies: a test of hypotheses. *Oecologia* 69:446–53.

RYTI, R. T. & CASE, T. J. (1988). The regeneration niche of desert ants: effects of establishing colonies. *Oecologia* 75:303–6.

SALISBURY, F. B. & ROSS, C. W. (1991). *Plant Physiology*, 4th edn. Wadsworth Publishing Company, Belmont, California.

SANBORN, S. R. (1972). Food habits of *Sauromalus obesus obesus* on the Nevada Test Site. *Journal of Herpetology* 6:142–4.

SANCHEZ-DIAZ, M. R. & MOONEY, H. A. (1979). Resistance to water transfer in desert shrubs native to Death Valley, California. *Physiologia Plantarum* 46:139–46.

SANTOS, P. F., DEPREE, E., & WHITFORD, W. G. (1978). Spatial distribution of litter and microarthropods in a Chihuahuan Desert ecosystem. *Journal of Arid Environments* 1:41–8.

SANTOS, P. F., ELKINS, N. Z., STEINBERGER, Y., & WHITFORD, W. G. (1984). A comparison of surface and buried *Larrea tridentata* leaf litter decomposition in North American hot deserts. *Ecology* 65:278–84.

SANTOS, P. F., PHILLIPS, J., & WHITFORD, W. G. (1981). The role of mites and nematodes in early stages of buried litter decomposition in a desert. *Ecology* 62:664–9.

SANTOS, P. F. & WHITFORD, W. G. (1981). The effects of microarthropods on litter decomposition in a Chihuahuan Desert ecosystem. *Ecology* 62:654–63.

SCHAEFFER, J. R. (1968). *Climatology of Tonopah Test Range, 1967. U.S. Atomic Energy Commission Report* SC-M-68-522.

SCHAMBERGER, M. L. & TURNER, F. B. (1986). The application of habitat modeling to the desert tortoise (*Gopherus agassizii*). *Herpetologica* 42:134–8.

SCHIMPER, A. F. W. (1903). *Plant-Geography upon a Physiological Basis.* Tr. by W. R. Fisher. Clarendon Press, Oxford.

SCHLESINGER, W. H. & HASEY, M. M. (1980). The nutrient content of precipitation dry fallout and intercepted aerosols in the chaparral of southern California. *American Midland Naturalist* 103:114–22.

SCHMIDT, P. J., SHERBROOKE, W. C., & SCHMIDT, J. O. (1989). The detoxification of ant (*Pogonomyrmex*) venom by a blood factor in horned lizards (*Phrynosoma*). *Copeia* 1989:603–7.

SCHMIDT-NIELSEN, K. (1964). *Desert Animals: Physiological Problems of Heat and Water.* Oxford University Press.

SCHMIDT-NIELSEN, K. (1972). *How Animals Work.* Cambridge University Press.

SCHMIDT-NIELSEN, K. (1975). Desert rodents: physiological problems of desert life, pp. 379–88. *In* I. Prakash & P. K. Ghosh (eds.), *Rodents in Desert Environments.* Monographiae Biologicae, vol. 28. Dr. W. Junk, The Hague.

SCHMIDT-NIELSEN, K. (1984). *Scaling: Why is Animal Size so Important?* Cambridge University Press, Cambridge.

SCHMIDT-NIELSEN, K. (1990). *Animal Physiology: Adaptation and Environment,* 4th edn. Cambridge University Press, Cambridge.

SCHMIDT-NIELSEN, K., DAWSON, T. J., HAMMEL, H. T., HINDS, D., & JACKSON, D. C. (1965). The jack rabbit – a study in its desert survival. *Hvalradets Skrifter* 48:125–42.

SCHMIDT-NIELSEN, K., HAINSWORTH, F. R., & MURRISH, D. E. (1970). Counter-current heat exchange in the respiratory passages: effect on heat and water balance. *Respiratory Physiology* 9:263–76.

SCHREIBER, U. & ARMOND, P. A. (1977). Heat-induced changes in chlorophyll fluorescence and related heat-damage at the pigment level. *Carnegie Institution of Washington Yearbook* 76:341–46.

SCHULTZ, J. C., OTTE, D., & ENDERS, F. (1977). *Larrea* as a habitat component for desert arthropods, pp.176–208. *In* T. J. Mabry, J. H. Hunziker, & D. R. DiFeo (eds.), *Creosote Bush: Biology and Chemistry of Larrea in New World Deserts.* Dowden, Hutchinson and Ross, Inc., Stroudsburg, Pennsylvania.

SCHULTZ, V. (1966). References on Nevada Test Site ecological research. *Great Basin Naturalist* 26:79–86.

SCHULZE, E.-D. & HALL, A. E. (1982). Stomatal responses, water loss and CO_2 assimilation rates of plants in contrasting environments, pp. 181–230. *In* O. L. Lange, P. S. Nobel, C. B. Osmond, & H. Ziegler (eds.), *Encyclopedia of Plant Physiology,* New Series vol. 12B. Springer-Verlag, Berlin.

SCHULZE, E.-D., LANGE, O. L., BUSCHBOM, U., KAPPEN, L., & EVENARI, M. (1972). Stomatal responses to changes in humidity in plants growing in the desert. *Planta* 108:259–70.

SEELY, M. K. & HAMILTON, W. J. III (1976). Fog catchment sand trenches constructed by tenebrionid beetles, *Lepidochora,* from the Namib Desert. *Science* 193:484–6.

SEEMANN, J. R., BERRY, J. A., & DOWNTON, W. J. S. (1984). Photosynthetic response and adaptation to high temperature in desert plants. A comparison of gas exchange and fluorescence methods for studies of thermal tolerance. *Plant Physiology* 75:364–8.

SERVENTY, D. L. (1971). Biology of desert birds, pp. 287–331. *In* D. S. Farber & J. R. King (eds.), *Avian Biology,* vol. 1. Academic Press, New York.

SHARIFI, M. R., MEINZER, F. C., NILSEN, E. T., RUNDEL, P. W., VIRGINIA, R. A., JARRELL, W. M., HERMAN, D. J., & CLARK, P. C. (1988). Effect of resource manipulation on the quantitative phenology of *Larrea tridentata* (creosote bush) in the Sonoran Desert of California. *American Journal of Botany* 75:1163–74.

SHARP, A. L., BOND, J. J., NEUBERGER, J. W., KULMAN, A. R., & LEWIS, J. K. (1964). Runoff as affected by intensity of grazing on rangeland. *Journal of Soil Water Conservation* 19:103–6.

SHEPPARD, J. M. (1968). Thirty-second breeding bird census. *Audubon Field Notes* 22:722–4.

SHEPPARD, J. M. (1970). A study of LeConte's Thrasher. *California Birds* 1:85–94.

SHIELDS, L. M. (1950). Leaf xeromorphy as related to physiological and structural influences. *Botanical Review* 16:399–447.

SHIELDS, L. M. (1951). Leaf xeromorphy in dicotyledonous species from a gypsum sand deposit. *American Journal of Botany* 38:175–90.

SHIELDS, L. M. (1957). Algal and lichen floras in relation to nitrogen content of certain volcanic and arid range soils. *Ecology* 38:661–3.

SHIELDS, L. M. & DROUET, F. (1962). Distribution of terrestrial algae within the Nevada Test Site. *American Journal of Botany* 49:547–54.

SHIELDS, L. M. & WELLS, P. V. (1962). Effects of nuclear testing on desert vegetation. *Science* 135:38–40.

SHIELDS, L. M. & WELLS, P. V. (1963). Recovery of vegetation on atomic target areas at the Nevada Test Site, pp. 307–10. *In* V. Schultz & N. W. Klement, jr. (eds.), *Radioecology.* Reinhold Publishing Corporation, New York.

SHIELDS, L. M., WELLS, P. V., & RICKARD, W. H. (1963). Vegetational recovery on atomic target areas in Nevada. *Ecology* 44:697–705.

SHOEMAKER, V. H. (1988). Physiological ecology of amphibians in arid environments. *Journal of Arid Environments* 14:145–53.

SHOEMAKER, V. H., NAGY, K. A., & COSTA, W. R. 1976. Energy utilization and temperature regulation by jackrabbits (*Lepus californicus*) in the Mojave Desert. *Physiological Zoology* 49:364–75.

SHREVE, F. (1931). Physical conditions in sun and shade. *Ecology* 12:96–104.

SHREVE, F. (1942). The desert vegetation of North America. *Botanical Review* 8:195–246.

SHREVE, F. (1951). *Vegetation of the Sonoran Desert.* Carnegie Institution of Washington Publication 591, Washington, D.C.

SHREVE, F. & HINCKLEY, A. L. (1937). Thirty years of change in desert vegetation. *Ecology* 18:463–78.

SIEGEL, S. L. (1956). *Nonparametric Statistics for the Behavioral Sciences.* McGraw-Hill, New York.

SIMANTON, J. R., JOHNSON, C. W., NYHAN, J. W., & ROMNEY, E. M. (1986). Rainfall simulation on rangeland erosion plots, pp. 11–17. *In* L. J. Lane (ed.), *Proceedings of the Rainfall Simulator Workshop.* Society for Range Management, Denver.

SINNOCK, S. (1982). *Geology of the Nevada Test Site and Nearby Areas, Southern Nevada.* Report 82–2207, Department of Energy, San Francisco Operations Office, San Francisco.

SKIDMORE, E. L. & WOODRUFF, N. P. (1968). *Wind Erosion Forces in the United States and their Use in Predicting Soil Loss. USDA Agricultural Handbook* 346, Washington, D.C.

SKUJINS, J. (1976). *Nitrogen Dynamics in Stands Dominated by Some Major Cool Desert Shrubs. V. Studies on Denitrification and Nitrogen Fixation: Comparison of Biological Processes in Western Deserts. US/IBP Desert Biome Research Memorandum 76–26.* Utah State University, Logan.

SKUJINS, J. (1977). *Comparison of Biological Processes in Western Deserts. US/IBP Desert Biome Research Memorandum 77–26.* Utah State University, Logan.

SKUJINS, J. (1981). Nitrogen cycling in arid ecosystems, pp. 477–91. *In* F. E. Clark & T. Rosswall (eds.), *Terrestrial Nitrogen Cycles.* Swedish Natural Science Research Council, Stockholm.

SKUJINS, J. & KLUBEK, B. (1978). Nitrogen fixation and cycling by blue-green algae-lichen crusts in arid rangeland soils. *Ecological Bulletin* (Stockholm) 33:477–91.

SLADE, N. A., HORTON, J. S., & MOONEY, H. A. (1975). Yearly variation in the phenology of California annuals. *American Midland Naturalist* 94:209–14.

SLEEPER, E. L. & MISPAGEL, M. E. (1975). Shrub-dwelling arthropods, pp. 35–43. *In* F. B. Turner (ed.), *Rock Valley Validation Site Report. US/IBP Desert Biome Research Memorandum 75–2.* Utah State University, Logan.

SMALL, A. (1974). *The Birds of California.* Winchester Press, New York.

SMITH, D. D., MEDICA, P. A., & SANBORN, S. R. (1987). Ecological comparison of sympatric populations of sand lizards (*Cophosaurus texanus* and *Callisaurus draconoides*). *Great Basin Naturalist* 47:175–85.

SMITH, S. D., HARTSOCK, T. L., & NOBEL, P. S. (1983). Ecophysiology of *Yucca brevifolia*, an arborescent monocot of the Mojave Desert. *Oecologia* 60:10–17.

SMITH, S. D. & NOBEL, P. S. (1986). Deserts, pp. 13–62. *In* N. R. Baker & S. P. Long (eds.), *Photosynthesis in Contrasting Environments.* Elsevier Scientific Publishers B. V., Amsterdam.

SMITH, S. D. & NOWAK, R. S. (1990). Ecophysiology of plants in the intermontane lowlands, pp. 179–211. *In* C. B. Osmond, L. F. Pitelka, & G. M. Hidy (eds.), *Plant Biology of the Basin and Range. Ecological Studies*, vol. 80. Springer-Verlag, New York.

SMITH, W. K. (1978). Temperatures of desert plants: another perspective on the adaptability of leaf size. *Science* 201:614–16.

SMITH, W. K. & NOBEL, P. S. (1977). Influences of seasonal changes in leaf morphology on water-use efficiency for three desert broadleaf shrubs. *Ecology* 58:1033–43.

SMITH, W. K. & NOBEL, P. S. (1978). Influence of irradiation, soil water potential, and leaf temperature on leaf morphology of a desert broadleaf, *Encelia farinosa* Gray (Compositae). *American Journal of Botany* 65:429–32.

SMYTH, M. & BARTHOLOMEW, G. A. (1966). The water economy of the black-throated sparrow and the rock wren. *The Condor* 68:447–58.

SNEDECOR, G. W. (1956). *Statistical Methods*, 5th edn. Iowa State University Press, Ames.

SOHOLT, L. F. (1977). Consumption of herbaceous vegetation and water during reproduction and development of Merriam's kangaroo rat, *Dipodomys merriami*. *American Midland Naturalist* 98:445–57.

SOLBRIG, O. T. (1979). Life forms and vegetation patterns in desert regions, pp. 82–95. *In* J. R. Goodin & D. F. Northington (eds.), *Arid Land Plant Resources.* Texas Tech University, Lubbeck.

SOLBRIG, O. T. (1982). Plant adaptations, pp. 419–32. *In* G. L. Bender (ed.), *Reference Handbook on the Deserts of North America.* Greenwood Press, Westport, Connecticut.

SOLBRIG, O. T. & ORIANS, G. H. (1977). The adaptive characteristics of desert plants. *American Scientist* 65:412–21.

SPARROW, A. H. & WOODWELL, G. M. (1963). Prediction of the sensitivity of plants to chronic gamma radiation, pp. 257–70. *In* V. Schultz & A. W. Klement, jr. (eds.), *Radioecology.* Reinhold Publ. Corp., New York.

STAVE, M. E. & SHIFF, C. J. (1981). Temporal segregation in North American desert Membracidae. *Oecologia* 51:408–11.

STEBBINS, R. C. (1954). *Amphibians and Reptiles of Western North America.* McGraw-Hill, New York.

STEBBINS, R. C. (1985). *A Field Guide to Western Amphibians and Reptiles*, 2nd. ed. Houghton Mifflin Co., Boston.

STERNBERG, L. (1976). Growth forms of *Larrea tridentata*. *Madroño* 23:347–52.

STEYN, P. L. & DELWICHE, C. C. (1970). Nitrogen fixation by nonsymbiotic microorganisms in some California soils. *Environmental Science and Technology* 4:1122–8.

STRAIN, B. R. (1969). Seasonal adaptations in photosynthesis and respiration in four desert shrubs growing *in situ*. *Ecology* 50:511–13.

STRAIN, B. R. & CHASE, V. C. (1966). Effect of past and prevailing temperatures on the carbon dioxide exchange capacities of some woody desert perennials. *Ecology* 47:1043–5.

STRANGE, L. A. (1970). Revision of the ant-lion tribe Brachynemurini of North America (Neuroptera: Myrmeleontidae). *University of California Publications in Entomology* 55:1–192.

STROJAN, C. L., RANDALL, D. C., & TURNER, F. B. (1987). Relationship of leaf litter decomposition rates to rainfall in the Mojave Desert. *Ecology* 68:741–4.

STROJAN, C. L., TURNER, F. B., & CASTETTER, R. (1979). Litterfall from shrubs in the northern Mojave Desert. *Ecology* 60:891–900.

STRONG, D. R., LAWTON, J. H., & SOUTHWOOD, R. (1984). *Insects on Plants*. Blackwell Scientific Publications, Oxford.

STRONG, D. R., SZYSKA, L. A., & SIMBERLOFF, D. (1979). Tests of community-wide character displacement against null hypotheses. *Evolution* 33:897–913.

STURGES, D. C. (1975). *Hydrologic Relationships on Undisturbed and Converted Big Sagebrush Lands: The Status of Our Knowledge*. USDA Forest Service Research Paper RM–140.

SWIFT, M. J., HEAL, O. W., & ANDERSON, J. M. (1979). *Decomposition in Terrestrial Ecosystems*. University of California Press, Berkeley.

SZAREK, S. R. & WOODHOUSE, R. M. (1976). Ecophysiological studies of Sonoran Desert plants. I. Diurnal photosynthesis patterns of *Ambrosia deltoidea* and *Olneya tesota*. *Oecologia* 26:225–34.

SZAREK, S. R. & WOODHOUSE, R. M. (1977). Ecophysiological studies of Sonoran Desert plants. II. Seasonal photosynthesis pattern and primary production of *Ambrosia deltoidea* and *Olneya tesota*. *Oecologia* 28:365–75.

SZAREK, S. R. & WOODHOUSE, R. M. (1978). Ecophysiological studies of Sonoran Desert plants. IV. Seasonal photosynthetic capacity of *Acacia greggii* and *Cercidium microphyllum*. *Oecologia* 37:221–9.

TAIZ, L. & ZEIGER, E. (1991). *Plant Physiology*. Benjamin/Cummings Publishing Company, Redwood City, California.

TAMURA, T. (1975). Physical and chemical characteristics of plutonium in contaminated soils and sediments, pp. 213–29. *In, Transuranium Nuclides in the Environment*. International Atomic Energy Association, Vienna.

TANNER, V. M. & PACKHAM, W. A. (1965). Tenebrionidae beetles of the Nevada Test Site. *Brigham Young University Science Bulletin, Biology Series* 6(1):1–44.

TANNER W. W. (1969). New records and distributional notes for reptiles of the Nevada Test Site, Mercury, Nevada. *Great Basin Naturalist* 29:31–4.

TANNER, W. W. & HOPKIN, J. M. (1972). Ecology of *Sceloporus occidentalis longipes* Baird and *Uta stansburiana* Baird and Girard on Ranier Mesa, Nevada Test Site, Nye County, Nevada. *Brigham Young University Science Bulletin, Biology Series* 15:1–39.

TANNER, W. W. & JORGENSEN, C. D. (1963). Reptiles of the Nevada Test Site. *Brigham Young University Science Bulletin, Biology Series* 3:1–31.

TANNER, W. W. & KROGH, J. E. (1973a). Ecology of *Sceloporus magister* at the Nevada Test Site, Nye County, Nevada. *Great Basin Naturalist* 33:133–46.

TANNER, W. W. & KROGH, J. E. (1973b). Ecology of *Phrynosoma platyrhinos* at the Nevada Test Site, Nye County, Nevada. *Herpetologica* 29:327–42.

TANNER, W. W. & KROGH, J. E. (1974a). Ecology of the leopard lizard, *Crotaphytus wislizeni* at the Nevada Test Site, Nye County, Nevada. *Herpetologica* 30:63–72.

TANNER, W. W. & KROGH, J. E. (1974b). Variations in activity as seen in four sympatric lizard species of southern Nevada. *Herpetologica* 30:303–8.

TANNER, W. W. & KROGH, J. E. (1975). Ecology of the zebra-tailed lizard, *Callisaurus draconoides*, at the Nevada Test Site. *Herpetologica* 31:302–16.

TAYLOR, C. R. (1969). Metabolism, respiratory changes, and water balance of an antelope, the eland. *American Journal of Physiology* 217:317–20.

TAYLOR, C. R. & LYMAN, C. P. (1972). Heat storage in running antelope: independence of brain and body temperatures. *American Journal of Physiology* 111:114–17.

TAYLOR, F. (1980). Timing in the life histories of insects. *Theoretical Population Biology* 18:112–24.

TEMPLETON, J. R. (1960). Respiration and water loss at the high temperatures in the desert iguana, *Dipsosaurus dorsalis*. *Physiological Zoology* 33:136–45.

TEMPLETON, J. R. (1964). Nasal salt excretion in terrestrial lizards. *Comparative Biochemistry and Physiology* 11:223–9.

THOMAS, D. B. & SLEEPER, E. L. (1977). The use of pit-fall traps for estimating the abundance of arthropods, with special reference to the Tenebrionidae

(Coleoptera). *Annals of the Entomological Society of America* 70:242–8.

THOMBULAK, S. C. & KENAGY, G. J. (1980). Effects of seed distribution and competitors on seed harvesting efficiency in heteromyid rodents. *Oecologia* 44:342–6.

THOMPSON, S. D. (1982a). Microhabitat utilization and foraging behavior of bipedal and quadripedal heteromyid rodents. *Ecology* 63:1303–12.

THOMPSON, S. D. (1982b). Structure and species composition of desert heteromyid rodent species assemblages: effects of a simple habitat manipulation. *Ecology* 63:1313–21.

THOMPSON, S. D. (1985). Bipedal hopping and seed-dispersion selection by heteromyid rodents: the role of locomotion energetics. *Ecology* 66:220–9.

THOMPSON, S. D. (1987). Resource availability and microhabitat use by Merriam's kangaroo rat, *Dipodomys merriami*, in the Mojave Desert. *Journal of Mammalogy* 68:256–65.

THORNE, R. F. (1976). California plant communities, pp. 1–31. *In* J. Latting (ed.), *Plant Communities of Southern California. California Native Plant Society Special Publication 2*, Berkeley, California.

THORNE, R. F. (1982). The desert and other transmontane plant communities of southern California. *Aliso* 10:219–57.

THORNE, R. F., PRIGGE, B. A., & HENRICKSON, J. (1981). A flora of the higher ranges and the Kelso Dunes of the eastern Mojave Desert in California. *Aliso* 10:71–186.

TILMAN, D. (1988). *Plant strategies and the Dynamics and Structure of Plant Communities.*

TING, I. P. (1985). Crassulacean acid metabolism. *Annual Review of Plant Physiology* 36:595–622.

TING, I. P. & GIBBS, M., eds. (1982). *Crassulacean Acid Metabolism.* American Society of Plant Physiologists, Rockville, Maryland.

TINKLE, D. W. (1967). The life and demography of the side-blotched lizard, *Uta stansburiana. Miscellaneous Publications of the Museum of Zoology, University of Michigan* 132:1–182.

TOWNER, J. W. (1965). The effect of radioactive fallout at the Nevada Test Site on the chromosomes of the pocket mouse. *Health Physics* 11:1569–71.

TRACY, C. R. (1982). Biophysical modeling in reptilian physiology and ecology, pp. 275–321. *In* C. Gans & F. H. Pough (eds.), *Biology of the Reptilia*, vol. 12. Academic Press, London.

TUCKER, V. A. (1965). The relation between the torpor cycle and heat exchange in the California pocket mouse, *Perognathus californicus. Journal of Cellular Comparative Physiology* 65:404–14.

TUCKER, V. A. (1966). Diurnal torpor and its relation to food consumption and weight change in the California pocket mouse, *Perognathus californicus. Ecology* 47:245–52.

TUCULESCU, R., TOPOFF, H., & WOLFE, S. (1975). Mechanisms of pit construction in antlion larvae. *Annals of the Entomological Society of America* 68:719–20.

TURNER, F. B. (1963). *Influence of a Cratering Device on Close-in Populations of Lizards. U.S. Atomic Energy Commission Report* PNE–224F.

TURNER, F. B. (1970). The ecological efficiency of consumer populations. *Ecology* 51:471–2.

TURNER, F. B., ed. (1972). *Rock Valley Validation Site Report. US/IBP Desert Biome Research Memorandum* 72–2. Utah State University, Logan.

TURNER, F. B., ed. (1973). *Rock Valley Validation Site Report. US/IBP Desert Biome Research Memorandum* 73–2. Utah State University, Logan.

TURNER, F. B. (1975). Effects of continuous irradiation on animal populations, pp. 83–144. *In* J. T. Lett & H. Adler (eds.), *Advances in Radiation Biology*, vol. 5. Academic Press, New York.

TURNER, F. B., ed. (1979). *Ecological Base Line Studies at the Site of the Barstow 10 MWe Pilot Solar Thermal Power System. U.S. Department of Energy Report* UCLA 12–1223.

TURNER, F. B., ed. (1981). *Ecological Observations during Construction of the Barstow 10 MWe Pilot STPS. U.S. Department of Energy Report* UCLA 12–1311.

TURNER, F. B., ed. (1982). *Ecological Observations during Early Testing of the Barstow 10 MWe Pilot STPS. U.S. Department of Energy Report* UCLA 12–1385.

TURNER, F. B. (1986). Forward. *Herpetologica* 42: 56–8.

TURNER, F. B. & BERRY, K. H. (1984). *Population Ecology of the Desert Tortoise at Goffs, California. Southern California Edison Company Annual Report* 84–RD–4:1–63, Rosemead, California.

TURNER, F. B. & BERRY, K. H. (1985). *Population Ecology of the Desert Tortoise at Goffs, California, in 1984. Southern California Edison Company Annual Report* 85–RD–63. Rosemead, California.

TURNER, F. B. & CHEW, R. M. (1981). Production by desert animals, pp. 199–260. *In* D. W. Goodall & R. E. Perry (eds.), *Arid Land Ecosystems: Structure, Functioning and Management*, vol. 2. Cambridge University Press.

TURNER, F. B. & EDNEY, E. B. (1977). *Ecological Effects of Coal Fly Ash on Desert Plants and Animals. Southern California Edison Research and Development Series Report* 77–RD–114.

TURNER, F. B., EDNEY, E. B., & VOLLMER, A. T. (1979). *Ecological Effects of Precipitator Ash on Desert Plants and Animals. Southern California Edison Research and Development Series Report* 79–RD–101.

TURNER, F. B. & GIST, C. S. (1965). Influences of a

thermonuclear cratering test on close-in populations of lizards. *Ecology* 46:845–52.

TURNER, F. B., HAYDEN, P., BURGE, B. L., & ROBERSON, J. B. (1986). Egg production by the desert tortoise (*Gopherus agassizii*) in California. *Herpetologica* 42:93–104.

TURNER, F. B., HODDENBACH, G. A., & LANNOM, J. R., Jr. (1965). Growth of lizards in natural populations exposed to gamma irradiation. *Health Physics* 11:1585–93.

TURNER, F. B., HODDENBACH, G. A., MEDICA, P. A., & LANNOM, J. R., Jr. (1970). The demography of the lizard, *Uta stansburiana* Baird and Girard, in southern Nevada. *Journal of Animal Ecology* 39:505–19.

TURNER, F. B. & LANNOM, J. R., Jr. (1968). Radiation doses sustained by lizards in a continuously irradiated natural enclosure. *Ecology* 49:548–51.

TURNER, F. B., LANNOM, J. R., Jr., KANIA, H. J., & KOWALESKY, B. W. (1967). Acute gamma irradiation experiments with the lizard *Uta stansburiana*. *Radiation Research* 31:27–35.

TURNER, F. B., LANNOM, J. R., Jr., MEDICA, P. A. & HODDENBACH, G. A. (1969). Density and composition of fenced populations of leopard lizards (*Crotaphytus wislizenii*) in southern Nevada. *Herpetologica* 25:247–57.

TURNER, F. B., LICHT, P., THRASHER, J. D., MEDICA, P. A., & LANNOM, J. R., Jr. (1973). Radiation-induced sterility in natural populations of lizards (*Crotaphytus wislizenii* and *Cnemidophorus tigris*), pp. 1131–43. *In* D. J. Nelson (ed.), *Radionuclides in Ecosystems*. Proceedings Third National Symposium on Radioecology. CONF–710501–P2, U.S. Atomic Energy Commission, National Technical Information Services, Springfield, Virginia.

TURNER, F. B. & McBRAYER, J. F., eds. (1974). *Rock Valley Validation Site Report. US/IBP Desert Biome Research Memorandum* 74–2. Utah State University, Logan.

TURNER, F. B. & MEDICA, P. A. (1977). Sterility among female lizards (*Uta stansburiana*) exposed to continuous Y irradiation. *Radiation Research* 70:154–63.

TURNER, F. B., MEDICA, P. A., BRIDGES, K. W., & JENNRICH, R. I. (1982). A population model of the lizard *Uta stansburiana* in southern Nevada. *Ecological Monographs* 52:243–59.

TURNER, F. B., MEDICA, P. A., & BURY, R. B. (1987). Age-size relationships of desert tortoises (*Gopherus agassizii*) in southern Nevada. *Copeia* 1987:974–9.

TURNER, F. B., MEDICA, P. A., & KOWALESKY, B. W. (1976). *Energy Utilization by a Desert Lizard,* Uta stansburiana. *US/IBP Desert Biome Research Memorandum* (No. 1, 57p). Utah State University, Logan.

TURNER, F. B., MEDICA, P. A., LANNOM, J. R., Jr., & HODDENBACH, G. A. (1969a). A demographic analysis of fenced populations of the whiptail lizard, *Cnemidophorus tigris*, in southern Nevada. *The Southwestern Naturalist* 14:189–202.

TURNER, F. B., MEDICA, P. A., LANNOM, J. R., Jr., & HODDENBACH, G. A. (1969b). A demographic analysis of continuously irradiated and nonirradiated populations of the lizard *Uta stansburiana*. *Radiation Resarch* 38:349–56.

TURNER, F. B., MEDICA, P. A., & LYONS, C. L. (1981). A comparison of populations of desert tortoise, *Gopherus agassizii*, in grazed and ungrazed areas in Ivanpah Valley, California, pp. 139–62. *In* K. A. Hashagen & E. St. Amant (eds.), *Proceedings of the 1981 Symposium*. Desert Tortoise Council, Long Beach, California.

TURNER, F. B., MEDICA, P. A., & LYONS, C. L. (1984). Reproduction and survival of the desert tortoise (*Scaptochelys agassizii*) in Ivanpah Valley, California. *Copeia* 1984:811–20.

TURNER, F. B., MEDICA, P. A., & SMITH, D. D. (1974). *Reproduction and Survivorship of the Lizard* Uta stansburiana, *and the Effects of Winter Rainfall, Density and Predation on these Processes. US/IBP Desert Biome Research Memorandum* 74–26. Utah State University, Logan.

TURNER, F. B. & RANDALL, D. C. (1987). The phenology of desert shrubs in southern Nevada. *Journal of Arid Environments* 13:119–28.

TURNER, F. B. & RANDALL, D. C. (1989). Net productivity by shrubs and winter annuals in southern Nevada. *Journal of Arid Environments* 17:23–6.

TURNER, F. B. & VOLLMER, A. T. (1980). *Ecological Effects of Precipitator Ash on Desert Plants. Southern California Edison Research and Development Series Report* 80–RD–106.

TURNER, F. B. & VOLLMER, A. T. (1982). *Ecological Effects of Precipitator Ash on Desert Plants. Southern California Edison Research and Development Report,* Laboratory of Biomedical and Environmental Sciences, University of California, Los Angeles

TURNER, F. B. & WAUER, R. H. (1963). Survey of the herpetofauna of the Death Valley area. *Great Basin Naturalist* 23:119–28.

TURNER, J. S. & LOMBARD, A. T. (1990). Body color and body temperature in white and black Namib Desert beetles. *Journal of Arid Environments* 19:303–15.

TURNER, N. C. & JONES, M. M. (1980). Turgor maintenance by osmotic adjustment: a review and evaluation, pp. 87–103. *In* N. C. Turner & P. J. Kramer (eds.), *Adaptation of Plants to Water and High Temperature Stress*. Wiley, New York.

TURNER, R. M. (1982a). 152.1 Great Basin desertscrub, pp. 145–55. *In* D. E. Brown (ed.), *Biotic communities of the American Southwest – United States and Mexico. Desert Plants 4.*

TURNER, R. M. (1982b). 153.1 Mohave desertscrub, pp. 157–68. In D. E. Brown (ed.), *Biotic Communities of the American Southwest – United States and Mexico*. *Desert Plants* 4.

TWISSELMAN, E. C. (1967). A flora of Kern County, California. *Wasmann Journal of Biology* 25:1–395.

VAN DE GRAAFF, K. M. & BALDA, R. P. (1973). Importance of green vegetation for reproduction in the kangaroo rat, *Dipodomys merriami merriami*. *Journal of Mammalogy* 54:509–12.

VAN DEVENDER, T. R., MOODIE, K. B., & HARRIS, A. H. (1976). The desert tortoise (*Gopherus agassizii*) in the Pleistocene of the northern Chihuahuan Desert. *Herpetologica* 32:298–304.

VASEK, F. C. (1979/80). Early successional stages in Mojave desert scrub vegetation. *Israel Journal of Botany* 28:133–48.

VASEK, F. C. (1980). Creosote bush: long-lived clones in the Mojave Desert. *American Journal of Botany* 67:246–55.

VASEK, F. C. (1983). Plant succession in the Mojave Desert. *Crossosoma* 9:1–23.

VASEK, F. C. & BARBOUR, M. G. (1977). Mojave desert scrub vegetation, pp. 835–67. In M. G. Barbour & J. Major (eds.), *Terrestrial Vegetation of California*. John Wiley and Sons, New York.

VASEK, F. C., JOHNSON, H. B., & BRUM, G. D. (1975). Effects of power transmission lines on vegetation of the Mojave Desert. *Madroño* 23:114–30.

VASEK, F. C., JOHNSON, H. B., & ESLINGER, D. H. (1975). Effects of pipeline construction on creosote bush scrub vegetation of the Mojave Desert. *Madroño* 23:1–13.

VIRGINIA, R. A., JARRELL, W. M., & FRANCO-VIZCAINO, E. (1982). Direct measurements of denitrification in a *Prosopis* (mesquite) dominated Sonoran Desert ecosystem. *Oecologia* 53:120–2.

VITT, L. J. (1991). Desert reptile communities, pp. 249–77. In G. A. Polis (ed.), *The Ecology of Desert Communities*. University of Arizona Press, Tucson.

VITT, L. J., CONGDON, J. D., & DICKSON, N. A. (1977). Adaptive strategies and energetics of tail autotomy in lizards. *Ecology* 58:326–37.

VITT, L. J. & OHMART, R. D. (1974). Reproduction and ecology of a Colorado River population of *Sceloporus magister* (Sauria: Iguanidae). *Herpetologica* 30:410–17.

VITT, L. J. & OHMART, R. D. (1977a). Ecology and reproduction of lower Colorado River lizards: I. *Callisaurus draconoides* (Iguandiae). *Herpetologica* 33:214–22.

VITT, L. J. & OHMART, R. D. (1977b). Ecology and reproduction of lower Colorado River lizards: II. *Cnemidophorus tigris* (Teiidae), with comparisons. *Herpetologica* 33:223–34.

VLECK, D. (1979). The energy cost of burrowing by the pocket gopher, *Thomomys bottae*. *Physiological Zoology* 52:122–36.

VLECK, D. (1981). Burrow structure and foraging costs in the fossorial rodent, *Thomomys bottae*. *Oecologia* 49:391–6.

VOGELSBERG, M. C. (1975). *Carbon, Nitrogen, and Algal Biomass in Cold Desert Soil Crusts*. M.S. thesis, Utah State University, Logan.

VOLLMER, A. T. & BAMBERG, S. A. (1975). Response of the desert shrub *Krameria parvifolia* after ten years of chronic gamma irradiation. *Radiation Botany* 15:405–9.

VOLLMER, A. T. & BAMBERG, S. A. (1977). Observations on the distribution of microorganisms in desert soil. *Great Basin Naturalist* 37:81–6.

VOLLMER, A. T., MAZA, B. G., MEDICA, P. A., TURNER, F. B., & BAMBERG, S. A. (1976). The impact of off–road vehicles on a desert ecosystem. *Environmental Management* 1:115–29.

VOLLMER, A. T., TURNER, F. B., STAUGHAN, I. R., & LYONS, C. L. (1982). Effects of coal precipitator ash on germination and early growth of desert annuals. *Environmental and Experimental Botany* 22:409–13.

VON WILLERT, D. J., ELLER, B. M., WERGER, W. J. A., BRINCKMANN, E., & IHLENFELDT, H.-D. (1992). *Life Strategies of Succulents in Deserts*. Cambridge University Press.

WALLACE, A., BAMBERG, S. A., & CHA, J. W. (1974). Quantitative studies of roots of perennial plants in the Mojave Desert. *Ecology* 55:1160–2.

WALLACE, A. & ROMNEY, E. M. (1972). *Radioecology and Ecophysiology of Desert Plants at the Nevada Test Site*. U.S. Atomic Energy Commission Report TID–25954, National Technical Information Service, U.S. Department of Commerce, Springfield, Virginia.

WALLACE, A., ROMNEY, E. M., & CHA, J. W. (1980). Depth distribution of roots of some perennial plants in the Nevada Test Site of the northern Mojave Desert. *Great Basin Naturalist Memoirs* (4):201–7.

WALLACE, A., ROMNEY, E. M., & HUNTER, R. B. (1978). Nitrogen cycle in the northern Mohave Desert: implications and predictions, pp. 207–18. In N. E. West & J. Skujins (eds.), *Nitrogen in Desert Ecosystems*. Dowden, Hutchinson and Ross, Inc., Stroudsburg, Pennsylvania.

WALLACE, A., ROMNEY, E. M., & HUNTER, R. B. (1980a). The challenge of a desert: revegetation of disturbed desert lands. *Great Basin Naturalist Memoirs* (4):216–25.

WALLACE, A., ROMNEY, E. M., & HUNTER, R. B. (1980b). Relationship of small washes to the distribution of *Lycium andersonii* and *Larrea tridentata* at a site in the northern Mojave Desert. *Great Basin Naturalist Memoirs* (4):94–7.

WALLACE, A., ROMNEY, E. M., & HUNTER, R. B.

(1990). Variability and diversity caused by environmental forces in the vegetation at the Nevada Test Site, pp. 209–20. *In* E. D. McArthur, E. M. Romney, S. D. Smith & P. T. Tueller (compilers), *Proceedings – Symposium on Cheatgrass Invasion, Shrub Die-off, and Other Aspects of Shrub Biology and Management.* General Technical Report INT–276. U.S. Department of Agriculture, Forest Service, Intermountain Research Station, Ogden, Utah.

WALLACE, A., ROMNEY, E. M., KLEINKOPF, G. E., & SOUFI, S. M. (1978). Uptake of mineral forms of nitrogen by desert plants, pp. 130–51. *In* N. E. West & J. Skujins (eds.), *Nitrogen in Desert Ecosystems.* Dowden, Hutchinson and Ross, Inc., Stroudsburg, Pennsylvania.

WALLACE, A., ROMNEY, E. M., & KINNEAR, J. E. (1980). Frequency distribution of three perennial plant species to nearest neighbor of the same species in the northern Mojave Desert. *Great Basin Naturalist Memoirs* (4):89–93.

WALLWORK, J. A. (1972). Distribution patterns and population dynamics of the micro–arthropods of a desert soil in southern California. *Journal of Animal Ecology* 41:291–310.

WARMING, E. (1925). *Oecology of Plants.* Oxford University Press.

WASBAUER, M. S. (1973). The male brachycistidine wasps of the Nevada Test Site (Hymenoptera: Tiphiidae). *Great Basin Naturalist* 33:109–12.

WATSON, A. (1989). Desert crusts and varnishes, pp. 22–55. *In* D. G. S. Thomas (ed.), *Arid Zone Geomorphology.* Halsted Press, New York.

WEATHERS, W. W. (1970). Physiological thermoregulation in the lizard *Dipsosaurus dorsalis. Copeia* 1970: 549–57.

WEATHERS, W. W. (1979). Climatic adaptation in avian standard metabolic rate. *Oecologia* 42:81–9.

WEBB, R. H. (1982). Off-road motorcycle effects on a desert soil. *Environmental Conservation* 9:197–208.

WEBB, R. H., STEIGER, J. W., & TURNER, R. M. (1987). Dynamics of Mojave Desert shrub assemblages in the Panamint Mountains, California. *Ecology* 68:478–90.

WEBB, R. H., STEIGER, J. W., & WILSHIRE, H. G. (1986). Recovery of compacted soils in Mojave Desert ghost towns. *Journal of the Soil Science Society of America* 50:1341–4.

WEBB, R. H. & STIELSTRA, S. S. (1980). Sheep grazing effects on Mojave Desert vegetation and soils. *Environmental Management* 3:517–29.

WEBB, R. H. & WILSHIRE, H. G. (1980). Recovery of soils and vegetation in a Mojave Desert ghost town, Nevada, U.S.A. *Journal of Arid Environments* 3: 291–303.

WEBB, R. H. & WILSHIRE, H. G. (1983). *Environmental Effects of Off-Road Vehicles.* Springer-Verlag, New York.

WEBB, W. L., LAUENROTH, W. K., SZAREK, S. R., & KINERSON, R. S. (1983). Primary production and abiotic controls in forests, grasslands, and desert ecosystems in the United States. *Ecology* 64:134–51.

WEBB, W. L., SZAREK, S., LAUENROTH, W., KINERSON, R., & SMITH, M. (1978). Primary productivity and water use in native forest, grassland, and desert ecosystems. *Ecology* 59:1239–47.

WEEKS, L. V. & RICHARDS, S. J. (1967). Soil-water properties computed from transient flow data. *Proceedings of the Soil Science Society of America* 31:721–5.

WELLS, P. V. (1961). Succession in desert vegetation on streets of a Nevada ghost town. *Science* 134:670–1.

WELSH, S. L., ATWOOD, N. D., GOODRICH, S., & HIGGINS, L. C. (1987). *A Utah Flora. Great Basin Naturalist Memoirs* 9:1–894.

WENT, F. W. (1948). Ecology of desert plants. I. Observations on germination in the Joshua Tree National Monument, California. *Ecology* 29:242–53.

WENT, F. W. (1949). Ecology of desert plants. II. The effect of rain and temperature on germination and growth. *Ecology* 30:1–13.

WENT, F. W. (1955). The ecology of desert plants. *Scientific American* 192(4):68–75.

WENT, F. W. & STARK, N. (1968). The biological and mechanical role of soil fungi. *Proceedings of the National Academy of Sciences, U.S.A.* 60:497–504.

WENT, F. W. & WESTERGAARD, M. (1949). Ecology of desert plants. III. Development of plants in the Death Valley National Monument, California. *Ecology* 30:26–38.

WEST, N. E. (1980). Formation, distribution and function of plant litter in desert ecosystems, pp. 647–60. *In* D. W. Goodall & R. A. Perry (eds.), *Arid Land Ecosystems: Structure, Functioning and Management,* vol. 1. Cambridge University Press.

WEST, N. E. (1983). Nutrient cycling in desert ecosystems, pp. 301–24. *In* N. E. West (ed.), *Temperate Deserts and Semi-deserts. Ecosystems of the World,* vol. 5. Elsevier, Amsterdam.

WEST, N. E. (1990). *Structure and Function of Soil Microphytic Crusts in Wildland Ecosystems of Arid and Semiarid Regions.* Advances in Ecological Research. Academic Press, London.

WEST, N. E. & SKUJINS, J. (1978). *Nitrogen in Desert Ecosystems.* Dowden, Hutchinson and Ross, Inc., Stroudsburg, Pennsylvania.

WESTERMAN, R. L. & TUCKER, T. C. (1978a). Factors affecting denitrification in a Sonoran Desert soil. *Proceedings of the Soil Science Society of America* 42:596–9.

WESTERMAN, R. L. & TUCKER, T. C. (1978b). Denitrification in desert soils, pp. 75–126. *In* N. E. West & J. Skujins (eds.), *Nitrogen in Desert Ecosystems.* Dowden, Hutchinson and Ross, Inc., Stroudsburg, Pennsylvania.

WHITE, L. D. & ALLRED, D. M. (1961). Range of kangaroo rats in areas affected by atomic detonations. *Proceedings of the Utah Academy of Science, Arts and Letters* 38:101–10.

WHITFORD, W. G. (1976). Temporal fluctuations in density and diversity of desert rodent populations. *Journal of Mammalogy* 57:351–69.

WHITFORD, W. G. (1978). Foraging in seed–harvesting ants *Pogonomyrmex* spp. *Ecology* 59:185–9.

WHITFORD, W. G. (1986). Decomposition and nutrient cycling in deserts, pp. 93–117. *In* W. G. Whitford (ed.), *Patterns and Processes in Desert Ecosystems*. University of New Mexico Press, Albuquerque.

WHITFORD, W. G. & CREUSERE, F. M. (1977). Seasonal and yearly fluctuations in Chihuahuan Desert lizard communities. *Herpetologica* 33:54–65.

WHITFORD, W. G., FRECKMAN, D. W., ELKINS, N. Z., PARKER, L. W., PERMALEE, P., PHILLIPS, J., & TUCKER, S. (1981). Diurnal migration and responses to simulated rainfall in desert soil microarthropods and nematodes. *Soil Biology and Biochemistry* 13:417–25.

WHITFORD, W. G., FRECKMAN, D. W., PARKER, L. W., SCHAEFER, D., SANTOS, P. F., & STEINBERGER, Y. (1981). The contributions of soil fauna to nutrient cycles in desert systems, pp.49–59. *In* P. Lebran, H. M. Andre, A. de Midts, C. Gregoire-Wilba, & G. Wavthy (eds.), *New Trends in Soil Biology*. Dieu-Brichard, Louvain-La-Neuve, Belgium.

WHITFORD, W. G., JOHNSON, P., & RAMIREZ, J. (1976). Comparative ecology of the harvester ants *Pogonomyrmex barbatus* (F. Smith) and *Pogonomyrmex rugosus* (Emery). *Insectes Sociaux* 23:117–32.

WHITFORD, W. G., REPASS, R., PARKER, L. W., & ELKINS, N. Z. (1982). Effects of initial litter accumulation and climate on litter disappearance in a desert ecosystem. *American Midland Naturalist* 108:105–10.

WHITFORD, W. G., STINNETT, K., & ANDERSON, J. (1988). Decomposition of roots in a Chihuahuan Desert ecosystem. *Oecologia* 75:8–11.

WIENS, J. A. (1991). The ecology desert birds, pp. 278–310. *In* G. A. Polis (ed.), *The Ecology of Desert Communities*. University of Arizona, Tucson.

WIENS, J. A. & ROTENBERRY, J. T. (1981). Habitat associations and community structure of birds in shrub-steppe environments. *Ecological Monographs* 51:21–41.

WILCOTT, J. C. (1973). *A Seed Demography Model for Finding Optimal Strategies for Desert Annuals*. Ph.D. dissertation, Utah State University, Logan.

WILLIAMS, A. B. (1936). The composition and dynamics of a beech–maple climax community. *Ecological Monographs* 6:317–408.

WILLIAMS, R. B. & BELL, K. L. (1981). Nitrogen allocation in Mojave Desert winter annuals. *Oecologia* 48:145–50.

WILSHIRE, H. G. & NAKATA, J. K. (1976). Off-road vehicle effects on California's Mojave Desert. *California Geologist* 29:123–32.

WILSON, R. T. 1989. *Ecophysiology of the Camelidae and Desert Ruminants*. Springer-Verlag, Berlin.

WINOGARD, I. J. (1962). Interbasin movement of ground water at the Nevada Test Site. *U.S. Geological Survey Professional Paper* 450–C:108–11.

WINOGARD, I. J. & THORDARSON, W. (1968). Structural control of ground water movement in miogeosynclinal rocks of south-central Nevada. *Memoirs of the Geological Society of America* 110:35–48.

WINOGARD, I. J. & THORDARSON, W (1975). *Hydrogeological and Hydrochemical Framework, South-Central Great Basin, Nevada-California, with Special Reference to the Nevada Test Site. U.S. Geological Survey Professional Paper* 712–C: 1–126.

WINTER, K. (1985). Crassulacean acid metabolism, pp. 329–87. *In* J. Barber & N. R. Baker (eds.), *Photosynthetic Mechanisms and the Environment*, vol. 6. Elsevier, Amsterdam.

WISCHMEIER, W. H. (1959). A rainfall erosion index for a universal erosion equation. *Proceedings of the Soil Science Society of America* 23:246–9.

WISCHMEIER, W. H. & SMITH, D. D. (1965). *Predicting Rainfall-Erosion Losses from Cropland East of the Rocky Mountains. USDA Agricultural Handbook* 282, Washington, D.C.

WOOD, T. G. (1971). The distribution and abundance of *Folsomides deserticola* (Collembola: Isotomidae) and other microarthropods in arid and semi–arid soils in southern Australia, with a note on nematode populations. *Pedobiologia* 11:446–68.

WOODBURY, A. M. & HARDY, R. (1940). The dens and behavior of the desert tortoise. *Science* 92:529.

WOODBURY, A. M. & HARDY, R. (1948). Studies on the desert tortoise, *Gopherus agassizii*. *Ecological Monographs* 18:145–200.

WOODELL, S. R. J., MOONEY, H. A., & HILL, A. J. (1969). The behavior of *Larrea divaricata* (creosote bush) in response to rainfall in California. *Journal of Ecology* 57:37–44.

WOODRUFF, N. P. & SIDDOWAY, F. H. (1965). A wind erosion equation. *Proceedings of the Soil Science Society of America* 29:602–8.

WOODWARD, S. L. (1976). *Feral Burros of the Chenehuevi Mountains, California: The Biogeography of a Feral Exotic*. Ph.D. dissertation, University of California, Los Angeles.

WOODWELL, G. M. (1962). Effects of ionizing radiation on terrestrial ecosystems. *Science* 138:572–7.

WOODWELL, G. M. (1963). The ecological effects of radiation. *Scientific American* 208(6):40–9.

WOODWELL, G. M., ed. (1965). *Ecological Effects of Nuclear War*. BNL 917 (C–43). Brookhaven National Laboratory, Upton, New York.

WOODWELL, G. M. (1967). Radiation and the patterns of nature. *Science* 156:461–70.

WOODWELL, G. M. & OOSTING, J. K. (1965). Effects of chronic gamma irradiation on the development of old field plant communities. *Radiation Botany* 5:205–22.

WOODWELL, G. M. & SPARROW, A. H. (1962). *Predicted and Observed Effects of Chronic Gamma Radiation on a Near-Climax Forest Ecosystem. U.S. Atomic Energy Commission Report* BNL–6968.

WOOTEN, R. C., CRAWFORD, C. S., & RIDDLE, W. A. (1975). Behavioral thermoregulation of *Orthoporus ornatus* (Diplopoda: Spirostreptidae) in three desert habitats. *Zoological Journal of the Linnean Society* 57:59–74.

YOUNG, J. A., EVANS, R. A., & MAJOR, J. (1977). Sagebrush steppe, pp. 763–96. *In* M. G. Barbour & J. Major (eds.), *Terrestrial Vegetation of California.* John Wiley and Sons, New York.

YOUNG, R. A. (1972). Water supply for the Nuclear Rocket Development Station, at the U.S. Atomic Energy Commission's Nevada Test Site. *U.S. Geologic Survey Water-Supply Paper* 1938:1–19.

ZOBACK, M. C. & ZOBACK, M. D. (1980). Faulting patterns in north-central Nevada and strength of the crust. *Journal of Geophysical Research* 85:275–84.

ZOHARY, M. (1961). On the hydro-ecological relations of the Near East desert vegetation, pp. 195–212. *In, Plant–water Relationship in Arid and Semi-arid Conditions.* Proceedings of the Madrid Symposium, *Arid Zone Research* (UNESCO) 16.

ZWEIFEL, R. G. & LOWE, C. H. (1966). The ecology of a population of *Xantusia vigilis*, the desert night lizard. *American Museum Novitatas* (2247):1–57.

Species Index

If there are no references to the subject in the text of a particular page, then the following convention is used: m = map: f = figure: t = table

Abronia villosa (sand verbena) 114
Acacia 67(f), 276
Acamptopappus shockleyi
 bajadas 17
 clump densities 93
 size class/frequency 95, 96(f)
 soil requirements 92
 summer dormancy 75
Acarina 249
Acarospora strigata 256–7
Achnatherum hymenoides (Indian rice grass) 17, 80
Acrobeles complexus 262
Actinomycetes 260, 261(t), 262
Allenrolfea occidentalis (iodinebush) 18
Amaranthaceae 63, 113
Amaranthus palmeri 81
Amaryllidacae 119(t)
Ambrosia deltoidea (triangle-leaved bursage) 75
Ambrosia dumosa (white bursage) 4(f)
 age 73
 arthropod habitat 236, 237(t), 238(t)
 arthropod sampling 216, 218
 bajadas distribution 13
 biomass 92, 97
 borrow pit recolonization 297, 298(t)
 canopy increase 96
 coppice mounds 35(f)
 diet of *Sauromalus obesus* 183, 184
 drought-deciduous features 73, 83
 habitat 84, 86
 IBP site composition 92
 ionizing radiation 308–9
 leaf beetles 242
 leaf initiation 108, 109, 111(t)
 leafing phenology 111
 litterfall 104(t), 105
 microarthropod sampling 260
 nematode sampling 260, 263
 net primary production 100–2
 nitrogen content 282
 precipitation growth response 103
 revegetation 293, 296
 root system 98, 99(f)
 sap-feeders 244–7
 seed reserve dynamics 266–8, 270(t)

 soil requirements 92
 trichome shield 59
 as water source 144
 weevil population 242, 240–1(t)
Ambrosia/Atriplex association 17
Ambrosia/Larrea association 17
Ammospermophilus leucurus (antelope ground squirrel) 155, 156(t), 157(f)
 activity pattern 133
 body weight 172(t)
 estivation 139
 huddling 139
 hyperthermia 140, 153, 160
 population density 167
 water source 160
Amphispiza belli (sage sparrow) 209, 211
Amphispiza bilineata (black-throated sparrow) 202, 203(f), 207
 breeding 209, 210
 electrolyte control 149
 production 211
 renal concentration 148
 water requirement 145
Amsinckia tessellata 125, 312
Anabaena 275
Anepsius brunneus 220
annelids 258
Anthocoridae 247
Anthus rubescens (American pipit) 209
Antilocapra americana (pronghorn antelope) 159
Antrozous pallidus (pallid bat) 159
Anuroctonus phaeodactylus 229
Aphelenchoides sp. 262
Aphelenchus avenae 262
Apiaceae 12
Appolophanes texanus 232
Aquila chrysaetos (golden eagle) 208
Araeoschizus sulcicollis 220, 221
Araneida 215, 222–5(t), 229(t), 230, 231(t), 232
Arctomecon 12
Arenivaga erratica (cockroach) 221, 254
Arenivaga investigata (desert cockroach) 146
Arinolus nevadae 224
Aristida spp. 113

Artemisia spp. 85(m)
 bird breeding habitat 209
Artemisia nova (black sagebrush) 87
Artemisia spinescens 17, 18
Artemisia tridentata (big sagebrush)
 desert regime characteristics 90
 fault line growth 24
 Great Basin Desert 6, 17, 87, 89(t)
 nesting site 211
 precipitation requirement 91
 summer growth 107
Asclepias erosa 58
Aspergillus 260
Asteraceae 12, 119(t)
Astragalus spp. 119(t), 120, 276
Astragalus lentiginosus 118, 119(t)
Atamisquia emarginata 71(t)
Athene cunicularia (burrowing owl) 208, 210(f)
Atriplex spp.
 Arenivaga investigata food 146
 C_4 photosynthesis 65
 Dipodomys microps food 149
 Great Basin 87, 89(t)
 soil requirements 92
Atriplex canescens (four-winged saltbush) 17
 associations 88, 89(t)
 Great Basin 87, 89(t)
 max/min temperature 91
 post-nuclear colonizer 307
 shadscale scrub 18
Atriplex confertifolia (shadscale) 17
 bajadas 13
 desert holly scrub 18
 desert regime characteristics 90
 Great Basin 87
 habitat 85(m)
 IBP site composition 92
 maximum temperature 91
 nesting site 210
 net primary production 100–2
 phenology 106
 rodent population 168
 sap-feeders 244–6
 shadscale scrub 18
 soil nitrogen levels 278–9

Atriplex hymenelytra (desert holly)
  desert holly scrub 18
  evergreen 71, 83
  habitat 72, 83
  as halophyte 73
  leaf orientation 58
  leaf reflectivity 73
  organic acid synthesis 73
  salt glands 73
  stomal control 72
  WUE 73
Atriplex lentiformis, acclimation 61
Atriplex polycarpa 18
Atriplex-Kochia scrub 197
Azotobacter 275–6

Baileya pleniradiata 118
Bebbia juncea (sweet bush) 86
Beleperone californica 67
Berberis (barberry) 71(t)
Boerhaavia spp. 113
Bombylliidae 216, 243
Bootettix argentatus 243
  on *Larrea* 251, 252, 253
  leaf consumption 105
  and *Mythicomyia* population 248
  population abundance 243, 255
  predators 243
  sampling inaccuracy 219
Boraginaceae 12, 119(t)
Bouteloua spp. 113
Braconidae 248
Brassicaceae 12, 119(t)
Bromus spp.
  effect on lizards 313–14
  fire risk 313
Bromus rubens 120, 298, 299(t), 313
  on ash soil 312
  biomass 125
  density 121–2
  Gopherus forage 192
  post-nuclear colonizer 307
  survivorship 121
Bromus tectorum (cheatgrass) 312–13
Bromus trinii 313
Bryophytes 257
Bufo punctatus (desert toad) 146

Cactaceae 12, 63
Callipepla gambelii (Gambel's quail) 203,
  205–7
  drought adaptation 206–7
  food/water requirement 145, 206
Callisaurus draconoides (zebra-tailed
  lizard) 174, 176(f)
  biomass 187
  characteristics 182
  coloration 136–7
  densities 185
  effect of vehicles 304
  enclosed density estimates 186
  size 187

superlight coloration 137
thermoregulation 182
Camelus dromedarius (dromedary) 145–6,
  147
Camissonia boothii, rosette 114(f)
Camissonia claviformis 183
Camissonia munzii 192, 283
Campanulaceae 119(t)
Campylorhynchus brunneicapillum
  (cactus wren) 147
Canis latrans (coyote)
  den 136
  density 162
  Gopherus predator 195
  water search 145
Carnegiea gigantea (saguaro) 11
Carpodacus mexicanus (house finch) 197,
  207, 208
  flocking 266
  water requirement 145
Castela emoryi (crucifixion thorn) 78
Caulanthus cooperi 125, 312
Centrioptera muricata 220
Centrodontus atlas 246, 253
Cephalobidae 262
Ceratoides lanata (winter fat) 17
  bajadas 13
  Great Basin 87, 89(t)
  microflora sampling 259–60
  net primary production 100–2
  semi-evergreen 77
  shadscale scrub 18
Cercidium floridum (palo verde) 11, 77,
  83
Ceuthophilus fossor (camel cricket) 221,
  254
Chaenactis spp. 119(t), 120, 121
Chaenactis carphoclinia 119(t), 283
Chaenactis fremontii 119(t), 120, 125, 304
Chalcidae 248
Chalcidoidea 248
Chamaesyce spp. 119(t), 120, 113
Chenopodiaceae 12, 63
Chilopoda (centipedes) 222–5(t), 231, 232
Chilopsis linearis (desert-willow) 19, 61,
  67
Chionactis occipitalis ocipitalis (western
  shovel-nosed snake) 12, 181, 191
Chorizanthe spp. 119(t), 120, 125
Chrysomelidae 242
Chrysothamnus spp. 61, 86, 87
Chrysothamnus viscidiflorus 293, 296
  ssp. *puberulus* 87
Cicadellidae (leafhoppers) 247, 253
Cleridae 247
Clostridium 275–6
Cnemidophorus tigris (whiptail lizard) 174
  biomass 187
  burrow 134
  body temperature 182
  characteristics 182
  clutch frequency 188

densities 185, 189
diet 182
effect of vehicles 304
enclosed density 185
radiation effect 310–11
size 187
survival rate 188
Coccinellidae 247
Coleogyne ramosissima (blackbrush)
  associations, transition desert 87–8,
    89(t)
  clonal propagation 77
  CO_2 levels 52
  desert regime characteristics 90
  drought-deciduous features 76–7
  evergreen 71(t)
  Great Basin Desert 6, 17
  habitat 85(m), 86(f)
  in Joshua tree woodland 17
  leaf structure 76, 77(f)
  southern Mojave 17
  spiders 230
Coleonyx variegatus (banded gecko) 174,
  179(f)
  biomass 187
  characteristics 183
  drinking 145
  enclosed densities 186
Coleophoridae 239, 242
coleoptera 222(t), 223(t), 224(t), 225(t),
  242
  density/biomass 234(t)
  predatory 248–9
Collema 275
Conomyrma insana 250, 251
Corvus corax (raven) 138, 208
Crematogaster depilis 250, 251
Cricetidae 155, 156(t)
Crossidium aberrans 257
Crotalus cerastes (sidewinder) 12, 181,
  191
  coloration 136–7
Crotalus mitchelli stephensi (Panamint
  rattlesnake) 12
Crotalus mitchellii (speckled rattlesnake)
  181, 191
Crotaphytus collaris (collared lizard) 144,
  182, 184
Crotaphytus wislizenii xvi, 310–11
Cryptantha spp. 119(t), 120, 121
Cryptogamae 256–7
Cryptoglossa verrucosa 220, 221
Cucurbita digitata 58, 67
Curculionidae (weevils) 216, 233, 242
Cuscutaceae 119(t)
cyanobacteria 271
Cyperaceae 12
Cyprinodon (desert pupfish) 12

Dalea spinosus (smoke tree) 67
Datura wrightii 58

Dedeckera 12
Dendroica coronata (yellow-rumped
   warbler) 209
Dermatocarpon lachneum 257
Diplopoda 222, 223(t), 224, 227(t), 228(t)
Dipodomys spp.
   body weight 172(t)
   distribution 168(f)
   foraging pattern 168(f)
   metabolic water 146, 160
   open space foraging 170
   population densities 162–4
   reproduction 164–6
Dipodomys deserti (desert kangaroo rat)
   155, 156(t)
   foraging 169
   habitat 168
   indeterminate population 162–3
   shrub utilization 168–9
Dipodomys merriami (Merriams' kangaroo
   rat) 155, 156(t), 157(f)
   food/reproduction correlation 153
   foraging interactions 170
   home range 171
   Larrea habitat 167–8
   metabolic water dependence 146
   seasonal activity 170–1
   shrub utilization 168–9
   water ingestion and reproduction 164
Dipodomys microps (Great Basin
   kangaroo rat) 155, 156(t)
   food 168
   habitat 167
   home range 171
   seasonal activity 170–1
   salt ingestion control 149
   urine concentration 148
   water source 144
Dipodomys ordii (Ord's kangaroo rat) 155,
   156(t)
Dipsosaurus dorsalis (desert iguana) 174,
   179(f)
   activity pattern 135
   burrow 134
   characteristics 184
   estivation 139
   home range 184
   hyperthermia 140
   panting 144
   respiratory surface evaporation 143
   salt glands 148
   superlight coloration 137
   thermal tolerance 184
   water source 144
Diptera 215, 248
Distichlis spicata 19
Ditylenchus sp. 262

Edrotes robus 220
Elaphonema sp. 262
Eleodes armata 221
Eleodes grandicollis 221

Encelia farinosa (brittlebush)
   drought-deciduous features 75
   leaf absorbtance 59
   leaf dimorphism 57
   root system 69
   temperature maintenance 61
   water potential 67
Encelia frutescens 293, 297, 298(t)
Encelia virginensis 86
Encryrtidae 248
Ephedra spp.
   aphyllous 77, 83
   as food 167
   gelechiid on 239
   root system 69
Ephedra funerea 17
Ephedra nevadensis
   bajadas 13
   dead biomass 98
   disturbance response 294
   flowering phenology 112
   IBP site biomass 92
   ionizing radiation 308–9
   Joshua tree woodland 17
   leaf initiation 108
   net primary production 100–2
   nitrogen content 282
   revegetation 292, 293
   root system 98, 99(f)
   sap-feeders 244–6
   seed reserve dynamics 266–8
   southern Mojave 17
   transition desert 87
Ephedra torreyana 17
Equus asinus (burro) 143, 313
Equus caballus (horse) 313
Eremophila alpestris (horned lark) 197,
   208–9
Eriogonum spp. 119(t), 120
Eriogonum fasciculatum (buckwheat) 17
Eriogonum fasciculatum spp. polifolium
   71(t)
Eriogonum inflatum 118
Eriogonum trichopes 125
Erioneuron pulchellum 118
Erodium cicutarium (storksbill) 120, 298,
   299(t), 300(f), 313
   effect of compaction 304
   germination 114
   grazing resistance 314
Eucyllus spp. 233
Eudorylaimus monohystera 262
Eulophidae 248
Eumeces skiltonianus (western skink) 174
Eupelmidae 248
Euphorbiaceae 12, 119(t)
Eurybunus riversi 232
Eusattus dubius 220, 221
   biomass 226, 227(t), 228(t)

Fabaceae 12, 119(t)
Falco mexicanus (prairie falcon) 208

Falco sparverius (American kestrel) 208
Felis concolor (mountain lion) 159
Filago depressum, rosette 114(f)
Formicidae: *see* ant
Fraxinus velutina var. *coriacea* (velvet
   ash) 19

gamasina mites 287
Gambelia wislizenii (long-nosed leopard
   lizard) 174, 177(f)
   characteristics 183
   clutch frequency 188
   densities 185
   size 187
   survival rate 188
Garrya flavescens var. *pallida* (silk-tassel)
   71(t)
gelechiid larvae 238–9
Geococcyx californicus (roadrunner)
   basking 138
   panting 144
   salt glands 148
   thermoregulation 207, 208(f)
geometrid larvae 242
Geraea canescens 115(f)
Geraniaceae 119(t)
Gilmania luteola 12, 115(f)
Gopherus agassizii (desert tortoise) 192–6
   ageing estimates 195
   body size 194
   burrow 134, 175, 181, 193
   death rate 195
   drinking 191
   egg production 195–5
   electrolyte control 149
   estivation 139, 191, 192
   feeding 193–4
   FMR during estivation 152
   foraging 192
   grazing affect 314
   growth rates 194–5
   hibernation 191, 192
   home range 193
   nitrogen content 284
   NTS 174
   osmoregulation 193
   osmotic tolerance 147
   overview 196
   population samples 194
   predators 195
   surface activity 191–3
   ticks 249
   water drinking 145, 146
   water source 144
Grayia spinosa (hopsage)
   arthropods 236, 237(t), 238(t)
   arthropod sampling 218
   bajadas 13
   dead biomass 98
   desert regime characteristics 90
   drainage basin 17
   drought-deciduous features 76

early-growth 108
flowering phenology 112
habitat 85(m)
high bajadas 17
IBP site biomass 92
ionizing radiation 308–9
leafing phenology 111
Lepidoptera 237–9, 240–1(t), 242
nesting site 210
net primary production 100–2
radiation damage 307
sap-feeders 244–7
soil requirements 92
Grayia–Lycium association 87, 89(t)
  centipedes 232
  Dipodomys microps habitat 167
  snakes 181
  Venezillo arizonicus 222
Grimmia sp. 257
Guillenia lasiophylla 125

Hadrurus spp. 229–30
Hadrurus arizonensis (giant desert hairy
  scorpion) 143, 229
Haplodrassus eunis 230
Haplopappus cooperi
  disturbance response 294
  habitat water potential variation 69
  high bajadas 17
  resin 61
  southern Mojave 17
Hecastocleis 12
Heloderma suspectum (banded/venomous
  gila monster) 148, 174
Hemerobiidae 247
Hemiptera 222(t), 226(t), 247
  population size 244(t), 245(t)
Herpyllus hesperolus 232
Heteromyidae 155, 156(t)
Hilaria rigida (bunch grass) 80, 83
  acclimation 61
  disturbance response 294
  Joshua tree woodland 17
  Kranz anatomy 63(f)
  physiological characteristics 81(T), 81–2
  southern Mojave 17
Homoptera 215, 244(t), 245(t)
Hydrophyllaceae 12, 119(t)
Hymenoclea salsola (cheesebush) 291,
  292, 293, 294
  bajadas 13
  C_3 features 75, 76(f)
  disturbance response 294
  habitat 86
  leaf size 58
  post-nuclear colonizer 307
  southern Mojave 17
  temperature maintenance 61
Hymenoptera 226(t), 248
  burrows 215
Hypsiglena torquata (night snake) 12, 181

Hyptis emoryi (desert lavendar) 67, 75

Ichneumonoidea 248
Iridomyrmex pruinosus 251
Iridomyrmex spp. 250
Isopoda 215, 222–4, 222–5(t), 227(t),
  228(t)

Juniperus spp. (junipers) 11
Juniperus osteosperma 87, 89(t)
Justicia californica (chuparosa) 11, 67

Kochia americana (green molly) 87, 89(t)
Krameria erecta (Pima rhatany)
  arthropod 236, 237(t), 238(t)
  arthropod sampling 216, 218
  bajadas 13
  biomass 92
  drought-deciduous features 74–5, 83
  flowering 108, 109(t), 111(t)
  habitat 84, 86
  IBP site composition 92
  ionizing radiation 308–9
  late-growth 108
  leaf beetles 242
  leaf initiation 108, 109, 111(t)
  leafing phenology 111
  Lepidoptera 238–9, 240–1(t)
  limestone bajadas 17
  microarthropod sampling 260
  microflora sampling 259–60
  nematode sampling 260, 263
  net primary production 100–2
  nitrogen content 282–3
  phenology 106
  precipitation growth response 103
  root parasitism 75
  root system 69
  sap-feeders 244–7
  seed mass 271
  seed reserve dynamics 266–8
  size class/frequency 95, 96(f)
  soil nitrogen levels 280
Lagurus curtatus (sagebush vole) 155,
  156(t)
Laminaceae 12
Lampropeltis getulus (common kingsnake)
  191
Langloisia setosissima 192
Lanius ludovicianus (loggerhead shrike)
  208
Larrea tridentata (creosote bush) 4(f)
  acclimation 61
  adaption to stress 72
  age 71–2
  Arenivaga investigata food source 146
  arthropods 236, 237(t)
    food source 146, 215
    habitat 238(t)
    habitat/host 251–4, 255
  arthropod sampling 216, 218
  associations 88(t), 89(t)

bajadas 13, 17
borrow pit recolonization 298(t)
canopy 96
clones 72
CO_2 uptake 72
coppice mounds 35(f), 36
dead biomass 97
desert regime characteristics 90
diet of *Sauromalus obesus* 184
Dipodomys merriami correlation 167–8
evergreen adaptive strategy 70–2, 83
fauna 70
flowering 108, 109(t), 111(t)
flowering phenology 112
habitat 70, 84, 85(m)
height 88, 89(t)
height/density correlation 90, 91(f)
herbivory (*Bootettix argentatus*) 105
IBP site composition 92
late-growth 108
leaf beetles 242
leaf discrimination, old/new 101
leaf initiation 108, 109
leaf orientation 58
leafing phenology 111
leaves 61
Lepidoptera 240–1(t)
litterfall variation 105
microarthropod sampling 260
net primary production 100–2
nitrogen content 282
orthopteran habitat 243
osmotic adjustment 67
PAR saturation 72
Phytocoris on 247
pollination 72, 247
precipitation/density correlation 72
precipitation growth response 103
precipitation tolerance 91
predators on 248
radiation/dust effect 308, 308–9
re-establishment 292, 296
resin 61
road edge 298, 299(t)
root system 69, 98, 99(f)
sap-feeders 244–7
seed mass 271
seed reserve dynamics 266–8, 270(t)
Semiothisa larreana larvae 242
soil compaction tolerance 293
soil nitrogen levels 278, 280–1
soil requirements 92
sugar availability 247
upper desert boundary 11
vapor pressure deficit maintenance 59,
  60(f)
water potential 67
water source 144
Larrea–Ambrosia communities
  beeflies 248
  bird diversity 197
  CO_2 levels 52

Larrea–Ambrosia communities (*cont.*)
  Dipodomys deserti habitat 168
  lizards 174, 182
  low bajadas 17
  snakes 181, 182
  spiders 230
  Tenebrionidae 220
  Venezillo arizonicus 222
Larrea–Atriplex association 17
Larrea–Coleogyne transition 17, 91
Larrea–grassland ecosystem decomposers 258
Lasthenia chrysostoma 114–15
Lepidoptera 215, 226(t), 237–9, 242
  density/biomass 234(t)
  families 239(t)
Leptonchus sp. 262
Lepus californicus (black-tailed jackrabbit) 156(t), 158(f), 159
  density 162
  electrolyte control 149
  energy budget 159
  estivation 139
  grazing effect 316
  hyperthermia 140
  nitrogen content 284
  water requirement 159
  water source for 144
Ligurotettix coquilletti 243, 252
Loasaceae 12, 119(t)
Loxosceles unicolor 222(t), 232
Lupinus 276
Lycium spp.
  drought-deciduous features 73–4
  early-growth 108
  flowering phenology 112
  germination 74
  leaf initiation 108, 109, 111(t)
  leafing phenology 111
  mesophyll 73–4
  root system 69
  Toxostoma lecontei nesting 210
  transition desert 87, 88, 89(t)
Lycium andersonii (desert thorn)
  arthropod 236, 237(t), 238(t)
  arthropod sampling 216, 218
  bajadas 13, 17
  biomass 92, 97
  canopy 96
  disturbance response 294
  drainage basin 17
  habitat 84, 86(f), 86
  ionizing radiation 308–9
  leaf beetles 242
  Lepidoptera 238–9, 240–1(t)
  litterfall 104
  microarthropod sampling 260
  microflora sampling 259–60
  nematode sampling 260, 263
  nesting site 210
  net primary production 100–2
  nitrogen content 282

Phrynosma platyrhinos food 183
  revegetation 290–1
  root system 98, 99(f)
  sap-feeders 244–7
  seed mass 271
  seed reserve dynamics 266–8, 270(t)
  size class/frequency 95, 96(f)
Lycium andersonii–Grayia association 87, 89(t)
  Uta stansburiana density 185
Lycium pallidum (box thorn)
  arthropod habitat 238(t)
  arthropod sampling 218
  associations 88, 89(t)
  biomass 92
  canopy 96
  desert regime characteristics 90
  habitat 85(m), 86(f)
  Lepidoptera 238–9, 240–1(t)
  limestone bajadas 17
  litterfall 104
  living biomass 97
  net primary production 100–2
  nitrogen content 282
  sap-feeders 244–6
  seed mass 271
  water source 144
Lycium shockleyi
  desert regime characteristics 90
  habitat 85(m)
  limestone bajadas 17
  Nevada endomic 17
Lygaeidae (lygaeid bugs) 247

macroarthropods 258
Malacothrix glabrata, rosette 114(f)
Marpissa californica 222(t), 232
Masticophis flagellum (coachwhip) 191
Melyridae 247, 248–9
Membracidae (treehoppers) 246–7, 253, 254(t)
Menodora spinescens
  aphyllous 78, 83
  bajadas 17
  Joshua tree woodland 17
  transition desert 88
Mentzelia albicaulis (stickleaf) 121, 307
Mentzelia obscura 283;
microarthropods 258
Miloderes mercuryensis 233
Mimus polyglottos (mockingbird) 209
Mirabilis bigelovii 183–4
Mirabilis pudica
  facultative life cycle 118
  post-nuclear colonizer 307
Miridae (mirid bugs) 216, 247
Mohavena breviflora 115(f)
Mollugo spp. 113
Monoptilon bellioides 115(f)
Mortonia (scabrella) utahensis (sandpaper bush)
  evergreen 70, 71(t), 83

leaves 61
Mucor 260
Multareis cornutus 246, 253
Multareoides bifurcatus 246–7, 253
Mutillidae 248
Myiarchus cinerascens (ash-throated flycatcher) 208
Myotis californicus (California myotis) 159
Myrmecocystus spp. 250, 251
Myrmeleontidae 247
Mythicomyia 243, 248, 253

Nebidae 247
nematodes 258
Neotoma lepida (packrats; woodrats) 155, 156(t)
  nests 136
  population density 166–7
  recolonization role 297
  Xantusia predation 183
Neuroptera (ant lions) 215
Nitrophila occidentalis 18
Nostoc 275
Notholeana parryi 61
Notisorex (desert shrew) 138
Novomessor 266
Nyctaginaceae 12
Nysius ericae 247

Odocoileus hemionus (mule deer) 159
Oenothera deltoides (evening primrose) 114
Olneya tesota (ironwood) 11, 67
Onagraceae 12, 119(t)
Onychomis torridus (grasshopper mouse) 155, 156(t)
  body weight 172(t)
  food/water requirements 160
  population density 166–7
Ophryastes varius 233
Opuntia spp. (succulent chollas) 78, 79(f)
  cladode orientation 59
  Gopherus forage 192
  southern Mojave 17
  water source 145
Opuntia basilaris 17, 78, 79(f)
Opuntia biglovii 297, 298(t)
Opuntia echinocarpa 92
Opuntia ramosissima 17
oribatid mites 258
Ornithodorus spp. 249
Orthichelus michelbacheri 224
Orthoporus ornatus (desert millipede) 222, 224, 254
  overwintering 136
Orthoptera 221–2, 222(t), 224(t), 225(t), 227(t), 228(t), 242–3
  density/biomass 234(t)
  on *Larrea* 252–4
Oryzopsis hymenoides 80
  bajadas 17
  Gopherus forage 192
  post-nuclear colonizer 307

Ovis canadensis (bighorn sheep) 143, 313
Oxystylis 12

Pectis papposa 113, 116
Pectocarya spp. 119(t), 120, 122
Penicillium 260
Pentatomidae 247
Perognathus spp. 168(f), 170
Perognathus formosus (pocket mouse)
  body weight 172(t)
  foraging interactions 170
  home range 171
  population densities 162–4
  radiation exposure xv–xvi, 309–10
  reproduction 165–6
  seed reserve dynamics 266–7
Perognathus longimembris (pocket mouse)
  155, 156(t), 157(f)
  foraging 169, 170
  home range 171
  population densities 162–4
  reproductive index 173
  shrub utilization 168–9
  torpor 138–9, 153, 160
  vehicle effects 304
Perognathus parcus (Great Basin pocket
  mouse) 155, 156(t)
Peromyscus crinitus 166–7
Peromyscus eremicus (cactus mouse) 155,
  156(t)
  estivation 139
Phacelia fremontii 121, 283
Phacelia vallis-mortae 283
Phalangida (harvestman) 225(t), 232
Pheidole spp. 250, 251, 266
Phloeothripidae 247
Phrynosoma platyrhinos (desert horned
  lizard) 174, 177(f), 178(f)
  ants as water source 144
  biomass 187
  clutch frequency 188
  countercurrent heat exchange 141
  densities 185, 186, 189
  food 183
  natural history 183
  radiation effect 310–11
  rainfall and juvenile density correlation
  189
  shade useage 136
  size 187
  survival rate 188
Phyllorhynchus decurtatus (spotted
  leaf-nosed snake) 191
Phymatidae 247
Phytocoris, on *Larrea* 247
Pinus monophylla (pinyon pine) 11, 87,
  89(t)
  ant species 250
Pipistrellus hesperus (Western pipistrelle)
  159
Pituophis melanoleucus (gopher snake)
  191

Plantaginaceae 119(t)
Plecotus townsendii (Townsend's
  big-eared bat) 159
Pluchea sericea (arroweed) 19
Poaceae 12, 119(t)
Pogonomyrmex californicus
  (seed-harvester ants) 26, 183, 250,
  251
Pogonomyrmex rugosus 251
Polemoniaceae 12, 119(t)
Polygonaceae 12, 119(t)
Populus fremontii (cottonwood) 19
Prosopis glandulosa (mesquite)
  arroyos 19
  denitirification 278
  nitrogen fixation 276
  playas 19
  root system 69
  water potential 67
Prosopis pubescens (screwbean) 19
Prostigmata 265(t)
protozoa 258
Prunus fasciulata (desert almond),
  habitat 86
Pseudococcus helianthi 245–6, 253, 255
Psilochorus utahensis 230, 232, 254
Psorothamnus aborescens (mojave indigo
  bush) 276
  Arenivaga investigata food source 146
  aphyllous 83
  bajadas 13
  habitat 86
  limestone bajadas 17
  summer dormancy 76
Psorothamnus spinosus (smoke tree) 19
  aphyllous 77
  water potential 67
Pteromalidae 248
Pungenus sp. 262
Purshia glandulosa 71(t)

Quercus (scrub oak) 71(t)

Ranunculaceae 119(t)
Reduviidae 247
Rhinocheilus lecontei (long-nosed snake)
  191
rhizobium nodules 276
Rosaceae 12
Rumex hymenosepalus 58

Salazaria mexicana (bladder sage)
  bajadas 13
  habitat 86
  Joshua tree woodland 17
  southern Mojave 17
Salicornia spp. 18
Salix spp. (willow) 19
Salpinctes obsoletus (rock wren) 208,
  211(f)
Salsola (Russian thistle) community 174,
  197, 314

Salsola kali (Russian kale) 307
Salticidae 247
Salvadora hexalepis (western patch-nosed
  snake) 191
Salvia columbariae, rosette 114(f)
Salzaria mexicana (bladder sage) 291
Sarcobatus vermiculatus (greasewood) 18,
  87
Sauromalus obesus (chuckwalla)
  basking 135
  characteristics 183–4
  clutch frequency 188
  crevice usage 136
  diet 183–4
  electrolyte control 149
  energy budget 153
  food/reproduction correlation 153
  foraging efficiency 152
  habitat 182
  hibernation 149
  home range 183
  juvenile/adult coloration 137
  NTS 174, 180(f)
  salt glands 148, 149
  water source 144
  water storage 146
Scaphiopus (spadefoot toad) 146
Scarabaeidae 221
Sceloporus graciosus (sagebrush lizard)
  174
Sceloporus magister (spiny desert lizard)
  174, 178(f)
  biomass 187
  characteristics 182–3
  crevice usage 136
  enclosed densities 186
Sceloporus occidentalis (western fence
  lizard) 174
Schismus barbatus 298, 299(t), 300(f), 313
  compaction effect 304, 305
  grazing resistance 314
Scolopendra michelbacheri 232
Scopulophila 12
Scorpionida 222–5(t), 229(t), 231(t), 254
Scrophulariaceae 12, 119(t)
Scyotonema 275
Semiothisa larreana 242, 253
Senna armata (spiny senna) 77–8, 83,
  86
Simmondsia chinensis (jojoba)
  evergreen 71(t), 83
  leaf orientation 58
  leaves 61
  water potential 67
Solanaceae 12
Solenopsis 266
Solpugida 215, 222–5(t), 229(t), 230,
  231(t), 254
Sonora semiannulata (ground snake) 191
Spanthus excavatus 247
Spermophilus tereticaudus (round-tailed
  ground squirrel) 155, 156(t), 167

Spermophilus townsemdii (Townsend's ground squirrel) 155, 156(t)
Sphaeralcea ambigua 184
  disturbance response 294
  facultative life cycle 118
  post-nuclear colonizer 307
Spizella breweri (Brewer's sparrow) 209
Spizella passerina (chipping sparrow) 209
Stephanomeria exigua 184
Stephanomeria pauciflora 297, 298(t)
Stipa speciosa (bunchgrass) 291, 292, 307
Streptanthella longirostris 125
Streptomyces 260
Suaeda spp. 18
Swallenia 12
Sylvilagus spp. 159, 162
Syspira eclectica 222(t), 230, 232

Tamarix ramosissima (saltcedar) 19, 313
Tantilla planiceps (black-headed snake) 181
Tarentula kochi 230
tarsonemid mites 287
Tenebrionidae (darkling beetles) 215, 216, 220, 221(t), 224(t), 227(t), 228(t), 254
  biomass 226, 227–8(t)
  density 218
  size 221
Tetradymia axillaris (horsebrush) 17
Thamnosma montana (turpentine bush) 291, 292
  aphyllous 78, 83
  habitat 86
Thomisidae 247
Thomomys (bottae) umbrinus (pocket gopher) 155, 156(t), 158(f)
  activity 138
  burrow system 167
  ecophysiology 159–60
  fencing ineffectiveness 315
  main food 167
  nitrogen content 284
  population 167
Thysanoptera (thrips) 215, 243–4, 245(t)
  on *Larrea* 253

predatory 247
  shrub habitat 236, 237(t)
Tidestromia oblongifolia 65, 118
Tiphiidae 248
Tortula bistratosa 257
Toxostoma lecontei (Le Conte's thrasher) 207, 208
  breeding 209, 210
  food 210
  production 211
Trimorphodon biscutatus (lyre snake) 181
Triorophus laevis 220
Trogloderus costatus 220, 221
Tylenchida 262
tydeid mites 287
Tylenchorhynchus spp. 262

Uma scoparia (fringe-toed lizard) 12
Uta stansburiana (side-blotched lizard)
  biomass 187
  burrow 134
  characteristics 182
  clutch frequency 187
  density 185, 189
  effect of vehicles 304
  egg production 187–8, 189
  enclosed density 185
  estivation 139
  foraging 182
  juvenile data 189–90
  NTS 174, 176(f)
  population model 188–9
  precipitation and egg laying 188, 189
  radiation effect 310–11
  reproductive energy requirement 153
  secondary production 190
  size 187
  solar heating coloration 137
  sterility 311
  survival rate 188

Vejovis spp. 229, 230
Venezillo arizonicus (sowbug) 222–4
Veromessor spp. 250, 266

Viscainoa geniculata, evergreen 71(t)
Vulpes macrotis (kit fox), den 136, 162, 195
Vulpia octoflora 120, 122, 123, 124(t), 125, 126

Washingtonia filifera (California fan palm) 19, 58

Xantusia vigilis (desert night lizard) 174, 180(f)
  activity pattern 133
  characteristics 183
  drinking 145
  habitat 182
  predators 183
Xylorhiza canescens 118

yponomeutid larvae 239, 242(t)
Yucca spp.
  age 79
  root system 69
Yucca brevifolia (Joshua tree)
  habitat 78, 83, 86
  high bajadas 17
  MDR diagnostic 5
  photosynthesis 79
  Sceloporus magister habitat 182–3
  shadscale scrub 18
  succession 294
  as succulent 78
Yucca schidigera (Mojave yucca)
  habitat 79, 83
  limestone bajadas 17
  Peromyscus crinitus location 167
  Sceloporus magister habitat 182–3
  as succulent 78–9

Zenaida macroura (mourning dove) 202, 204(f), 207
  drinking water 144, 145
Zonotrichia leucophrys (white-crowned sparrow) 209

Main Index

If there are no references to the subject in the text of a particular page, then the following convention is used: m = map: f = figure: t = table

absorptance, leaf 58–9
absorptivity, soil 39(t)
abundance index 216, 224(t)
acclimation 61, 80, 81(t); *see also* temperature
activity, thermoregulatory 133
activity pattern, seasonal 133
adaptive stategies (animals)
  coloration 136–8
  ectothermic thermoregulation 131, 132
  endothermic thermoregualtion 131, 133
  heat balance 130–41
  heat exchange 140–1
  hyperthermia 140, 153, 160
  osmoregulation 147–50
  water balance 131, 141–7, 149–50
  water drinking 131, 144, 145–6
  see also burrow; estivation; cooling; hibernation; shade; water, metabolic; energy budget
adaptive strategies (plants) 55
  acclimation 61
  leaf anatomy 61–2, 63
  leaf orientation 58–9
  leaf reflectance 58–9
  models 55–6
  photosynthetic pathway 62–6
  water conservation 62–3
  WUE 66–7
AEC ix; *see also* ERDA
AET (actual evapotranspiration) 103
age
  plant 71–2
  white bursage 73
  Yucca spp. 79
  see also life span
ageing, and torpor 139, 170
ageing estimates, tortoise 195
air spaces, intercellular 62
algae 256, 276, 288
Algodones Dunes, vehicles 306
alkali sink 18
alkaline basin 18
allogenic drainage 7
allscale-alkali scrub 18
alluvial soil 24
  bajadas 28, 29

NTS 30
  Rock Valley 25
Amargosa River system 7
Amargosa Valley 18, 23–4, 85(m)
American desert scorpion: *see Hadrurus arizonensis*
American kestrel: *see Falco sparverius*
American pipit: *see Anthus rubescens*
ammonia volatilization 278
ammonium 274–5, 281
ammonium pool, Rock Valley 280–1
amphibians, water requirements 146
amphistomaty 61, 62, 82
angiosperms 81(t)
annuals 80–1, 83, 119(t)
  adaptive strategy 83
  biomass, plant parts 127(t)
  compaction 304–5
  carbon gain 55–6
  density and habitat 128(t)
  dry mass 126(t)
  facultative perennials 118
  germination 113–16, 129
  grazing 314
  growth 116, 118
  microhabitat distribution 127, 129
  nitrogen concentration 283–4
  NTS 113
  physiology 81(t), 81–2
  seedling count 121
  size variance 125
  species diversity 123
  survivorship 116, 118(t), 120–1, 129
  water stress 56
annuals, winter 11, 114–15
  biomass 123–7
  density 120(t)
  diversity 120
  IBP site 118–20
  lizard egg production 188
  net productivity 120(t)
  population densities 122–3
  precipitation dependence 115–16
  road edge 298–300
  Rock Valley 120
  rodent density fluctuation 163–4, 165
  rodent reproduction 166

total annual productivity 124(t)
  vehicle damage 303–4
ant 216, 249–51
  foraging pattern 249–50
  lizard food 182, 183
  nests, beetle association 220
  nitrogen source 206
  NTS species 250–1
  resource partitioning 249–50
  and sap-feeding insects 243
  size grading 251
  territoriality 250
  water source 144
  see also harvester ant
ant consumption, quail 206
ant lions: see Neuroptera
ant venom tolerance 144
antelope ground squirrel: *see Ammospermophilus leucurus*
antidiuretic hormone 206–7
aphyllous shrubs 77–8, 81(t), 81–2
aqueducts, succession 296, 297(t)
arachnid species diversity 229(t)
arboreal lizards 182–3
Argillic horizon 29
Aridisols 29
arroweed: *see Pluchea sericea*
arroyo, plant community 19
arthropod
  biomass 219
  bird food source 210
  concealment 252
  density 222(t), 223(t), 252–3
    analysis 219
    peaks 235
  feeding guild 224(t)
  feeding habits 218(t)
  field studies 216, 218–19
  fog moisture uptake 146
  food resource 215
  life-history strategy 215
  nitrogen content 284, 285(t)
  nocturnal activity pattern 133
  NTS survey 216, 217(t), 218(t)
  population, *Larrea* 252–3
  predators/parasitoids 247–9
  predatory, on *Larrea* 252

anthropod (*cont.*)
  radiation effects xiii
  respiration 220
  sap-feeding, on *Larrea* 253–4
  shrub-dwelling 218
  spatial patterning 235
  surveys 215–19
  temperatures, lethal 130, 131(f); *see also* ectotherms
  tortoise diet 194
  trophic specializations 214–15
  water loss 143
arthropod, foliage 235–49
  Coleoptera 242
  defoliating insects 236–7, 239–42(t)
  leafhoppers 247, 253
  Lepidoptera 237–9, 240–1(t), 242
  mealybugs 245–6, 253
  Orthoptera 242–3
  pollinating insects 247
  sap-feeding 243
  types 236
  see also Membracidae; Thysanoptera
arthropod, ground-dwelling 219–35, 222–6(t)
  biomass (detritivorous) 225(t), 226, 227(t), 229
  biomass (predatory) 231(t), 233
  density 227(t)
  detritivorous beetles 220–1, 254
  detritivorous Orthoptera 221–2
  isopods 222–4
  see also specific species
ash, landfill 311–12
ash flow 25
Ash Meadows
  groundwater system 26, 28
  plant communities 19
  saltbush scrub 19
ash-throated flycatcher: *see Myiarchus cinerascens*
association model (Beatley) 108
associations, vegetational 91–2, 95(f)
atmospheric nitrogen inputs 274–5; *see also* dryfall; wetfall
atmospheric testing xii-xiii
auxilliary roots 98

bacteria 261(t), 261–2
  in decomposition process 287
  free-living 288
  nitrogen fixation 275–6
bajada 4
  alluvial soils 29
  basin soil 28
  calcareous, vegetation 84
  Grayia–Lycium andersonii association 87
  lichens 256–7
  moisture retention 39
  plant communities 17
  precipitation 13

slopes, soils 29
soil composition 32(t)
vegetation patterns 13, 17
weathering deposits 24
Yucca spp. 79
banded gecko: *see Coleonyx variegatus*
banded gila monster: *see Heloderma suspectum*
Banded Mountain 23(f), 24, 85(m)
basal metabolic rate (BMR) 152
basins 23–4
  hydrologic 26
  soils 28–9, 29
  structure 3–4
  temperature inversion 89–90
  see also Frenchman Flat; Jackass Flats; Yucca Flat
basking 132, 135
  coloring changes 137
  roadrunner 138
bat population 159
Beatley, J. C.
  annual plant phenology model 116, 117
  annual survivorship 116, 118
  Mojave site monitoring 88–91
beavertail cactus: *see Opuntia basilaris*
beeflies 248, 253; *see also* Bombylliidae
bees, as pollenators 247
beetles 220–1
  biomass 221, 226–9
  coloration 137–8
  fog moisture uptake 146
  mass 221, 225–6(t)
  predatory 247, 248–9
  seasonal activity 220–1
  soil preference 220
Belted Range 27(f), 85(m)
Belted Range tuff 25, 26(t)
berm 294, 296
beta radiation damage, plant 307
big sagebrush: *see Artemisia tridentata*
bighorn sheep: *see Ovis canadensis*
biomass 55
  aboveground 96–8
  animal, nitrogen 284–5
  annuals 126(t), 129
  aqueduct line 296, 297(t)
  arthropod 219, 231(t), 233, 244(t)
  belowground 98–100
  birds 213(t)
  dead 98
  herb 124–5
  IBP site 92
  living/dead 97
  lizards 185, 186(t), 187
  millipede 224
  nematodes 263
  nitrogen reserve 274
  NTS shrubs 112
  orthopteran 221–2, 228(t), 243
  plant, nitrogen 281–4
  plant parts 127

reproductive allocation 127
  and root/shoot ratio 100(t)
  seed 271–2
  snake 191–2
  solpugids 230
  spiders 232(t), 232
  winter annuals 123–7
biomass size classes 95
biomass/net production equivalence 124
biota 10–11
birds
  activity pattern 133
  breeding 209–10
  densities 211–13
  excretion 148
  FMR and environment 152
  habitat partitioning 210–11
  hypothermia 140
  lethal temperatures 130, 131(f)
  migratory 197
  natural history review 203
  nest-building 136
  NTS breeding status 206(t)
  panting 144
  predatory 203
  production 211–13
  residency 206(t), 208–9
  species summary 12, 197–201
  thermoregulatory activity pattern 133, 203
  transient 197
  water loss predictions 141–2
  see also endotherms; passarine birds
black coloration 137–8
black sagebrush: *see Artemisia nova*
black-headed snake: *see Tantilla planiceps*
black-tailed jackrabbit: *see Lepus californicus*
black-throated sparrow: *see Amphispiza bilineata*
blackbrush: *see Coleogyne ramosissima*
bladder sage: *see Salzaria mexicana*
blood shunting, thermoregulatory 140–1
BMR (basal metabolic rate) 152
body mass: *see* mass
borrow pit, recolonization 296–8
box thorn: *see Lycium pallidum*
breeding, birds 209–10
  density 211–13
  status 206(t)
  variables 210
Brewer's sparrow: *see Spizella breweri*
brittlebush: *see Encelia farinosa*
broad-leafed plants 58
Brookhaven radiation experiments xiii
buckwheat: *see Eriogonum fasciculatum*
bunchgrass: *see Hilaria rigida; Stipa speciosa*
bundle sheath cells 64
burro: *see Equus asinus*
burrow 133–5, 150(t)
  food storage 139

foraging behavior (rodents) 169
foraging limit 160
gopher 160, 167
heat unloading 140
humidity 135
hibernation 139
Hymenoptera 215
lizards 135
microclimate 134–5
rabbit heat avoidance strategy 159
spadefoot toad 146
tortoise 175, 181, 193
burrowing owl: see Athene cunicularia
bursage: see Ambrosia dumosa
butterflies 144, 247

C₃ metabolism 82
annuals 81
bunchgrasses 80
in C₄ pathway 64
cheesebush 75, 76(f)
IBP site 92
photorespiration 65
summer dormancy 75–6
C₄ metabolism 63–4
annuals 81, 120
bunchgrasses 80
facultative life cycle 118
high capacity 65
IBP site 92
photorespiration 65
temperature-tolerant 65
water use efficiency 65
cacti 78, 79
limestone bajadas 17
tortoise consumption 194
water source 145
see also succulents
cactus mouse: see Peromyscus eremicus
cactus wren: see Campylorhynchus
brunneicapillum
calcareous (CaCO₂) undercoating 30, 31(f)
calcareous soils 29
calcrete, pedogenic 29, 279
caldera 25
caliche hardpan 17
California fan palm: see Washingtonia
filifera
California myotis: see Myotis californicus
Calvin-Benson cycle: see C₃ metabolism
CAM (Crassulacean acid metabolism) 64
energy requirements 65–6
IBP site 92
succulents 78, 79
temperature-dependent efficiency 66
water conservation 65, 82
Cambrian rocks 24
camel cricket: see Ceuthophilus fossor
Camp Clipper 293
Cane Spring 25, 26
canopy 95–6
annual density 128(t), 129

arthropod density 215
coverage 95–6
environmental influence 96
NTS site 112
moisture 53
Rock Valley shrubs 95–6
seed density 268–9
shade 136
soil drying rate 52
temperature 47, 48(t)
transition site 91
see also coppice mound
canopy volume, and seed density 269
carbon, and denitrification 278
carbon dioxide measurement 151–2
carbon dioxide soil level 52
carbon dioxide uptake
annuals 81, 83
C₃/C₄ species 64
CAM 64
creosote bush 72
and NAAP 103
carbon gain
creosote bush 72
influence on growth form 56
strategy 55–6
carbonates, water flow 26
carnivores 159, 208
electrolyte level control 148
population 162
salt glands 148
casebearer larvae 239
cation exchange, soil 32–3
Cenozoic 21
centipedes: see Chilopoda
cheatgrass: see Bromus tectorum
cheesebush: see Hymenoclea salsola
chemical potential energy budget 151
chiggers 249
Chihuahuan Desert 1
arachnid species diversity 229(t)
decomposition studies 286–7
precipitation 8
temperature 9
chipping sparrow: see Spizella passerina
chloroplast density 62
chlorosis, radiation-provoked 309
chollas: see Opuntia
chromosomal volume, and radiation
damage 309
chubasco 8, 9
chubascos 40
chuckwalla: see Sauromalus obesus
chunascos 40
chuparosa: see Justicia californica
cities, MDR 2–3
cladode, orientation 59
clastic rocks, water table 26
clay 29, 32, 34, 38(t)
climate 40
and BMR 142
vegetation type 88–91
see also precipitation; temperature

climatological station 44
clone 72, 76
clutch, tortoise 195–6
clutch frequency, and egg production
187–8
clutch size 187, 211–12
coachwhip: see Masticophis flagellum
coal ash deposition 311–12
cockroach: see Arenivaga erratica; A.
investigata
cold air drainage 10
collared lizard: see Crotaphytus collaris
collenchyma 61
colluvial deposits 24, 25
colonizing perennials 293
Colorado River woodland 19
coloration 136–8
common kingsnake: see Lampropeltis
getulus
community, vegetational 91–2, 95(f)
community assembly (rodents) 171–2
community composition, altered by
grazing 314
community-level root biomass 100
community litterfall 105
community occurence, environmental
control 91
community partitioning, by mass 160–1
community similarity analysis 123, 294
compaction 316
infiltration rate 305
perennial re-establishment 293
roadedge 298, 300
vehicular 304–5
competition
bunchgrasses 80
foraging behavior 169
plants 70
rodent communities 160–1
concealement sites, Larrea 252
conductivity 38(t), 39
convectional storm 8
Convergent Ecosystem program 247
cooling
broad leaves 57, 57–8
evaporative 132, 133, 140, 141, 143–4,
153
reptiles 133
coppice mound 35(f), 36, 277; see also
canopy
coprovores 214
cottonwood: see Populus fremontii
coyote: see canis latrans
CP Hills 23(f), 24, 25, 85(m)
Crassulacean acid metabolism: see
CAM
Crater Flat 85(m)
creosote bush: see Larrea tridentata
crickets 242–3; see also orthoptera
crop, annuals 129
crops estimate 124
crucifixion thorn: see Castela emoryi

crust 256
  algal 288
  altered by grazing 314
  deformation 21, 23
  modulus of rupture 302
  nitrogen fixation 275–6
cryptogamic plants 256–7
crystalline rocks, water table 26
Curlew Valley, runoff 277
cuticle 61, 62
cyanobacteria 256, 275

D-vac sampling 217(t), 218, 236, 237
  limitations 219
  spiders 230
daddy-long-legs 232
darkling beetles: see Tenebrionidae
death rate, tortoise 195
Death Valley 2–3(m), 4
  annuals 114–15
  reptiles 174
  saltbush scrub 18
  temperature range 10
decomposition 257, 258, 272
  nitrogen flux 286–7
  soil organsisms 256
defoliating insects 236–43
  Larrea population 253
  shrub proportion 240–1(t)
  weevil 233
  see also Coleoptera; Lepidoptera;
    Orthoptera
defoliation, dust 214, 308
dehydration tolerance, dromedary 145–6
denitrification 277–8
density
  annuals
    rodent population 163–4
    habitat 128(t), 129
    vehicle impact 303(t)
  arthropod 219, 220(t), 235
    on *Larrea* 252–3
    soil arthropod 222(t), 223(t)
  birds 211–13
  grasshopper 243
  lepidopteran larvae 238
  lizard 186(t), 189(t), 314
  orthopterans (detritivorous) 221
  plant height 90, 90(f)
  scorpions 230
  seasonal, treehoppers 253
  secondary production, Uta 190
  sowbug 223–4
  tenebrionid beetles 218
  treehoppers 246
  weevils 233
  see also biomass; *specific species*
density analysis, arthropods 219
density-dependent population 165, 229
density study undercounts 185
desert almond: see *Prunus fasciculata*
desert bighorn sheep: see *Ovis canadensis*

Desert Biome Program ix, xvi
desert cockroach: *see Arenivaga
    investigata*
desert cottontail: *see Sylvilagus audubonii*
desert environments 1
desert fern: *see Notholeana parryi*
desert holly: *see Atriplex hymenelytra*
desert holly scrub 18
desert horned lizard: *see Phrynosoma
    platyrhinos*
desert iguana: *see Dipsosaurus dorsalis*
desert kangaroo rat: *see Dipodomys
    deserti*
desert lavender: *see Hyptis emoryi*
desert night lizard: *see Xantusia vigilis*
desert night snake: *see Hypsiglena
    torquata deserticola*
desert oasis woodland 19
desert pavement 30, 31(f), 300
desert shrew: *see Notisorex*
desert thorn: *see Lycium andersonii*
desert toad: *see Bufo punctatus*
desert tortoise: *see Gopherus agassizii*
desert transitions 17
desert wash community 13, 19
desert willow: *see Chilopsis linearis*
desert wood rat: *see Neotoma lepida*
detritivores 214, 220–9
  beetles 220–1
  biomass 226–9
  density 220
  enclosed plot presence 227(t)
  facultative 226
diapause 138
diet, and electrolyte fluctuation 148–9
diversity curves 123
dolomite
  Banded Mountain 24
  hydrologic basin 26
  origin 21
  vegetation 85
dominance-diversity curves 123
dormancy
  spadefoot toad 146
  thermoregulatory 132
  winter 138
  see also estivation; hibernation, torpor
dosimeter, gamma radiation study xv
dove: *see Zenaida macroura*
drainage 3, 7
drainage basins, plant communities 17
drinking water 144, 145–6
dromedary: *see Camelus dromedarius*
drought 40
  rosette survival 116
  solar tracking 59
  stomatal opening 68
drought adaptation 68, 206–7
drought avoidance 67–8, 197, 207
drought-decicuous plants 67
  shrubs 73–7, 81(t), 81–2, 83
  Sonoran Desert 75

dryfall 274
dune: see sand dune
dust storms 51

ecostem disturbance persistence 290
ectotherm
  FMR 152
  thermoregulation 131, 132
ectotherms/endotherms, metabolic differ-
    ences 152
edge effect 298
egg production
  environment (lizards) 187–8
  Le Conte's thrasher 210
  precipitation 188, 189
  tortoise 195–6
electrolyte levels 148–9
elevation
  precipitation 9, 107
  vegetation distribution 85
enclosure, NTS plots xiv, xv(f)
endotherm
  FMR 152
  thermoregulation 131, 133
endotherms/ectotherms, metabolic differ-
    ences 152
energetics 130–1, 150–3
energy balance, leaf form dependence 57
energy balance maintenance 150(t)
energy budget 150–3, 159
energy content, food 152
energy development projects 311–12
energy flow
  lizards 190
  nematodes 263–4
  see also respiration
Entisols 29
environment
  BMR and FMR differences 152
  community occurence 91
environmental phenological triggers 117(f)
environmental stress 55–6
ephemerals 56, 114–15
epicuticular lipids 143
epidermis, evaporative cooling 143
equipotential flow 28(f)
ERDA (formerly AEC) ix
erosion 289
  nitrogen loss 276–7
  wind 302
erosion rate 300–1
estivation 138–9, 139, 150(t)
  daily (pocket mouse) 139
  reptilian FMR 153
  tortoise 192
Eureka quartzites 24
evaporation, respiratory surface 143–4
evaporative cooling: *see under* cooling
evapotranspiration, actual (AET) 103
evening primrose: see *Oenothera deltoides*
evergreen shrubs 70–3, 81(t), 81–2, 83

leaf orientation 58
semi-evergreen 77
excretion, energy budget 151
experimental plots: *see* plot

fallout xi–xiii, 306
fault zones 24, 25
faunal diversity, and biomass 55
feeding guild, arthropods 224(t)
fence design, NTS plots xiv, xv(f)
fencing 315
field metabolic rate (FMR) 151–2
finch: *see Carpodacus mexicanus*
fire 312, 313, 315
flash floods 301–2
flatrock areas 29
flocculants, water movement restriction 30
flocking, finches 266
flood damage 301–2
flooding, plant preferences 92
floral tissue, weevil larvae 242
floristic surveys 11
flower
    annuals 80
    biomass 127(t)
    nitrogen content 282
flower initiation 108, 109(t)
    prediction 110, 111(t)
flower inhibition, temperature 108
flower/leaf production date relationship 110
FMR (field metabolic rate) 151–2, 153, 154
fog moisture uptake 146
folding, NTS 21, 23
foliage, dust/radiation 307–8
food
    arthropod 218(t), 251–4
    breeding onset 153, 210
    dry, water stress from 144–5
    metabolizable energy content 152
    rodent 160
    water source 144
food cache, burrow 135
foraging
    ant 249–50
    birds 210–11
    lizards 182
    and moonlight 170
    quail 205
    rodent 167–71, 173
    seed reserve 169
    shrub cover 168–9
foraging behavior 153, 169
foraging efficiency 152–3
forest, plant communities 14–16(t); *see also* woodland
form, and carbon gain strategy 56
Forty-mile Canyon 23, 87
four-winged saltbush: *see Atriplex canescens*
French, N. R. xiii
Frenchman Flat
    Callisaurus densities 186

clay concentration 32
geological structures 25
lizard population studies 184
nuclear testing xii
phenological data 105
precipitation record 41
soils 29
    salinity 34
temperature inversion 89
vegetation type 85(m)
fringe-toed lizard: *see Uma scoparia*
fruit
    biomass 127(t)
    inhibition, radiative 309
    initiation 108, 109(t)
    nitrogen content 282
fungi 257, 258, 260, 261(t), 263, 287

Gambel's quail: *see Callipepla gambelii*
gamma radiation ix, 316
    animals 184, 309–11
    Brookhaven experiments xiii
    plants 308–9
    plot B experiments xiv
gas exchange analysis 103
gas pipeline, succession 294–5, 295, 297(t)
gecko: *see Coleonyx variegatus*
gelechiid larvae 238, 238–9
geological history, NTS 21–6
geological structures, NTS 25(f)
geometrid larvae 242
geomorphic landform, NTS 23(f)
germination
    annuals 80, 113–16, 129
    Lycium 74
    precipitation timing 121
    temperature 113
germination threshold 272
giant desert hairy scorpion: *see Hadrurus arizonensis*
gila monster: *see Heloderma suspectum*
gnaphosid spiders 232
golden eagle: *see Aquila chrysaetos*
gopher snake: *see Pituophis melanoleucus*
gourds, water potential 67
gradients 24
granivores
    electrolyte control 149
    metabolic water dependence 146
grass foraging (tortoise) 192
grasses 17, 19
    facultative perennials 118
    see also bunchgrasses
grasshopper mouse (southern): *see Onychomys torridus*
grasshoppers 242–3; *see also* orthoptera
gravel cover, and erosion 300–1
gravelly soil, scorpion preference 229
grazing 159, 314, 315
greasewood: *see Sarcobatus vermiculatus*
Great Basin Desert 1
    communities 87, 91

creosote bush diminishment 17
desert transition 17
MDR boundary 6
precipitation 9
shrubs 17, 89(t)
Great Basin–Mojave transition 90
Great Basin kangaroo rat: *see Dipodomys microps*
Great Basin pocket mouse: *see Perognathus parcus*
green molly: *see Kochia americana*
ground snake: *see Sonora semiannulata*
ground temperature, snake activity 133
groundwater supply, leaf-form 58
groundwater systems 26, 28
growth
    Mojave–Great Basin comparison 106–7
    and nitrogen pool 288
    precipitation enhanced 102
    reduction by radiation 308–9
    summer 106
growth competition, bunchgrasses 80
growth rate, tortoise 194–5
growth rings 71–2
gular flutter 144, 154, 203
gymnosperms, aphyllous 77
gypsophilous species 19
gypsum 19

habitat
    classification, arthropod 215
    paritioning, birds 210–11
    water potential variation 69
    see also microhabitat
Halfpint Range 23(f), 24, 85(m)
halophytic saltbush scrub 18
hardpan
    calcrete soils 29
    high bajadas 17
    root penetration 98
Harrisburg revegetation 292–3
harvester ants (graniverous) 144, 249–50, 255, 266
harvestman 232
heat 39, 131–2
    metabolic 131, 133
heat balance 130–2
heat budget 150
heat dissipation, hyperthermic 140
heat exchange, countercurrent 140–1
heat product, density specific 39(t)
heat reflection, trichomes 59
heat stress avoidance
    bird 203
    mammals 159–60
    quail 205
height, plant 88, 89(t), 90
herb biomass 124–5, 125, 129
    and rodent reporduction 166
herbivores 105
    arthropod 214

herbivores (*cont.*)
  electrolyte control 149
  perennial plant comsumption 144–5
herbs, Rock Valley 120; *see also* annuals
herpetofauna: *see* reptiles
hibernation 138–9, 150(t)
  tortoise 192, 193
  see also estivation; torpor
home range
  Dipodomys dorsalis 184
  male/female differences 171
  rodent 170–1
  tortoise 193
homeostasis control: *see* electrolyte; temperature control
honeydew 243, 247, 250, 251
hopsage: *see Grayia spinosa*
horned lark: *see Eremophila alpestris*
horse: *see Equus caballus*
horsebrush: *see Tetradymia axillaris*
house finch: *see Carpodacus mexicanus*
huddling 139, 150(t)
humidity 53–4
  burrow 135
  and precipitation 48–9
  Rock Valley 47–9
  temperature 48–9
hummocks: *see* coppice mounds
hurricane 8
hydrologic pattern 301
hydrologic year 102–3
hydrology 26–8
hyperthermia 140, 153, 160
hypothermia 138

Ibis Mountain camps 293
IBP (International Biological Program)
  desert sites ix, xvi
  perennial densities 92–3, 94(t)
  precipitation records 41–2
  soil profile 36–9
  species composition 92–105
infiltration rate 305
infra-red reflection, trichomes 59
insects
  flying, thermoregulation 132
  metabolizable energy content 152
  nest thermoregulation 132
  as water source 160
International Biological Program: *see* IBP
intershrub space
  foraging (rodent) 169
  foraging pattern 170
  lizard foraging 182
  nitrogen pool 279–80
  seed density 169(t), 268, 269(t), 270(t)
  seed distribution 169
introduced species 119(t), 120, 291, 312–14
invertebrates
  life-history strategy 215
  microhabitat 146
  nitrogen content 285(t)

Inyo Mountains 6, 11, 20
iodinebush: *see Allenrolfea occidentalis*
iodinebush alkali scrub 18
ion distribution, soil 31–3, 34
ion exchange 32–3, 36(t)
ionizing radiation: *see* beta *and* gamma
  radiation
Iron Mountain camps 293
ironwood: *see Olnea tesota*
irradiation 49, 58–9
'islands of fertility' 274
isopods 222–4

Jackass Flats 23, 24
  ammonium pool 281
  ash deposition 312
  geological structures 25
  soils 29, 32(t)
  species density 93
  vegetation type 85(m), 87–8
  winter annual survivorship 121
jojoba: *see Simmondsia chinensis*
Joshua tree: *see Yucca brevifolia*
Joshua Tree National Monument, reptiles
  174
Joshua tree woodland
  plant associates 17, 18(f)
  succession 294
juniper: *see Juniperus* spp.

kangaroo rat: *see Dipodomys* spp.
Kelso Dunes 6
kidney efficiency 148
kit fox: *see Vulpes macrotis*
Kranz anatomy 63, 81

lactation, rodent 165
lagomorphs 159
lake, dry: *see* playa
lake, temporary 26
Lake Manly 7
landfill, coal ash 311–12
landforms 3–4
larvae
  lepidopteran 237–9, 239–40(t), 242
  as water source 144
Las Vegas, irradiance 49(t)
lateral roots 69, 98, 99
Le Conte's thrasher: *see Toxostoma*
  lecontei
leaching, and germination 113
leaf
  absorptance 58–9
  anatomy 61–2, 63
  biomass 127(t)
  broad
    overheating 82
    water stress adaptation 58
  dimorphic 57
  evergreen 61
  effect of fallout 307–8
  narrow, advantages 57

nitrogen content 282
non-succulent 70, 71
orientation 58–9, 82
polymorphism 82
reflectance 58–9, 73
resins 61
solar tracking 59
structure, blackbrush 76
temperature, and size 57
water-loss minimization 58, 82
  as water source 144
water vapor concentration 57
winter annuals 80
see also drought-deciduous plants
leaf area index 75
leaf beetles 242; *see also* Coleoptera
leaf initiation 108, 109(t)
  prediction 110, 111(t)
leaf succulents 78–9, 81(t), 81–2
leaf/flower production date 110
leaf/stem photosynthesis 75
leafhoppers 247, 253
legumes 77, 276; *see also Acacia; Prosopis*
lichens 256–7, 272, 288
  distribution 258(t)
  nitrogen fixation 275
life-span
  arthropods 215
  ionizing radiation 310
light, and breeding onset 210; *see also*
  moonlight
lime 32(t), 38(t)
limestone 21, 24, 26
limestone mountain plants 70
limestone soil 19, 33(t), 36, 38; *see also* calcareous soils
limestone vegetation 85
line transect sampling 88
lipids, epicuticular 143
litter, decomposition process 287
litter habitat, arthropod 214–15
litter nitrogen 284
litter traps 104
litterfall 104–5
  community 105
  nitrogen flux 286
  seasonal patterns 104
livestock, and introduced species 312
livestock grazing 314, 315(t)
lizard 196
  activity patterns 185–6
  arboreal 136, 182–3
  and brome grass introduction 313
  burrows 134, 135
  characteristics 182–4
  clutch frequency 187–8
  coloring changes 137
  countercurrent heat exchange 141
  densities 185, 186(t)
  dry weights 185, 186(t)
  egg production and environment 187–8
  energy flow 190

FMR and environment 152
foraging efficiency 152
habitats 182–4
huddling 139
hyperthermia 140
juvenile/adult coloration 137
mortality 190
natural history 182
nitrogen content 284, 285(t)
panting 144
predation 183, 187
productivity, and precipitation 188–90
radiation xvi, 184, 310–11
respiratory surface evaporation 143
salt glands 148
size relationships 187
spring density 189
survivorship 188
thermoregulation 133, 135–6
vehicle effect (dunes) 306
lizard population studies 184
effect of grazing 314
enclosed 185–6
lizard density determination 185
radiation effect xvi, 307, 310–11
sampling 184–5
undercounts 185
lizard/substrate reflectance 137
loess, NTS 32
loggerhead shrike: see Lanius ludovicianus
long shoot–short shoot organization 74
long-nosed leopard lizard: see Gambelia
    wislizenii
long-nosed snake: see Rhinocheilus lecontei
Lygaeid bugs: see Lygaeidae
lymph sac water storage 146
lyre snake: see Trimorphodon biscutatus

malate, in CAM 64
male/female differences, home range 171
mammals
    foraging efficiency 152
    lethal temperatures 130, 131(f); see also
        endotherms
    population densities 156(t), 162–7
    species, NTS 155, 156(t), 157–8(f), 159
    water loss predictions 141–2
mass
    arthropod 219
    basking 135
    clutch size 187
        tortoise 196
    community assembly 171–2
    ecological niche (rodents) 160–1
    heat exchange 132
    population density (lizards) 187
    resource partitioning (rodents) 160–1
    rodent 172(t)
    seed 271
    snakes 191
    soil arthropods 225–6(t)
    tortoise 194

water flux 141–2, 143
weevils 233
MDR (Mojave Desert Region)
    arachnid species diversity 229(t)
    area 1, 3
    boundaries 4–6
    drainage 3
    lakes 7
    lizards 174
    map 2–3
    modelling 272
    precipitation 8–9, 10(f)
    rivers 7
    shrub data 89(t), 90(t)
    situation 1–3
    southern, perennials 17
    topography 6–7
    as transitional community 19
MDR/Sonoran Desert, faunal similarity 12
mealybugs 245–6, 253
mean precipitation 42
mean volume (shrub) 97
measuring worm 253
melyrid beetles 247, 248–9
Mercury Valley
    ammonium pool 281
    phenological data 105
    soil composition 32(t)
    species densities 92–3
    vegetation type 85(m)
Merriams' kangaroo rat: see Dipodomys
    merriami
Mesic family soil 28
mesophyll 61, 62, 73–4
Mesozoic, NTS 21
mesquite: see Prosopis spp.
mesquite woodland, denitrification 278
metabolic rate 142, 151–2; see also BMR;
    FMR
metabolic water: see water, metabolic
metabolism, components 151
metabolism control, thermoregulatory 131,
    131–2
microarthropods
    decomposition process 287
    density 265(t)
    sampling 260
    taxonomic listing 264
microclimate 153
microhabitat 127–9, 136
micronutrients
    coppice mound 36(t)
    cycling 34
    saline soil 33–4
    soil 32(t), 33(t)
microphylly 57
millipedes 222, 224, 226
mineralization 258, 287, 288, 289
Miocene, faulting 24
mirid bugs 216, 247
mites 232–3, 249, 264
    debris fragmentation 258

decomposition process 287
on Larrea 253
mockingbird: see Mimus polyglottos
moisture
    denitrification 278
    facultative life cycle 118
    gravimetric measurement 51
    microorganism population 262
    nematode activity 263
    see also water
moisture depletion (soil) 301
moisture dynamics, and depth 52
moisture potential 52
moisture recharge 301, 302(t)
Mojave Desert Region: see MDR
Mojave indigo bush: see Psorothamnus
    arborescens
Mojave River 7
Mojave site monitoring (Beatley) 88–91
Monotony Valley tuff 25, 26(t)
moonlight, foraging behavior 135, 170, 171
mormon tea: see Ephedra nevadensis
mortality, lizard 190
moths 144, 247
motorcycle, compaction 304–5
mountain lion: see Felis concolor
mountain systems, MDR 6–7
mourning dove: see Zenaida macroura
mule deer: see Odocoileus hemionus
multivoltine life history 215
muscle vibration, thermoregulatory 132

NAAP (net annual aboveground
    production) 102(t), 105, 112, 283(t)
NAAP equations 102–3
nasal drinking 145, 192–3
nasal mucosa, thermoregulation 143
nasal water resorption 147
natural springs, plant communities 19
nectar 152, 247
Negev Desert ix
nematodes 262–4, 273
    decomposition process 287
    energy consumption 263–4
    prey of mites 233
    sampling 260
    seasonal distribution 263
    spatial distribution 262
nest thermoregulation 132
nesting production 211–12
nesting sites 210–11
net annual aboveground production: see
    NAAP
net primary production
    annuals 113, 123–7, 129
    and ant density 249
    data estimation, perennials 101
    dry mass, perennials 102(t)
    loss 105
    NTS 112
    perennials 100–2
    precipitation correlation 127

net primary productions (*cont.*)
    predictive equations 125
    shrubs, arthropod abundance 236
    winter annuals 120(t), 123–7
    see also NAAP; production
neuropterans, predatory 247
Nevada Test Site: *see* NTS
night lizard: *see Xantusia vigilis*
night snake: *see Hypsiglena torquata*
nitrate, playa soil 34
nitrate input 274–5
nitrate nitrogen 280–1
nitrogen
    from annuals 113
    atmospheric 274–5
    coppice mounds 277
    cycling 272
    distribution 274
    erosional loss 289
    flux 285–7, 289
    input estimates (fixation) 275
    intake, ant consumption 206
    mineralization 258, 287, 288
    termite activity 258–9
    translocation 285
    uptake 285–6, 289
    wetfall 274
    see also ammonium; denitrification;
        nitrate
nitrogen cycle 287–8, 289
nitrogen fertilizer 315
nitrogen fixation 275–6
nitrogen pool 278–85, 286(t), 289
    animal 284–5
    beneath shrubs 279(t), 284
    equilibrium 288
    intershrub spaces 279–80
    plant 281–4
    Rock Valley 279–80
    soil nitrogen 278–9
    vertical gradient 278
nocturnal activity pattern 133, 135, 170,
    171
NTS (Nevada Test Site) iv
    dominant shrubs 112
    geological history 21–6
    history ix
    1950s studies xi-xii
    plot establishment xiii-xiv
    reptiles 174–81
    topography 21–6
nuclear testing
    aboveground 306–7
    atmospheric xi
    NTS areas xii
    site choice xi
    see also beta *and* gamma radiation
nutrient cycling 258
Nuttall's cottontail: *see Sylvilagus nuttallii*
nymphal development, treehoppers 246

OAA (oxaloacetate) 64

ochric epipedons 29
off-road vehicles: *see* vehicles
open space: *see* intershrub space
Operation Jangle xii
Ord's kangaroo rat: *see Dipodomys ordii*
ordination techniques 91
organic acid synthesis 73
organic matter
    IBP validation site 38(t)
    arthropod concentration 235
    soil 32(t), 34
oribatid mites 287
orientation, thermoregulatory 131, 135
orthopteran biomass 221–2, 228(t)
osmoregulation 68, 147–50
osmotic concentration tolerance 147
osmotic potential lowering 68
ovarian tissue loss 310
overwintering, treehoppers 246, 247
oviposition, on *Larrea* 251–2
oxaloacetate (OAA) 64
oxidative water: *see* water, metabolic

packrats: *see Neotoma*
Pahute Mesa 23(f), 24
    groundwater system 26, 28
    phenological data 106, 107
    vegetation type 85(m)
Paintbrush tuff 25, 26(t)
Paleozoic, NTS 21
palisade cells 61–2, 73
pallid bat: *see Antrozous pallidus*
palo verde: *see Cercidium floridum*
pan, cemented 39
Panamint rattlesnake: *see Crotalus mitch-
    elli stephensi*
panting 143–4
PAR 59, 72
parasites, arthropods 247–9
passarine birds 197, 203, 207; *see also
    Amphispiza; Carpodacus*
pavement disturbance 300
pedogenic calcretes 29
PEP (phosphoenolpyruvate) 64
perennial production: *see* net primary
    production
perennials
    arthropod food resource 215
    carbon gain 55–6
    colonizing 293
    consumption by herbivores 144–5
    facultative 118
    Harrisburg 292–3
    IBP site 92–105
    mean size 96
    nitrogen concentration 281–3
    nitrogen pool increase 288
    NTS 84–92
        map 85
    re-establishment 290–2, 293, 296, 297(t)
    Rock Valley 119(t)
    southern Mojave Desert 17

Wahomonie 290–2, 292(t)
water storage 67
water stress avoiders 56
permeability, soil 38(t)
PET (potential evapotranspiration) 52–3,
    54
PGA (phosphoglycerate) 63
pH
    ammonia volatilization 278
    ash landfill 311
    changes in CAM plants 64–5
    plant preferences 92
    saline soils 34(t)
    soil 32(t), 33, 38(t)
    vegetation 36
phenology
    annual plants (Beatley) 116, 117
    coloring differences 137
    quantitative analysis 108–12
    shrub 105–12
    triggers 117(f)
    Turner and Randell model 110,
        111–12
    vegetative/reproductive 106(f), 107(f)
phenophase, treehopper 246
phenophase identification 105–6
phenophase initiation 108, 110
phloem, primary 61
phloem-feeding 243
phosphoenolpyruvate (PEP) 64
phosphoglycerate (PGA) 63
photoperiod change, diapause 138
photorespiration 65
photosynthesis
    angiosperms 81(t), 81–2
    C_3 63
    C_4 63–4
    cheesebush 75
    drought-deciduous plants 75
    ionizing radiation effect 308–9
    maximisation, and leaf anatomy 62
    restricted, low temperature 77
    stem 75, 77–8
    temperature optima 81, 82; *see also*
        acclimation
    temperature tolerant 65
    thermal stress 56
    transpiration 66–7
    Yucca brevifolia 79
    see also CAM; C_3 metabolism; C_4
        metabolism
photosynthetic symbiots 275
phreatophytes 19, 69
phytophages 220, 251–4
    orthoptera 221
    soil arthropods 233, 234(t)
pickleweed: *see Allenrolfea occidentalis*
Pima rhatany: *see Krameria eracta*
pinyon pine: *see Pinus monophylla*
pinyon–juniper woodland
    ants 250
    bird species diversity 197

lizard population 174
mammal species 159
pioneer species 291, 312; *see also* introduced species
pitfall traps 191, 216, 222, 224, 230
plant community
    classifications 13, 14–16(t)
    similarity analysis 123
plant densities, rodent population 163–4
plant growth, nitrogen pool 288
plant population soil seed reserve 271(t), 272(t)
plasma electrolyte 148–9
plastron length 194–5
platyopuntias: *see Opuntia*
playa 4
    clay 34
    saline soils 33–4
    silt 34
    soils 29, 33–4
    vegetation 18–19, 87, 88(t)
    water-filled 5(f)
playa lakes 26
pleistocene, drainage 7
plot, experimental xiii–xiv, 37(f)
plowed field succession 294
pocket gopher: *see Thomomys umbrinus*
pocket mouse: *see Perognathus longimembris*
pollinators 214, 247
population
    arthropods 219, 220(t)
    body size (lizards) 187
    creosote bush 72
    density-dependent, scorpion 229
    rodent 163(t), 165, 172
    seasonal, on *Larrea* 252–3
    spiders 232
    tortoise 194
    Uta 188–9
    winter annuals 122–3
    see also density
population studies, and predation 166
potassium excretion 148
potential evapotranspiration 52–3, 54
power transmission line, succession 295
prairie falcon: *see Falco mexicanus*
precambrian, NTS 21
precipitation 7–9, 18, 40
    aboveground net production 125
    AET 103
    bajada 13
    differential growth characteristics 103
    elevation 9, 40
    fall (autumn), shrub arthropod index 236
    fall (autumn), arthropod abundance 239(f)
    germination of annuals (Beatley) 116, 117
    growth enhancement measurement 97–8
    growth renewal 102
    humidity 48–9
    intensity 42, 43–6(t)

litter disappearance 287
lizard density 189(t), 190
lizard egg production 187, 188, 189
long-term records 10(f), 40–1, 41–2
MDS 20
mealybug nutrition 246
millipede activity 224
net annual production correlation 127
nitrogen loss 276, 277
nitrogen wetfall 274–5
pattern 8, 42, 43–6(t), 53
    in phenology model 110, 112, 117(f)
    temperature correlation 111
plant height 90
revegetation 316
scrub composition 90(t)
seasonal 41–2
    Rock Valley 41–2, 43–6(t)
    summer 8, 9, 113
    winter 8, 9, 187, 195
simulation experiments 300
site species composition 88–9, 90(t)
species intolerance 91
summer growth 106
temperature 47, 111
tortoise growth rate 195
transition sites 91
variability 9, 10(f)
    see also productivity
precipitation gradient, site elevation 107
precipitation timing
    germination, annuals 113, 115, 116
    survivorship 121
    community structure 123
predation
    beetles 248–9
    birds 203, 208
    black coloration 137
    insect 247–9
    of lizards 183
    population density studies 166
predatory arthropods 229–33
    biomass 233
    density 220
    on *Larrea* 252
    see also Chilopoda; mites; Phalangida; scorpions; Solpugida; spiders
preformed water 144, 149, 160
primary production: *see* net primary production
production, secondary
    birds 211–13
    lizards 190–1
    mammalian 172
productivity, and precipitation
    lizard 188–90
    plant 265
profile: see soil profile
pronghorn antelope *see Antilopcapra americana*
Project Sedan 307, 308

pupfish, desert: *see Cyprinodon*
pylons, succession 295

quail: *see Callipepla gambelii*
quartzites 24
Quaternary deposits 24

radiation xiii, 306–11, 316
    effect on lizards xvi, 309–11
    human exposure xi
    long-term studies xiii
    plant damage 308–9
    toxicity levels 306
rain shadow 7–8; *see also* precipitation
rain-year 53
Rainer Mesa 23(f), 24, 85(m)
rainfall intensity: *see* mean precipitation intensity
ratany: *see Krameria*
rattlesnakes 181
raven: *see Corvus corax*
recharge, soil water 28(f), 39
reflectance 58–9, 137
relative abundance index 216, 224(t)
relative humidity: *see* humidity
reproduction, density-dependent 165
reproduction index 165–6, 173
reproductive energy requirements 153
reproductive success, pocket mice 165
reptiles
    heat gain 133
    lethal temperatures 130
    metabolic water 147
    NTS 174–81
    torpor 138
    water drinking 145
    water loss predictions 141–2
    see also ectotherms
residual soils 29
resin 61, 70, 72
resource competition 161
resource partitioning 249–50
respiration 220, 263–4; *see also* energy
respiratory resorption of water 147
respiratory surface evaporation 143–4, 154
revegetation
    colonizing perennials 293
    compaction 292
    military camps 293
    NTS 315–16
    species changes 290–2, 292–3
    utility corridors 294–6
    see also succession
rivers, MDR 3
road edge vegetation 298–300
roadrunner: *see Geococcyx californicus*
rock
    cricetid rodent 167
    lichen 257
    sorting 29
    undercoating 30
rock crevice, thermoregulatory 136

Rock Valley iv
　algal crust 276
　ammonia volatilization 278
　calcrete layer 279
　choice xiii
　denitirification 278
　geological structure 25–6
　history of studies ix-x
　irradiance 49
　lizard population 184
　mammals 156(t), 162–7
　nitrate/ammonium nitrogen 280–1
　nitrogen fixation estimates 276
　nitrogen losses 277
　nitrogen pool size 279–80
　precipitation events 41–4, 45(t), 46(t)
　research (1976-　) xvii
　soils 29
　　composition 32(t)
　　profile 36–9
　topography 22(m)
　as validation site xvi
　vegetation type 85(m)
　wind 49–51
　see also IBP validation site
rock wren: see Salpinctes obsoletus
rocky terrain 182, 184
rodent
　burrows 134
　cricetid
　　diet 266
　　population densities 166–7
　density fluctuation 164
　density manipulation 165
　distribution pattern 155
　dry body weight 172(t)
　food preferences 160
　foraging interactions 169–70
　foraging, and moonlight 135
　heteromyid
　　densities 162–4
　　foraging 167–71
　　nitrogen content 284, 285(t)
　　population fluctuation 155
　　seed reserves 266
　　survey methods 162
　home range 170–1
　mass, and community membership 160
　nests (woodrats) 136
　nocturnal activity patterns 133, 135
　population density 172
　population fluctuation 155, 163(t), 164, 265–6
　as prey 160
　respiratory water resorption 147
　seasonal activity 170–1
　shrub utilisation 168–9
　vehicles 303–4
rodent community (guild) 171–2
rodent competition 161
rodent diversity (dunes) 306
rodent reproduction 164–6, 265–6

root
　biomass 100, 127(t)
　nitrogen content 281, 282(t)
　mass estimation 98–9
root parasites 75, 233
root/shoot ratio 69, 98–100
　annuals 125, 127
　and nitrogen vertical gradient 278
　NTS shrubs 112
root system 69–70, 98–100
　aphyllous gymnosperms 77
　blackbrush 76
　Tidestromia oblongifolia 118
rosette 80, 116, 114(f), 129, 193
round-tailed ground squirrel: see Spermo-
　　philus tereticaudus
Rubisco 63
running
　Dipsosaurus dorsalis 184
　quail 205
　rodents 169
runoff 276, 277
　nitrogen loss 276–7
　NTS 30
runoff rate 301, 302(t)
rupture modulus, soil 302
Russian thistle: see Salsola

sage sparrow: see Amiphispiza belli
sagebrush: see Artemisia tridentata
sagebrush lizard: see Sceloporus graciosus
sagebush vole: see Lagurus curtatus
saguaro, giant: see Carnegiea gigantea
saline soil 33–4, 70
saline-sodic desert pavement 30
salivary gland water uptake 146
salt balance adaptation 150(t)
salt concentration 18, 34
salt glands 142(t), 148, 207
salt ingestion avoidance 149
salt tolerance 18, 73
saltbush scrub 18
saltcedar: see Tamarix ramosissima
salts 4, 19
sampling
　arthropods 216, 218
　　limitations 219
　bird studies 207–8
　soil microflora 259–60
sand
　bajadas soil composition 32(t)
　calcrete overlay 29–30
　eolian 25–6
　IBP validation site 38(t)
　scorpion preference 229
　snake habitat 191
　vegetation 87
sand dune 160, 306
sand verbena: see Abronia villosa
sandpaper bush: see Mortonia utahensis;
　　M. scabrella

sap-feeding insects 243, 253–4; see also
　　Hemiptera; leafhoppers; mealybugs;
　　Thysanoptera; treehoppers
scarab beetles 221
scorpions 229–30
　habitat 215
　moonlight foraging 135
　seasonal activity 225(t), 229–30
　water loss 143
screwbean: see Prosopis pubescens
scrub
　CO_2 soil level 52
　NTS 84, 86
　plant communities 14–16(t)
scrub oak: see Quercus
scute annuli 195
seasonal activity
　lizards 136
　scorpions 225(t), 229–30
　solpugids 230
secondary growth production 100–1
secondary plant succession 290–302, 316;
　　see also revegetation
seed
　arthropod food 214
　biomass 271–2
　demography 272
　density 169, 269–71
　distribution 169–70, 266–7
　dormancy 272
　germination threshold 272
　mass 271
　metabolizable energy content 152
　predation, ants 249–50
　quail diet 206
　radiation 308–9
　rodent community structure 171
　tortoise comsumption 193
seed reserves 264–72, 273
　annual variation 269–71
　dynamics 266–7
　persistence 272
　population density 266, 271(t), 272(t)
　rodent foraging 169
　sampling 266–7
　shrub canopy volume 269
　uses 265–6
seepage zone 26; see also springs
shade 136
　heat dissipation 140
　lizards 182
　quail 205
　rabbits 159
shadscale: see Atriplex confertifolia
shadscale scrub 18
sheep 313, 314, 315(t)
Sheep Range 3–4(m), 6–7
sheet flow 39, 277
sheetwash, NTS 30
shovel-nosed snake: see Chionactis occipi-
　　talis occipitalis
shrew, desert: see Notisorex

shrub
  ammonium pool 280–1
  arthropods 218, 235, 236, 237(t), 238(t)
  ash deposition 312
  Coleoptera habitat 242
  defoliating insects 239–42(t)
  early/late growth 108
  frequency distribution 93–5
  height 88, 89(t)
  lepidopteran larvae densities 238,
    240–1(t)
  litterfall 104–5
  mean volume 97
  Mojave Desert sites 88
  nematode population 260, 263
  net primary production 100–2
  new growth nitrogen 283
  nitrate pool 280
  nitrogen content 274, 281–2, 282(t)
  NTS 112
  protective (rodents) 168
  seed density 169(t)
  soil nitrogen concentration 279(t),
    279–80
  species variety 89(t)
  tortoise consumption 194
  see also aphyllous shrubs; drought-
    deciduous plants; evergreen shrubs
shrub canopy: *see* canopy
shrub arthropod index 236
shrub cover 90
  Coleogyne occurence 87
  Dipodomys distribution 168(f)
  sheep grazing 314
  Uta density 185
side-blotched lizard: *see Uta*
    stansburiana
sidewinder: *see Crotalus cerastes*
silica rock undercoating 30, 31(f)
silicate deposits 21
silk-tassel: *see Garrya flavescens* var.
    pallida
silt 29, 32(t), 34
Silverbell (Arizona) site studies 251–2
similarity coefficient 294
site elevation, precipitation and tempera-
    ture gradients 107
size 132
size grading, ants 251
skin, evaporative cooling 143
Skull Mountain 23(f), 24, 85(m)
Small Boy 308
smoke tree: *see Dalea spinosus; Psoro-*
    thamnus spinosus
snake 191–2, 196
  activity pattern 133, 135–6, 191
  excretion 148
  huddling 139
  Mojave species 12
  nitrogen content 284, 285(t)
  NTS 181
snout-to-ventral length: *see* SVL

snow 9
Soda Lake 5(f), 7
sodic desert pavement 30
sodium excretion 148
soil
  arthropod habitat 214–15
  biologically active 279
  biomass, nematodes 263
  biotas 258–9
  carbonates 29–30
  denning potential 175
  drying rate, soil canopy 52
  environment 256
  erosion rate variables 300
  horizons, and vegetation 34, 36
  hydrologic pattern 301
  ion distribution 31–3
  microclimate 51–2
  micronutrients 32(t), 33(t), 33–4
  nitrogen 274, 278–9, 280
    uptake 285–6
  particle size 31–3
  physical attributes 39(t)
  recovery times 292
  restoration 315
  Rock Valley site 54 profile 33(t)
  rodent preference 162
  root system development 98–100
  salinity 33–4
  sorting 29
  structure (NTS) 30
  temperature, burrow profile 135
  tortoise burrows 175
  types 28–9
  vegetation type 91–2
  weed colonization 313
soil compaction 316
  perennial re-establishment 293
  roadedge 298, 300
  townsite 291, 292
  vehicular 304–5
soil crevice habitat 215
soil crusts 256, 275–6, 314
soil depth
  microarthropods 264
  microorganisms 261–2
  moisture dynamics 52
soil formation 34–6
soil loss equation 277
soil microoorganisms 257–64
  Actinomycetes 260, 261, 262
  bacteria 261(t)
  decomposition 257, 258
  fungi 257, 260, 261(t), 263
  invertebrates 257–8
  microarthropod 260, 264, 265(t)
  microflora 260–2
  nematodes 260, 262–4
  nutrient cycling 258
  sampling 259–60
  spatial distribution 260–1
  trophic relationships 258, 259(f)

soil moisture
  facultative life cycle 118
  grasshopper abundance 243
  gravimetric measurement 51
soil preference, scorpion 229
soil profile 36–9
soil profile development, and vegetation
  34
soil water 39–40, 52–3
soil water potential 39–40
solar heating, adaptive coloration 137
solar irradiance: *see* irradiance
solar power stations 311
solar tracking 59, 82
Sonoran Desert 1, 3
  aphyllous shrubs 77
  arachnid species diversity 229(t)
  creosote bush scrub 13
  faunal similarity 12
  precipitation 8
  temperature 9
  see also Silverbell
sorting, soil particles 29
sowbug: *see Venezillo arizonicus*
spadefoot toad: *see Scaphiopus*
sparrow: *see Amphispiza bileneata*
species, introduced 119(t), 120, 291,
  312–14
species composition, IBP 92–105
species differences, revegetation 290–2,
  292–3
species diversity 123
speckled rattlesnake: *see Crotalus*
    mitchellii
Specter Range 23(f), 25–6, 85(m)
spiders: 230, 232, 247; *see also* Araneida
spine formation 74
spiny desert lizard: *see Sceloporus magister*
spiny senna: *see Senna armata*
spiracles 144
spotted leaf-nosed snake: *see*
    Phyllorhynchus decurtatus
Spotted Range 23(f), 24, 85(m)
Spring Mountains 3–4(m), 6–7
springs 26
standing crop 124, 129
stem
  biomass 127(t)
  nitrogen content 281, 282(t)
  photosynthetic 59, 75, 77–8; *see also*
    cladode
stem succulents 78–9, 81(t), 81–2
sterility, radiation-induced xvi, 310, 311
stomata 61, 66–8
stones, crevice habitat 215
storksbill: *see Erodium cicutarium*
storm 8, 20, 40, 41
stormworm: *see Orthoporus ornatus*
subshrub 73
succession 290–302, 316
  pioneers 291, 312
  plowed fields 294

succession (*cont.*)
  utility corridors 294–6
  see also revegetation
succulents 78–9, 83
  physiological characteristics 81(t), 81–2
  re-establishment 294
  see also Opuntia
suffrutescents (subshrubs) 73
sugar 247
sulfates (sulphates), soil 33, 34
summer annuals: *see* annuals
sunning: *see* basking
sunspiders 230
superlight coloration 137
surface area, heat exchange 132
surface temperature, desert 130
surface water, permanent 26
survey methods, heteromyid rodents 162
survivorship 116, 118, 120–1, 129
SVL (snout-to-ventral length) 187, 191(t)
swallows 203
sweat glands 143
sweating 143
sweet bush: *see Bebbia juncea*
swifts 203

tap roots 69, 98, 99
tarantula 232
tarsonemid mites 287
temperature 9–10
  arthropod activity 215
  angiosperm optimum 81(t)
  burrow profile 135
  day/night, scrub composition 90(t)
  denitrification 278
  desert surface 130
  diapause 138
  germination of annuals 115
  grasshopper abundance 243
  Great Basin 91
  humidity correlation 48–9
  lizard 182
  low 44, 46(t), 47
    flowering inhibition 108
    nocturnal, and *Lycium* 17
    restricted photosynthesis 77
    species presence 90–1
    vegetation communities 87
  max/min 10, 44, 46(t), 47
  maximum, NTS 91
  millipede activity 224
  minimum, basins 89–90
  Mojave Desert 44, 47(t)
  nematode population 263
  phenology 110, 111, 112
  photosynthetic adaptation 56
  photosynthetic C_4 pathway 65
  photosynthetic optima 81, 82–3
  precipitation 47, 111
  rosette survival 116
  shrub canopy modulation 47, 48(t)
  site species composition 88–9, 90(t)

torpor (pocket mouse) 139
  unpredictability 44, 46(t)
  winter, and shrub arthropod index 236
  see also acclimation; water stress
temperature gradient, site elevation 107
temperature inversion 10, 89–90
temperature tolerance, birds 197
tenebrionid beetles: *see* Tenebrionidae
termites 182, 249, 258–9, 287
territoriality, ants 250
test ban treaty xii
thermal conductivity 39(t)
thermal tolerance 184, 191
Thermic family soil 28
thermoregulation 133
  during hibernation 138
  ectotherms 132
  endotherms 133
  quail 206
  torpor 138–9
  zebra-tailed lizard 182
  see also burrows; coloration
Thirsty Canyon tuff 25, 26(t)
thrips: *see* Thysanoptera
thrust faults 24
ticks 249
Timber Mountain 23(f), 24
  geological structures 25
  vegetation type 85(m)
  volcanic centre development 27(f)
time budget 151, 205(f)
toad: *see Bufo punctatus; Scaphiopus*
toe clipping 184–5
topography, NTS 21–6
Topopah Valley 85(m)
torpor 150(t), 153, 160, 170
  circannual rhythm 138–9
  winter 138
  see also estivation; hibernation
Townsend's big-eared bat: *see Plecotus townsendii*
Townsend's ground squirrel: *see Spermophilus townsendii*
transects 162, 185
transient heat models 39
transition desert 90
  canopy coverage 91
  communities 87–8
  lizard population 174
  Mojave as 19
  orthoptera 243
  precipitation 91
  shrub data 89(t), 90(t)
transit speeds: *see* running
transpiration
  broad leaves 57, 57–8
  CAM 65, 79
  narrow leaves 57
  photosynthesis 66–7
  trichomes 59
  water loss 82
  water potential 68

trapping
  rodents 162
  snakes 191
  reliability of 167, 170
treehoppers 246–7, 253; *see also* Membracidae
trees, breeding habitat 209
trenching 294, 296
triangle-leaved bursage: *see Ambrosia deltoidea*
trichomes 59, 72, 73, 82
trophic specialization 214–15
tuffs 25, 26, 27(f)
tuft, annuals 80
turgor maintenance 68
Turner, F. B. x, xvi-xvii
turpentine bush: *see Thamnosma montana*
tydeid mites 233

undercount compensation 185
univoltine life-history 215
uplifting, NTS 21, 23
upper respiratory disease syndrome (URDS) 195
uric acid 148
urinary bladder 145, 146, 148, 149, 193
urine
  concentration 148
  quail 206–7
  rodent granivores 149, 160
utility corridors 294–6

vapor phase transport 40
vapor pressure deficit 59
varnish (rock) 31(f)
vascular plants 11–12
vegetation
  erosion reduction 301
  Great Basin Desert 87
  lizard consumption 182
  rodent requirement 165, 166
  soil formation 34–6
  soil profile 34, 39
  soil water potential 301
  water intake (tortoise) 192
vegetation association 8(t), 91–2, 95(f)
  transition desert 87–8
vegetation patterns 13, 17, 18–19
vegetation type
  climate 88–91
  NTS 85(m)
  soil 91–2
vegetative rosette: *see* rosette
vehicle damaging effect 302–6, 316
velvet ash: *see Fraxinus velutina* var. *coriacea*
venomous gila monster: *see Heloderma suspectum*
vertebrates 12, 285(t); *see also specific species*
volcanic centre development 27(f)
volcanic formations 21, 25, 26(t)

VPD (vapor pressure deficit) 59

Wahomonie revegetation 290–2
washes
  Lycium spp. 73
  plant community 13
  vegetation 86
  water potential 67, 69
wasps 248
water
  drinking 144, 145–6
  fog 146
  free 133
  labelled 151–2
  lack in Mojave Desert 40
  metabolic 146–7, 154
    kangaroo rat 160
    passarine birds 207
    quail 206
    rodent granivores 149
  NTS 39–40
  osmotically active 68
  oxidative 142(t)
  plant species competition 70
  preformed 144, 149, 160
  and soil nitrogen fixation 275
  sources 144
  as spatially variable resource 83
  see also groundwater system; hydrologic
    basin; water table
water availability
  plants 83
  termite activity 258–9
water balance 131, 141–7, 149–50
water conservation
  CAM 79
  leaf 56, 62–3
  see also transpiration
water economy index (WEI) 142–3
water flow system 26, 28
water flux 141–2, 143
water gains (animals) 142(t)
water loss
  animals 142(t)
  excretory/fecal 148
  physiological effects 141
  spiracular 144
  transpiration 82
  control 150(t), 153–4
  limitation strategies 82, 147
    cuticle 61, 62
    torpor 139
  see also evaporation; salt glands; urine
water movement 29–30
water permeability, NTS soils 30

water potential 67
  angiosperms 81(t)
  blackbrush 76
  drought-deciduous plants 75
  transpiration relationship 68
water potential change, and vegetative
  cover 301
water relations (plant) 66–70
  floristic surveys 11
  metabolizable energy content 152
  spatial variation 69
  vascular 11–12
  as water source 144
water relations
  passarine birds 207
  quail 206
  rodent, and population variation 265–6
  terrestrial mammals 159–60
water resorption, respiratory 147
water runoff: *see* runoff
water storage
  chuckwalla 146
  plants 67
water stress
  adaptation strategy 56, 141–54
  and dry food comsuption 144–5
  and spine formation 74
  stem/leaf photosynthesis 75
water stress gradient 69
water table 26
water use, NAAP correlation 103
water use efficiency: *see* WUE
water use optimization, plants 56
water vapor concentration 57, 63
water vapor loss 141
water vapor movement, soil 40
wax 59, 143
weather stations, S. Nevada 40–1
weathering 34
weeds 312, 313
weevils 216, 233, 242; *see also* Coleoptera
WEI (water economy index) 142–3
western fence lizard: *see Sceloporus
  occidentalis*
western patch-nosed snake: *see Salvadora
  hexalepis*
western pipistrelle: *see Pipistrellus
  hesperus*
western shovel-nosed snake: *see Chionactis
  occipitalis*
western skink: *see Eumeces skiltonianus*
western whiptail lizard: *see Cnemidophorus tigris*
wet bulb depression (PET studies) 52
wetfall 274
whiptail lizard: *see Cnemidophorus rigris*

white bursage: *see Ambrosia dumosa*
White Rock Springs 26
white-crowned sparrow: *see Zonotrichia
  leucophrys*
white-tailed antelope ground squirrel: *see
  Ammospermophilus leucurus*
Williams spot mapping census technique
  207
willow: *see Salix* spp.
wind 7–8, 49–51, 54
wind erosion
  maximal 277
  nitrogen loss 276–7
  and soil damage 302
wind run 51
wind speed, and leaf temperature 57(f)
windscorpions 230
winter, hibernation/diapause 138–9
winter annuals: *see* annuals, winter
winter fat: *see Ceratoides lanata*
winter feeding ground (birds) 197
winter-rainfall deserts ix
woodland
  Colorado River 19
  Great Basin 87
  plant communties 14–16(t)
  see also pinyon–juniper
woodrats: *see Neotoma*
WUE (water-use efficiency) 66–7
  angiosperms 81(t)
  C_4 pathway 65, 77
  CAM 65
  desert holly 73
  plant longevity 69
  water stress gradient 69

xerophytic saltbush scrub 18

year, hydrologic 102–3
yellow-rumped warbler: *see Dendroica
  coronata*
yponomeutid larvae 239, 242(t)
Yucca dry lake 25
Yucca Flat 23, 24
  lizard population studies 184
  nuclear testing xii
  phenological data 106
  soils 29
  temperature inversion 89–90
  vegetation 85(m), 87
Yucca Mountain 23(f), 24, 85(m)
Yuma desert pavement 30

zebra-tailed lizard: *see Callisaurus
  draconoides*